The Psychology of Argument
Cognitive Approaches to Argumentation and Persuasion

Edited by
Fabio Paglieri,
Laura Bonelli
and
Silvia Felletti

© Individual authors and College Publications 2016
All rights reserved.

ISBN 978-1-84890-195-7

College Publications
Scientific Director: Dov Gabbay
Managing Director: Jane Spurr

http://www.collegepublications.co.uk

Original cover design by Orchid Creative   www.orchidcreative.co.uk
Printed by Lightning Source, Milton Keynes, UK

All rights reserved. No part of this publication may be reproduced, stored in a retrieval system or transmitted in any form, or by any means, electronic, mechanical, photocopying, recording or otherwise without prior permission, in writing, from the publisher.

Studies in Logic
Studies in Logic and Argumentation
Volume 59

The Psychology of Argument
Cognitive Approaches to Argumentation and Persuasion

Volume 48
Trends in Belief Revision and Argumentation Dynamics
Eduardo L. Fermé, Dov M. Gabbay, and Guillermo R. Simari

Volume 49
Introduction to Propositional Satisfiability
Victor Marek

Volume 50
Intuitionistic Set Theory
John L. Bell

Volume 51
Metalogical Contributions to the Nonmonotonic Theory of Abstract Argumentation
Ringo Baumann

Volume 52
Inconsistency Robustness
Carl Hewitt and John Woods, eds.

Volume 53
Aristotle's Earlier Logic
John Woods

Volume 54
Proof Theory of N4-related Paraconsistent Logics
Norihiro Kamide and Heinrich Wansing

Volume 55
All about Proofs, Proofs for All
Bruno Woltzenlogel Paleo and David Delahaye, eds

Volume 56
Dualities for Structures of Applied Logics
Ewa Orłowska, Anna Maria Radzikowska and Ingrid Rewitzky

Volume 57
Proof-theoretic Semantics
Nissim Francez

Volume 58
Handbook of Mathematical Fuzzy Logic, Volume 3
Petr Cintula, Petr Hajek and Carles Noguera, eds.

Volume 59
The Psychology of Argument. Cognitive Approaches to Argumentation and Persuasion
Fabio Paglieri, Laura Bonelli and Silvia Felletti, eds

Studies in Logic Series Editor
Dov Gabbay   dov.gabbay@kcl.ac.uk

Contents

Introduction: A walk on the cognitive side i

Part I: Models

1. Reasoning about arguments from consequences 1
Jean-François Bonnefon

2. Commitment attribution and the reconstruction of arguments 17
Steve Oswald

3. Erotetic problem solving: From real data to formal models. An analysis of solutions to Erotetic Reasoning Test task 33
Mariusz Urbański, Katarzyna Paluszkiewicz and Joanna Urbańska

4. Initiating argument in social confrontation: Development of a behavioral complexity model to understand variations in initiation types 47
Susan Kline and Wen Song

Part II: Rationality

5. Pushing the bounds of rationality: Argumentation and extended cognition .. 67
David Godden

6. Don't blame the norm. On the challenge of ecological rationality 85
Maarten Boudry, Michael Vlerick and Ryan McKay

7. The fragility of argument 99
John Woods

8. Arguments and their sources 129
Ulrike Hahn and Peter Collins

Part III: Biases and Fallacies

9. Don't worry, be gappy! On the unproblematic gappiness of alleged fallacies . 153
Fabio Paglieri

10. Reliable debiasing techniques in legal contexts? Weak signals from a darker corner of the social science universe 173
Frank Zenker, Christian Dahlman and Farhan Sarwar

11. The biased use of argument evaluation criteria in motivated reasoning: Does argument quality depend on the evaluators' standpoint? 197
Hans Hoeken and Mariecke van Vugt

12. Evidence quality variations and claim acceptance: An experimental investigation of the role of distraction and dilution 211
Jos Hornikx

13. Does expertise favour the detection of the metaphoric fallacy? 223
Francesca Ervas, Antonio Ledda and Antonio Pierro

14. Face the consequences! Strategic maneuvering with the *argumentum ad consequentiam* 245
Bart Garssen

Part IV: Communication and Persuasion

15. The psychological approach to interpersonal argumentation in the U.S. argumentation community 257
Dale Hample

16. Ethos, familiars and micro-cultures 275
Michael Gilbert

17. Multimodal persuasion in judicial debates standpoint? 287
Francesca D'Errico and Antonella Bellon

18. Finding Mussolini's charisma in his multimodal discourse 305
Isabella Poggi and Francesca D'Errico

19. Persuasive lexicon extraction from political speeches 327
Marco Guerini, Gözde Özbal and Carlo Strapparava

Part V: Learning and Development

20. The trajectory of argumentation and its multifaceted functions 347
Matthew Fisher and Frank Keil

21. Arguing your way out of confusion 363
Blair Lehman and Art Graesser

22. The importance of multi-modality in mathematical argumentation 387
Baruch Schwarz and Naomi Prusak

23. The psychology of far transfer from classroom argumentation 407
Michael Nussbaum and Christa Asterhan

List of Contributors

Christa S. C. Asterhan
Hebrew University of Jerusalem, Jerusalem, Israel
asterhan@huji.ac.il

Antonella Bellon
TIM S.p.A., Legal Office, Milan, Italy
antonundici@gmail.com

Laura Bonelli
Istituto di Scienze e Tecnologie della Cognizione, Consiglio Nazionale delle Ricerche, Roma, Italy; "La Sapienza" University of Rome
laura.bonelli@istc.cnr.it

Jean-François Bonnefon
Centre for Research in Management, Toulouse School of Economics, Toulouse, France
jfbonnefon@gmail.com

Maarten Boudry
Department of Philosophy & Moral Sciences, Ghent University, Ghent, Belgium
maartenboudry@gmail.com;

Peter Collins
Department of Psychological Sciences, Birkbeck, University of London

Christian Dahlman
Law Faculty, Lund University, Sweden
christian.dahlman@jur.lu.se

Francesca D'Errico
Faculty of Psychology, Uninettuno University, Rome, Italy
f.derrico@uninettunouniversity.net

Francesca Ervas
University of Cagliari, Cagliari, Italy
ervas@unica.it

Silvia Felletti
Istituto di Scienze e Tecnologie della Cognizione, Consiglio Nazionale delle Ricerche, Roma, Italy; "La Sapienza" University of Rome
silvia.felletti@istc.cnr.it

Matthew Fisher
Yale University, New Haven, USA
matthew.fisher@yale.edu

Bart Garssen
Speech Communication, Argumentation Theory and Rhetoric, University of Amsterdam, Amsterdam, the Netherlands
b.j.garssen@uva.nl

Michael A. Gilbert
Department of Philosophy, York University, Toronto, Canada
gilbert@yorku.ca
David Godden
Michigan State University, East Lansing, Michigan, USA
dgodden@msu.edu

Art Graesser
University of Memphis, Memphis, USA
graesser@memphis.edu

Marco Guerini
FBK-Irst - Povo, I-38100 Trento
guerini@fbk.eu

Ulrike Hahn
Department of Psychological Sciences, Birkbeck, University of London
u.hahn@bbk.ac.uk

Dale Hample
Department of Communication, University of Maryland, College Park MD, USA
dhample@umd.edu.

Hans Hoeken
Utrecht Institute for Linguistics OTS, Utrecht University, Utrecht, The Netherlands
j.a.l.hoeken@uu.nl

Jos Hornikx
Centre for Language Studies, Radboud University, Nijmegen, the Netherlands
j.hornikx@let.ru.nl

Frank C. Keil
Yale University, New Haven, USA
frank.keil@yale.edu

Susan L. Kline
School of Communication, Ohio State University, Columbus OH, USA
kline.48@osu.edu

Antonio Ledda
University of Cagliari, Cagliari, Italy
antonio.ledda@unica.it

Blair Lehman
Educational Testing Service, Princeton, USA
blehman@ets.org

Ryan McKay
ARC Centre of Excellence in Cognition and its Disorders, Department of Psychology, Royal Holloway, University of London, Egham, Surrey, United Kingdom

E. Michael Nussbaum
University of Nevada, Las Vegas, Nevada, USA
nussbaum@unlv.nevada.edu

Steve Oswald
University of Fribourg, Switzerland
steve.oswald@unifr.ch

Gözde Özbal
FBK-Irst - Povo, I-38100 Trento
g.ozbal@fbk.eu

Fabio Paglieri
Istituto di Scienze e Tecnologie della Cognizione, Consiglio Nazionale delle Ricerche, Roma, Italy
fabio.paglieri@istc.cnr.it

Katarzyna Paluszkiewicz
Institute of Psychology, Adam Mickiewicz University, Poznań, Poland
k.paluszkiewicz@amu.edu.pl

Antonio Pierro
University of Cagliari, Cagliari, Italy
antonio.pierro@gmail.com

Isabella Poggi
Dipartimento di Filosofia Comunicazione e Spettacolo, Università Roma Tre
isabella.poggi@uniroma3.it

Naomi Prusak
The Hebrew University of Jerusalem, Israel
inlrap12@netvision.net.il

Baruch Schwarz
The Hebrew University of Jerusalem, Israel
baruch.schwarz@mail.huji.ac.il

Wen Song
College of Literature and Journalism, Sichuan University, Cheng du, Sichuan Province, P.R. China
wen.song17@outlook.com.

Carlo Strapparava
FBK-Irst - Povo, I-38100 Trento
strappa@fbk.eu

Joanna Urbańska
Institute of Psychology, Adam Mickiewicz University, Poznań, Poland
joanna.urbanska@amu.edu.pl

Mariusz Urbański
Institute of Psychology, Adam Mickiewicz University, Poznań, Poland
e-mail: mariusz.urbanski@amu.edu.pl

Mariecke van Vugt
Utrecht Institute for Linguistics OTS, Utrecht University, Utrecht, The Netherlands

Michael Vlerick
Department of Philosophy, University of Johannesburg, Johannesburg, South Africa

John Woods
Director, Abductive Systems Group, Department of Philosophy, University of British Columbia, Vancouver, BC, Canada
john.woods@ubc.ca; www.johnwoods.ca.

Frank Zenker
Department of Philosophy & Cognitive Science, Lund University, Sweden
frank.zenker@fil.lu.se

Introduction: A Walk on the Cognitive Side

Fabio Paglieri[1], Laura Bonelli[2] & Silvia Felletti[3]

[1,2,3] Istituto di Scienze e Tecnologie della Cognizione, Consiglio Nazionale delle Ricerche, Roma, Italy;
[2,3] "La Sapienza" University of Rome;
[1] fabio.paglieri@istc.cnr.it; [2] laura.bonelli@istc.cnr.it; [3] silvia.felletti@istc.cnr.it

Compared to other disciplines, so far psychology has contributed only marginally to the interdisciplinary enterprise known as "argumentation theory": this is apparent in the scant number of psychologists taking part in the main conferences on argumentation and the relatively small percentage of psychological contributions appearing in leading journals on argumentation studies, such as *Argumentation, Informal Logic, Argument and Computation, Argumentation and Advocacy*, and the *Journal of Argumentation in Context*. Granted, there are many excellent scholars working on the psychology of argument (Dale Hample, Ulrike Hahn, Mike Oaksford, Lance Rips, Jean-François Bonnefon are names that come to mind, among others), but in the argumentation theory community they remain exceptions, albeit very notable ones, as compared to the legions of philosophers, linguists, rhetoricians, and computer scientists that populate the field. As this volume endeavors to demonstrate, the problem is not lack of interest, but rather lack of cross-disciplinary integration: there is plenty of outstanding research being conducted on the psychology of argumentation across the world (and there has been for some time, actually), but most of it happens to take place in areas that have not yet been fully integrated in argumentation theory proper. The aim of this collection is to foster such an integration, by bringing into the fold of argumentation theory contributions typically categorized under different labels, e.g. psychology of reasoning, psychology of education, communication studies, persuasion research, and so on.

The fact that there is still a need for this integrative effort is somehow surprising, for two reasons: first, the relevance of the cognitive dimension of argumentation to its study is hard to miss, not only (albeit obviously) for descriptive projects, but also for normative approaches (consider for instance the role of bounded rationality in discussions on normative models of inference); second, argumentation theory is already a highly interdisciplinary endeavor, so there is certainly no hostility from the community to cross-fertilization across disciplines. Nonetheless, psychology so far remained on the outskirts of that endeavor, probably due to some historical accident. Whatever the reason, it is now high time to revert the trend: the present volume is unlikely to manage it on its own, but it is certainly meant to be a step in that direction. As a consequence of this mission, the roster of authors include both "usual suspects" of argumentation theory (that is, well-known scholars in the community) and authoritative newcomers (that is, people that are certainly well-established in their own fields, but are less known to most argumentation scholars). Moreover, their contributions manage to cover a broad swath of topics that are relevant to argumentation theory, demonstrating that the psychology of argument is not a quaint niche curiosity, but rather a promising and powerful approach, worthy of becoming mainstream in the field – the sooner, the better.

The variety of ways in which argumentation theory can benefit from psychological contributions is mirrored in the internal structure of the volume, which is organized in five parts. The first section includes essays that discuss how insights from the psychology of reasoning can inform models of argument. Jean François Bonnefon's contribution, "Predicting behavior on the basis of arguments from consequences", nicely exemplifies this approach. Bonnefon starts by pointing out how arguments from consequences are characterized by a dual nature, since they can be used to predict other people's future behavior as well as to persuade them to undertake a course of action. Through the use of utility grids representing the agents' actions and consequences, he shows how folk axioms of decisions can be used to predict the inferences afforded by conditional statements in which the consequences of an action are beneficial or detrimental for one or more agents. The author gives an insight of his studies on the predictive use of arguments from consequences, and presents interesting results about the interpretation of different kinds of ambiguity, the prediction of others' behavior on the basis of material or hedonic utility, and the processing of utility violations, outlining a link between argumentation research and theories of mind and reasoning.

The second contribution, "Commitment attribution and the reconstruction of arguments" by Steve Oswald, deals with the identification of missing or unexpressed premises in argumentative discourse. Addressing the issue from the perspective of the addressee, the paper focuses on commitment as a specific attitude that the speaker expresses – more or less overtly – in her utterances, and that bears a strong relationship with meaning: attributing a meaning to the utterance is also attributing a commitment to the speaker. But argumentation in natural settings does not always allow to easily reconstruct the speaker's commitment, for example when argumentation is incomplete. The author assumes that while pragma-dialectics can assist the analyst in the normative reconstruction of arguments, that approach has little to contribute to commitment attribution at the level of naïve interpretation. Based on the assumptions of relevance theory, he proposes a two-step model to reconstruct plausible representations of real argumentative exchanges, that gives a deeper insight on commitment attribution at the level of naïve interpretation.

The third contribution is also an effort to account for people's actual reasoning processes in difficult deductive tasks involving conditionals. In their "Erotetic problem solving: From real data to formal models. An analysis of solutions to erotetic reasoning test task", Mariusz Urbanski, Katarzyna Paluszkiewicz and Joanna Urbanska examine how people deals with erotetic inferences, i.e., inferences which involve questions as conclusions, or as premises and conclusions. According to Inferential Erotetic Logic (IEL), the validity of erotetic inferences depends on two primary conditions: the transmission of truth/soundness of the premises into soundness of the conclusions, and cognitive usefulness. The authors examine the results of a study with a task (the Erotetic Reasoning Test) they designed to assess people's fluency with erotetic inferences, opening an interesting descriptive window on the formal framework of IEL.

The last contribution in this section, "Initiating argument in social confrontation: Development of a behavioral complexity model to understand variations in initiation types" by Susan Kline and Weng Song, shifts the analysis on conflictual argumentation. In their article, they develop a framework for understanding variations in argumenta initiations in the social confrontation episode. They review the literature on argument initiations, and then use interpersonal communication literatures on conflict orientation and multiple goal perspectives to develop a behavioral complexity model of argument initiations, offering evidence to support their model.

In the second section of the volume, the topic of how the theories of rationality can meet, contrast or be aided by those of argumentation is seen from up close. Among the many goals of argumentation theories, one is certainly to provide a satisfactory explanation of how people evaluate arguments in a rational way and according to specific norms and standards. As David Godden neatly describes in the opening of his chapter, "these can be understood as rational norms, where the core idea of rationality is that we rightly respond to reasons by according the credence we attach to our doxastic and conversational commitments with the probative strength of the reasons we have for them" (p. 67). Godden's "Pushing the bounds of rationality: Argumentation and extended cognition" opens this section with relevant reflections on how both argumentation and argumentation theories can extend cognition in such a way that individuals can better meet the rational norms and standards classically put forward by logic and probability theories – a series of standards which is idealized, and quite hard to meet by mankind in real life. After briefly summarizing the principles of logic and presenting the "prescriptivity gap" between the rational norms provided by logic and how humans fail to follow them in daily reasoning, Godden explains how theories of bounded rationality possibly widened the aforementioned gap and how, on the other hand, argumentative rationality might move from a perspective where humans rightly respond to reasons and possibly cooperate in doing so. According to Godden, five kinds of assumptions are contained within argumentative rationality: the normativity assumption, deontological assumptions, structuralist assumptions, internalist assumptions, and the assumption of reflective stability. After discussing each of them, Godden considers the effects of bounded rationality on argumentative rationality and shows how argumentation and critical thinking can extend cognitive abilities such that individuals can extend their responsibilities to adhere to logical standards and norms.

On a similar vein – but holding quite different conclusions – Maarten Boudry, Michael Vlerick and Ryan McKay argue against the criticism of the classical norms of rationality sustained by ecological rationalists, by advocating that these norms remain valid nonetheless. In their chapter "Don't blame the norm. On the challenge of ecological rationality", the authors explain how ecological rationality presents two apparent challenges to the traditional canons of rationality and how, in both cases, they fail at refuting the classical norms of rationality. In the first challenge posited by ecological rationality's advocates, the norms which appear violated re-emerge at the level of evolutionary adaptation. In the second challenge, the norms turn out to be not applicable, and therefore no actual challenge takes place. The authors also highlight how advocates of ecological rationality still use the traditional norms of rationality as a benchmark. Even though the findings of ecological rationalists (which are briefly summarized in the first paragraphs of the chapter) are warmly welcomed by the authors, Boudry, Vlerick and McKay claim that what stands between logical and ecological rationality is simply a false opposition with a thought provoking argumentation against the "anti-norms" rhetoric.

The third chapter of this section is "The fragility of argument" by John Woods, a complex and very well-articulated work on the necessary relations between human inference and argument theory. The chapter is divided in twelve distinct sections, the first two of which describe the importance of a distinction, in the logics of argument, between a proposition's consequences, the individual who simply recognizes them, and a reasoner who draws them – and how this distinction can be already detected in Aristotle's theory. Subsequently, Woods presents other two distinctions, this time between implication and inference, and between inference and argument. Moreover, addressing a topic that has been touched in the previous chapters of this book, he argues how human inference may not find a good model in a theory of argument. Following a line similar to the one addressed by Godden, the fourth section discusses the normativity problem: the author argues that normativity presumptions of idealized rationality still have to find a satisfactory justification. A further development of the topic treated in the third section is found in section five, where the author draws a distinction between conscious and unconscious cognitive processes. The following three paragraphs deal with several vulnerabilities within argument theory: the fragilities of the more formal precincts of argument theory within the scope of the normativity issue and the vulnerability of the non-monotonic premise-conclusion links if compared to "the openness of the world" (p. 100). The author further suggests that non-monotonic consequence relations should be considered epistemic relations, another issue that stresses the implication-inference divide. Woods' rich chapter continues with the *pars construens* of argument theory: the author emphasizes the importance of real life facts for the logic of argument and proposes a tentative solution to the normativity problem, explaining how the way reasoning plays out in the conditions of real life is the way it should play out in the premise-conclusion compound. Woods further addresses the distinction between arguing and meta-arguing, and cites the difficulties inherent in the latter as one of the reasons for the relative infrequency of face-to-face combat arguments. Finally, he discusses the distortive influences of paradigmatic theories in argument theory and he concludes his dense and multifaceted work with a praise of naturalism as an antidote to such risky dogmatisms.

The final chapter of this section deals with "Arguments and their sources", as the title chosen by Peter Collins and Ulrike Hahn clearly states. As argumentation theory has moved away from classical logic as a standard, sources have played an increasingly important role in the psychology of argumentation. Collins and Hahn's chapter reunites and presents together different perspectives in the psychology of argumentation: procedural rules, pragmatics, argumentation schemes and Bayesian Argumentation, which all offer distinct points of view on the characteristics of argument sources. The authors argue for a reunified framework that reconciles these three different approaches around a probabilistic notion of relevance, by addressing how exactly each of them positively contributes to argument theory.

The third part of the volume deals with the analysis of reasoning errors. Historically, this is one of the most important areas of overlap between the psychology of reasoning and argumentation theory, since the notion of bias and the concept of fallacy, while not identical, are certainly intertwined to one another. In his contribution "Don't worry, be gappy! On the unproblematic gappiness of alleged fallacies", Fabio Paglieri puts in contact these two research areas and argues against the standard conception of fallacies as attractive and universal errors that are hard to eradicate. In particular, Paglieri discusses several recent attempts to deal with "non-fallacious fallacies", that is, arguments that fit the bill of one of the traditional fallacies but are actually respectable enough to be used legitimately in the right context. The fact that reinterpreting alleged fallacies as non-fallacious arguments requires supplementing the textual material with something else, e.g. probability distributions, pragmatic considerations, dialogical context, makes them *gappy*, yet this gappiness, according to Paglieri, is typically unproblematic, and thus no reason for concern. This, in turn, calls into question the usefulness of the very notion of fallacy.

In a similar vein, Frank Zenker, Christian Dahlman and Farhan Sarwar, in their paper "Reliable debiasing techniques in legal contexts? Weak signals from a dark corner of the social science universe", take stock on the effectiveness, or lack thereof, of debiasing techniques in legal matters. These techniques aim to impact positively on personal or procedural features of reasoning and decision-making: unfortunately, the empirical record on their effectiveness in legal contexts, as well as in other areas, is mixed. This paper discusses theoretical and empirical considerations that help explaining such puzzling pattern of results, using a historical perspective and emphasizing the rise and popularization of research on human biases and heuristics. Along the way, the authors also provide abundant examples of such techniques and suggestions for methodological improvements to the *status quo*.

The chapter "The biased use of argument evaluation criteria in motivated reasoning: Does argument quality depend on the evaluators' standpoint?", by Hans Hoeken and Mariecke van Vugt, investigates one of the many ways in which biased reasoning interacts with argument evaluation. Research on motivated reasoning suggests that people take a more critical stance toward arguments that go against their opinions compared to arguments that are in accordance with these opinions: this may lead to distortions in how they distinguish strong from weak arguments. In this study, Hoeken and van Vugt told participants that they would take part in a debate and would have to defend a pre-assigned claim: all participants received sixteen (strong or weak) arguments and were asked to prepare themselves for the debate while thinking aloud. Analysis of the think aloud protocols showed that people almost exclusively used criteria to boost the quality of arguments supporting their claim while disqualifying arguments that went against it, thus providing a powerful illustration of the role of confirmation bias in argument assessment.

Even when people demonstrate sensibility to factors that affect the epistemic value of an argument, e.g. evidence quality, this is often subjected to systematic distortions: the challenge is thus to identify the cognitive mechanisms responsible for such biases. In his paper "Evidence quality variations and claim acceptance: An experimental investigation of the role of distraction and dilution", Jos Hornikx critically discusses a previous finding, namely, the fact that claims supported by high-quality evidence are better accepted than claims supported by low-quality evidence only when short texts are considered, whereas adding information unrelated to the evidence at the end of the text undermines sensibility to evidence quality. This effect of text length could be explained either by distraction (the additional text at the end distracts the reader) or by dilution (the additional text makes the fragment less diagnostic for claim evaluation). In a novel study, Hornikx found support for the distraction explanation, since an effect of evidence quality on claim acceptance was observed in two conditions: in short texts, and in longer texts with additional information placed at the start. The fact that the effect disappears only when the irrelevant information is added at the end suggests that the crucial factor is not text length, but rather the distracting effect of reading it after the relevant evidence had already been presented.

Another important aspect in fallacy studies is the role of expertise: several studies have found experts to be as vulnerable to biases as laypeople, yet this may not hold universally. Francesca Ervas, Antonio Ledda and Antonio Pierro, in their "Does expertise favor the detection of the metaphoric fallacy?", aims at clarifying whether and to what extent expertise plays a role in the detection of ambiguity fallacies, such as *quaternio terminorum*, where a metaphor is the middle term in one of the premises (metaphoric fallacy). They used systematic training in philosophical logic as the discriminating criterion between experts and non-experts, and then asked participants to evaluate a series of verbally presented arguments, having the structure of *quaternio terminorum* and containing either a lexical ambiguous or a metaphorical middle term. Their experimental results show that non-experts tend to assess as sound instances of *quaternio terminorum* with lexicalized metaphors as middle terms, when the conclusion of the argument is far from being patently false. Interestingly, metaphorical middle terms have an effect also on experts' intuitions on fallacious argument with plausible conclusion, albeit to a minor degree. Thus in the case of the metaphoric fallacy expertise acts as a mitigating factor, rather than a universal antidote.

Bart Garssen, in his "Face the consequences! Strategic maneuvering with the argumentum *ad consequentiam*", tackles another alleged fallacy, the argument from consequences (already discussed by Bonnefon in his contribution, albeit from a very different perspective), and argues that it mimics other reasonable types of argumentation: pragmatic argumentation and ad absurdum argumentation. Garssen shows how different variants of the *ad consequentiam* fallacy manage to exploit its resemblance to reasonable argument forms, and suggests that this mimicry is precisely what makes such arguments appealing, in spite of their flaws. Along the way, Garssen also clarifies the crucial difference between the pragma-dialectical understanding of the argument from consequences and other approaches to this type of reasoning: in a nutshell, the traditional concept of an argument from consequences involves advocating a certain course of action based on its consequences, which most often does not involve any fallacy; what is fallacious, however, is drawing a conclusion on the truth of a statement based on the desirability of that truth – and this is the type of fallacious reasoning that Garssen strives to analyze.

In the fourth section of the volume, the focus shifts to the role of arguments in communication and persuasion: this has been a thriving area of research in communication studies, unfortunately with very little cross fertilization with argumentation theory proper – a state of affair that the following papers strive to remedy. Thus it is fitting that this section is opened by Dale Hample, one of the few scholars active across both communities: in his "The psychological approach to interpersonal argumentation in the U.S. argumentation community", Hample offers a broad-spectrum analysis of the development of the United States' approach to studying interpersonal arguing. By adopting an historical perspective, the author presents the intellectual climate that led into the main researches of the 1970's, by especially focusing on Joseph Wenzel's influence at the University of Illinois and the relevance of his interests in non-traditional perspectives, such as cognitive and other psychological approaches to interpersonal communication and argumentation. Hample's work continues with a systematic and captivating conceptual inventory of the American cognitive and psychological research on interpersonal argumentation and presents its major lines of inquiry, which include the arguers' attitudes (e.g.: verbal and argumentative aggressiveness) and metacognition (e.g.: arguers' understanding of arguing), the effects of arguing (e.g.: persuasive effects), the age of the arguers (e.g.: children vs adults), the arguing context (e.g.: where arguments are situated and how situations affect them), and cross-cultural research perspectives. Although there has never been anything such as a "unified American school of interpersonal argumentation" (p. 262), Hample outlines both historical and current trends in the scholarly tradition of interpersonal argumentation in a clear and engaging way, and draws attention to the need of more works, perspectives and inquiries in this research scope.

By focusing the discourse on the speakers' attitude in interpersonal argumentation previously pinpointed by Hample, Michael Gilbert examines how arguers attribute ethos to others and try or expect to have it attributed to them depending on trust levels and degrees of acceptance of their opinions and statements (or "ethotic ratings"). In his chapter "Ethos, familiars and micro-cultures", Gilbert suggests that the concepts, which are usually employed with respect to public individuals, can also be applied on a micro-level between individuals that are particularly familiar with each other and these "ethotic standings" posit specific consequences and effects in interpersonal arguments. Whereas "ethotic ratings" of experts are made by a hearer towards a speaker – and not the other way around – Gilbert explains that they are a result of relational work when the context is an interaction with someone familiar. Even though in these cases speakers may try to influence their ethotic ratings by performing different face-saving acts (which may or may not succeed), their ability to judge and be judged is greater than the case when they are confronted with public speakers or experts. Gilbert hypothesizes that speakers who are familiar with each other work together on their interactional ethotic standings. In familiar contexts, ratings are more interactive and bound to different norms and to context-specific maxims – an observation which leads the author to conclude that we cannot simply rely on traditional judgments and models that apply primarily to experts in analyzing the role of ethos in arguments. Gilbert finally invites to further explore how these dynamics work in situated, micro-cultural contexts.

In their chapter "Multimodal persuasion in judicial debates", Francesca D'Errico and Antonella Bellon rather focus on the interplay of persuasive effects and verbal aggression in three Italian judicial debates, analyzed in terms of multimodal persuasion. The authors explore the different levels of judicial persuasion by analyzing the three, most frequent types of communicative acts: discrediting, argumentative and dominant moves. Whereas discrediting and dominant moves are characterized by a strategic intensification of speech tones and gestures ("peripheral" strategies), D'Errico and Bellon notice that argumentative moves are more focused on strategic verbal and rhetorical choices ("central" strategies), and the interplay of these two macro-strategies attempts to trigger different persuasive effects. Discredits, argumentative and dominant moves explain different levels of persuasiveness: interestingly, the more the moves are aggressive, the less they seem to be persuasive.

Another thought-provoking analysis in the framework of multimodal persuasion is Isabella Poggi and Francesca D'Errico's chapter "Finding Mussolini's charisma in his multimodal discourse". In this work, the authors present a theoretical model of charisma and explore in detail the specific features of Mussolini's communicative behaviors. In a first, qualitative analysis the authors identify the multimodal, communicative aspects that typically identify the dictator's charisma. This first analysis is focused on a small corpus of discourses, from which Poggi and D'Errico isolate verbal features (e.g.: *"nicewordism"* and *"strongwordism"*), illocutionary aspects (e.g.: orders or threats), emotive attitudes (e.g.: attempts of empathetic attunement) and other sets of argumentative moves and phonetic and kinesic features, of which they explore the distribution and organizational patterns. In a second moment, the authors test how these charismatic features are represented in a wider sample of Mussolini's public speeches by means of a quantitative lexicographic analysis. Results show how the specific, "charismatic" features they detected seem to occur more often when they are organized in the co-text of speeches addressed to big crowds rather than those addressed to particular or smaller audiences.

Marco Guerini, Gözde Özbal, and Carlo Strapparava continue to explore communication and persuasion in public political discourse in a chapter called "Persuasive lexicon extraction from political speeches". The authors worked on a wide corpus of speeches tagged with audience reactions and applied different distributional semantic approaches (i.e.: tf-idf, weighted tf-idf, LSA, and LDA) to isolate and extract the lexica of persuasive words, each of them being associated with a specific score of persuasive impact. Subsequently, the authors present a series of experiments on pairs of persuasive / non persuasive sentences aimed at assessing the different distributional approaches and show how some methodologies are clearly more effective than others. Guerini, Özbal, and Strapparava's work also points out how computational methodologies can be fruitfully employed to depict the most efficacious lexical choices in persuasive communication.

The fifth and final section of the volume deals with the role that argumentation plays in empowering domains such as learning and development. As recent developments in the debate on critical thinking education suggest, this is really an area in which a stronger alliance between psychologists, educators and argumentation scholars is of primary importance. In an individual's span of life, the way arguments are constructed, their primary functions in their context of occurrence and their specific aims vary significantly. Matthew Fisher and Frank Keil's chapter "The trajectory of argumentation and its multifaceted functions" opens this section and focuses on how early learners engage in proto-argumentation in a "learning mode" that leaves aside any competitive feature.

Arguments appear to be used by children as a means to increase their knowledge about the world. As early learners develop more sophisticated reasoning abilities, their approach shifts to an "argue-to-win" style, which has the promotion of one's own point of view and the devaluation of others as its primary goal. The simplest form of this mode of argument is *ad hominem*-style attacks that ignore the actual structure of the argument; but once the structure of the argument has been internalized by the young interactants, they would rather re-adopt a learning mode equipped with more powerful tools for discovery. Moreover, Fisher and Keil argue that strategies like "considering the opposite" can influence reasoning and enable productive exchanges of ideas because they may expose gaps in understanding and probe the understanding of others. The authors address the need of further research to determine how to best promote learning-oriented arguments for firmly held topics, as well as the need to identify the conditions under which people willingly use arguments to learn and how they resist influences that push them towards "argue-to-win" mindsets.

In their "Arguing your way out of confusion", Blair Lehman and Art Graesser shift the focus on the role of confusion in learning processes, showing how precious cognitive efforts such as problem solving and deliberation intervene in the construction of arguments. Arguing provides learners – especially students – with the opportunity to compare and contrast opposing views, as well as consider why one perspective is correct rather than incorrect. Lehman and Graesser present an experimental environment where contradictory information is communicated to students by two animated agents during the discussion of the scientific merits of a research case study. The students had to construct their arguments either with or without additional information about the presented topic. The authors assessed the quality and the structure of the students' arguments according to three different criteria: the presence of claims and evidence, the semantic match score to ideal arguments, and particular linguistic features such as cohesion and syntactic complexity. Lehman and Graesser also provide a very interesting discussion about the impact that argument construction has on both confusion resolution and learning, and highlight the factors that facilitate confusion resolution: this leads them to some fascinating concluding remarks on how arguing has a beneficial role in promoting learning at deeper levels.

Shifting the focus to the specific case of mathematical reasoning, Baruch Schwarz and Naomi Prusak explain how non-verbal argumentative moves prove to be crucial in mathematical argumentation, in a chapter entitled "The importance of multi-modality in mathematical argumentation". By presenting different examples from a context of collaborative mathematical problem-solving among third-grade students, the authors show how gestures and actions such as drawing or folding may not simply accompany verbal utterances in a folkloristic fashion. Instead, they may function as independent argumentative moves communicated on a kinesic, non-verbal channel, which interweave with verbal peer argumentation in order to create sound mathematical arguments. Schwarz and Prusak discuss how multimodal argumentation led the young students to fundamental insights in elementary geometry and suggest several roles of non-verbal actions in learning through mathematical argumentation, especially the proleptic role of gestures and of inscriptions in argumentation in geometry.

Finally, Michael Nussbaum and Christa Asterhan conclude this fifth section discussing "The psychology of far transfer from classroom argumentation". School lessons that consistently engage students in argumentation have been shown to provide learning benefits in other types of lessons with different programs and teachers. For example, argumentation-rich teaching in science classes or mathematics has resulted in higher student achievement in English Language Arts. In their work, the authors review previous explanations for these effects rooted in theories of development, such as argumentation schema theory, ACT-R theory, motivation theory, and situativity theory. They subsequently extend these accounts by proposing that in argumentation-rich school programs, students discover and practice "proactive executive control strategies" that involve the activation or inhibition of cognitive processes such as protection from interference, as well as the generation of general production rules and the acquisition of conceptual agency through participation in conversations that matter.

Taken together, these twenty-three chapters are meant to convey both the diversity and the complementarity of the cognitive approaches to argumentation and persuasion, as well as their overall relevance to the broader community of argumentation scholars. This collection is far from complete, in terms of both topics and notable contributors: yet we trust it is comprehensive and authoritative enough to invite greater attention to the psychological analysis of argumentative processes, hopefully helping to inspire other like-minded initiatives in the future. Such result would have been impossible to achieve, without the generous efforts of our authors and of the many anonymous reviewers who helped improve their contributions: to all these colleagues we offer our most sincere gratitude. Equally instrumental to the making of this collection were John Woods, managing editor of the *Studies in Logic and Argumentation* series, to which this volume proudly belongs, and Jane Spurr, managing director of our publisher, College Publications: we thank both of them for the care, competence and patience with which they supported us throughout the whole editorial process. Needless to say, any remaining flaw in the editing of the volume remains our sole responsibility. Whatever these flaws may be, we do hope readers will enjoy this walk on the cognitive side of arguments.

Part I
Models

Chapter 1

Predicting Behavior on the Basis of Arguments From Consequences

Jean-François Bonnefon

Centre for Research in Management, Toulouse School of Economics, Toulouse, France, jfbonnefon@gmail.com

Abstract. People reason about the preferences of others when they seek to win them to a course of action, or when they seek to predict their behavior. In order to win others to a course of action, one has to design arguments from consequences, showing others how their preferences will be satisfied. Since this persuasive use of preferences is already well described in the argumentation literature, this chapter offers a survey of the other kind of reasoning about preferences, which occurs when reasoners use arguments from consequences to predict the behavior of others. I will describe what inferences people derive from arguments from consequences, how they fit ambiguous arguments into templates, how they resolve ambiguity in the causal structure of these arguments, how they react to behavior that violates their predictions, and how they use expected affect to predict behavior based on complex arguments from consequences which impact several agents.

1. Introduction

Arguments from consequences attempt to persuade an agent to engage in or refrain from a behavior, by stating desirable and/or desirable consequences of this behavior for the agent — and assuming that the agent will make a consequentialist decision to engage in (resp., refrain from) the behavior if its consequences are, on the whole, positive (resp., negative).

For example, (1a) states the consequence of eating fresh fruits everyday, (1b) makes it explicit that that this consequence is desirable to the recipient, and (1c) concludes the argument by an injunction to eat fresh fruits everyday. Accordingly, Example (1ac) is a fully explicit argument from consequences, which states the consequences of a behavior, the desirability of these consequences, and the conclusion that the recipients ought to engage in the behavior.

(1)
a. Eating fresh fruits everyday will improve your health;
b. Improving your health is desirable;
c. Therefore, you should eat fresh fruits everyday.

Arguments from consequences are ubiquitous in persuasion attempts. For example, Schellens and De Jong (2004) observed that arguments from consequences were present in each and every government brochure they analyzed, that aimed at behavioral change. Given the prevalence and importance of arguments from consequences, it is perhaps unsurprising that they have been a prime target for investigation in both the argumentation and the reasoning literatures (e.g., Corner, Hahn, & Oaksford, 2011; Evans, Neilens, Handley, & Over, 2008; Feteris, 2002; Hoeken, Timmers, & Schellens, 2012; Thompson, Evans, & Handley, 2005).

Importantly for the present chapter, there are two sides to arguments from consequences. They can be used to persuade people to undertake a course of action, but they can also be used to predict individual behavior. Consider first that, in many instances, only the first part of the argument is made explicitly (Schellens & De Jong, 2004), leaving it to the agent to realize that the consequences are (un)desirable, and that they should (not) engage in the behavior as a result. For example, one may simply state (2a), and leave it to the recipient to fill in (2b) in order to come to the conclusion (2c).

(2)
a. Maxing out your credit cards hurts your credit score;
b. It is not desirable to hurt your credit score;
c. Therefore, you should not max out your credit cards.

Sentence (2a) is the *compact form* of the argument from consequences (2ac). An interesting characteristic of compact arguments from consequences is that they can be expressed as conditional statements, as in (3a) or (3b).

(3)
a. If you max out your credit cards, you will hurt your credit score.
b. If Andrew sends his invoice tomorrow, he will receive $3,000.

This is an interesting characteristic because it makes it easier to connect argumentation to the psychology of reasoning, which always had a special interest for conditional statements. Note in that respect that Sentence (3b) can be used as an argument to convince Andrew to send his invoice the day after, but it also provides a reason to infer that Andrew, as a rational agent, will indeed send his invoice the day after. More generally, the compact form of an argument from consequences (i.e., action a will lead to a consequence valued by an agent A) can be used as a basis for persuasion or prediction. When used as a basis for persuasion, they attempt to convince agent A to take or not to take action a; when used as a basis for prediction, they lead to the inference that agent A will or will not take action a.

My intention in this chapter is to provide an overview of the work I have conducted over the years on the predictive (rather than persuasive) use of compact arguments from consequences, when they are expressed as conditionals. In the rest of the chapter, we will consider the compact form of arguments from consequences illustrated in Sentence (3b). This compact form consists of a conditional statement 'if p then q' where p is an action of an individual, whose consequence q has utility for this or other individuals. First, I will introduce the theoretical framework dealing with these utility conditionals (Bonnefon, 2009). Next, we will see that people have expectations about the structure of such statements, which led them to apply utility templates when an argument from consequences is ambiguous about how agents value some states of the world. We will come to know that the underlying causal structure of these statements impose strong constraints on the inferences they afford, even when the surface structure of the argument does not change. We will consider data showing that expected affect drives the inferences based on arguments from consequences, in complex situations where the actions of an agent impact other individuals besides herself. Finally, we will examine how people process stories in which characters violate expectations based on arguments from consequences: How do people explain why a character failed to take the action suggested by an argument from consequences, and what can their eye movements tell us about their online processing of such situations?

2. Utility Conditionals

Utility conditionals are statements of the form 'if p then q' where p is an action of an individual, whose consequence q has utility for this or other individuals (Bonnefon, 2009). They can be unpacked as: if agent x takes action p which has utility u for agent y, then agent x' will take action q which has utility u' for agent y'. This information can be represented in the *utility grid* of the conditional. The role of the utility grid is to provide a standard representation of who can do what to which consequences for whom, so that formal principles (the folk axioms of decision introduced later in this section) can be applied to the grid in order to systematically predict the inferences afforded by the conditional statement.

$$\begin{Bmatrix} x & u & y \\ x' & u' & y' \end{Bmatrix}$$

The first row of the grid contains the information related to the if-clause of the conditional. That is, it displays the agent who can potentially take action p (left column), and the utility of this action (central column) for a given target (right column). The second row of the grid contains the corresponding information with respect to the then-clause of the conditional.

The set of all agents is denoted by A. By convention, the agent who states the conditional is denoted by s (for 'speaker'), the agent at whom the conditional is directed is denoted by h (for 'hearer'), and e (for 'someone else') denotes an agent who is neither the speaker nor the hearer. When p or q is not an action that can be taken by an intentional agent but is rather an event or a state of the world, it is noted as being undertaken by a special, neutral agent ω. The agent ω can be thought as 'the world' or the body of laws that govern the world. Finally, utility is represented in the grid by its sign: u and u' take their values from $\{-, 0, +\}$, where $-$ and $+$ respectively stand for any significantly negative and positive values. Note that $u = 0$ means that action p is not known to have any utility for any agent. By convention, such an action has the whole set of agents A as a target.

A statement like 'If Andrew sends his invoice tomorrow, he will receive $3,000' would typically receive the following utility grid:

$$\begin{Bmatrix} e & 0 & A \\ \omega & + & e \end{Bmatrix}$$

This grid means that if agent e (here, Andrew) takes action p which does not have any obvious intrinsic utility, an event q will occur which has positive utility for the same agent e. To illustrate the power of the utility grid notation, let us consider a diverse set of arguments from consequences, expressed as utility conditionals:

(4)
a. If you testify against me, you will have an accident.
b. If I study instead of partying, I will get good grades.
c. If she moves to Paris, I'll be unhappy.
d. If I let you go away with this, my boss will fire me.

These four examples would receive these four utility grids:

$$\begin{Bmatrix} h & - & s \\ \omega & - & h \end{Bmatrix}$$

$$\begin{Bmatrix} s & - & s \\ \omega & + & s \end{Bmatrix}$$

$$\begin{Bmatrix} e & 0 & A \\ \omega & - & s \end{Bmatrix}$$

$$\begin{Bmatrix} s & + & h \\ e & - & s \end{Bmatrix}$$

Once a utility conditional has been translated to an utility grid, the theory makes use of folk axioms of decision to predict the inferences it affords. These folk axioms correspond to heuristics that people use to predict the behavior of other individuals on the basis of arguments from consequences. For example, the folk axiom of self-interested behavior states that agents tend to take actions that increase their utility, and do not take actions that decrease their utility.

Accordingly, the following grids (where the black dot stands for any legitimate value of the parameter) afford the inference that agent x will take action p:

$$\begin{Bmatrix} x & + & x \\ . & . & . \end{Bmatrix}$$

$$\begin{Bmatrix} x & . & . \\ . & + & x \end{Bmatrix}$$

whereas the following grids afford the inference that agent x will not do p:

$$\begin{Bmatrix} x & - & x \\ . & . & . \end{Bmatrix}$$

$$\begin{Bmatrix} x & . & . \\ . & - & x \end{Bmatrix}$$

To give another example, the folk axiom of limited altruism states that people take actions that increase the utility of other agents, as long as these actions do not decrease their own utility (and *mutatis mutandis*, they do not take actions that would decrease the utility of other agents, unless these actions would increase their own utility). Accordingly, the following grid is one among others which affords the inference that agent x will take action p:

$$\begin{Bmatrix} x & 0 & A \\ \omega & + & y \end{Bmatrix}$$

Looking back at Examples (4-a-d), the folk axioms of self-interested behavior and of limited altruism, applied to the four utility grids of the four statements would predict the following inferences:

(5)
a. The hearer will not testify.
b. The speaker will study unless he or she finds positive utility in partying.
c. She will not move to Paris.
d. The speaker will not let the hearer go away.

The theory of utility conditionals thus relies on a representational tool (the utility grid) on which folk axioms of decisions are applied to predict the inferences that people will derive from an argument from consequences. The theory accounts for the results of reasoning experiments featuring arguments from consequences (Bonnefon & Hilton, 2004; Corner et al., 2011; Evans et al., 2008; Ohm & Thompson, 2004, 2006), and it has been further tested on randomly generated statements whose utility grids were not found in previous experiments (Bonnefon, 2012). Consider for example the following statement, which is arguably weird but nevertheless corresponds to a well-formed utility grid:

(6)
If I do this, you will hurt yourself.

$$\left\{ \begin{matrix} s & 0 & A \\ h & - & h \end{matrix} \right\}$$

The folk axiom of limited altruism applies to this grid and predicts the inference that the speaker will not do 'this'. And indeed, reasoners gave a -2.7 rating (on a scale from -5 to $+5$) for the likelihood that the speaker would do 'this' (Bonnefon, 2012). Other statements left reasoners puzzled, for example:

(7)
If you hurt me I will help Luis.

Example (7) seems to have some sort of Necker cube quality, flipping between two possible interpretations depending on how utility are assigned to the various agents. Some people seem to interpret it as a threat (You don't want me to help your enemy Luis, do you? Then don't hurt me), whereas other people seem to interpret it as a promise (I want you to hurt me so I'm bribing you with a promise to help your pal Luis). This anecdotal observation is interesting because it suggests that people have a tendency to rearrange utilities so that unusual arguments from consequences fall within familiar categories. In the next section, I will consider evidence for this claim, and two examples of the utility templates that seem to guide the interpretation of ambiguous arguments from consequences.

3. Utility Templates

Bonnefon, Haigh, and Stewart (2013) hypothesized that some utility grids would act as templates for the interpretation of ambiguous or incomplete utility conditionals. These templates would correspond to special configurations of the grid that guide and constrain the interpretation of conditional statements. The first candidate for this template status is the social contract grid:

$$\begin{Bmatrix} x & u & y \\ y & u & x \end{Bmatrix}$$

This grid denotes that if agent x takes action p with utility u to agent y, then y will take action q which has the same utility grid (or more generally utility of the same sign) for agent x. For example:

(8)
a. If you vote for me, I will reward you.
b. If he hurts her, she will take revenge on him.

Specialists of reasoning have long observed that individuals are exquisitely sensitive to these social contracts and the inferences the afford (Haigh, Stewart, Wood, & Connell, 2011; Hilton, Kemmelmeier, & Bonnefon, 2005; Legrenzi, Politzer, & Girotto, 1996; Perham & Oaksford, 2005; Politzer & Nguyen-Xuan, 1992). Various authors have attributed this sensitivity either to repeated exposure (Cheng & Holyoak, 1985) or to an innate cheater detection algorithm (Cosmides, Barrett, & Tooby, 2010). As a result of this sensitivity, Bonnefon *et al.* (2013) suggested that reasoners would give priority to the social contract grid when interpreting or re-interpreting conditionals whose grid is close enough to that of a social contract.

Consider for example the following statement, which features an unknown verb whose valence is ambiguous:

(9)
If Peyton votes for Jesse, then Jesse will yorb Peyton.

Example (9) would correspond to the following utility grid, where the utility of the then-clause is unknown:

$$\begin{Bmatrix} x & + & y \\ y & ? & x \end{Bmatrix}$$

If reasoners apply the social contract template to the ambiguous statement (9), they should infer that there is positive utility for Peyton to be yorbed by Jesse, or, ore informally, that Peyton likes to be yorbed by Jesse. The same should apply to premises such as (10), featuring two nonverbs of which one is disambiguated in the premises:

(10)
If Peyton tymps Jesse, then Jesse will yorb Peyton. Jesse likes to be tymped.

Mutatis mutandis, similar predictions can be derived from statements in which one verb has negative valence, semantically or by disambiguation:

(11)
a. If Peyton harms Jesse, then Jesse will yorb Peyton.
b. If Peyton tymps Jesse, then Jesse will yorb Peyton. Jesse dislikes to be tymped.

In both cases, reasoners who apply the social contract template should infer that Peyton dislikes to be yorbed. And indeed, two experiments offered support for all these predictions.

An additional and stronger prediction is that reasoners might reinterpret the valence of known verbs, if a non-verb is disambiguated so that the resulting grid violates the social contract template. For example:

(12)
a. If Peyton harms Jesse, Jesse will zim Peyton.
b. Peyton likes to be zimmed.

If 'harming' is understood as carrying negative utility, as it typically is, then the utility grid of (12-a) should be the following, in line with the information in (12-b):

$$\left\{ \begin{matrix} x & - & y \\ y & + & x \end{matrix} \right\}$$

If reasoners are drawn to the social contract template, they should attempt to reinterpret (12) in a way that would fit a social contract. Given that the interpretation of 'zimming' is fixed by (12-b), the only degree of freedom left for shoehorning (12-a) in a social contract template is to reinterpret 'being harmed' as having positive utility for Jesse. And again, this is what was observed in an additional study by Bonnefon et al. (2013).

In addition to the social contract template, Bonnefon et al. (2013) identified (and verified) three other templates, suggested by a completion study in which participants had to fill conditional fragments such as:

(13)
If Peyton harms Jesse, then . . .

Participants often completed such fragments in such a way to make them social contracts (11% of completions), but other completions were frequent enough to be considered as potential templates. One of these completions, which was dubbed the Justice template (13% of completions), correspond to the following grid:

$$\left\{ \begin{matrix} x & u & y \\ \omega & u & x \end{matrix} \right\}$$

which would typically be expressed as:

(14)
a. If Peyton harms Jesse, then Peyton will be punished.
b. If Peyton helps Jesse, then Peyton will be rewarded.

The justice template seems to fit with a Just World intuition, according to which good things happen to people who do good, and bad things happen to people who do harm (Hafer & Begue, 2005). The justice template was tested just as the social contract was tested, and these tests revealed comparable interpretation and reinterpretation effects.

To sum up what we have covered so far: arguments from consequences can be framed as utility conditionals, and thus represented as utility grids. Reasoners hold utility templates that guide their interpretation toward some salient utility grid, when the utility of the if-clause or the then-clause is ambiguous. Once reasoners have settled on a utility grid, they apply folk axioms of decision (i.e., the folk axiom of self-interested behavior, or the folk axiom of limited altruism) to predict the behavior of the agents featured in the grid. An important twist on these results, though, is that utility conditionals can be ambiguous beyond the utility of their clauses: Another ambiguity has to do with their underlying causal structure, a problem that I address in the next section.

4. Utility and Causal Structure

Even when utilities can be unambiguously assigned to the clauses of an argument from consequences, another factor can complicate its interpretation, and affect the inferences it affords. Specifically, arguments from consequences expressed as utility conditionals can be ambiguous with respect to their underlying causal structure. Although conditional statements can support various causal structures (Dancygier, 1998; Declerck & Reed, 2001), we are chiefly interested in the contrast between causal and diagnostic conditionals. Consider for example:

(15)
a. If there is fire then there is smoke.
b. If there is smoke then there is fire.

Example (15-a) is a causal conditional, in which the if-clause is a cause of the then-clause. In contrast, Example (15-b) is a diagnostic conditional, in which the if-clause is diagnostic of the then-clause. Inferences made from causal and diagnostic conditionals generally track their underlying causal structure (e.g., Ali, Chater, & Oaksford, 2011; Politzer & Bonnefon, 2006), and causal structure can also affect the inferences afforded by an argument from consequences, when this argument is expressed as a conditional. Consider for example:

(16)
If Corey accepts this deal, then he's rich.

This conditional would receive the following utility grid:

$$\begin{Bmatrix} x & 0 & A \\ \omega & + & x \end{Bmatrix}$$

This grid expresses the fact that if Corey takes an action which has no apparent intrinsic utility, a state of affairs will occur that has positive utility for Corey. Applied to that grid, the folk axiom of self-interested behavior allows the prediction that Corey will accept the deal. But note that this application of the folk axiom of self-interested behavior assumes that Example (16) expressed a causal relation between accepting the deal and becoming rich: presumably, accepting the deal would cause Corey to become rich because it is a very good deal.

Example (16), though, may very well express a diagnostic relation. Accord- ing to this diagnostic interpretation, Corey would have to be rich to accept the deal (say, because the entry cost in the deal is huge), and thus the fact that Corey would accept the deal would indicate that he's rich, rather than causing him to become rich. In that case, the folk axiom of self-interested behavior is silent about whether Corey will accept the deal, because there is no clear advance in self-interest linked to accepting the deal. Note that a similar analysis applies to Example (17), which features a negative utility then-clause:

(17)
If Corey accepts this deal, then he's ruined.

Once again, accepting the deal may cause the ruin of Corey (because the deal is very bad), or it may simply be diagnostic of the ruin of Corey (say, because the deal only appeals to desperate persons). And once more, the folk axiom of self-interested behavior applies under the first interpretation (affording the prediction that Corey will not accept the deal) – but it is silent under the second interpretation.

In sum, some utility conditionals are ambiguous with respect to their underlying causal structure: They can be interpreted causally, making them genuine arguments from consequences, or they can be interpreted diagnostically, in which case it is not clear what reasoners will do. Will reasoners track the causal structure of a utility conditional, and refrain from drawing conclusion when it is diagnostic? Or will they rely on the surface structure of the conditional, and use it as any other argument from consequences, regardless of its underlying causal structure?

In several experiments, Bonnefon and Sloman (2013) investigated this question and provided evidence that reasoners were sensitive to the underlying causal structure of utility conditionals. These experiments featured ambiguous conditionals such as (16) or (17), and used different disambiguating manipulations. For example:

(18)
a. If he accepts this deal, then he's ruined. (Because it's a very bad deal.) Is he going to accept the deal?
b. If he accepts this deal, then he's ruined. (Because the deal only appeal to desperate persons.) Is he going to accept the deal?

Or:

(19)
a. The fact that he buys this house would make him rich as a result. In other words: If he buys this house, then he is rich. Is he going to buy the house?
b. The fact that he buys this house would indicate that he is rich. In other words: If he buys this house, then he is rich. Is he going to buy the house?

Responses to these questions tracked the underlying causal structure of the causal conditional, and not its surface structure. That is, reasoners were highly likely to agree with the conclusion for (18-a) and (19-a), and much less likely to do so for (18-b) and (19-b). Accordingly, inferences derived from utility conditionals cannot be predicted on the sole basis of the utilities of the if-clause and the then-clause: Causal structure needs be incorporated in the utility grid. Bonnefon and Sloman (2013) suggested one such modification to the utility grid format, which would make the top and bottom rows refer to the cause and the effect, rather than to the if-clause and the then clause. Consider again the example:

(20)
If Corey accepts this deal, then he's ruined.

The original version of the utility grid theory (Bonnefon, 2009) was insensitive to causal structure, and would have assigned the following grid to (20), whatever its underlying causal structure:

$$\begin{Bmatrix} x & 0 & A \\ \omega & - & x \end{Bmatrix}$$

The updated version of the theory would assign one of two possible grids to (20), depending on whether it is causal (top) or diagnostic (bottom):

$$\begin{Bmatrix} x & 0 & A \\ \omega & - & x \end{Bmatrix}$$

$$\begin{Bmatrix} \omega & - & x \\ x & 0 & A \end{Bmatrix}$$

Folk axioms of decision can then be applied to the retained grid, just as they were in the original version of the theory. The folk axiom of self-interested behavior applies to the grid on the left (affording the conclusion that Corey will not accept the deal), but not to the grid on the right (thus not affording any conclusion). This modification to the utility grid format can thus adequately address situations in which the meaning of a sentence can be that of a genuine argument from consequences, or that of a merely diagnostic observation.

Now that we have addressed the various problems that can complicate the interpretation of an argument from consequences, we can turn to situations in which the argument is clearly spelled and readily affords an inference. Most of these situations will fall under the basic framework already summarized in the Section Utility Conditionals. Some situations, though, call for closer scrutiny. In the rest of this chapter, I will discuss two such special situations. First, we will consider how people process situations in which a character's actions have consequences for other individuals besides himself or herself, which can conflict with her own personal utility. Second, we will look at how people process and explain situations in which a character acts against her best-interest.

5. Conflicting Utilities

So far we only considered situations in which a single agent acts on the basis of her personal preferences. A more complicated situation arises when the action of one agent have consequences for other people besides herself, especially when consequences for the self and consequences for others have the same valence.

Let us consider a situation in which the actor can decide to take an action which has either positive or negative consequences for herself, and at the same time positive or negative consequences for a recipient. If reasoners expect people to act on the basis of their material self-interest, then they should predict the actor to take the action that maximizes positive consequences for herself, regardless of the consequences for the recipient. Although this view has had its proponents (Kruger & Gilovich, 1999; Miller, 1999), it is at odds with one of the basic results of behavioral economics, according to which people care about the material interests of others, and expect others to feel the same (Cooper & Kagel, in press).

Another possibility formulated by De Vito and Bonnefon (2014) is that people expect others to maximize hedonic rather than material utility. That is, they expect others to take actions that maximize their positive affect, rather than their material benefits. If they expect another individual to feel bad about a selfish action, to the extent that these negative feelings surpass the positive feelings produced by material benefits, then they should expect this actor not to take the selfish action. This analysis extends to all other situations – for example, if a reasoner expect an actor to feel good about being altruist, to the extent that these positive feelings exceed the negative feelings produced by the material cost of altruism, then they should expect the actor to take the altruist action. Note that this is only selfishness under another guise. Indeed, according to this account, actors do not care about the feelings of the recipient: they simply maximize their own positive feelings, regardless of the feelings of the recipient.

To test this hypothesis, De Vito and Bonnefon (2014) presented reasoners with four types of vignettes in which an actor could take an action that had either positive or negative material consequences for herself, and either positive or negative material consequences for the recipient. Just like in the other experiments reviewed so far, reasoners had to predict whether the actor would take the action. Critically, they then responded to an exhaustive inventory of questions about the emotions that the actor would feel after taking or not taking the action, and the emotions that the recipient would feel if the action was taken or not taken. Based on these 32 questions about emotions, it was possible to compute for each situations, the net expected hedonic utility of the action for the actor, and the net expected hedonic utility of the action for the recipient.

Results wholly supported the predictions. The behavior of the actor could not be predicted by simple material utility maximization, but was accurately predicted by the net expected affect of the action for the actor. Furthermore, the behavior of the actor was uniquely predicted by her own affect, independently of the expected affect of the recipient. In other words, when an argument from consequences features both an actor and another individual (the recipient), people expect the actor to selfishly maximize her own positive emotions, which may or may not coincide with the material utility of the action, regardless of the material or affective consequences of the action for the recipient.

6. Utility Violations

The research I reviewed so far firmly established that reasoners draw inferences from arguments from consequences, in particular when they are expressed as utility conditionals. For example, given (21), reasoners infer that Alice, as a self-interested agent, will file her taxes online:

(21)
If Alice files her taxes online, she will save 300 euros.

But how do they react if they discover that Alice did not in fact file her taxes online? That is, how do people process information which contradicts the expectations they formed from an argument from consequences?

There are two ways to address these questions, which I'll elaborate in this section. Firstly, we can simply ask people for their explanation of Alice's behavior. This I will call the 'offline' method, because it does not track the cognitive processes at work, only their output. Secondly, we can monitor cues to cognitive processing (and specifically, eye movements) at the very moment people discover Alice's unexpected behavior – this I will call the 'online' method.

Bonnefon, Girotto, and Legrenzi (2012) used the offline method to investigate how reasoners would explain an action that contradicted the inferences they had derived from a statement like (21). They focused on what appeared to be the two most plausible explanations, one based on the beliefs of the agent, and the other based on the preferences of the agent. More specifically, they argued that to infer that 'Alice will file her taxes online' from (21) required two assumptions: (a) that Alice knows about the opportunity to save 300 euros, and (b) that Alice cares about saving 300 euros. If one of this assumptions is incorrect, then there is no ground to predict that Alice will file her taxes online. Accordingly, they expected reasoners to revise at least one of these two assumptions when learning that Alice did not file her taxes online.

To test this hypothesis, Bonnefon *et al.* (2012) assigned one control group of participants to a standard utility conditional reasoning task, and another group to a modified task that included information about the unexpected behavior. That is, participants in the control group read conditionals similar to (21), were asked whether Alice was going to file her taxes online, and rated how probable it was that Alice knew about this opportunity and how probable it was that she cared. The conditionals featured various actions and the money that could be saved ranged from 40 to 320 euros. Participants in the other group read the same conditional statements, but were informed that Alice did not take the action that would have saved her money. They then rated how probable it was that Alice knew about the opportunity, and how probable it was that she cared.

The key question in this experiment was how participants in the second group would perceive the knowledge and preferences of Alice, as compared to the participants in the first group. The response depends in part on how much Alice could save. For the minimal amount (40 euros), participants typically considered that Alice knew she could save money, but could not be bothered to do so. For any other amount, participants either revised their belief that Alice knew about the opportunity, or their belief that she cared. There was a significant negative correlation ($r = -.54$) between the two revisions, suggesting that the more participants adopted one of these explanations, the less they adopted the other – as would be predicted by a strategy of minimal belief revision.

In sum, an experiment using the 'offline' method (i.e., simply asking participants about their explanation of Alice's nonconsequential behavior) suggested that people do not have a clear preference for either explanations based on beliefs, or explanations based on desire – although they may have a preference for the latter when the utility involved is small. To achieve a more subtle understanding of how people respond to utility violations, it is necessary to adopt an 'online' method, which tracks the cognitive processing of individuals when they first read about the nonconsequential behavior. This is the strategy adopted by Haigh and Bonnefon (2015), who recorded the eye movements of participants reading stories of nonconsequential behavior.

Haigh and Bonnefon (2015) presented participants with vignettes in which a character did not take a beneficial action, systematically manipulating whether the character knew or cared about this opportunity. For example, in the following vignette, Alice knows she can save money by renewing her insurance policy online, and cares about saving that money (emphasis and structure added):

(22)

a. Alice needed to renew her car insurance before it expired.
b. She knew that if she renewed over the Internet she would save £100.
c. Such a saving was important as she was struggling financially and desperately needed to save money.
d. After gathering together the relevant documents she renewed her policy over the phone.
e. The call lasted nearly half an hour.

In other versions of the vignette, Alice did not know about this opportunity: (22-b) was replaced with 'She didn't know that if she renewed over the Internet she would save £100.' Still in other vignettes, Alice did not care about this opportunity: (22-c) was replaced with: 'However, because she was very wealthy such a saving was not important to her.' Accordingly, four different versions were constructed for each vignette.

The key idea in this study was to record eye movements in order to capture any disruptions of reading when participants reached the critical, emphasized portion of text in (22-d), which describes the nonconsequential behavior of Alice. The conditions in which reading is disrupted are informative with respect to whether rapid inferences about the behavior of Alice are driven by her desires, her beliefs, or both. For example, if reading was disrupted in conditions where Alice knows she can save money, regardless of her desire to do so, we would be in a position to conclude that online inferences are mostly driven by belief, rather than desire.

Results suggested that inferences were initially driven by desire alone, and only then moderated by belief. When readers reached the critical region in (22-d), the likelihood that they would look back to an earlier region of text (a disruption measure known as First Pass Regression Out) was significantly impacted by whether Alice desired to save money, but not by whether she knew she could. However, the total time readers spent on the critical region before moving on (a disruption measure known as Regression Path Time) was impacted by the interaction of belief and desire. That is, participants took longer to move on when Alice's behavior contradicted her preferences, but only when she knew about the consequences of her behavior. These two effects suggest that readers made quick online inferences about the behavior of Alice based on her desires alone, but were then able to incorporate the beliefs of Alice in their processing, if they had to make sense of her nonconsequential behavior.

Overall, offline and online studies of reasoners' reaction to utility violations suggest that the utility component of an argument from consequences (is the action beneficial to an agent?) is more cognitively salient than its knowledge component (does the agent know about the beneficial consequences?). Not only is the utility component processed first (as per the online results), but it can also weigh heavier in explaining nonconsequential behavior (as per the offline results). These results resonate with findings in both the reasoning and the persuasion literature (Corner *et al.*, 2011; Evans *et al.*, 2008; Hoeken *et al.*, 2012) suggesting that when presented with an argument from consequences, people are more sensitive to differences in the desirability of consequences, than to differences in their likelihood.

7. Conclusion

People use arguments from consequences in order to convince others to take a course of action, but also to predict which course of action others will take. In this chapter, I have reviewed the work that my colleagues and I have conducted on this latter, predictive use of arguments from consequences. I believe this work to open new perspectives for argumentation research, by establishing a link between argumentation and theory of mind (i.e., mentalizing, or reasoning about the beliefs, desires and intentions of other individuals). Indeed, the arguments (from consequences) we can use to argue for a course of action, can also be used to predict that an individual will take this course of action, or to explain that an individual took this course of action. This dual nature of arguments from consequences naturally places them at the intersection of argumentation, reasoning, and theory of mind.

References

Ali, N., Chater, N., & Oaksford, M. (2011). The mental representation of causal conditional reasoning: mental models or causal models. *Cognition, 119*, 403–418.

Bonnefon, J. F. (2009). A theory of utility conditionals: Paralogical reasoning from decision-theoretic leakage. *Psychological Review, 116*, 888–907.

Bonnefon, J. F. (2012). Utility conditionals as consequential arguments: A random sampling experiment. *Thinking and Reasoning, 18*, 379–393.

Bonnefon, J. F., Girotto, V., & Legrenzi, P. (2012). The psychology of reasoning about preferences and unconsequential decisions. *Synthese, 187*, 27-41.

Bonnefon, J. F., Haigh, M., & Stewart, A. J. (2013). Utility templates for the interpretation of conditional statements. *Journal of Memory & Language, 68*, 350–361.

Bonnefon, J. F., & Hilton, D. J. (2004). Consequential conditionals: Invited and suppressed inferences from valued outcomes. *Journal of Experimental Psychology: Learning, Memory, and Cognition, 30*, 28–37.

Bonnefon, J. F., & Sloman, S. A. (2013). The causal structure of utility conditionals. *Cognitive Science, 37*, 193–209.

Cheng, P. W., & Holyoak, K. J. (1985). Pragmatic reasoning schemas. *Cognitive Psychology, 17*, 391–416.

Cooper, D., & Kagel, J. (in press). Other regarding preferences: A survey of experimental results. In J. Kagel & A. Roth (Eds.), *The handbook of experimental economics*. Princeton, NJ: Princeton University Press.

Corner, A., Hahn, U., & Oaksford, M. (2011). The psychological mechanism of the slippery slope argument. *Journal of Memory and Language, 64*, 153-170.

Cosmides, L., Barrett, H. C., & Tooby, J. (2010). Adaptive specializations, social exchange, and the evolution of human intelligence. *Proceedings of the National Academy of Science of the USA, 107*, 9007–9014.

Dancygier, B. (1998). *Conditionals and prediction. Time, knowledge, and causation in conditional constructions*. Cambridge: Cambridge University Press.

De Vito, S., & Bonnefon, J. F. (2014). People believe each other to be selfish hedonic maximizers. *Psychonomic Bulletin & Review, 21*, 1331–1338.

Declerck, R., & Reed, S. (2001). *Conditionals. A comprehensive empirical analysis*. Berlin and New York: Mouton de Gruyter.

Evans, J. S. B. T., Neilens, H., Handley, S. J., & Over, D. E. (2008). When can we say 'if'? *Cognition, 108*, 100–116.

Feteris, E. T. (2002). A pragma-dialectical approach of the analysis and evaluation of pragmatic argumentation in a legal context. *Argumentation, 16*, 349–367.

Hafer, C. L., & Begue, L. (2005). Experimental research on just-world theory: Problems, developments, and future challenges. *Psychological Bulletin, 131*, 128–167.

Haigh, M., & Bonnefon, J. F. (2015). Eye movements reveal how readers infer intentions from the beliefs and desires of others. *Experimental Psychology, 62*, 206-213.

Haigh, M., Stewart, A. J., Wood, J. S., & Connell, L. (2011). Conditional advice and inducements: Are readers sensitive to implicit speech acts during comprehension? *Acta Psychologica, 136*, 419–424.

Hilton, D. J., Kemmelmeier, M., & Bonnefon, J. F. (2005). Putting ifs to work: Goal-based relevance in conditional directives. *Journal of Experimental Psychology: General, 135*, 388–405.

Hoeken, H., Timmers, R., & Schellens, P. J. (2012). Arguing about desirable consequences: What constitutes a convincing argument? *Thinking and Reasoning, 18*, 394–416.

Kruger, J., & Gilovich, T. (1999). 'Naive cynicism' in everyday theories of responsibility assessment: On biased assumptions of bias. *Journal of Personality and Social Psychology, 76*, 743–753.

Legrenzi, P., Politzer, G., & Girotto, V. (1996). Contract proposals: a sketch of a grammar. *Theory & Psychology, 6*, 247–265.

Miller, D. T. (1999). The norm of self-interest. *American Psychologist, 54*, 1053–1060.

Ohm, E., & Thompson, V. (2004). Everyday reasoning with inducements and advice. *Thinking and Reasoning, 10*, 241–272.

Ohm, E., & Thompson, V. (2006). Conditional probability and pragmatic conditionals: Dissociating truth and effectiveness. *Thinking and Reasoning, 12*, 257–280.

Perham, N. R., & Oaksford, M. (2005). Deontic reasoning with emotional content: Evolutionary psychology or decision theory? *Cognitive Science, 29*, 681–718.

Politzer, G., & Bonnefon, J. F. (2006). Two varieties of conditionals and two kinds of defeaters help reveal two fundamental types of reasoning. *Mind and Language, 21*, 484–503.

Politzer, G., & Nguyen-Xuan, A. (1992). Reasoning about conditional promises and warnings: Darwinian algorithms, mental models, relevance judgements or pragmatic schemas? *Quarterly Journal of Experimental Psychology, 44*, 401–412.

Schellens, P. J., & De Jong, M. (2004). Argumentation schemes in persuasive brochures. *Argumentation, 18*, 295-323.

Thompson, V. A., Evans, J. S. B. T., & Handley, S. J. (2005). Persuading and dissuading by conditional argument. *Journal of Memory and Language, 53*, 238–257.

Chapter 2

Commitment Attribution and the Reconstruction of Arguments[1]

Steve Oswald

University of Fribourg, Fribourg, Switzerland, steve.oswald@unifr.ch

Abstract. The notion of commitment has been shown to be pivotal in the study of argumentation (see e.g. Hamblin 1970, Walton & Krabbe 1995), in particular as far as argumentative reconstruction is concerned. In this chapter I will consider how the notion of *commitment attribution*, as developed in cognitive pragmatic research (Morency *et al.* 2008, Saussure & Oswald, 2008, 2009), can contribute clear criteria to be used in an analytical task argumentation theorists regularly have to undertake, namely the identification of missing or unexpressed premises.

1. Introduction

The notion of commitment is one of the fundamental concepts argumentation theorists rely on as they engage in the reconstruction of argumentative exchanges. Although definitional consensus has not been reached by linguists, argumentation theorists and philosophers of language across scientific communities (see Boulat, 2014), one of the recurring features of commitment that emerges seems to be its attitudinal import. That is, commitment denotes a specific attitude of the speaker's towards the content of what she has uttered. Extant discussions of commitment touch upon parallel and neighboring notions such as endorsement and involvement (see e.g. Katriel & Dascal, 1989), responsibility (Nølke *et al.*, 2004), or, in the francophone tradition, *prise en charge* (Culioli 1971). While these different notions have been distinguished and to a fair extent theorized (see Dendale & Coltier, 2011), they all seem to converge on one specific point, namely a construal of commitment in terms of – or in relation to – belief and truth (see Boulat, 2014).

In this chapter I adopt a restricted and moderately informal definition of commitment and construe it, following Katriel & Dascal, as a content "which the speaker can be said to have 'taken for granted' in making his or her utterance" (Katriel & Dascal, 1989, p. 286)[2].

---

[1] I would like to thank Michael Baumtrog for his valuable input on form- and content-related issues on a previous version of this work. Remaining mistakes are my own.

[2] An anonymous reviewer points out that this definition seems to leave out commitment to explicit contents, in the sense that one could hardly consider that a speaker takes for granted what he utters, to the extent that he needs to utter it. It would indeed make sense to broaden the definition so that it can also include explicit contents. In this paper, I will thus consider that what speakers take for granted include the explicit meaning of their utterances, and would accordingly adapt Katriel & Dascal's definition to stipulate that commitment is what speakers take for

Here the intimate and private nature of commitment as a psychological notion will matter less than its public appraisal in verbal exchanges – and the cognitive mechanisms of interpretation underlying this appraisal. In other words, I will be concerned with the public facet of commitment in discourse, which will lead me to consider the issue from *the perspective of the addressee*.

Commitment is fundamental in argumentative exchanges because it allows conversational participants to keep track of each other's arguments, positions, standpoints – i.e., of each other's performance of relevant speech acts (cf. Hamblin's *commitment stores*, 1970). This means that at the level of argumentative conversation, commitment scrutiny is both what allows participants to keep track of what they have previously said (or implied) and, as a consequence, endorsed, and what allows them to proceed with the exchange. Since I target these very processes, it is only natural for me to be interested in processes of *commitment attribution* (Morency et al., 2008) which are taken to characterize the *naïve* or *pre-theoretical* appraisal of commitment by conversational participants. In a nutshell, I will be concerned with what speakers *mean* when they argue and predominantly with *how* their addressees come to recognize what speakers mean as they utter it, which is a prerequisite for the mere possibility of conducting an argumentative exchange in the first place.

The reason these mechanisms should be of prime concern for argumentation theorists lies in the necessity of being able to assess the reality – in terms of actual meaningfulness – of a given argumentative exchange with clear-cut criteria, in order to minimize the chances of over-, or misinterpretation, as much as possible. This chapter accordingly tackles the naïve more than the normative interpretation of argumentative discourse and provides methodological guidelines to achieve plausible argumentative reconstructions. In particular, the hypothesis that will be defended is that the selection difficulties posed by the reconstruction of implicit premises can be overcome by relying on a cognitive model of human communication with clear comparative criteria.

2. Commitment, Meaning and Commitment Attribution

Walton and Krabbe characterize commitments as follows: "one's commitments are personal – that is, indexed to a distinct person, or individual – and they may even be, in some cases, private and only partially accessible to others" (1995, p. 14). For Walton and Krabbe, the personal nature of commitments does not make them psychological in nature (on this point, see also Paglieri, 2010); besides, even if we were to construe commitment as a psychological state denoting a relationship between an attitude of the speaker's and a propositional content, its assessment in communication would still remain an important feature of the success of communicative exchanges. By this I mean that commitment bears a strong relationship with (speaker) meaning. If, like Grice (1989), we construe communication as a successful exchange of meaning, then we also need to accept that communication is successful only when the addressee has understood speaker meaning. When an addressee has understood the speaker's utterance, he will by default consider that the output of his comprehension procedure – i.e. the proposition(s) he identified as speaker meaning – corresponds to what the speaker has actually meant by her utterance. In other words, understanding comes with the usually implicit and intuitive judgement, on behalf of the addressee, that the speaker has meant precisely what he has understood. I would like to argue that such judgement is the attribution of commitment: when an addressee pairs the

---

granted in making their utterances and which they cannot retract without causing semantic of pragmatic inconsistencies.

content he derives with speaker meaning, he is ipso facto attributing the speaker a commitment to that content. This is why comprehension straightforwardly involves commitment attribution. Of course, this does not mean that addressees never fail in the process: misunderstandings do occur for a variety of reasons. Interestingly, such cases can also be characterized in terms of commitment misattribution, in which a mismatch occurs between what the speaker means and what the addressee takes her to mean. If an addressee fails to properly recover speaker meaning, he will take the speaker to be committed to a set of propositions she is not actually committed to. Assuming the parallel between interpretation/misinterpretation and commitment attribution/misattribution is relevant, we can hypothesize that commitment is intimately linked to meaning. In principle, therefore, this would warrant the inclusion of a pragmatic theory of interpretation, concerned with the study of contextualized meaning, in any argumentative theory interested in commitment.

As they engage in communicative exchanges, conversational participants formulate utterances loaded with a variety of indicators meant to facilitate the addressee's recovery of speaker meaning. The semantic (relative) stability of the linguistic code is trivially a highly reliable tool for meaning encoding, as literal and explicit meaning is usually unequivocal – provided, of course, the speaker has opted for the linguistic units that best convey her communicative and informative intentions. In turn, the relative transparency of explicit and literal meaning allows for straightforward and unproblematic commitment attribution. The problems raised by Moorean utterances such as (1) are a good indication that such is the case, since their inherent contradiction seems to rest precisely on the impossibility of asserting a state of affairs and simultaneously denying being committed to the truth of the propositional content used to express that state of affairs:

(1)   ?? Laszlo is home but I do not believe that Laszlo is home.

Understanding assertions thus goes hand in hand with the possibility of attributing the speaker a commitment to the propositional content.[3] While we could thus say commitment attribution to explicit contents is unproblematic, safe and automatic (judging by the difficulties linked to the retraction of commitment from an assertion as illustrated above), the story might be different for commitment attribution to implicit contents, to which human communication very often resorts (e.g., humor, politeness, irony, metaphor, and figurative language more broadly). Although commitment attribution may be trickier in those cases, if only because the addressee is largely responsible for the identification of implicit contextual assumptions that are necessary for the relevant interpretation of speaker meaning, it remains a necessary step of the process. When the addressee infers irony or humor, usually he does so assuming that it was meant as such by the speaker. In other words, implicit contents also come with commitment. From the perspective of the usefulness of communication in the species, it would indeed be pointless for implicit mechanisms of communication to exist in the absence of the assumption that these also serve to intentionally convey meaning. Commitment also goes through with implicatures, for instance, even if the addressee's grounds for attributing commitment are necessarily weaker than the grounds he has for assessing the speaker's commitment to explicit contents (but this does not mean that commitment is weaker, see Morency *et al.* (2008) for a discussion). Further, just because implicatures are typically cancellable does not mean that commitment is too. If a speaker cancels an implicature, she is signaling that the implicature should not have been drawn in the first place and therefore that she is not committed to its

---

[3] There are different types of commitments, even if the view adopted here is moderately reductionist. For a detailed account, see issue 22 of the *Belgian Journal of Linguistics* (2008) and section 2 of the chapter by Morency *et al.* (2008, pp. 199-204).

content, which can be assimilated to a form of mistake acknowledgement/correction on her behalf. However, if she does not cancel the implicature, she can *ipso facto* be taken to intend the addressee to infer it, along with her commitment to its content and to her intention of having the addressee recognizing it.

From this brief discussion, it appears that even if commitment is a psychological state belonging to a set of directly unobservable objects, it can be inferred and follows from the identification of speaker meaning, regardless of whether the latter is explicit or implicit.[4] This is a direct consequence of the intentional nature of communication: speakers engaging in a communicative exchange not only intend to inform their audience of something, they also intend their audiences to recognize that they are intentionally attempting to inform them of something.[5]

If we now consider what happens in the mind of conversational participants, a corollary of these considerations is that the meaning derivation procedure provides both an interpretation of speaker meaning and a judgement on her commitment. That is, the cognitive operations that are responsible for the identification of speaker meaning are also (perhaps indirectly) responsible for the identification of speaker commitment by virtue of the intentional nature of communication. This means, from an analytical perspective, that we can gain access to mechanisms of commitment attribution by looking at meaning and at the cognitive machinery involved in its interpretation.

3. The Analysis of Argumentation

3.1. Problems in Argumentative Reconstruction

The disorderly and rather untidy format of argumentation as it occurs in natural settings can be a great obstacle for transparent and thorough analysis. For instance, arguers often fail to explicitly provide all components of their argumentations and many times leave it up to their addressees to infer the set of contextual information needed to establish the connections between explicit premises and conclusion (as is the case with enthymemes, for instance). Thus, the justificatory link is often only incompletely provided by speakers. Now, for arguers, this is not a source of great concern, as their cognitive environment contains information that is relevant to the argumentative situation they find themselves in. This, in addition to the obvious possibility of asking for more information to make sure they have understood what was meant, allows them to effortlessly and efficiently fill in the blanks or select the missing information required to make sense of the arguments offered by the speaker. For analysts, however, this may be problematic in many respects, mainly because they do not have the possibility of interacting with the speakers when the identification of speaker commitment turns out to be problematic or when they lack crucial contextual information that was easily available when the original interaction took place.

The difficulty involved in attributing commitment on the basis of incomplete argumentation is in turn problematic for at least two (interrelated) reasons: (i) analysts may fail to do justice to argumentative reality by misinterpreting arguers' utterances, and (ii) as

---

[4] Walton (1993) adopts this perspective when he positively answers the question of whether commitment is or not "an inference to be drawn from what you say and how you act when you are interacting with another participant in a social situation" (1993, p. 93). Crucially, thus, whether speakers are sincere or not is a separate question; the notion of commitment attribution as will be used here allows us to leave the question aside and to focus on the communicative features of commitment, that is, on communicated commitment.

[5] Note that this idea echoes Sperber & Wilson's distinction between informative and communicative intention (Sperber & Wilson, 1995).

a consequence, analysts may end up 'wrongly' – or at least illegitimately – evaluating the arguments they analyze, thus running the risk of committing the straw man fallacy.

One of the tasks of argumentation theorists has thus been to formulate clear-cut criteria to reconstruct plausible representations of the argumentative exchange as it has been carried out by actual participants, so that the margin of error in the reconstruction can be reduced to a minimum. One way of tackling the issue consists in finding the right balance between naïve and normative interpretations of argumentative exchanges (see Lewiński, 2012; Lewiński & Oswald, 2013; and Oswald & Lewiński, 2014). Crucially, the question of reconstructing argumentative exchanges could be formulated as follows (see also Paglieri, 2007): when there is doubt as to which contents should be identified as adequate/correct/intended in order to reconstruct the argument provided in an incomplete form (as with enthymemes), should we go for those that seem to be the ones that participants have made use of (i.e. should we find reliable ways of identifying the speakers' commitments), or should we go for those that logically make sense in the argument, with no particular regard to whether they correspond to what the arguers effectively had in mind? That is, should we consider arguments as the result of a specific intention of arguers that needs to be contextually specified, or should we consider arguments as inferences needing to comply with some normative standard? The answer to this question boils down to a choice of perspective: either (i) we decide to address what we think the arguers meant, and thus the naïve level of interpretation or (ii) we focus on the argument's value and merit and strive to make the most of it, even if we risk overlooking some crucial pragmatic aspects of argumentation. Here I will try to assess how far we can go into the development of option (i).[6]

Both logical and pragmatic considerations have been put forth, for instance by pragma-dialecticians, to provide a sound method of argumentative analysis – and reconstruction in particular: "[t]o establish precisely what someone who has advanced argumentation can be held to if the argumentative discourse is analyzed as a critical discussion, an analysis must be carried out both at a pragmatic and at a logical level" (van Eemeren & Grootendorst, 1992, p. 60). Both perspectives become complementary in the analysis and reconstruction of arguments because each of them targets a distinct aspect of argumentation: pragmatic tools (inspired mainly from Speech Act Theory, à la Austin, 1962 and Searle, 1969) allow us to come up with a sensible and plausible reconstruction meant to provide a reliable representation of arguers' respective argumentative moves (i.e. an interpretation of the linguistic material), while logical norms provide the means to reconstruct the line of reasoning of the argumentation being analyzed. According to pragma-dialecticians, the pragmatic and the logical levels of analysis are related in such a way that "[i]n practice, the logical analysis is instrumental for the pragmatic analysis" (van Eemeren & Grootendorst 1992, p. 60). In the analysis of unexpressed premises, for instance, the identification of implicit material will first apply a criterion of logical validity to single out possible candidates for the implicit premise, which will then be assessed pragmatically to single out the optimal candidate among them.

Moreover, a criterion of charity has been put forth to provide a way of settling the question when the balance between logic and pragmatics falls short of providing the optimal solution. This basically amounts to choosing, among possible alternatives, the one that favors best the arguer in terms of argumentative strength (see e.g. Snoeck Henkemans, 1992). Following Lewiński (2012) and Lewiński & Oswald (2013), however, it appears that charity should not be blind, but attuned to contextual constraints dictating the appropriate

---

[6] See Walton & Reed (2005), Lewiński (2012) and Lewiński & Oswald (2013) for alternative accounts which cater for both options. In this chapter I am however only concerned with the first layer of analysis, centered on naïve interpretation.

amount of charity to be adopted by analysts in their reconstruction. This line of argument will not be further explored here, as extant discussions on the principle of charity abound (see Lewiński 2012 for a comprehensive review) and are not directly concerned with the level of naïve interpretation I am focusing on here.

3.2. The Pragmatic Optimum

Reliance on pragmatic research in order to deal with the identification of commitments, and those concerning missing premises in particular, is not a new idea in argumentation studies (see e.g. van Eemeren & Grootendorst, 1992; Gerritsen, 2001; Becker, 2012). Pragma-dialectics has specifically elaborated a detailed procedure meant to identify unexpressed premises, namely the procedure to determine the *pragmatic optimum*.

As seen in the previous sub-section, according to pragma-dialectics, a meaningful reconstruction of arguments that have been presented in an incomplete form should be conducted in such a way as to make them logically valid. In this sense, analysts should strive for the *logical minimum* first, defined as "the premise that consists of the 'if...then' sentence that has as its antecedent the explicit premise and as its consequent the conclusion of the explicit argument" (van Eemeren & Grootendorst 1992, p. 64). In many cases, however, this yields reconstructions that do not seem to do justice to the original format of the argument or that leave out some relevant aspects of meaning (see van Eemeren & Grootendorst, 1992, pp. 60-72 for a discussion). For this reason, pragma-dialecticians have introduced the notion of *pragmatic optimum*, which they define as "the premise that makes the argument valid and also prevents a violation of (...) any other rule[s] of communication", adding that "[p]redominantly, this is a matter of generalizing the logical minimum" (van Eemeren & Grootendorst, 1992, p. 64). Assessing the pragmatic optimum in an argument, as indicated by its name, involves considering pragmatic aspects of meaning that should help the analyst figure out the contents that seem to best fit the context in which the argument took place. Also, it allows the analyst to identify unexpressed constituents of the argument that are optimally formulated to fulfil their argumentative function.

A 5-step procedure for identifying the pragmatic optimum is accordingly formulated as follows (van Eemeren & Grootendorst 1992, pp. 66-67):

(i) Determine what the argumentation is in which a premise has been left unexpressed.
(ii) Determine how well-defined the context is in which the argumentation occurs.
(iii) Determine which added premises could validate the argument underlying the argumentation.
(iv) Determine which of these added premises may, in the context at hand, be considered to be part of the commitments of the speaker.
(v) Determine which of the added premises to which the speaker is committed is the most informative in the context at hand.

Steps (i), (ii) and (iii) involve decoding and inferring the meaning of the argumentative material that has been explicitly uttered and listing the argumentative material that has potentially been left by the speaker for the addressee to infer. Step (iii) may consist in an exhaustive survey of every possible proposition plausibly related to the argument and plausibly fulfilling the role of premise, following a criterion of logical validity meant to identify all candidates that are instrumental to making the argument logically valid. Steps (iv) and (v) are the most interesting for my purposes in that they explicitly target the identification of commitments (step (iv)) and the criteria that allow us to select which ones are appropriate among those available (step (v), criterion of informativeness).

While the last two steps of the procedure tell us what to do to precisely identify the pragmatic optimum, much more can be said about *how* it should be identified, beyond the requirement of non-violation of any "rule of communication", mentioned by van Eemeren & Grootendorst (1992, p. 64), and presented as a condition for the identification of implicit contents. The rules of communication alluded to by pragma-dialecticians refer to Searle's theory of indirect speech acts (1969) and Grice's conversational maxims (Grice, 1989, p. 26-27). Yet, in their pragma-dialectical interpretation, these are presented as normative rules enjoining speakers to contribute utterances that comply with rational conversational conduct – itself instrumental to the resolution of a difference of opinion.[7] While these rules undoubtedly have clear descriptive and normative value (they provide a clear-cut norm of rationality envisaged in terms of their instrumentality to fulfil a specific goal), whether they constitute solid guidelines for the reconstruction of argumentative *meaning* remains to be seen.

The psychological reality of these rules of communication is also controversial: Grice himself did neither claim that his model was psychologically plausible nor that his 'working out schema' reflected the actual cognitive computation of implicatures. His system of maxims, together with the principle of cooperation, allows us to identify the contents that *can count as* conversational implicatures, but this does not *ipso facto* mean that conversational participants actually use these principles (Grice, 1989, p. 31). While the rules of communication as formulated in pragma-dialectics can assist the analyst in the normative reconstruction of arguments, they have little to contribute to commitment attribution at the level of naïve interpretation. Pragma-dialectics does not say much about the latter, which seems to suggest that the criteria to be used rest somewhere in our intuitions about meaning. Yet, stage (iv) in the determination of the pragmatic optimum seems to be targeted at assessing speaker meaning (i.e., prior to a normative reconstruction); the question of how this is done is left open. Additionally, since all judgements about commitment, as stipulated in stage (v), have to be contextually grounded, the method to identify commitments needs to involve some sort of operation meant to assess the adequacy of commitments relatively to the context in which they are identified. Pragma-dialectics mentions in this respect a criterion of informativeness, but what is meant by 'informativeness' remains vague and intuitive in the theory. The following proposal is accordingly meant to cognitively ground steps (iv) and (v) above.

4. A Cognitive Pragmatic Take on Interpretation – and Commitment Attribution

While pragma-dialectics gives us insights on normative reconstructions, it only provides limited insights on naïve reconstructions. The perspective defended here is thus that a meaning-informed take on commitment attribution can be instrumental to the reconstruction of arguments. Relevance Theory (RT) (Sperber & Wilson, 1995; Carston, 2002; Wilson & Sperber, 2012) represents a solid theoretical choice in this respect because it formulates precise criteria defining what it means for a language user to contextually understand any given utterance. This cognitive account of communication considers that

---

[7] These rules are formulated in order to enforce a general principle (see van Eemeren & Grootendorst, 1992, p. 50): "Be clear, honest, efficient and to the point".

understanding is identifying the information meant to be communicated by the speaker. In turn, this is defined as identifying the contextual relevance of what is communicated.

Given the resource-boundedness of our cognitive systems, we have neither unlimited time nor unlimited resources as we process verbal messages to take into account all available information in order to reach an interpretation of a speaker's utterance. We take shortcuts, mobilizing only the most relevant information. Sperber & Wilson postulate that we are naturally equipped with cost-effective means to do that. That is, whenever we interpret speaker meaning, we go straight for the interpretation that seems to be the most relevant within the context in which the utterance was uttered. A *cognitive principle of relevance* has been formulated to reflect cost-effectiveness under such constraints:

(2) Cognitive principle of relevance: "Human cognition tends to be geared to the maximization of relevance." (Wilson & Sperber, 2002, p. 254)

According to this principle, our mind is guided by considerations of relevance which determine how and to what extent resources should be allocated to the processing of any given stimulus. Applied to the case of information processing within communicative contexts, this principle is accompanied by the *communicative principle of relevance*, which is formulated as follows:

(3) Communicative principle of relevance: "Every ostensive stimulus conveys a presumption of its own optimal relevance." (Wilson & Sperber, 2002, p. 256)

Following this principle, ostensive stimuli in communication carry a presumption of relevance: if you recognize that the speaker has uttered something and moreover that she meant for you to recognize that her utterance was meant to be recognized by you as an intentional stimulus, this constitutes a trigger for you to infer whatever needs to be inferred from the utterance. Relevance, thus, is a property of utterances; the recognition of the intentional character of utterances gives your cognitive system an indication that effort should be spent in the search of the utterance's contextual relevance.

The capital contribution of RT to the study of meaning can be said to lay in its technical and precise characterization of what relevance is in cognitive terms. Relevant information (for a cognitive system) is defined along two conditions called the *extent conditions of relevance* (Sperber & Wilson, 1995, p. 125):

(4) "Extent condition 1: an assumption is relevant in a context to the extent that its contextual effects in this context are large.
(5) Extent condition 2: an assumption is relevant in a context to the extent that the effort required to process it in this context is small."

Contextual effectiveness (4) is defined in epistemic terms, through three effects generally discussed in RT: the addition of new reliable information, the suppression of old and unreliable information, and the revision of old but uncertain information already stored in the system. This means that contextual effectiveness as referred to in (4) can be characterized as instrumentality to secure a more adequate and reliable cognitive environment (the cognitive environment being defined as the set of all assumptions that are manifest to an individual, which is made of the set of assumptions that she takes to be true or probably true, see Sperber & Wilson, 1995, p. 39). In parallel, the second extent condition of relevance targets processing effort and specifies that relevant information is information that requires little processing effort to be represented. In other words, the second condition of relevance stipulates that relevant information is information that is easily accessible in the context of interpretation. Such a cost-effective construal of the cognitive mechanisms at play in interpretation yields the general assumption that the most relevant of all pieces of information are those that best satisfy the effort-effect ratio.

RT assumes *optimal* relevance to obtain when communication is successful. Given the constraints in effort and effect induced by the utterance the speaker has chosen and the communicative principle of relevance (3), the addressee is entitled to expect that the way the speaker has phrased her utterance is the best of all possible ways of phrasing it to convey exactly what she means. It follows that successful communication can be described as a situation in which the output of the comprehension procedure of a speaker's utterance, this output being defined as the set of assumptions that the utterance has contributed to the cognitive environment of the addressee, resembles the set of assumptions that the speaker meant to convey in formulating her utterance. This has strong implications for the analysis of argumentative material: it means that the contents the analyst is likely to identify, comparatively, as the most contextually relevant contents, are probably the ones the speaker has intended to go through – and also the ones the addressee considers that the speaker has intended to go through.[8]

Looking at actual interactions and taking into account their context of occurrence – context being here construed as the sets of relevant information both speaker and addressee mobilize to make sense of each other's utterances – the RT model of information processing gives us tools to plausibly assess (i) what speakers mean as they communicate and (ii) what each is likely to consider that the other has meant. In other words, once we have identified speaker meaning by evaluating which interpretations are contextually optimal along the effort-effect dimensions, we will be able to identify plausible interpretations of their utterances. To the extent that those interpretations can be taken to correspond to speaker meaning, they are *ipso facto* good candidates to represent the conversational participants' commitments.

In what follows I try to illustrate how RT can be used in argumentative reconstructions.[9] I also believe that such a model can be used for other tasks in the study of argumentation – but this merely follows from the fact that RT is a general theory of cognition and communication in particular, and that argumentation is one particular instance of communication. I have argued elsewhere (Oswald, forth.) that RT can tentatively be used for rhetorical analysis, trying to elaborate on Paglieri's intuition that "rhetorical persuasion is partially dependent on our cognitive limitations in assessing rational criteria for argument evaluation" (Paglieri, 2007, p. 5). In what follows, however, I will restrict my contribution to methodological considerations and illustrate them with a clear example in which the

---

[8] An anonymous reviewer points out that a similar argument could be made for pragma-dialectics, as it could be claimed that the contents that the pragma-dialectical analyst is likely to identify, namely those that are optimal in persuading the addressee on reasonable grounds, are probably the ones the speaker has intended to go through. While this proposal has merit, I do not think the parallel would hold, mainly due to the difference in scope of both theories. RT targets actual cognitive mechanisms, while Pragma-Dialectics neither needs nor wants to postulate the cognitive reality of the phenomena it tackles (see this idea explicitly formulated in van Eemeren & Grootendorst, 2004, p. 74). Pragma-dialectical reconstructions are tailored to fit an externalised model of argumentative interaction and inherit their plausibility from agreement with the theory's normative claims, while RT-based reconstructions, which are based on a naturalistic account of cognitive processes, inherit their plausibility from a cognitively-grounded model of human cognition which lends itself to experimental testing (see e.g. van der Henst & Sperber, 2004). While disagreement is possible with both accounts, its nature and scope will therefore differ.

[9] There have been a few attempts to integrate the insights of RT into argumentation theory in the past: Tindale (1992) argues that RT's notion of cognitive environment can provide "a framework for assessing candidates for the hidden premise" (1992, p. 185). Paglieri (2007) and Paglieri & Woods (2011) try to go beyond what RT has to offer as they focus on the virtues of parsimony to model arguers' behaviour in argumentative practices. Woods (1992) critically discusses RT's notion of relevance while recognising the merits and limitations of the theory.

choice of unexpressed premises is highly problematic for subsequent argumentative evaluation.

5. Analyzing Arguments and Determining Unexpressed Premises

5.1. A Procedure to Identify Unexpressed Premises

One might hold that the reconstruction of naïve interpretations might yield biased representations of the argumentative moves being analyzed, since intentions and commitments are private and only indirectly accessible. There is a simple way of replying to this objection, which consists in acknowledging that the analyst's results are based on her own hypotheses about meaning (i.e., on the output of her own psychological processes), and that as such they are at least as plausible as the actual arguer's own hypotheses about the particular meanings that are being exchanged in conversation. It is assumed that the analyst is a competent language user just as any actual arguer, meaning that both of them are in principle equipped with the same cognitive information processing devices. What the analyst is doing is taking an extra reflective step, which, if properly guided by a cognitive theory of communication, will allow her to assess meaning. To make an analogy with reasoning (as per Mercier & Sperber, 2009), what the analyst concerned with meaning assessment is doing is soliciting the same cognitive mechanisms as the arguer, only *reflectively*, which allows her to go beyond the intuitive representations yielded by the comprehension mechanism, whereas actual arguers usually do not need to take that step (comprehension is automatic) and thus remain within the bounds of intuitive inference. Accordingly, there is no reason to suspect that the analyst is less competent than any actual arguer in meaning assessment. In fact, analysts have more time and more resources to devote to the cognitive operations underlying comprehension, without this making them 'different' interpreters. This line of argument, which to my knowledge was first defended by Saussure (2005), legitimates recourse to a cognitive pragmatic theory in the analysis of discourse.

At this point I am ready to propose a rather simple procedure to identify unexpressed premises. A simple two-step procedure, which incorporates the assessment of relevance as defined in section 4, can be formulated to identify implicit argumentative material:

(6) Procedure to identify unexpressed premises:
    a) focus on meaning at the level of the arguers' management and exchange of meaning
    b) based on considerations of cognitive effort and effect, identify speaker meaning that is contextually relevant
        b1) take into account different candidates for the unexpressed premise and assess which one yields the best ratio between cognitive effort and contextual effect in the context of the utterance
        b2) select the optimally relevant candidate as the one corresponding to speaker meaning

(6) states that analysts should be something like 'informed arguers': their own competence as language users, together with enabling circumstances such as increased time and cognitive resources available for reflective processes, should allow them to reach plausible assumptions on speaker meaning, i.e., on what the speaker has meant and what the addressee assumes the speaker has meant. In turn, this should mirror the addressee's

commitment attribution processes. Once this is done, the next stage of analysis can be considered, namely evaluation, which is where the normative component of argumentation analysis comes into play.

5.2. Identifying an Unexpressed Premise: An Example

The data I will be analyzing to illustrate how the abovementioned framework can be used to reconstruct unexpressed premises comes from an article published in the Swiss tabloid *20minutes* in November 2011.[10] In the article, the journalist reports that Hollywood actor Ashton Kutcher had given marital advice in a men's magazine at a point where his wife, Hollywood actress Demi Moore, was actually filing for divorce, thereby calling into question Kutcher's credibility as a marital counsellor. (7) below is a literal translation of the title and (8) of the headline:

(7) Ashton's love lessons leave much to be desired.
Les leçons d'amour d'Ashton laissent à désirer.
(8) Ashton Kutcher has given marital counselling in the press. Demi Moore has however just filed for divorce.
Ashton Kutcher a donné des conseils matrimoniaux dans la presse. Demi Moore vient pourtant de demander le divorce.

The argumentative nature of the example is given away by the presence of the connective *pourtant* ('however'), which, in the terms of Anscombre (2002), is used to fulfil a counter-argumentative function: *pourtant* introduces a piece of information whose argumentative orientation runs contrary to the argumentative orientation of the sequence (or of its implications) that precedes the connective. Moreover, if we think about the fact that *20minutes* is a tabloid and that tabloids are known for regularly making a business out of exposing (and many times mocking) the life of celebrities, an argumentative relationship between the title and the headline is not hard to infer based on genre considerations.

The standpoint is explicitly mentioned in (7). One explicit (minor) premise is explicitly available in the second half of the headline in (8), which I will number below as (9) for exposition purposes:

(9) Demi Moore has just filed for divorce.

In order to evaluate the argument, we need to make explicit a major premise that connects (7) and (9). The logical minimum (see section 3.2 above) would yield something like (10):

(10) If Demi Moore has just filed for divorce, then Ashton's love lessons leave much to be desired.

(10) would render the argument logically valid, but does not quite do the job in terms of meaning, to the extent that it seems overly specific: furthermore, there seems to be something missing, for the connection between being a poor marital counsellor and going through a divorce is not yet explicit enough.

The tabloid is manifestly trying to make fun of Ashton Kutcher by exposing the irony arising out of a mismatch between his status as someone who has been asked by a magazine

---

[10] The article can be found online at http://www.20min.ch/ro/entertainment/people/story/10899919, last accessed 27.03.2015.

to give love advice (which should say something about his credibility as a marital counsellor) and his personal life (in terms of marital situation) which indicates an inability to maintain a marriage. The tabloid can therefore be taken to communicate (and defend) that Kutcher's personal circumstances are an obstacle to his credibility. To reflect this idea, I suggest that the standpoint in (7) should be reformulated in the following way:

(11) Ashton Kutcher is a poor marital counsellor.

The support for (11) given in the headline hints at Kutcher's own unsuccessful marriage through the mention of Demi Moore's action in view of divorcing; the argumentative pivot here rests on the connection between getting a divorce and being a credible marital counsellor. The explicit premise in (9) could thus be rephrased as follows, granted we assume that in our culture divorce is a symptom of an unsuccessful marriage:

(12) Ashton Kutcher's marriage is not successful.

The missing (major) premise should thus be something that connects (11) and (12) in a relevant way. Quite a few parameters render the search for the missing premise arduous, since theoretically speaking, and regardless of the context, many candidates are in principle possible. (11) and (12) could both be taken to be either the antecedent or the consequent of the conditional premise we are looking for, which already yields two options. The presence of negation, drawn from the fact that we are considering an unsuccessful marriage, might further complicate things, for a positive or negative formulation may multiply candidates. Moreover, the degree of generalization to be expected in the premise is unspecified: should it be about Ashton Kutcher, or about the larger set of people with unsuccessful marriages? And finally, some *topoi* (e.g., (21), (22), and (23)) might be considered as well, to the extent that they seem to straightforwardly relate to the issue under discussion in the article. Below is a list of propositions that could act as a potential major premise:[11]

(13) If AK is a poor marital counsellor, then (it would be expected that) AK's marriage is not successful.
(14) If AK is a credible marital counsellor, then (it would be expected that) AK's marriage is successful.
(15) If AK's marriage is successful, then AK is a credible marital counsellor.
(16) If AK's marriage is not successful, then AK is a poor marital counsellor.
(17) If X is a poor marital counsellor, then X's marriage is not successful.
(18) If X is a credible marital counsellor, then X's marriage is successful.
(19) If X's marriage is successful, then (it would be expected that) X is a credible marital counsellor.
(20) If X's marriage is not successful, then (it would be expected that) X is a poor counsellor.
(21) Happily married people give good love advice.
(22) If you preach what you practice, you advice is credible.
(23) Don't preach what you don't practice.

---

[11] I am not saying that all of these are equally plausible, but merely envisaging theoretical possibilities to show how problematic the identification of implicit premises can turn out to be. Items numbered (13) to (23) below represent a subset of the set of all possible implicit premises an analyst might come up with and do not consequently constitute an exhaustive closed list.

In light of all these possibilities, a crucial question arises, since some possibilities would make the argument formally fallacious, while some others would not: how do we determine whether the journalist is guilty of providing a formally fallacious inference? Consider the contrast between (15)/(19) on the one hand and (16)/(20) on the other – respectively represented in (24) and (25) below:

(24) If AK's/X's marriage is successful, then AK/X is a credible marital counsellor [(15)/(19)]
AK's/X's marriage is not successful [(12)]
Therefore, AK/X is not a credible marital counsellor [(11)]

(25) If AK's/X's marriage is not successful, then AK/X is a poor marital counsellor [(16)/(20)]
AK's/X's marriage is not successful [(12)]
Therefore, AK/X is not a credible marital counsellor [(11)]

(24) is an instance of denying the antecedent, while (25) is a canonical instance of the *modus ponens*. The example discussed here is interesting precisely because our choice of unexpressed premise in the argumentative reconstruction will itself determine our evaluation of the argument in terms of (non)fallaciousness.

I claim that the conditions of relevance discussed earlier can help with the argumentative reconstruction of unexpressed premises and thus prevent us from misattributing the responsibility of a fallacy to the speaker. This means that in order to have a plausible and complete interpretation of the journalist's intended argument, we need to assess which of all possible options seems to be the optimal one in terms of processing effort and cognitive effect. For reasons of space and because I see them as representative of the type of problems an analyst might encounter in the reconstruction of arguments, I will only compare assumptions (19) and (20).[12]

From the perspective of cognitive effect, (19) and (20) could be considered to be rather equivalent: there is *a priori* no reason to assume that the two sides of the story (i.e. a successful marriage boosting the credibility of a person in terms of love counselling, or an unsuccessful marriage weakening their credibility in the same domain) yield any significant difference in terms of the consequences of adding either of the two representations to an individual's cognitive environment. Both (19) and (20) are about the connection between marital success and love counselling credibility. Knowing that if someone's marriage is unsuccessful, it usually means that they are not credible love counsellors seems *prima facie* to be equivalent to knowing that if someone's marriage is successful, it also usually means that they can be credible love counsellors. It is quite hard to imagine that one would yield more cognitive benefits than the other. Furthermore, if we recall Geis & Zwicky's (1971) work on *invited inferences* and conditional perfection, it appears that many times conditional statements are interpreted as triggering a biconditional reading, where 'if P, then Q' is interpreted as also meaning 'if ¬P, then ¬Q'.[13] Following this line of research, it would appear that (19) is also interpreted as (20), which is another argument in support of the claim that from the perspective of cognitive effect, (19) and (20) do not significantly differ: if (19) is uttered, (20) will also be very likely to be inferred and vice-versa. It

---

[12] Though in principle all assumptions mentioned from (13) to (23) could be equally assessed with the tools drawn from RT.
[13] Geis & Zwicky's original example, 'If you mow the lawn, I will give you 5$' is pretty straightforward, and it is not difficult to imagine that the addressees of that utterance will also infer that if they do not mow the lawn, they will not get 5$.

therefore seems that the extent condition related to cognitive effect does not play a role in our example.

Now what about processing effort? Is (19) easier to process than (20)? The piece of information explicitly provided in (9) tells us that Demi Moore just filed for divorce, and represents a strong indication of Ashton Kutcher's unsuccessful marriage [(12)]. We can argue that assumptions that are consonant with (12), which is already present in the cognitive environment of the reader of the news article, will be easier to process than assumptions that are not, in particular assertions about the opposite state of affairs. So, given that the reader is already entertaining (12), we can hypothesize that (20) will be easier to process than (19), since (20) is precisely about *un*successful marriages, while (19) is about successful marriages. We could in fact hypothesize that (9) primes (20) but not (19), and that as a consequence (20) is easier to process than (19). Another way of looking at this is to assume that within a cognitive environment, incoming information that contradicts some assumptions that are already present and active in the cognitive environment require more processing effort because some adjustment needs to be made to accommodate the contradiction. This is not the case with information that is not contradictory or problematic – i.e. this is not likely to happen with (20).

Summing up, processing effort is lower for (20) than for (19), while cognitive effect is equivalent for both. The optimally relevant candidate for the unexpressed premise is thus (20). Assuming that the human cognitive system, in communication, is geared towards the production and the identification of optimally relevant representations, we can conclude that the intended unexpressed premise, i.e. the one the journalist/arguer is committed to, is (20). Subsequently, the argumentative inference presented in the title and headline of the article can be evaluated as formally and deductively valid, for it turns out to be a *modus ponens*. The reconstruction of the argument thus looks as follows:

(26) If X's marriage is not successful, then X is not a credible marital counsellor
AK's marriage is not successful
Therefore, AK is not a credible marital counsellor

One clarification is in order as to the psychological reality of this intricate reconstruction. While this type of analysis lays down a step-by-step procedure which justifies reconstruction by resorting to well-defined comparative cognitive criteria, I do not claim that this procedure, in particular its comparative dimension, corresponds to what went on in the journalist's mind as he wrote his article. The comprehension procedure is said to follow this effort/effect dynamics, but this does not necessitate a comparative assessment, as, at least in theory, the addressee is supposed to get to the most relevant one right away. However, we can *a posteriori* model this procedure to explain why (20) is the most contextually relevant premise, and thus the one both the journalist and the reader are likely to have gone for.

6. Conclusion

The main assumption the framework for argumentative analysis presented here builds upon is that analysts are competent language users and that as such they are legitimated to rely on their own comprehension mechanisms, although perhaps only reflectively. In other words, analysts are just like addressees: they understand verbal stimuli with the same cognitive mechanisms, with the difference that they have more time and processing resources. They can spend as much time as they want in figuring out and precisely assessing what speakers mean, they can make some inquiries about context to have a better grasp of the communicative situation, and they have the material possibility of comparing competing candidates to identify unexpressed premises.

While I do not challenge the idea that argumentative evaluation ultimately has to rely on normative criteria, I have tried to show that prior to evaluation, argumentative reconstructions could be carried out on the assumption that it is possible to work out plausible interpretations of the argumentative data, even in cases involving unexpressed premises.

The type of analysis performed here involves an assessment of the relative weight of two cognitive parameters regulating verbal information processing, namely processing effort and cognitive effect. While it is probably impossible (and it would make little sense anyway) to determine in absolute terms whether a given piece of information is easy to process or able to trigger significant cognitive effects, a *comparative* assessment can be performed, which enthymemes afford due to their inherent incomplete form. Within such a cognitive pragmatic framework, the reconstruction of unexpressed premises will become more likely to correspond to the actual argumentative data. The methodology described here, which builds on cognitive modelling of spontaneous information processing mechanisms, can therefore benefit from psychologically-grounded assumptions.

References

Anscombre, J.-C. (2002). Mais/pourtant dans la contre-argumentation directe: raisonnement, généricité, et lexique. *LINX, 46*, 115-131.

Austin, J.L. (1962). *How to do things with words*. Cambridge: Harvard University Press.

Becker, T. (2012). The pragmatics of argumentation. Commitment to implicit premises. In A. Schalley (Ed.), *Practical theories and empirical practice: a linguistic perspective* (pp. 257-272). Amsterdam, Benjamins.

Boulat, K. (2014). Are you committed? A pragmatic model of commitment. Paper delivered at The 8th Days of Swiss Linguistics, 19-21 June 2014, Zurich. Available at http://www.linguistics-phd.uzh.ch/zling/index.html?page=archive.

Carston, R. (2002). *Thoughts and utterances. The pragmatics of explicit communication*. Oxford: Blackwell.

Dendale, P. & Coltier, D. (Eds.). (2011). *La prise en charge énonciative: études théoriques et empiriques*. Paris, Bruxelles: De Boeck, Duculot.

Eemeren, F.H. van, & Grootendorst, R. (1992). *Argumentation, communication, and fallacies: A pragma-dialectical perspective*. Hillsdale, NJ: Lawrence Erlbaum.

Eemeren, F.H. van & Grootendorst, R. (2004). *A systematic theory of argumentation. The pragma-dialectical approach*. Cambridge: Cambridge University Press.

Geis, M. & Zwicky, A. (1971). On invited inferences. *Linguistic Inquiry, 2*, 561-566.

Gerritsen, S. (2001). Unexpressed premises. In F. H. van Eemeren (Ed.), *Crucial Concepts in Argumentation Theory* (pp. 51-79). Amsterdam: Amsterdam University Press.

Grice, H.P. (1989). *Studies in the way of words*. Cambridge: Harvard University Press.

Hamblin, C. (1970). *Fallacies*. London: Methuen.

Henst, J.-B- van der & Sperber, D. (2004). Testing the cognitive and communicative principles of relevance. In I. Noveck & D. Sperber (Eds.), *Experimental Pragmatics* (pp. 141-171). Basingstoke: Palgrave.

Katriel, T. & Dascal, M. (1989). Speaker's commitment and involvement in discourse. In Y. Tobin (Ed.), *From Sign to Text: A Semiotic View of Communication* (pp. 275-295). Amsterdam/Philadelphia: John Benjamins.

Lewiński, M. (2012). The paradox of charity. *Informal Logic, 32(4)*, 403-439.

Lewiński, M. & Oswald, S. (2013). When and how do we deal with straw men? A normative and cognitive pragmatic account. *Journal of Pragmatics, 59B*, 164-177.

Mercier, H. & Sperber, D. (2009). Intuitive and reflective inferences. In J. S. B. T. Evans & K. Frankish (Eds.), *In Two Minds* (pp. 149-170). New York: Oxford University Press.

Morency, P., Oswald, S. & Saussure, L. de. (2008). Explicitness, implicitness and commitment attribution: a cognitive pragmatic account. *Belgian Journal of Linguistics, 22,* 197-219.

Nølke, H. Fløttum, K. & Norén, C. (2004). *ScaPoLine. La théorie scandinave de la polyphonie linguistique*. Paris: Kimé.

Oswald, S. & Lewiński, M. (2014). Pragmatics, cognitive heuristics and the straw man fallacy. In T. Herman & S. Oswald (Eds.). *Rhétorique et cognition: perspectives théoriques et stratégies persuasives / Rhetoric & Cognition: theoretical perspectives and persuasive strategies* (pp. 313-343). Bern: Peter Lang.

Paglieri, F. (2007). No more charity, please! Enthymematic parsimony and the pitfall of benevolence. In H.V. Hansen et. al. (Eds.), *Dissensus and the Search for Common Ground, CD-ROM* (pp. 1-26). Windsor, ON: OSSA.

Paglieri, F. (2010). Committed to argue: On the cognitive roots of dialogical commitments. In C. Reed, C. Tindale (Eds.), *Dialectics, Dialogue and Argumentation: An Examination of Douglas Walton's Theories of Reasoning and Argument* (pp. 59-71). London: College Publications.

Paglieri, F. & Woods, J. (2011). Enthymematic parsimony. *Synthese, 178,* 461-501.

Saussure, L. de (2005). Pragmatique procédurale et discours. *Revue de sémantique et de pragmatique, 17,* 101-125.

Searle, J. (1969). *Speech acts: An Essay in the Philosophy of Language*. Cambridge: Cambridge University Press.

Snoeck Henkemans, A. F. (1992). *Analysing complex argumentation: The reconstruction of multiple and coordinatively compound argumentation in a critical discussion*. Amsterdam: SicSat.

Sperber, D. & Wilson, D. (1995). *Relevance: Communication and Cognition*, (2nd ed). Cambridge, MA: Blackwell Publishers.

Tindale, C. (1992). Audiences, relevance, and cognitive environments. *Argumentation, 6(2),* 177-188.

Walton, D. (1993). Commitment, Types of Dialogue, and Fallacies. *Informal Logic, 14(2&3),* 93-103.

Walton, D. & Krabbe, E. (1995). *Commitment in dialogue. Basic Concepts of Interpersonal Reasoning*. Albany, NY: SUNY Press.

Walton, D., & Reed, C. (2005). Argument Schemes and Enthymemes. *Synthese, 145,* 339-370.

Wilson, D. & Sperber, D. (2002). Relevance theory. *UCL Working Papers in Linguistics, 14,* 249-287.

Wilson, D. & Sperber, D. (2012) *Meaning and relevance*. Cambridge: Cambridge University Press.

Woods, J. (1992). Apocalyptic relevance. *Argumentation, 6(2),* 189-202.

Chapter 3

Erotetic Problem Solving: From Real Data to Formal Models. An Analysis of Solutions to Erotetic Reasoning Test Task

Mariusz Urbański[1], Katarzyna Paluszkiewicz[2] & Joanna Urbańska[3]

Institute of Psychology, Adam Mickiewicz University, Poznań, Poland; [1] mariusz.urbanski@amu.edu.pl; [2] k.paluszkiewicz@amu.edu.pl; [3] joanna.urbanska@amu.edu.pl

Abstract. In this paper we model solutions to one of the Erotetic Reasoning Test tasks. Logical framework for our research is set up by one of the most influential current paradigms in the logic of questions – Inferential Erotetic Logic. We propose a weakened version of the relation of erotetic implication, which allows for adequate account on suboptimal solutions to the task in question.

1. Introduction

The aim of this paper is to offer formal models of solutions to one of the Erotetic Reasoning Test (ER) tasks (Urbański *et al.*, 2013). ER was designed in order to operationalize fluency in solving difficult deductive tasks involving conditionals, thus extending the framework for research on deduction set up by the Wason selection task (see Stenning and van Lambalgen, 2008).

Erotetic inferences are inferences which involve questions as conclusions or as premises and conclusions. They form good representations of some techniques of problem solving, either by reduction of an initial problem to a simpler one(s), or by identifying missing information which is needed in order to solve the initial problem (Urbański & Łupkowski, 2010). By analogy to the tripartite division of reasoning rules (Stanovich, 1999) we shall show that such inferences can be modeled not only from the point of view of a normative yardstick of assumed logical background. Our formal framework allows for modeling such inferences from a prescriptive perspective, as well as for an adequate descriptive account on slips and errors made by reasoning subjects.

We start with outlining logical basis of our research, which is one of the most influential current paradigms in the logic of questions – Inferential Erotetic Logic (section 2). Then we present the ER task we are interested in and its normatively correct solution (section 3). Subsequently, drawing on the subjects' justifications to suboptimal solutions (section 4), we introduce the concept of weak erotetic implication (section 5), in terms of which solutions described earlier can be modeled.

2. Logical Framework: Inferential Erotetic Logic

In order to define validity of erotetic inferences a logic of questions is needed, which allows to define semantic properties of and relations between questions (or interrogatives). However, there are many possible models of validity of erotetic inferences.

Logical framework for this research was set up by Inferential Erotetic Logic (IEL; Wiśniewski, 1995, 2013). This choice is justified by a couple of reasons. Firstly, IEL is flexible: it is not tied up to any specific logic of declaratives. IEL-style logic of questions can be based on any such logic which satisfies some simple syntactic and semantic conditions; examples include many-valued logics (Urbański, 2002), modal logics (Leszczyńska-Jasion, 2007, 2008), paraconsistent logics (Wiśniewski, Vanackere, & Leszczyńska, 2005; Chlebowski & Leszczyńska-Jasion, 2015), intuitionistic logic (Skura, 2005). Consequently, IEL allows for formalization of erotetic reasoning at different levels of complexity of specification of considered verbal representations and with different underlying semantic requirements.

Secondly, formal representation of questions in IEL is friendly to the user, as in representing questions IEL follows so-called set-of-answers methodology (Harrah, 2002; see Peliš, 2016, for a comprehensive introduction). The idea stems from Hamblin's (1958, p. 162) postulate that "Knowing what counts as an answer is equivalent to knowing the question". Thus in IEL a representation for the question 'Is John the good, the bad, or the ugly?' is $?\{p, q, r\}$ (where p stands for 'John is the good', q for 'John is the bad' and r for 'John is the ugly'). In general, for any question Q of the form $?\{A_1, ..., A_n\}$ the formulas $A_1, ..., A_n$ are considered as direct answers to Q (the assumption is that $A_1, ..., A_n$ are pairwise syntactically distinct formulas). As a result, while advocating for a non-reductionist account on questions as having the meaning on their own (see Belnap, 1983; Wiśniewski, 1995, pp. 37-42), IEL offers some straightforward tools for modeling erotetic inferences. It should be noted, however, that we are not entering into the details of the lively discussion concerning relations between questions and interrogatives (see for example Wiśniewski, 2015).

Thirdly, in general IEL is not limited to an analysis of any specific class of questions (although in this paper we are going to consider only questions with finite number of answers). In particular, IEL defies another Hamblin's (1958, p. 163) postulate which states that "The possible answers to a question are an exhaustive set of mutually exclusive possibilities". Such questions (which elsewhere we call maximally informative questions; see Paluszkiewicz and Urbański, 2016) are ubiquitous in natural language; nevertheless, as humans do reason with questions lacking this property, we need a framework which is able to cope with such inferences.

We were interested in inferences in which conclusion and one premise are questions and other premises – if there are other premises at all – are declarative sentences. In IEL validity of such inferences is defined in terms of semantic relation of erotetic implication (e-implication for short), which meets the following conditions:

1. transmission of truth/soundness into soundness: if the question-premise is sound (i.e., there exists a true direct answer to this question) and all the declarative premises (if there are any) are true, then the question-conclusion is sound as well;
2. cognitive usefulness: each answer to the question-conclusion is useful in answering the question-premise (each answer to the question-conclusion narrows down the class of possible answers to the question-premise), provided that all the declarative premises (if there are any) are true.

Consider a simple example (Urbański & Łupkowski, 2010, p. 68). Suppose that our problem is expressed by the initial question:

(Q) *Who stole the tarts?*

Suppose also that we managed to establish the following evidence:

(E_1) *It is one of the courtiers of the Queen of Hearts attending the afternoon tea-party who stole the tarts.*

Thus the initial question together with the evidence erotetically implies the question:

(Q*) *Which of the Queen of Hearts' courtiers attended the afternoon tea-party?*

It is intuitively justified to ask for the list of courtiers – participants of the afternoon tea-party (Q*) in order to solve the problem (Q), in view of the established evidence (E_1). This justification can be expressed in exact terms by fulfilment of both conditions of e-implication. First, if somebody really stole the tarts and if it is true, that the culprit is one of the courtiers of the Queen of Hearts attending the afternoon tea-party (that is, if Q is sound and if E_1 is true), then some of the courtiers must have attended the party (that is, Q is sound as well). Second, each non-empty list of courtiers – participants of the party narrows down the class of suspects, provided that it is really one of the courtiers of the Queen of Hearts attending the afternoon tea-party who stole the tarts (that is, each direct answer to Q narrows down the class of possible answers to Q, in view of E_1).

If moreover we know that:

(E_2) *Queen of Hearts invites for a tea-party only these courtiers who made her laugh the previous day.*

then Q* and E_2 erotetically imply the question:

(Q**) *Which courtiers made the Queen of Hearts laugh the previous day?*

It is easy to check that in this case both conditions of e-implication are fulfilled as well.

Formal definition of e-implication offers precise explication for conditions of transmission of truth/soundness into soundness and of cognitive usefulness (Definition 1; see Wiśniewski, 2013). For the sake of simplicity we consider only the case of questions with finite sets of direct answers.

Definition 1. *A question Q e-implies a question Q_1 on the basis of a set X of declaratives (**Im(Q; X; Q_1)**) iff:*

1. *transmission of truth/soundness into soundness: for each direct answer A to the question Q: X∪{A} entails the disjunction of all the direct answers to the question Q_1, and*
2. *cognitive usefulness: for each direct answer B to the question Q_1 there exists a non-empty proper subset Y of the set of direct answers to the question Q such that X∪{B} entails the disjunction of all the elements of Y.*

Such erotetic inferences clearly involve deductive reasoning (especially in view of the first condition imposed on e-implication). However, the presence of verbal representations different than the usual declaratives, and the presence of an additional condition of cognitive usefulness suggest that carrying out erotetic reasoning may be a task both more difficult than carrying out simple syllogistic reasoning, widely used in deductive reasoning tests (Stenning & van Lambalgen, 2008) and more comprehensive with respect to deduction than solving conditionals of Wason selection task. These claims were confirmed in the research in which tests based on syllogistic reasoning as well as on erotetic reasoning were used (Urbański et al., 2013).

3. Erotetic Reasoning Test: An Exemplary Task and Its Normatively Correct Solution

Erotetic Reasoning Test (ER), designed by the authors and carried out in Polish, contains 3 items (time limit 30 min). Each item consists of a detective-like story in which the initial problem and the evidence gained are indicated. The task is to pick a question (one out of four), each answer to which will lead to some solution to the initial problem. The subjects are asked to justify their choices.

All three stories describe some investigation and they invoke search for a solution to an initial problem by means of posing further (auxiliary) questions. The stories are set up in such a way that the impact of previous content-related experience of the subjects on the choice of solution is minimized. All the relevant information is explicitly listed and the subjects are asked to solve each task (i.e., to pick a correct question) solely on the basis of what is given.

There is only one correct answer in each item. However, the criterion of correctness is somewhat complicated. What matters is not only correct choice of the question-solution but also a proper justification of the choice, based on two conditions of validity of erotetic implication: transmission of truth/soundness into soundness and cognitive usefulness. Thus assessment of overall correctness of an answer in each ER task is based on the presence of all these three elements.

Our research were carried out between February and May, 2012 and between January and March, 2015. 137 subjects were recruited (M=21.69, SD=1.44, 111 women), students at Adam Mickiewicz University in Poznań, who volunteered to participate in these research. Reliability of the ER test turned out to be acceptable (Cronbach's α = .74). Interested reader can find detailed results of this study (including more psychometric data) and discussion of their implications in (Urbański et al., 2016).

We shall elaborate our formal model on the example of one of the ER tasks, 'The Bomb' (the phenomena we are going to address were also observed in case of the remaining two tasks). It runs as follows:

In the capital of a certain country someone planted a bomb in the palace of the king. The best royal engineer, who arrived immediately, established the following evidence:

1. There are three wires in the bomb: green, red and orange.
2. To disarm the bomb either the green or the red wire must be cut. Cutting the wrong wire will cause an explosion.
3. If the bomb has been planted by Steve, cutting the green wire will disarm it.
4. If the bomb has been planted by John, cutting the red wire will disarm it.

Moreover, no one but John would have used the red wire.
5. If the bomb has not been planted on an even day of the month, the culprit is Steve.
6. The bomb has been planted either by Steve, or by John, or by someone else.

Each of the following questions below can be answered either 'yes' or 'no'. Mark the question to which the answer (regardless of it being 'yes' or 'no') will allow you to establish in the shortest possible time which wire should be cut in order to disarm the bomb:

- ☐ Was the bomb planted on an even day of the month?
- ☐ Was the bomb planted by Steve?
- ☐ Was the bomb planted by John?
- ☐ Was the bomb planted by someone else than Steve or John?

Justify your choice.

The normative yardstick for correctness of solving ER tasks was determined by the aforementioned concept of erotetic implication (which we shall call further on canonical e-implication).

Reasoning underlying solutions to ER tasks can be represented concisely in terms of Erotetic Search Scenarios (ESSs; Wiśniewski, 2013, p. 103–126). ESSs provide a formal account on the Erotetic Decomposition Principle:

Transform a principal question into auxiliary questions in such a way that: (a) consecutive auxiliary questions are dependent upon previous questions and, possibly, answers to previous auxiliary questions, and (b) once auxiliary questions are resolved, the principal question is resolved as well.

A scenario is a tree-like structure consisting of branches. Each branch satisfies the following conditions (Wiśniewski, 2013, p. 106–107):

1. It begins with the principal question and ends with a direct answer to it;
2. Each declarative sentence involved:
   (a) is an initial premise, or
   (b) is a direct answer to an auxiliary question that immediately precedes it on the branch, or
   (c) is entailed by some declarative sentence(s) which occur(s) earlier on the branch.
3. Each auxiliary question involved is e-implied, in the sense of IEL, by some question and declarative sentence(s) that occur earlier on the branch.

Some auxiliary questions in a scenario are queries: these are questions which are immediately followed by all their direct answers (and thus queries are the only branching points of a scenario). Notice, that not all auxiliary questions need to be answered in a scenario; we shall come back to this issue in the section 5.

From a pragmatic point of view, a scenario offers conditional instructions on what questions should be asked and when they should be asked in order to solve an initial problem. Moreover, a scenario shows where to go if such-and-such a direct answer to a query appears to be acceptable and does so with respect to any direct answer to each query (Wiśniewski, 2003, p. 122).

A scenario underlying the normatively correct solution to the task is presented in Figure 1, expressed in the language which is an erotetic extension of Classical Propositional Calculus (see Wiśniewski, 2013, p. 18–20); $\neg$ stands for negation, $\wedge$ for conjunction, $\vee$ for disjunction, $\perp$ for exclusive disjunction, $\rightarrow$ for implication and $\leftrightarrow$ for equivalence. Propositional variables represent the following sentences:

p – Cutting the green wire disarms the bomb.
q – Cutting the red wire disarms the bomb.
v – Cutting the orange wire disarms the bomb.
s – The bomb has been planted by Steve.
r – The bomb has been planted by John.
t – The bomb has been planted on an even day of the month.
u – The bomb has been planted by someone else than Steve or John.

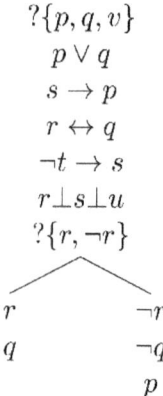

$?\{p, q, v\}$
$p \vee q$
$s \rightarrow p$
$r \leftrightarrow q$
$\neg t \rightarrow s$
$r \perp s \perp u$
$?\{r, \neg r\}$

r $\neg r$
q $\neg q$
       p

Figure 1. A scenario for normatively correct solution

In the scenario presented in Figure 1, the question $?\{r, \neg r\}$ ("Was the bomb planted by John?"), which is the correct solution to the considered task, is canonically e-implied by the initial question, $?\{p, q, v\}$, on the basis of the following declarative premises: $p \vee q$, $r \leftrightarrow q$.

4. Subjects' Justifications to ER Task Solutions

In ER tasks the subjects were asked to provide justifications to their solutions. This way some insight into reasoning leading to these solutions was obtained. Some subjects offered justifications which exactly fit normatively correct solution given in the previous section. These justifications complied with the two conditions of canonical e-implication (Definition 1), as in the example 1.

Example 1. [Subject A100, solution: Was the bomb planted by John?] I considered all the possibilities. If it is John, then one should cut the red one. If not, then the green one, because only these two disarm the bomb. And only John would have used the red one, so anybody else would have used the green one. This is the only question answer to which gives clear solution. Using any other there is a risk that one will need to answer some further questions.

This is exactly our normatively correct solution, represented by the scenario in the previous section. We shall focus, however, on suboptimal solutions, which nevertheless are somewhat justified in view of the informational goal being pursued.

A significant number of participants did not comply with the requirements imposed by the canonical e-implication. In particular, they violated the cognitive usefulness condition, expressed in the instruction as "to choose that question to which the answer (regardless of it being 'yes' or 'no') will allow to establish solution" to the initial problem. Nevertheless, most choices were motivated by the possibility of gaining relevant information; the subjects often found it justified to choose a question only *some* answer to which will lead to the solution to the initial problem. Let us clarify this issue by means of further examples.

Example 2. [Subject A81, solution: Was the bomb planted on an even day of the month?] This answer, because if the bomb wasn't planted on an even day of the month, then Steve did this. And it will be known who did this, you only need to pay attention to whether it was an odd or an even day of the month.

In this case the subject has chosen a solution which is useful, but only partially. The negative answer to it leads to the solution to the initial problem. However, the affirmative answer is of no use. The solution can be modeled by means of the scenario presented in Figure 2.

Example 3. [Subject A78, solution: Was the bomb planted on an even day of the month?] Because if it was an even day of the month, then the culprit is known.

This solution can be modeled by means of the same scenario as the previous one, albeit with focus on different branch of it. It points out at the branch containing affirmative answer, which in fact does not lead to any solution to the initial problem. Most probably this was caused by misreading the content of the task ("the bomb has been planted on an even day of the month", instead of "has not been planted"). Still, the justification refers to partial usefulness of the chosen question.

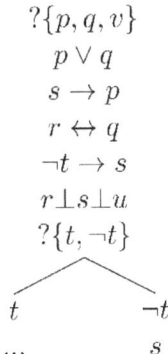

Figure 2. A scenario for examples 2 and 3.

Example 4. [Subject A101, solution: Was the bomb planted on an even day of the month?] No matter which question we'll ask it will exclude one suspect and it will be necessary to ask another question which will decide who planted the bomb.

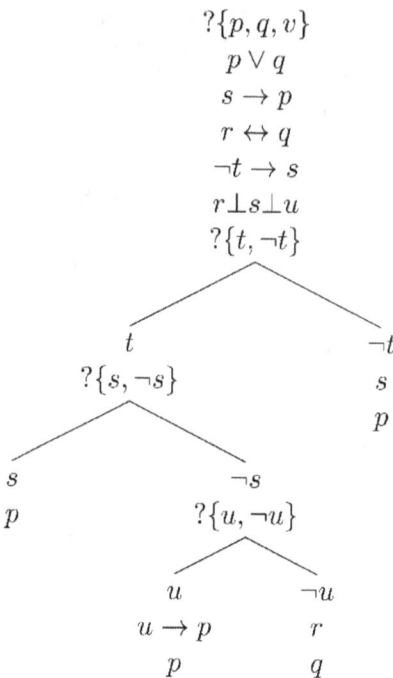

Figure 3. A scenario for example 4

This is an interesting solution. The subject's claim is to some extent justified: although it is not the case that all the questions offered as solutions are of only partial use, asking two partially useful questions will lead to the solution. In this case, after asking the question concerning a day of planting the bomb one might ask the question "Was the bomb planted by Steve?" and then "Was the bomb planted by someone else than Steve or John?" (in whichever order, in fact). This way one would obtain a scenario for solving the initial problem. The scenario in question is represented in Figure 3. Notice, that in order to generate such a scenario an additional premise is needed ($u \to p$, i.e. "If the bomb has been planted by someone else than Steve or John, then cutting the green wire disarms it"). This premise is justified in view of the information provided, so it can be interpreted as an enthymematic one. It is important, however, at which stage this premise is added. If it precedes the query $?\{u, \neg u\}$, this query becomes canonically e-implied by the initial question and declarative premises; if not, introduction of the query does not meet the usefulness condition.

Another possibility, consistent with the subject's justification, would be to combine one of the above questions with the question $?\{r, \neg r\}$ (although this would mean that the subject did not consider both answers to this question as useful).

Example 5. [Subject A82, solution: Was the bomb planted on an even day of the month?] We don't know if the bomb exploded at all and we do not know who planted the bomb.

The subject is not able to infer any solution, so the choice of a question can be viewed as just a random one.

Example 6. [Subject A91, solution: Was the bomb planted by Steve?] Because if the bomb was planted by Steve, then cutting the green wire will disarm it.

This solution is similar to the one of example 2, as witnessed by the scenario presented in Figure 4. The subject has chosen a question which is partially useful and did not consider the branch containing negative answer to the query.

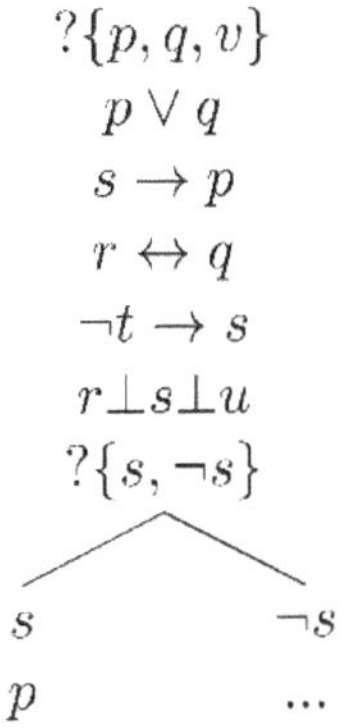

$$?\{p, q, v\}$$
$$p \vee q$$
$$s \rightarrow p$$
$$r \leftrightarrow q$$
$$\neg t \rightarrow s$$
$$r \perp s \perp u$$
$$?\{s, \neg s\}$$

```
        s           ¬s
        p           ...
```

Figure 4. A scenario for example 6

Example 7. [Subject A103, solution: Was the bomb planted by Steve?] If we'll get an affirmative answer to this question, then we'll know that the green wire needs to be cut. If a negative one, then there will be only one possibility left – the red wire, and additionally we'll know that the culprit is John.

Here the subject interprets the conditional s→p as a biconditional, which is quite common in everyday reasoning (cf. Stenning and van Lambalgen, 2008 on Wason selection task). As a result, inferences marked by an asterix (Figure 5) are not valid.

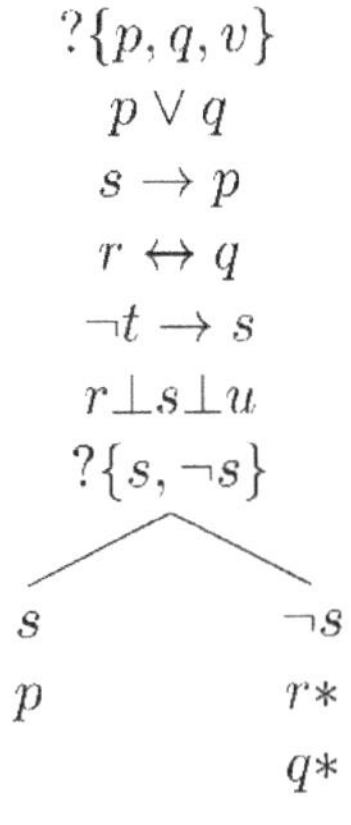

$$?\{p, q, v\}$$
$$p \vee q$$
$$s \rightarrow p$$
$$r \leftrightarrow q$$
$$\neg t \rightarrow s$$
$$r \perp s \perp u$$
$$?\{s, \neg s\}$$

```
        s           ¬s
        p           r*
                    q*
```

Figure 5. A scenario for example 7

5. Weak Erotetic Implication

Reasoning by which choices of suboptimal solutions to the ER task were justified do not comply with the normative requirements imposed by canonical e-implication. Nevertheless, as we have shown in the previous section, some of them are not just incorrect: from a prescriptive point of view there is some decent rationality involved in chosing partially useful solutions. Such reasoning can be modeled in terms of a relation weaker than canonical e-implication; we shall call it weak erotetic implication.

Definition 2. *A question Q weakly e-implies a question Q_1 on the basis of a set X of declaratives (**$Im_w(Q; X; Q_1)$**) iff:*

1. *Transmission of truth/soundness into soundness: for each direct answer A to the question Q: $X \cup \{A\}$ entails the disjunction of all the direct answers to the question Q_1, and*
2. *Partial cognitive usefulness: for some direct answer B to the question Q_1 there exists a non-empty proper subset Y of the set of direct answers to the question Q such that $X \cup \{B\}$ entails the disjunction of all the elements of Y.*

As we have seen in the justification of their choices, there are two main reasons for which subjects find it reasonable to indicate these partially useful solutions. The first one can be concisely summarized by a quotation from one of the subjects: because by asking and answering such a question "we know at least something". This resembles somewhat the 'seizing' phase of information gathering in terms of the need for cognitive closure (Kruglanski, 2004). The second reason is that some subjects found it difficult to point at a single question, each answer to which would allow to solve an initial problem, but what they claimed was that the task can be solved by means of asking a sequence of questions. From this point of view choices of partially useful weakly implied questions form a working example of what Hintikka calls strategic rules; such rules make sense not in move-by-move terms, only in terms of complete strategies for solving problems (Hintikka, 1999, p. 97–98).

As the relation of weak e-implication is weaker than the canonical one, there is an obvious relationship between the two: if a question Q canonically e-implies a question Q_1 on the basis of a set X of declaratives, then Q also weakly e-implies Q_1 on the basis of X, and not the other way around. Erotetic implications on which our models of exemplary solutions to the ER task are based are summarized in Table 1 (in each case $Q = ?\{p, q, v\}$).

Examples	$Q; X; Q_1$	E-implication
1	$Q; p \lor q, r \leftrightarrow q; ?\{r, \neg r\}$	Im, Im_w
2, 3, 4	$Q; \neg t \to s, s \to p; ?\{t, \neg t\}$	Im_w
4, 6, 7	$Q; s \to p; ?\{s, \neg s\}$	Im
4	$Q; p \lor q, r \perp s \perp u, r \leftrightarrow q; ?\{u, \neg u\}$	Im_w
4	$Q; p \lor q, r \perp s \perp u, r \leftrightarrow q, \neg s; ?\{u, \neg u\}$	Im, Im_w

Table 1. Weak and canonical e-implications

Now we are in a position to present a more general overview of the solutions to the considered ER task. In Table 2 a statistics of the solutions chosen by the subjects is given.

The solution	% of choices
Was the bomb planted on an even day of the month?	13.7
Was the bomb planted by Steve?	9.3
Was the bomb planted by John?	66.2
Was the bomb planted by someone else than Steve or John?	10.8

Table 2. Solutions chosen by the subjects

As we mentioned above, criterion of correctness of each of ER tasks consists not only of correct choice of the question-solution but also of a proper justification to the choice, based on two conditions of validity of erotetic implication. Some of the subjects who have chosen the normatively correct solution (i.e., the question "Was the bomb planted by John?") did not meet requirements for its proper justification: 9.8% of them have given justifications modeled by weak e-implication while 3.3% have given solutions for which we were unable to find a formal model within our framework. Of all the subjects, 57.5% have given the normatively correct justification modeled by canonical e-implication, 33.1% have given justifications modeled by weak e-implication. The remaining 9.4% have given solutions for which we were unable to find a formal model (see example 5).

Models based on weak erotetic implication account well for reasoning of subjects who solve ER tasks choosing partially useful questions. However, these solutions can be represented in terms of canonical e-implication as well, in a way emulating its weaker version.

We shall elaborate this issue on an example of asking a sequence of questions in order to solve the task. Some of the subjects pointed out that a viable solution would be to ask if the bomb has been planted by Steve and then, in case the answers is negative, if the bomb has been planted by someone else than Steve or John (this solution is somewhat similar to that of example 4). Clearly, the question about Steve ($?\{s, \neg s\}$) is only partially useful in solving the initial problem and thus it is only weakly e-implied in this context. The second question is also only partially useful and also only weakly e-implied by the initial question and declarative premises. The relevant scenario is presented in Figure 6.

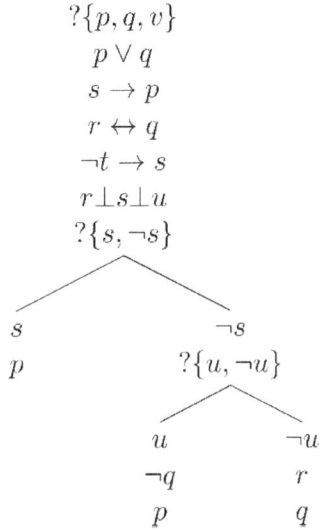

Figure 6. A scenario with a sequence of queries

This scenario can be interpreted as representing a strategy for solving the initial problem in the sense proposed by Hintikka. Not all inferential steps are 'safe' here (from the normative point of view of canonical e-implication), but combined they lead to a valid solution. Notice also, that if $\neg s$ is added to the declarative premises, the question $?\{u, \neg u\}$ becomes canonically e-implied by the initial problem and extended set of declarative premises (the same holds in the case of the scenario given in example 4).

The fact that both queries of the above scenario are not canonically e-implied by the initial question and declarative premises stems from a particular property of canonical e-implication. Namely, this relation is not transitive: the facts that Q_1 e-implies Q_2 on the basis of X and that Q_2 e-implies Q_3 on the basis of X does not warrant that Q_1 e-implies Q_3 on the basis of X.

Consider the next scenario (Figure 7) which differs from the previous one with respect to one additional auxiliary question: $?\{s \wedge u, s \wedge \neg u, \neg s \wedge u, \neg s \wedge \neg u\}$.

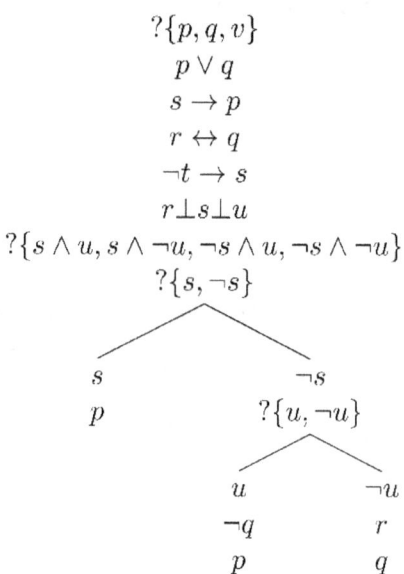

Figure 7. A scenario emulating weak e-implication, version 1

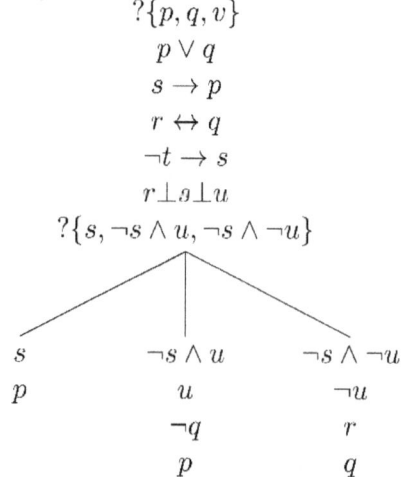

Figure 8. A scenario emulating weak e-implication, version 2

Let us label it by Q_c, as it is a conjunctive question (see Urbański, 2001, p. 76). Q_c is canonically e-implied by the initial question on the basis of the declarative premises. Also, Q_c canonically e-implies both ?$\{s, \neg s\}$ and ?$\{u, \neg u\}$ (no declaratives needed). However, as we mentioned above, neither of these two questions is canonically e-implied by the initial one and declarative premises.

Q_c is not a query of this scenario, as it is not answered in it. Its only role is to provide a bridge over an e-implicational gap between the initial question and the queries of the scenario. From the formal point of view, Q_c is just processed in order to arrive at the queries. The same holds in case of the question ?$\{s, \neg s \wedge u, \neg s \wedge \neg u\}$ in the next scenario (Figure 8).

Such 'bridging' questions are very useful in modeling erotetic reasoning by means of canonical e-implication. However, they are not present in subjects' justifications. Thus modeling the solutions by means of weak e-implication introduces an important descriptive factor into the formal framework of IEL.

6. Conclusion

There is only a little exaggeration in saying that fluency in erotetic inferences is a landmark of intelligence. Extended framework of IEL allows for modeling of such inferences from all the three interesting perspectives: normative, prescriptive and descriptive. In particular, weak erotetic implication is a useful tool for formally addressing the issue of strategic component of reasoning with questions. We need to acknowledge that, even with weakened condition of cognitive usefulness, our analysis is constrained by the condition of transmission of truth/soundness into soundness. As we aimed here at modeling certain class of inferences, forming erotetic counterparts to deductive problem solving, this is not a major limitation in the context of reasoning underlying solutions to Erotetic Reasoning Test task. Nevertheless, further research on even weaker versions of e-implication are needed in order to account for erotetic inferences which do not involve deductive reasoning.

Acknowledgments

Research for this paper were supported by the National Science Centre, Poland (DEC-2012/04/A/HS1/00715). The authors would like to thank Monika Zawadzka for her assistance in translation of the test materials.

References

Belnap, N. D. (1983). Approaches to the semantics of questions in natural language. In: R. Bäuerle et al. (Eds.), *Meaning, Use, and Interpretation of Language* (pp. 21-29). Berlin: de Gruyter.

Chlebowski, S., & Leszczyńska-Jasion, D. (2015). Dual Erotetic Calculi and the Minimal LFI. *Studia Logica*, published on-line 03.06.2015.

Hamblin, C. (1958). Questions, Australasian Journal of Philosophy, 36(3), 159–168.

Harrah, D. (2002). The logic of questions. In D. Gabbay & F. Guenthner (Eds.), *Handbook of Philosophical Logic*, (Vol. 8, pp. 1–60). Dordrecht: Kluwer Academic Publishers.

Hintikka, J. (1999). What is abduction? The fundamental problem of contemporary epistemology. In *Inquiry as inquiry: A logic of scientific discovery* (pp. 91-113).

Springer Netherlands.
Kruglanski, A. (2004). *The Psychology of Closed Mindedness*. New York: Psychology Press.
Leszczyńska-Jasion D. (2007). *The Method of Socratic Proofs for Normal Modal Propositional Logics*. Poznań: Adam Mickiewicz University Press.
Leszczyńska-Jasion, D. (2008). The Method of Socratic Proofs for Modal Propositional Logics: K5, S4.2, S4.3, S4M, S4F, S4R and G. *Studia Logica*, 89(3), 371–405.
Paluszkiewicz, K., & Urbański, M. (2016). Generalized conjunctive questions. The case of Classical Propositional Calculus and some normal modal logics. Submitted.
Peliš, M. (2016). *Inferences with Ignorance: Logics of Questions*. The Prague: Karolinum.
Skura T. F. (2005). Intuitionistic Socratic procedures. *Journal of Applied Non-Classical Logics*, 15(4), 453–464.
Stanovich, K. E. (1999). *Who Is Rational? Studies of Individual Differences in Reasoning*. Mahwah, NJ: Lawrence Erlbaum.
Stenning, K., & van Lambalgen, M. (2008). *Human Reasoning*. The MIT Press.
Urbański, M. (2001). Synthetic tableaux and erotetic search scenarios: Extension and extraction. *Logique et Analyse*, 173, 175, 69–91.
Urbański, M. (2002). Synthetic tableaux for Łukasiewicz's calculus Ł3. *Logique et Analyse*, 45(177-178), 155–173.
Urbański, M., & Łupkowski, P. (2010). Erotetic Search Scenarios: Revealing Interrogator's Hidden Agenda. In: P. Łupkowski, M. Purver (Eds.), *Semantics and Pragmatics of Dialogue* (pp. 67–74). Poznań: Polskie Towarzystwo Kognitywistyczne.
Urbański, M., Paluszkiewicz, K., & Urbańska, J. (2013). *Deductive Reasoning and Learning: a Cross-Curricular Study*. Research report.
Urbański, M., Paluszkiewicz, K., & Urbańska, J. (2016). Different types of deductive abilities: the case of easy and difficult deduction tasks. Submitted.
Wiśniewski, A. (1995). *The Posing of Questions. Logical Foundations of Erotetic Inferences*. Dordrecht: Kluwer Academic Publishers.
Wiśniewski, A. (2003). Erotetic Search Scenarios. *Synthese*, 134(3), 389–427.
Wiśniewski, A. (2013). *Questions, Inferences, and Scenarios*. London: College Publications.
Wiśniewski, A. (2015). Semantics of questions. In: S. Lappin & Ch. Fox (Eds.), *The Handbook of Contemporary Semantic Theory* (2nd Ed.) (pp. 273–313). Wiley-Blackwell.
Wiśniewski A., Vanackere G., Leszczyńska D. (2005). Socratic proofs and paraconsistency: A case study. *Studia Logica*, 80(2–3), 433–468.

Chapter 4

Initiating Argument in Social Confrontation: Development of a Behavioral Complexity Model to Understand Variations in Initiation Types

Susan L. Kline[1] & Wen Song[2]

[1] School of Communication, Ohio State University, Columbus OH, USA 43210, kline.48@osu.edu; [2] College of Literature and Journalism, Sichuan University, Cheng du, Sichuan Province, P.R. China, wen.song17@outlook.com

Abstract. The purpose of this essay is to develop a framework for understanding variations in argument initiations in the social confrontation episode. We review literature on argument initiations, and then use interpersonal communication literatures on conflict orientations and multiple goal perspectives to develop a behavioral complexity model of argument initiations. Evidence supporting the model is summarized.

1. Introduction

Argument scholars have focused much of their attention to understanding reasoning in public discourse, with less attention paid to the everyday argumentation that occurs between friends, families, neighbors and coworkers. Interpersonal scholars, however, have conducted numerous studies on argument and conflict (Caughlin, Vanghaelisti, & Mikucki-Enyart, 2013; Roloff, 2014; Sillars & Canary, 2013), and found that effective conflict management is important for maintaining relationship and individual well-being (Caughlin & Vangelisti, 2006; Roloff & Soule, 2002). Those who have studied conversational argument or conflict have focused either on general styles and behaviors to understand their overall effectiveness, or focused on conversation practices that characterize disagreement. Fewer researchers have focused on the moves arguers engage in to initiate argumentation in interpersonal contexts.

To address these gaps, our essay is focused on everyday argument that occurs in personal relationships. Interpersonal arguments can involve disagreement over ideas, conduct, or decision-making, but our focus here is on disagreements over behavioral conduct. Our essay is focused on the initial moves expressed in social confrontation, when an actor for the first time engages another actor about a perceived disagreement. Such interactions are "focused" in that actors express "moves" within turn exchanges (Goffman, 1967, p. 20). These initial moves have been referred as an "opening" (Benoit & Benoit, 1990), "initiation" (Newell & Stutman, 1988), and the "confrontation stage" (van Eemeren & Grootendorst, 1984).

The study of argument initiations has value, because the initiating phase can influence actors' beliefs that the disagreement can be resolved (Benoit & Benoit, 1990; Reznik & Roloff, 2011). The importance of argument initiation is also shown by Carrere and Gottman's (1999) research on newlyweds' conflict interactions. Carrere and Gottman had 124 newlywed couples discuss a recurring problem. The interactions were coded for the number of positive (interest, validation, affection) and negative (domineering, whining, disgust) affect states expressed. Six years later, Carrere and Gottman found that the couples who had divorced, compared to those who remained married, had expressed significantly more negative emotions and fewer positive emotions during the first three minutes of the discussion they had experienced years earlier.

Despite the significance of argument initiations, however, the research literature is not extensive and is fragmented across topics and approaches. Thus, our essay has four objectives. First, we review our conception of the social confrontation episode and the role of argument within social confrontation. Second, we review approaches taken to study argument initiations. Third, we selectively review interpersonal research on conflict styles and multiple goal perspectives, and assess the evidence for one multiple goals perspective called the behavioral complexity model. Finally, we use behavioral complexity to advance our conceptual framework for argument initiation moves, and offer preliminary evidence for the framework.

2. The Social Confrontation Episode

A role for argument initiation moves can be seen in Newell and Stutman's model of the social confrontation episode (1988). Drawing from the language-action tradition, they conceptualized social confrontation as a type of communication that occurs when expectations about general social norms or relationship rules are violated and a social situation becomes problematic. As a basic unit of interaction, the social confrontation episode begins with an initiation and ends with some type of remedy or resolution. Newell and Stutman analyzed undergraduates' role-played confrontations and found that social confrontation begins when a confronter signals to the other that his/her behavior is not appropriate. Moves may involve the confronter expressing disagreement or dissatisfaction, giving reasons to warrant his/her claims, and making attributions about causes or responsibilities, all of which are hearable as *reproaches* or *complaints*.

Newell and Stutman found that argument occurs within the social confrontation episode around distinct stock issues. Social actors may disagree over whether a legitimate rule for appropriate conduct has been violated, whether there is a superseding rule that would take precedence in the situation, whether the other person performed the behavior, whether the behavior violated a rule, and whether the other person is responsible.

Within this view, social confrontations are just one type of interaction used to manage disagreement or conflict within relationships. Conflict is generally seen as disagreement between two people who are interdependent and who have incompatible goals (Hocker & Wilmot, 1998). Conflict is further seen as inevitable and potentially beneficial for relationships, but characterized by various myths, such as that conflict is polarizing. Rather, conflict interactions can assume various forms, from antagonistic exchanges to civilized reflective discussion (Paglieri, 2015).

3. Argument Initiation Research

As discussed, the social confrontation episode represents a particular type of conversation that focuses on disagreements over behavioral conduct. We now turn to research on the way social confrontations begin and arguments within the episode are initiated. To date, four approaches have been used to study argument initiations: discourse analytic, strategic, social-cognitive, and normative.

Several groups of conversation and interpersonal scholars have focused on the discursive acts involved in the initiation of disagreements. For instance, Jacobs and Jackson (1982) have shown that argumentation is the expansion of a disagreement relevant adjacency pair, with lines of disagreement organized by the felicity conditions associated with the speech act that is the focus of the disagreement. Often these speech acts express "disapproval for the actions and attributes of speaker or hearer – accusations, insults, criticisms, complaints, challenges, self-deprecation" (p. 227). Cody and McLaughlin (1985) have focused on reproaching in arguments, which are considered the speech acts that raise questions about the reasonableness or worth of another's actions. They found that reproaches typically assume four types: rebukes (e.g., "You shouldn't have borrowed my car without asking"), requesting an account (e.g., "Why did you leave your job?"), implying superiority ("I just don't have time to watch TV"), and expressing surprise or disgust ("That's gross!"; Tracy & Robles, 2013, p. 93). Benoit and Benoit (1990) have also found that four reproaches are typical in conflicts between roommates and romantic partners: insults, commands, accusations, and refusals of requests.

Argument initiations can also include complaints, which Alberts (1988) has analyzed in couples' interactions. She found five types: complaints about actions done or not done; performance complaints about how an action was done; personal characteristic complaints; complaints about physical appearance; and meta-complaints, or complaints about complaining. Overall, Tracy and Robles (2013) have found that reproaches include three general types of speech acts: expressing dissatisfaction, describing problematic actions, and trying to get the other to change (i.e., expressive, representative, and directive speech acts, respectively).

Besides discourse analyses, strategic analyses of argument initiations have focused on ways of performing initiations to increase their effectiveness. Rawlins (1992) has observed that reproaches often involve a behavioral dilemma as part of the dialectic of judgment and acceptance embedded in friendship. While friends and couples are expected to accept and support one another, they also expect to advise, criticize, and help each other improve. Because of this dilemma, expressing a reproach can be complex. Benoit and Benoit (1990) used politeness theory to compare aggravated to mitigated (face-attentive) conflict openings, and found that aggravated openings were viewed as less appropriate and effective than their mitigated versions. Alberts (1988) has similarly found that satisfied couples were more likely to use behavioral complaints that were face-attentive, while dissatisfied couples used personal characteristic complaints that engaged in face attack.

A final strategic approach to argument initiations is Remer and de Mesquita's (1990) stage model of confrontation skill. Three stages are relevant to argument initiations: preparation, lead-in, and confrontation. The preparation stage involves the confronter's cognitive activities (e.g., appraisal of the situation) followed by the lead-in stage in which the confronter's statements usher the confrontee into the confrontation conversation. Lead-in statements may include affirming the relationship and suggesting how the interaction be structured, moves that can reduce resistance and facilitate conflict resolution. The confrontation stage is most effective when statements are specific and take the form: "When you (confrontee) *(description of behavior)*, I (confronter) feel *(feeling state(s))*, because (*consequence to the confronter*) (Remer & de Mesquita, 1990, p. 239).

Unlike pragmatic and strategic studies of argument initiations, social cognitive approaches have focused on the scripts and subjective expectations people have about how argumentative interactions begin and whether people will engage in argument. For instance, Miller (1991) has examined scripts for common friendship conflicts, and found that people expect that confrontations over broken promises, cumulative annoyances, criticism, and illegitimate demands will begin with a question, but for rebuffs the confrontation should begin with an accusation. Fehr, Baldwin, Collins, Patterson and Benditt (1999) examined scripts for anger in conflicts among dating couples, and found that anger is expected the most from conflicts involving betrayals of trust, followed by rebuffs, criticism, negligence and cumulative annoyances. Fehr et al also found that people expected that their partners would reciprocate with positive/constructive actions if they themselves engaged in positive behaviors.

A more extensive program of research on argument engagement is being conducted by Hample, Paglieri and their colleagues (e.g., Hample, Paglieri, & Na, 2012; Paglieri, 2013). They have found that the decision to engage in an argument is a choice that can be modeled as an expectancy value calculation based on people's analysis of the costs and benefits of arguing. In particular, the intention to argue is predicted by the perceived likelihood of winning and the appropriateness of arguing in the situation.

A final approach to understanding argument initiation has been the pragma-dialectic theory developed by Frans van Eemeren and Rob Grootendorst (1984, 2004). Their normative theory identifies four stages for resolving differences of opinion, with the first stage designed for the expression of disagreement. The confrontation stage is ideally composed of arguers expressing their standpoints, accepting or rejecting those standpoints, and asking for or providing definitions, amplifications, and explications (called *usage declaratives*). Usage declaratives help determine if the disagreement is spurious.

Taken together, these lines of research suggest that in social confrontation arguers express directly or indirectly that the other's behavior is inappropriate, arguers provide reasons to warrant their claims, they may ask about the other's standpoint, and they try to get the other to change. Arguers face the dilemma between expressing and legitimizing that the other has broken a social rule with the need to show support and acceptance toward the other. While balancing this dilemma is often achieved with mitigated discourse forms, less is known about how arguers' relational aims and identities are integrated into argument initiation moves. Less is known, too, about how different argument initiations affect arguers' perceptions and responses.

4. The Role of Interpersonal Conflict Orientations and Multiple Goals

We next review research on interpersonal conflict and multiple goals and their implications for understanding argument initiation. For many scholars conflict management involves broad styles or strategies, along with specific patterns and acts (Sillars & Canary, 2013). Conflict styles embody design logics about how to use argument and reasoning to produce desired outcomes in contexts of opposition.

4.1 Conflict Styles Research

Interpersonal researchers have produced typologies of conflict styles from both dual concern models and conceptions of negotiation processes (Putnam, 2013). Both Thomas's (1976) five styles that vary along assertiveness and cooperativeness (competing, collaborating, avoiding, accommodating, compromising), and Rahim's (1983) fives styles that vary along concern for self and concern for others (dominating, integrating, avoiding, obliging, and compromising) have been used to categorize interpersonal conflict behaviors. Researchers have also focused on negotiation processes. *Distributive* negotiation involves fixed values, with the negotiators' aim to maximize individual gains and capture the "largest share of a 'fixed pie'" (Canary, 2003; Putnam, 2013, p. 11). *Integrative* negotiation involves negotiators having a joint problem, shared interests and shared resources to create value that maximizes gain and achieves mutual goals. To this literature conflict scholars have added the avoidant style (Sillars, Coletti, Parry, & Rogers, 1982), reduced five conflict styles to three or four (such as by absorbing compromising into integrating), and added indirect fighting to the list (Donohue & Cai, 2014). Nevertheless, Putnam (2013) observes that distributive and integrative negotiation processes have served as a "foundation for communication strategies and tactics in…interpersonal conflict" (p. 11).

Whether or not particular conflict styles and behaviors facilitate conflict resolution and relationships has been a consistent question for interpersonal scholars (Canary, Cupach, & Messman, 1995; Canary & Spitzberg, 1989; Caughlin & Vangelisti, 2006; Sillars, 1980). The competing style has generally be seen as the least effective in meeting arguers' goals and producing satisfying relationships; by contrast, the collaborative style generally produces constructive conflict resolution and relational outcomes, with integrative tactics associated with relational trust, intimacy, and satisfaction. The collaborative style is viewed as effective and appropriate because it gives arguers the opportunity to understand each other's viewpoint, which can help them define the problem and design a solution (Tutzauer & Roloff, 1988). The compromise style has been associated with moderate levels of effectiveness and relational satisfaction, but conflict avoidance has been found to be ineffective and negatively associated with relational satisfaction (Caughlin & Vangelisti, 2006), except in certain conditions (Roloff & Ifert, 2000).

Other notable evidence of negative conflict behaviors has been John Gottman's (1994) conflict pattern, the Four Horsemen of the Apocalypse. This destructive pattern consists of criticism, contempt, defensiveness, and stonewalling; Gottman's research on married couples has found that complaints and criticism among couples can lead to contempt, defensiveness and stonewalling, which can produce increasingly negative attributions, withdrawal, and even divorce. Longitudinal studies have found that negativity in married couples' arguments (e.g., cruelty/intensity, lacking in mutual appreciation) and the lack of positive affect and expressions can predict divorce (Gottman & Levenson, 2000; McGonagle, Kessler, & Gotlib, 1993).

4.2. Conflict Style Revisions

While the conflict styles literature has produced substantial insights about interpersonal conflict, over the years issues with these schemes have resulted in modifications. Two issues with conflict styles are relevant for studying argument initiations. One issue has been that the popular distributive conflict style actually confounds different types of reasoning that make it difficult to understand how particular conflict tactics are successful. At least four types of reasoning can be identified within the distributive style: a conventional process in which one appeals to social obligations and commitments; a persuasion process in which one uses arguments and persuasion to advocate; a negotiation process in which one maximizes self-gain by seeking concessions, and an expressive process in which accommodation is gained through threats or verbal aggression (Roloff, 2014; Roloff, Putnam, & Anastasio, 2003; Walton & Krabbe, 1995). By differentiating between these reasoning processes distinct argument initiation practices can be identified.

A second issue with styles research has been that particular conflict strategies can play multiple roles in conflict management (Putnam & Wilson, 1989; Roloff, Tutzauer, & Dailey, 1989; Sillars & Canary, 2013). The styles approach only examines general behavioral tendencies instead of specific message practices, so it becomes difficult to address identity and relational dimensions that characterize conflict situations, and fails to account for how conflict messages can address multiple aims (Putnam, 2006).

Given these issues, the study of argument initiation in social confrontation calls for understanding the different orientations people take to resolve their conflicts, as well as understanding how people orient to identity and relationship goals in conflict situations. A framework for differentiating conflict orientations and interaction goals can be found in particular multiple goal perspectives employed in interpersonal communication research.

4.3. Multiple Goals in Interpersonal Argumentation

Over the last few decades one approach to the study of interpersonal communication has been the multiple goals perspective, which has sought to identify differences in messages by identifying goal structures resident in interaction contexts, messages and message producers. Multiple goals perspectives include both psychological and discourse-level perspectives (Craig, 1986; Tracy & Coupland, 1990; Wilson, 2007) that have produced multiple lines of research, such as Dillard's (2004) goals-plans-action model; cognitive editing (Hample & Dallinger, 1987); conversational constraints (Kellerman, 2004); constructivist and message design logic theory (O'Keefe & Delia, 1982; O'Keefe, 1988, 1997); identity implications theory (Wilson, Aleman, & Leatham, 1998); conversation and interactional dilemmas (Tracy, 1997), and goal formation and assembly (Greene, 1997). Multiple goals perspectives have been applied to argument-focused interactions involving conflict management (Keck & Samp, 2007), compliance-gaining (Wilson, 2002), persuasion (Dillard, 1990), serial arguments (Bevan, Finan, & Kamisky, 2008), and dyadic conflict patterns (Caughlin & Scott, 2010).

Most multiple goal perspectives assume that communication is goal-directed, that individuals' communicative efforts reflect multiple aims, and that interaction goals can conflict or compete with each other (Berger, 1997; O'Keefe & Delia, 1982). Thus, communicative behavior can be understood to be a strategic response to the multiple aims existing in the socio-cultural and interactional context. This reasoning has specifically been used to examine arguments in regulative messages calling for behavior change (Kline, 1991; O'Keefe, 1988).

Most multiple goal perspectives can be traced to a particular constructivist theoretical statement by Clark and Delia (1979). Integrating tenets from symbolic interactionist, communication, and cognitive-developmental theories, Clark and Delia argued that any interaction involves participants defining the situation in terms of at least an instrumental objective, interpersonal objectives that create or sustain desired relationships, and identity aims that sustain desired identities for the interactants (Weinstein, 1969). Identity aims include general face wants (Brown & Levinson, 1978) as well as situated identity aims (McCall & Simmons, 1973). These situational definitions and aims form the working consensus in an interaction, with beliefs about tasks, identities and relationships shaping communicators' strategies and perceived obstacles, and socio-cultural definitions and roles producing constraints and resources for what lines of action are best to use in the interaction.

Building from Clark and Delia's (1979) essay, O'Keefe and Delia (1982) further theorized that multiple goals can be managed along varying levels of behavioral complexity. Specifically, they argued that "messages can be seen as the product of multiple communicative intentions and message design as the product of reconciling multiple objectives in performance" (p. 52). They proposed that when faced with complex tasks, people manage their messages with one of three goal management strategies: *selection* (giving priority and expression to one single goal and ignoring other aims); *separation* (pursuing competing goals in temporally or behaviorally separated aspects of a message); or *integration* (reconciling competing goals through message designs that simultaneously accomplish task, identity, and relational goals). Behaviorally complex messages (i.e., those with goal integration or separation) are theorized to be more effective and appropriate than less behaviorally complex messages (i.e., those with goal selection).

4.4. Evidence for the Behavioral Complexity Model

Over the years there have been at least six types of evidence produced to corroborate a behavioral complexity model of multiple goal management.

A first test of the behavioral complexity account comes from several lines of research which show that messages which enact task, identity and relationship aims, or arguers who approach the interaction with these multiple goals, will be more effective. Dillard, Hardin and Segrin (1989) have shown that positive identity and relationship goals are associated with perceived persuasiveness in compliance gaining messages. Bevan et al. (2008) have also shown that serial arguers' positive and mutual goals are associated with integrative conflict tactics. Finally, Hample and his colleagues have shown that people can have relationship and identity motivations for engaging in argument besides the motive to secure agreement, such as for asserting dominance or engaging in play (e.g., Hample & Dallinger, 1995; Hample Han, & Payne, 2010). These aims are pursued with others in different ways, using blurting, degree of cooperation, and degree of civility (e.g, Hample & Benoit, 1999; Hample, Richards, & Skubisz, 2012).

A second and more rigorous test of behavioral complexity has examined the effectiveness of the integrative goal management strategy. O'Keefe and Shepherd (1987) analyzed dyadic discussions undergraduates had over policy conflicts. Their analysis found that compared to the selection strategy, the integrative goal management strategy was positively associated with overall liking for the discussion partner. More recently, Scott and Caughlin (2014) analyzed the discussions of adult children and their parents about end-of-life decisions, and found that discussion moves that collectively enacted identity, relationship and instrumental aims were particularly effective in reaching mutually satisfying decisions.

A third test of the behavioral complexity model is embedded in O'Keefe's message design logic theory (1988). Extending O'Keefe and Shepherd's theorizing (1987), O'Keefe asserts that messages differ in three ways: (a) their complexity of goal structure (that is, whether messages employ a minimal, unifunctional, or multifunctional structure); (b) the way goals are expressed in messages (such as by employing an expressive, conventional or rhetorical message design logic); and (c) the type of multiple goal management strategy used (selection, separation, or integration). People use different message design logics or ways of reasoning about situations, with circumstances calling for an expressive design logic (inducing change by expressing one's unedited thoughts and affect), a conventional design logic (by appealing to social roles, norms, obligations, commitments), or a rhetorical design logic (creating desired identities and a new situation).

Tests of message design logic theory have found that rhetorical and conventional design logics in messages that initiate social confrontation are more effective than messages containing an expressive design logic (Bingham & Burleson, 1989; O'Keefe & McCornack, 1987). Bingham and Burleson (1989) also found that messages that accomplished multiple goals (i.e., task, identity, relationship) were more relationally effective than messages that were either only task focused or messages that did not address a clear goal. A similar test of the behavioral complexity model has been Keck and Samp's (2007) sequential analysis of multiple goals in conflict interactions. Keck and Samp analyzed the conflict interactions of close friends and dating partners, and found that arguers with other identity or relationship goals were associated with integrative conflict tactics, while arguers with self-focused or instrumental goals were associated with distributive conflict tactics.

A related of test of multiple goals has come from identity implications theory (Wilson & Kunkel, 2000). Wilson et al (1999) reasoned that influence goals are often difficult to pursue because of inherently conflicting face threats for the message producer and recipient that are linked to the influence goal. In support of their theory, Wilson and Kunkel (2000) analyzed young adults' recalled conversations about giving advice and asking favors, and found that different face threats were associated with distinct influence goals (such as appearing nosy when giving advice or too lazy when asking for a favor). Not managing these threatened identities could be seen as obstacles to obtaining compliance-gaining success.

A fifth test of the behavioral complexity model has been to determine if speakers' use of behaviorally complex goals or the integrative goal management strategy are related to the development of their social cognitive systems. Those with complex cognitive systems should have a basis for designing collaborative and goal-integrating messages. Consistent with these expectations, O'Keefe and Shepherd (1987) found that the cognitive complexity of interactants was positively associated with the use of the integrative goal management strategy, and to the separation strategy in initiating lines of action. In other studies, cognitively complex communicators have been found to be more likely to pursue identity and relational aims, design messages with person-centered arguments, and produce more listener adapted message plans and strategies in conflict interactions (Kline, 1991, 2007; O'Keefe, 1988; Waldron & Applegate, 1991).

A final test of the behavioral complexity model comes from health persuasion studies that have identified ways of expressing multiple goals in messages. Goldsmith and her colleagues (Goldsmith, Lindholm, & Bute, 2006; Goldsmith, Bute, & Lindholm, 2012) have studied conversational dilemmas between spouses regarding lifestyle changes needed by a partner who had experienced a cardiac event. They discovered that spouses managed their conversation dilemmas in ways that mapped onto the selection, separation, and integration goal management strategies posited by O'Keefe and Delia (1982). Specific practices included rationing or reducing the amount of talk, saying it nicely to protect identities and emphasize rapport, and framing the talk cooperatively so that the talk no longer threatened identities or relational definitions.

In sum, the findings from these six types of tests provide a basis for creating a behavioral complexity framework to characterize argument initiation moves in social confrontation. The next section describes our framework and presents initial evidence for its validity.

5. Conceptualizing Argument Initiation Moves in Social Confrontation

Our framework delineates argument initiations along two dimensions: the conflict orientation, or argument design frame construed as needed by an arguer, and the type of goal structure embedded in the arguer's initiation moves. We believe that there are at least three conflict orientations relevant to argument initiations that reflect differences in message design logics and dialogue types.

When facing an interpersonal disagreement with one's friend or romantic partner (such as when a dating partner violates a relationship norm), a first decision is whether to remain silent or engage the partner to resolve the issue. Such decisions reflect means-end reasoning (Brown & Levinson, 1978). When a partner decides to engage in confrontation, initiating moves can reflect one of several argument, or conflict-solving orientations, such as an *eristic* orientation, a *conventional confrontation* orientation, or a *collaborative* orientation.

5.1 Argument Orientations

The *eristic orientation* approaches interpersonal conflict with antagonistic reactivity and a desired outcome the gaining of immediate compliance, accommodation, or temporary settlement (Walton & Krabbe, 1995). The design principle is to express what you think and feel straightforwardly, with initial moves expressing emotions, unedited thoughts, and recycled blaming. This orientation could be considered a belligerent form of the distributive style (Canary, et al., 1995), with moves characterized by accusations, demands, expressions of intense negative emotions, and threats. The orientation shares propositions with Sillars and Canary's (2013) *direct* and *face-attacking* conflict strategy and O'Keefe's (1988) expressive design logic. Eristic initiations would likely contain negative affect and judgments, as illustrated in this example:

> I can't believe it happened again! You have been a jerk tonight. You made so many mean jokes about me in front of your friends. You keep putting me down in front of your friends. You and your friends are so rude, getting a good laugh at my sacrifice! Don't you ever embarrass me like that again!

The *conventional confrontation orientation* sees behavioral conflict as conduct that prevents the achievement of culturally recognized goals, or that violates sociocultural rules or expectations, as discovered by Newell and Stutman (1988) in their description of the social confrontation episode. Some objectives are culturally and conventionally relevant to situations, with a person expected to employ communicative acts that are appropriate and relevant to the aims and expectations of the situation (Wilson, 2014). Thus, the conventional confrontation orientation sees that the task is to indicate that the other has violated a rule, and to obligate the other to change his or her behavior so that social expectations are fulfilled.

This orientation employs a conventional message design logic (O'Keefe, 1988) in seeing communication as "played cooperatively according to socially conventional rules and procedures" (p. 86). The use of persuasion is similar to Wilson and Krabbe's (1995) *persuasion dialogue* orientation. Conventional argument initiations likely express dissatisfaction about the partner's conduct by citing the social rule the behavior violates, as in this example:

> Some of the jokes you and your friends made tonight were annoying. You might think they're funny, but I actually found them offensive. I felt really embarrassed when you made jokes about me like that in front of your friends. And I don't even know these people. As my girlfriend/boyfriend, you should think more about my feelings. I'd rather you didn't make fun of me in front of your friends.

A third way to approach argument initiations in conduct disagreements is through a *collaborative orientation*, which aims to find a working definition of the problem, possible solutions, and dialogue to promote mutual understanding and trust. A collaborative orientation conceives of ways to link both parties' objectives together, and promotes mutual action to transform the situation. Arguers may ask questions about each other's goals to begin the collaborative process. This conflict orientation embodies O'Keefe's (1988) rhetorical design logic in that it regards communication as a way to define or re-define the situation and its identities that transcend opposing ideas; the orientation also shares propositions with Walton and Krabbe's (1995) dialogue framework, with arguers using information-seeking to learn others' perspectives, deliberation to engage in means-end discussion, and persuasion to advance reasoning. Strategic maneuvering within these initial moves may include discussing shared interests and using individuated reasoning. Collaborative initiations express dissatisfaction but encourage the partner to discuss her/his interests, as illustrated below:

> You and your friends seem to enjoy teasing each other for fun. Am I right? While that's ok, I'm upset. I don't have the same sense of humor as you guys, and I feel uncomfortable especially when those jokes are about me. I'm not saying you should stop cracking jokes. I understand that it's how you and your friends connect. So can we talk about it? Let's figure out a way that both of us can feel comfortable.

5.2. Goal Structures in Argument Initiations

Besides adopting a particular orientation toward designing an argument initiation in social confrontation, we also assume that conflict messages reflect people's efforts to satisfy at least three goal types: instrumental goals, identity goals, and relationship goals. In support of this view Keck and Samp (2007) have shown that people strive to achieve task, identity and relationship goals in conflict discussions. Stutman and Newell (1990) have also confirmed that confronters strive to achieve primary instrumental goals such as advancing positions, and secondary goals such as appearing competent and likeable and preserving the relationship with the partner. Thus, a confronter faces the challenge of managing potentially conflicting goals when designing the opening moves of a confrontation. Consistent with the literature and following O'Keefe (1988), we contend that an argument initiation can be identified by the number and type of goals it is designed to serve. Argument initiations can reflect at least three different levels of goal structures: minimal, unifunctional, and multifunctional.

Minimal messages refer to those in which the speaker pursues no clear goal, such as when the speaker is "confused, ambivalent" or is only communicating "phatically" (O'Keefe, 1988, p. 90). Variations in minimal goal initiations are illustrated below by messages that confront a partner about prioritizing friends over the confronter:

> *Eristic minimal*: I CAN'T BELIEVE IT HAPPENED AGAIN! I've had it. You think everything and everyone is more important than me!

> *Conventional minimal*: I'm really upset! You spend more time with your friends than with me! I can't believe you are putting your friends before me, after all the sacrifices I have made for you. It's really frustrating.

> *Collaborative minimal*: I'm really upset that we won't be seeing each other today. It's been a while since we got together. I know we all need friends. But I can't see how we can keep things this way. Can we talk about it?

Unifunctional initiations are those in which the confronter pursues one dominant goal (O'Keefe, 1988). Eristic unifunctional initiations may pursue the instrumental goal by expressing negative emotions or accusations; conventional unifunctional initiations may assert how the interactional partner has violated a relationship norm or expectation; and collaborative unifunctional initiations may ask the other about why he/she has behaved in a particular way. Such variations in unifunctional goal initiations are illustrated below:

Eristic unifunctional: I can't believe it happened again! It's always work, friends, work, friends. You never spend time with me. You are my girlfriend/boyfriend, but all you do is ignore me. I've had it. Everyone is more important than me. You better do something different.

Conventional unifunctional: Look, you spend more time with your friends than with me. After all the space I gave you when you were really busy with work, you are still putting your friends before me. Why be my girlfriend/boyfriend if you don't even want to spend time with me? You should set aside more time for us to be together.

Collaborative unifunctional: I'm upset that we won't spend the day together. I feel we don't see each other enough. Do you feel the same? Although we text and talk on the phone, they're different from spending time face-to-face. What do you think? I'm not saying you need to give up your friendships. I'm sure we can figure out a way to get together more often and still hang out with friends.

Finally, *multifunctional argument initiations* are those in which the confronter addresses identity or relationship aims in addition to the instrumental aim. The arguer not only conveys his/her dissatisfaction and the need to resolve the issue, but also likely affirms positive identities and the relationship with the partner. A collaborative multifunctional initiation likely redefines the situation to improve it and boost the identities and relationship of the parties. A conventional multifunctional initiation may fulfill the primary goal of confronting a partner's behavior while using politeness strategies to preserve positive face and autonomy. An eristic multifunctional message initiates confrontation by expressing dissatisfaction but not be as strongly toned. These variations are illustrated below:

Eristic multifunctional: I can't take it anymore. I'm so upset. Most people can't handle a person who always puts everyone else first but her/his partner. It happens again and again. I'm losing my patience. Why does my girlfriend/boyfriend think everyone else is more important than me? I just don't get it!

Conventional multifunctional: I'm not sure if you've noticed, but you spend more time with your friends than with me. I'm actually upset. I understand there're not enough hours in a day to make time for everyone, but I feel left out when you always make plans with your friends instead of with me. So could you consider setting aside more time for us to be together?

Collaborative multifunctional: It's great that you have so many friends to hang out with. You've always been popular. I'm just sad that we won't be spending time together today. I feel we don't see each other enough. Do you feel the same? It's been great between us. I'm afraid that we would drift apart. What do you think? I'm sure we can figure out a way that's best for us and our friendships.

Since conflict orientations and goal structures are orthogonal dimensions, argument initiations can reflect an eristic, conventional or collaborative conflict orientation that may further have a minimal, unifunctional, or multifunctional goal structure. By crossing the three conflict orientations with three types of goal structure, a behavioral complexity model of nine types of initiating moves in confrontation results. Multifunctional goal management strategies adopt a selection form within the eristic conflict orientation, a separation form within the conventional confrontation orientation, and an integrative form within the collaborative conflict orientation.

The most complete test of this model of argument initiations was recently completed by Song (2015). Song developed two scenarios that reflected disputes commonly reported by young dating couples--one in which a dating partner expresses dissatisfaction about the partner choosing to spend leisure time with friends rather than with the confronter, and the other in which the confronter expresses dissatisfaction about the partner making inappropriate jokes about the confronter at a party. For each scenario, Song developed realistic argument initiations of generally equal length that reflected combinations of the conflict orientations and goal structures. Participants were randomly assigned to imagine themselves being confronted by their dating partner who used one of the nine argument initiations in response to one of the two scenarios. Participants then responded to a series of scales measuring their perceptions of the effectiveness and outcomes of the argument initiation.

Song found substantial support for the behavioral complexity model of argument initiations. As predicted, eristic, conventional, and collaborative argument initiations were progressively higher in perceived message effectiveness, conflict resolvability, and relationship enhancement. Multifunctional argument initiations were progressively higher in perceived message effectiveness and effectiveness at accomplishing identity and relationship goals than unifunctional or minimal structured argument initiations.

When combined, argument initiations with the highest level of sophistication, defined as having more a complex type of conflict-solving orientation and more complex goal structures (i.e., collaborative multifunctional; collaborative unifunctional; conventional multifunctional) had higher message effectiveness, perceived conflict resolvability, relationship enhancement, and likelihood of attaining multiple goals than argument initiations that combined lower types of conflict-solving orientation and simpler types of goal structure.

6. Research Directions and Conclusions

Our framework for understanding argument initiations in social confrontations produced several insights about argument initiations in personal relationships. Argument initiations that conveyed no clear goal or that expressed an eristic orientation were perceived to be less effective than initiations that had a clear goal or used a conventional or collaborative conflict orientation. Argument initiations that managed multiple goals with a conventional or collaborative orientation were perceived as most effective. The separation goal management strategy was as effective as the integrative goal management strategy in argument initiations.

Given the importance of argument initiations in social confrontations, we believe that they should be studied further. A number of avenues could be pursued. Learning how argument initiation moves influence the impressions and subsequent moves of interactants over specific interactions would expand our knowledge of how initiations shape reasoning practices and argument elaboration. Learning more ways people skillfully initiate arguments, such as understanding how arguers manage their emotions, would expand our knowledge of the methods available to arguers. Variations in initiations may meaningfully differ as a function of the topic, lines of argument, and choices of their expression, currently understood as strategic maneuvering. Understanding the influence of media affordances on the design and outcome of argument initiations is also needed, given the multiplexity of choices available to arguers to initiate disagreement. The framework summarized here utilized several, but not all of the dialogues proposed by Walton and Krabbe (1995). Future work could examine how argument initiations commence in other types of dialogue, such as reaching a settlement with bargaining (negotiation), agreeing to a plan of action (deliberation), and sharing information (information-seeking dialogue).

Finally, understanding how argument initiations operate will necessitate linking them to the decisions of arguers to engage in argument, as conceived by Hample, Paglieri and his colleagues (Paglieri, 2013). Whether or not people's analysis of the costs and benefits of arguing is associated with particular argument initiation moves could help explicate how arguers become involved in particularly destructive or constructive patterns of argument.

In conclusion, this essay differentiates between eristic, conventional-confrontation, and collaborative conflict orientations, and employs a complete test of behavioral goal complexity theory to provide support for a rational model of argument initiation design. Conflict in relationships is inevitable, and confronting a partner's relationship transgression may not be easy. The framework offered here displays variations available to arguers and suggests ways to manage the complexity by constructing argument initiations that 1) engage the partner to facilitate mutual understanding about the situation, and 2) affirm each other's identity and the relationship in the process.

References

Alberts, J. K. (1988). An analysis of couples' conversational complaints. *Communication Monographs, 55*, 184-197.
Benoit, W. L., & Benoit, P. J. (1990). Aggravated and mitigated opening utterances. *Argumentation, 4*, 171-183.
Berger, C. R. (1997). *Planning strategic interaction: Attaining goals through communicative action.* Mahwah, NJ: Lawrence Erlbaum.
Bevan, J. L., Finan, A., & Kaminsky, A. (2008). Modeling serial arguments in close relationships: The serial argument process model. *Human Communication Research, 34*, 600-624.
Bingham, S. G., & Burleson, B. R. (1989). Multiple effects of messages with multiple goals: Some perceived outcomes of responses to sexual harassment. *Human Communication Research, 16*, 184-216.
Brown, P., & Levinson, S. (1978). Universals in language usage: Politeness phenomena. In E. Goody (Ed.), *Questions and politeness: Strategies in social interaction* (pp. 56-310). Cambridge: Cambridge Univ. Press.
Canary, D. J. (2003). Managing interpersonal conflict: A model of events related to strategic choices. In J.O. Greene & B.R. Burleson (Eds.), *Handbook of communication and social interaction skills* (pp. 515-550). Mahwah, NJ: Lawrence Erlbaum.
Canary, D. J., & Spitzberg, B. H. (1989). A model of the perceived competence of conflict strategies. *Human Communication Research, 15*, 630-649.
Canary, D.J., Cupach, W.R., & Messman, S.J. (1995). *Relationship conflict.* Thousand Oaks, CA: Sage Publications.

Carrere, S., & Gottman, J.M. (1999). Predicting divorce among newlyweds from the first three minutes of a marital conflict discussion. *Family Process, 38*, 293-301.

Caughlin, J. P., & Scott, A. M. (2010). Toward a communication theory of the demand/withdraw pattern of interaction in interpersonal relationships. In S. W. Smith, & S. R. Wilson (Eds.), *New directions in interpersonal communication research* (pp. 180-200). Beverly Hills, CA: Sage.

Caughlin, J.P. & Vangelisti, A. L. (2006). Conflict in dating and marital relationships. In J.G. Oetzel & S. Ting-Toomey (Eds.), *The Sage handbook of conflict communication: Integrating theory, research, and practice* (pp. 129-158). Thousand Oaks, CA: Sage Publications.

Caughlin, J.P., Vangelisti, A. L. & Mikucki-Enyart, S.L. (2013). Conflict in dating and marital relationships. In J.G. Oetzel & Stella Ting-Toomey (Eds.), *The Sage handbook of conflict communication* (2nd ed., pp. 161-186). Thousand Oaks, CA: Sage Publications.

Clark, R. A., & Delia, J. G. (1979). Topoi and rhetorical competence. *Quarterly Journal of Speech, 65*(2), 187-206.

Cody, M.J., & McLaughlin, M.L. (1985). Models for the sequential construction of accounting episodes: Situational and interactional constraints on message selection and evaluation. In R. L. Street, & J. N. Cappella (Eds.), *Sequence and pattern in communicative behavior* (pp. 50-69). London: Arnold.

Craig, R. T. (1986). Goals in discourse. In D. Ellis (Ed.), *Contemporary issues in language and discourse processes* (pp. 257-273). Hillsdale, NJ: Lawrence Erlbaum.

Dillard, J. P. (1990). The nature and substance of goals in tactical communication. In M. J. Cody & M. L. McLaughlin (Eds.), *The psychology of tactical communication* (pp. 70-90). Clevedon, UK: Multilingual Matters.

Dillard, J. P. (2004). The goals-plans-action model of interpersonal influence. In J. S. Seiter & R. H. Gass (Eds.), *Perspectives on persuasion, social influence, and compliance gaining* (pp. 185-206). Boston, MA: Allyn & Bacon.

Dillard, J. P., Segrin, C., & Harden, J. M. (1989). Primary and secondary goals in the production of interpersonal influence messages. *Communication Monographs, 56*, 19–38.

Donohue, W.A., & Cai, D.A. (2014). Interpersonal conflict: An overview. In N.A. Burrell, M. Allen & B.M. Gayle (Eds.), *Managing interpersonal conflict: Advances through meta-analysis* (pp. 22-41). New York: Routledge.

Fehr, B., Baldwin, M., Collins, L., Patterson, S., & Benditt, R. (1999). Anger in close relationships: An interpersonal script analysis. *Journal of Personality & Social Pyschology, 25*, 299-312.

Goffman, E. (1967). *Interaction ritual: Essays in face to face behavior.* New York: Anchor Books.

Goldsmith, D. J., Lindholm, K. A., & Bute, J. J. (2006). Dilemmas of talking about lifestyle changes among couples coping with a cardiac event. *Social Science & Medicine, 63*(8), 2079-2090.

Goldsmith, D.J., Bute, J.J., & Lindholm, K.A. (2012). Patient and partner strategies for talking about lifestyle change following a cardiac event. *Journal of Applied Communication Research, 40*(1), 65-86.

Gottman, J.M. (1994). *What predicts divorce? The relationship between marital processes and marital outcomes.* Hillsdale, NJ: Lawrence Erlbaum.

Gottman, J. M., & Levenson, R. W. (2000). The timing of divorce: Predicting when a couple will divorce over a 14-year period. *Journal of Marriage and Family, 62*, 737-745.

Greene, J. O. (1997). A second generation action assembly theory. In J. O. Greene (Ed.), *Message production: Advances in communication theory* (pp. 151-170). Mahwah, NJ: Lawrence Erlbaum.

Hample, D., & Benoit, P. J. (1999). Must arguments be explicit and violent: A study of naive social actors' understandings. In F. H. van Eemeren, R. Grootendorst, J. A. Blair, & C. A. Willard (Eds.), *Proceedings of the fourth international conference of the International Society for the Study of Argumentation* (pp. 306-310). Amsterdam, the Netherlands: SICSAT.

Hample, D., & Dallinger, J. M. (1987). Individual differences in cognitive editing standards. *Human Communication Research*, 14(2), 123-144.

Hample, D., Han, B., & Payne, D. (2010). The aggressiveness of playful arguments. *Argumentation, 24*, 405-421.

Hample, D., Paglieri, F., & Na, L. (2012). The costs and benefits of arguing: Predicting the decision whether to engage or not. In F. H. van Eemeren & B. Garssen (Eds.), *Topical themes in argumentation theory: Twenty exploratory studies* (pp. 307-322). New York NY: Springer.

Hample, D., Richards, A. S., & Skubisz, C. (2013). Blurting. *Communication Monographs, 80*, 503-532.

Hocker, J.L. & Wilmot, W.W. (1998). *Interpersonal conflict* (5thd ed.). Dubuque, IA: William C. Brown Publishers.

Johnson, K.L., & Roloff, M.E. (1998). Serial arguing and relational quality: Determinants and consequences. *Communication Research, 25,* 327-343.

Keck, K. L., & Samp, J. A. (2007). The dynamic nature of goals and message production as revealed in a sequential analysis of conflict interactions. *Human Communication Research, 33,* 27-47.

Kellermann, K. (2004). A goal-directed approach to gaining compliance: Relating differences among goals to differences in behaviors. *Communication Research*, 31(4), 397-445.

Kline, S. L. (1991). Construct differentiation and person-centered regulative messages. *Journal of Language and Social Psychology*, 10(1), 1-27.

Kline, S. L. (2007). Displaying reasonableness: Two argument practices and their developmentally-based differences. In F. Van Eemeren, J. A. Blair, C.A. Willard & B. Garssen (Eds.). *Selected papers from the Sixth Conference of the International Society for the Study of Argumentation* (pp. 767-774). Dordrecht, Netherlands: Foris Publications.

Liu, M., & Wilson, S. R. (2011). The effects of interaction goals on negotiation tactics and outcomes: A dyad-level analysis across two cultures. *Communication Research*, 38(2), 248-277.

McCall, G.J., & Simmons, J.L. (1973). *Identities and interaction, Rev. Ed.* New York: The Free Press.

McGonagle, K. A., Kessler, R. C., & Gotlib, I. H. (1993). The effects of marital disagreement style, frequency, and outcome on marital disruption. *Journal of Social and Personal Relationships, 10,* 385-404.

Miller, J.B. (1991). Women's and men's scripts for interpersonal conflict. *Psychology of Women's Quarterly, 15,* 15-29.

Newell, S. E. & Stutman, R.K. (1988). The social confrontation episode. *Communication Monographs, 55,* 266-285.

O'Keefe, B. J. (1988). The logic of message design: Individual differences in reasoning about communication. *Communications Monographs*, 55(1), 80-103.

O'Keefe, B. J. (1997). Variation, adaptation, and functional explanation in the study of message design. In T. L. Albrecht & G. Philipsen (Eds.), *Developing communication theories* (pp. 85-118). Albany, NY: SUNY Press.

O'Keefe, B. J., & Delia, J. G. (1982). Impression formation and message production. In M. E. Roloff & C. R. Berger (Eds.), *Social cognition and communication* (pp. 33-72). Thousand Oaks, CA: Sage Publications.

O'Keefe, B. J., & McCornack, S. A. (1987). Message design logic and message goal structure: Effects on perceptions of message quality in regulative communication situations. *Human Communication Research, 14*(1), 68-92.

O'Keefe, B. J., & Shepherd, G. J. (1987). The pursuit of multiple objectives in face-to-face persuasive interactions: Effects of construct differentiation on message organization. *Communications Monographs, 54*(4), 396-419.

Paglieri, F. (2013). Choosing to argue: Towards a theory of argumentative decisions. *Journal of Pragmatics, 59*, 154-163.

Paglieri, F. (2015). Arguments, conflict, and decisions. In F. D'Errico, I. Poggi, A. Vinciarelli, & L. Vincze (Eds.), *Conflict and multimodal communication: Social research and machine intelligence* (pp. 117-136). Switzerland: Springer International.

Putnam, L.L., & Wilson, S.R. (1989). Argumentation and bargaining strategies as discriminators of integrative outcomes. In M.A. Rahim (Ed.), *Managing conflict: An interdisciplinary approach* (pp. 121-144). New York: Praeger.

Putnam, L. L. (2006). Definitions and approaches to conflict and communication. In J. Oetzel & S. Ting-Toomey (Eds.), *The SAGE handbook of conflict communication: Integrating theory, research, and practice* (pp. 1-32). Thousand Oaks, CA: Sage Publications.

Putnam, L. L. (2013). Definitions and approaches to conflict and communication. In J.G. Oetzel & S. Ting-Toomey (Eds.), *The Sage handbook of conflict communication: Integrating theory, research, and practice* (2nd ed., pp. 1-40). Thousand Oaks, CA: Sage Publications.

Rahim, M. A. (1983). A measure of styles of handling interpersonal conflict. *Academy of Management Journal, 26*, 368-376.

Rawlins, W.K. (1992). *Friendship matters: Communication, dialectics and the life course.* Hawthrone, NY: de Gruyter.

Remer, R. & de Mesquita, P. (1990), Teaching and learning the skills of interpersonal confrontation. In D.D. Cahn (Ed.), *Intimates in conflict: A communication perspective* (pp. 225-252). Hillsdale, NJ: Lawrence Erlbaum.

Reznik, R. M., & Roloff, M. E. (2011). Getting off to a bad start: The relationship between communication during an initial episode of a serial argument and argument frequency. *Communication Studies, 62*, 291-306.

Roloff, M. E. (1987). Communication and conflict. In C. R. Berger, & S. H. Chaffee (Eds.), *Handbook of communication science* (pp. 484-534). Newbury Park, CA: Sage.

Roloff, M.R. (2014). Conflict and communication: A roadmap through the literature. In N.A. Burrell, M. Allen, & B.M. Gayle (Eds), *Managing interpersonal conflict: Advances through meta-analysis* (pp. 42-58). New York: Routledge.

Roloff, M.E., & Ifert, D.E. (2000). Conflict management through avoidance: Withholding complaints, suppressing arguments, and declaring topics taboo. In S. Petronio (Ed.), *Balancing the secrets of private disclosures* (pp. 151-163). Mahwah, NJ: Lawrence Erlbaum.

Roloff, M.E., Putnam, L. L., & Anastasiou, L. (2003). Negotiation skills. In J.O. Greene & B.R. Burleson (Eds.), *Handbook of communication and social interaction skills* (pp. 801-833). Mahwah, NJ: Lawrence Erlbaum.

Roloff, M.E., & Soule, K.P. (2002). Interpersonal conflict: A review. In M.L. Knapp & J.A. Daly (Eds.), *Handbook of interpersonal communication* (3rd ed., pp. 475-528). Thousand Oaks, CA: Sage Publications.

Roloff, M.E., Tutzauer, F.E., & Dailey, W.O. (1989). The role of argumentation in distributive and integrative bargaining contexts: Seeking relative advantage but at what cost? In M.A. Rahim (Ed.), *Managing conflict: An interdisciplinary approach* (pp. 109-120). New York: Praeger.

Scott, A. M., & Caughlin, J. P. (2014). Enacted goal attention in family conversations about end-of-life health decisions. *Communication Monographs, 81*, 261-284.

Sillars, A. L. (1980). The sequential and distributional structure of conflict interactions as a function of attributions concerning the locus of responsibility and stability of conflicts. In D. Nimmo (Ed.) *Communication Yearbook 4*. New Brunswick: Transaction Books.

Sillars, A.L., Coletti, F.S.F., Parry, D., & Rogers, M.A. (1982). Coding verbal conflicts: Non-verbal and perceptual correlates of the "avoidance-distributive-integrative" distinction. *Human Communication Research, 9*, 83-95.

Sillars, A.L., & Canary, D.J. (2013). Conflict and relational quality in families. In A.L. Vangelisti (Ed.), *The Routledge handbook of family communication* (pp. 338-358). New York: Routledge.

Stutman, R. K., & Newell, S. E. (1990). Rehearsing for confrontation. *Argumentation, 4*, 185-198.

Song, W. (2015). *Conflict-solving orientation and goal management: Effectiveness of opening messages in interpersonal conflict*. Unpublished doctoral dissertation, Ohio State University.

Thomas, K. W. (1976). Conflict and negotiation management. In M. D. Dunnette (Ed.), *Handbook of industrial and organizational psychology* (pp. 889-935). Chicago: Rand McNally.

Tutzauer, F. & Rooff, M.E. (1988). Communication processes leading to integrative agreements: Three paths to joint benefits. *Communication Research, 15*, 360-380.

Tracy, K. (1997). *Colloquium: Dilemmas of academic discourse*. Norwood, NJ: Ablex.

Tracy, K., & Coupland, N. (1990). Multiple goals in discourse: An overview of issues. *Journal of Language and Social Psychology, 9*(1-2), 1-13.

Tracy, K., & Robles, J.S. (2013). *Everyday talk: Building and reflecting identities* (2nd ed). New York: Guilford Press.

Van Eemeren, F.H., & Grootendorst, R. (1984). *Speech acts in argumentative discussions: A theoretical model for the analysis of discussions directing towards solving conflicts of opinion.* Berlin: Foris/Mouton de Gruyter.

Van Eemeren, F.H., & Grootendorst, R. (2004). *A systematic theory of argumentation. The pragma-dialectical approach.* Cambridge, England: Cambridge University Press.

Waldron, V.R., & Applegate, J. L. (1994). Interpersonal construct differentiation and conversation planning: An examination of two cognitive accounts for the production of competent verbal disagreements tactics. *Human Communication Research, 21*, 3-35.

Walton, D.N., & Krabbe, E.C.W. (1995). *Commitment in dialogue: Basic concepts of interpersonal reasoning.* Albany, NY: State University of New York.

Wilson, S. R. (2002). *Seeking and resisting compliance: Why people say what they do when trying to influence others.* Thousand Oaks, CA: Sage.

Wilson, S. R. (2007). Communication theory and the concept of "goal." In B. B. Whaley, & W. Samter (Eds.), *Explaining communication: Contemporary theories and exemplars* (pp. 77-111). Mahwah, NJ: Lawrence Erlbaum.

Wilson, S. R. (2014). Conventional and personal goals during conflict: A commentary on managing interpersonal conflict. In N.A. Burrell, M. Allen, & B.M. Gayle (Eds.), *Managing interpersonal conflict: Advances through meta-analysis* (pp. 59-74). New York: Routledge.

Wilson, S. R., Aleman, C. G., & Leatham, G. B. (1998). Identity implications of influence goals: A revised analysis of face-threatening acts and application to seeking compliance

with same-sex friends. *Human Communication Research, 25*(1), 64-96.
Weinstein, E.A. (1969). The development of interpersonal competence. In E.A. Goslin (Ed.), *Handbook of socialization theory and research* (pp. 753-775). Chicago: Rand-McNally.

Part II
Rationality

Chapter 5

Pushing the Bounds of Rationality: Argumentation and Extended Cognition

David Godden

Michigan State University, East Lansing, Michigan, USA, dgodden@msu.edu

Abstract. One of the central tasks of a theory of argumentation is to supply a theory of appraisal: a set of standards and norms according to which argumentation, and the reasoning involved in it, is properly evaluated. In their most general form, these can be understood as rational norms, where the core idea of rationality is that we rightly respond to reasons by according the credence we attach to our doxastic and conversational commitments with the probative strength of the reasons we have for them. Certain kinds of rational failings are so because they are manifestly illogical—for example, maintaining overtly contradictory commitments, violating deductive closure by refusing to accept the logical consequences of one's present commitments, or failing to track basing relations by not updating one's commitments in view of new, defeating information. Yet, according to the internal and empirical critiques, logic and probability theory fail to supply a fit set of norms for human reasoning and argument. Particularly, theories of bounded rationality have put pressure on argumentation theory to lower the normative standards of rationality for reasoners and arguers on the grounds that we are bounded, finite, and fallible agents incapable of meeting idealized standards. This paper explores the idea that argumentation, as a set of practices, together with the procedures and technologies of argumentation theory, is able to extend cognition such that we are better able to meet these idealized logical standards, thereby extending our responsibilities to adhere to idealized rational norms.

1. Logic and Reasoning: The Prescriptivity Gap

1.1. The Standard Picture and the Path to Prescriptivity

Logic, Frege tells us in his 1918 essay "The Thought", describes the laws of truth, from which prescriptions for asserting, thinking, judging, and inferring follow. Until recently, logic's 'path to prescriptivity' for reasoning, and thereby argument, was taken to be relatively straightforward. (Let us call the correct, prescriptive norms for thinking (i.e., reasoning and inference), judgment, assertion, and argument, whatever these turn out to be, 'rational norms,' or 'norms of rationality'). According to the *standard picture* (Stein, 1996, p. 4), logic provides rational norms.[1]

---

[1] The norm of rationality can be applied both to reasoners, when evaluating their rational doings (e.g., changes in view, inferences, and arguings), and to items in their 'web of belief' (e.g., propositional attitudes), when evaluating the cogency of the reasons on the basis of which the attitude is held. As such claims can be rationally held when doing so is permitted by the norm of rationality, just as changes in view can be made rationally when

> According to this picture, to be rational is to reason in accordance with principles of reasoning that are based on rules of logic, probability theory and so forth. If the standard picture of reasoning is right, principles of reasoning that are based on such rules are normative principles of reasoning, namely they are principles we ought to reason in accordance with.

Today, by contrast, the prevailing view is that the path from logic to rational norms is, at best, indirect.

1.2. The Prescriptivity Gap

As Harman (2002, p. 171; cf. 1984, 1986, 1995) states, "Inference and implication are very different things and the relation between them is rather obscure." For instance, adapting Harman's (1995, p. 184) example, the two sentences:

(1) Some set of premises, P, implies some conclusion, C, and

(2) If you believe P you should (or may) infer (or conclude) C,

say quite different things. As Harman notes (1995, p. 184), firstly, they are about entirely different things: (1) is about implication while (2) is about inference. "Inference and reasoning are psychological processes leading to possible changes in belief (theoretical reasoning) or possible changes in plans and intentions (practical reasoning). Implication is most directly a relation among propositions."

This difference in subject matter has an important consequence when understanding the descriptive relationship between logic and reasoning. According to Harman, logic does not describe reasoning processes—logic is not properly understood as representing some internal, psychological reasoning process. Considering something he calls the "deductive model of inference," according to which when we reason rightly we do so in accordance with logical rules, Harman objects that the reasoning processes involved in proof construction are neither accurately nor appropriately described in terms of the logical steps in the constructed proof.

> Except in the simplest cases, the best strategy is not to start with the premises, figure out the first intermediate step of the proof, then the second, and so on until the conclusion is reached. Often it is useful to start from the proposition to be proved and work backward. It is also useful to consider what intermediate results might be useful. (2002, p. 178)

The reasoning processes involved in proof construction are a kind of problem solving, whereby one attempts to figure out what sequence of logical rules applied to stated premises or derivable theorems will produce the required conclusion. While this reasoning makes use of logical rules by applying them, it is not accurately described by them, nor does it proceed according to them. "The so-called deductive rules of inference are not rules that you follow in constructing the proof. They are rules that the proof must satisfy in order to be a proof" (Harman, 1995, p. 193).

---

similarly permitted. Derivatively, reasoners can be said to be 'rational' when they generally behave (e.g., believe and reason) in ways permitted by the norm of rationality.

Not only does logic lack a descriptive relationship to reasoning, it lacks a straightforwardly prescriptive one as well. As Harman (1995, p. 184) observes, logic, in and of itself, is not prescriptive. For example, (1) does not "say anything normative about what anyone 'should' or 'may' do." And indeed, (2) is not a consequence of (1). Supposing (1) to be true, even if someone believed P, they might not thereby be justified in either believing or concluding C without explicitly drawing the inference from P to C and thereby recognizing the warranting relation of consequence obtaining between them. Whereas logical rules constitute relations of consequence among propositions, rational norms are epistemic or prudential in nature, and pertain to warranting or justificatory relations among reasons and claims.

This relationship between logic and reasoning creates a *prescriptivity gap* such that, at the very least, rational norms do not follow directly from standard logic.

1.3. Principles of Logic and Rational Norms

This prescriptivity gap is widened when it is asked whether logic provides the basis for fit rational norms at all. Consider, for example, the two basic logical relations of consistency and consequence, and some standard principles of deduction that follow from them. First is the principle of *non-contradiction*: that for any well-formed expression, it and its negation cannot both be true, or that standard logical systems are consistent such that logical falsehoods (contradictions) are excluded. Second is the principle of *deductive closure*: that consistent logical systems are closed under deduction, or that all logical consequences of stated premises and axioms are valid theorems of the logical system. Consider now those same principles interpreted or applied as rational norms:

Consistency: one's beliefs should be mutually consistent;

Closure: one's beliefs should be closed under deduction such that one believes all of the logical consequences of one's present beliefs.

And, in order to satisfy to those norms, one must also adhere to some rational norm like:

Tracking: one should keep track of the logical relations between one's beliefs, such that one can maintain consistency and closure among them.

Harman (1984, 1986, 1995, 2002) argues that none of these are suitable rational principles. Consider consistency: As before, Harman begins by noting (1995, p. 185) that the principles of deduction say very different things than their rational-norm counterparts. "It is one thing to say, 'Propositions A, B, C are inconsistent with each other.' It is quite another to say, 'It is irrational (or unreasonable) to believe A, B, C.'" And, the deductive principle can be true while the rational norm false (1995, p. 185). For example, A, B, and C might be equally well-supported by one's evidence and yet inconsistent, leaving one in the position of having no reason to abandon any one of them over the others. Further, Harman notes (1984, p. 109), consistency is not a fit rational principle *for us* since "a rational fallible person ought to believe the at least one of his or her beliefs is false," and believing this will make our entire belief set inconsistent. It would seem that rational yet fallible agents *ought* to have inconsistent beliefs!

Similarly with closure: It is one thing to say that 'C is a consequence of A and B,' and quite another to say that 'One ought to believe C (merely) because it is a consequence of one's present beliefs A and B.' Firstly, upon discovering an unacceptable consequence of one's present beliefs, it might be more rational to revise one's present beliefs, by giving one of them up, rather than to accept the unacceptable consequence. Moreover, there is an infinitude of trivial (e.g., just apply the rule of disjunction introduction repeatedly) or irrelevant (claims of no practical interest or significance) logical consequences of one's present beliefs; surely it is not rational to occupy oneself with these. Indeed, Harman proposes the rational norm of "clutter avoidance" (1986, p. 12; 1995, p. 186) to preempt such patently irrational (in this case, imprudent) behavior.

Thus far, we have seen that "ordinary rationality requires neither deductive closure nor consistency" (Harman 1995, p. 187). Johnson and Blair identify this line of reasoning as the *internal critique*, and describe it as follows: "formal deductive logic is inadequate as a normative theory of argument, supplying neither necessary nor sufficient conditions for logically good argument" (2002, p. 347).

2. Bounded Rationality: Widening the Prescriptivity Gap

2.1. The Empirical Critique

The internal critique is compounded by the *empirical critique* (Johnson and Blair 2002, p. 351), according to which not only does ordinary rationality not require adherence to norms based solely on logical principles, but ordinary rational agents are incapable of adhering to those norms.

The standard picture, on which logic provides rational norms, is committed to the *ambitious claim*, which Perkins identifies and defines as follows:

> By and large, people can, should (in the sense of adaptation), and do reason according to standard logic. (2002, p. 189)

The ambitious claim is built upon the principle that *ought implies can*: that the prescriptive force norms have over us presupposes our ability to meet, satisfy, or adhere to those norms. The ambitious claim simply attributes to us the capacity or ability to adhere to rational norms provided by logic.

Problematically, experimental results in the psychology of reasoning indicate that, in predictable and systematic ways, we do not, by and large, adhere to many of the rational norms provided by logic (Perkins 2002). To cite but two well-known examples, Wason's (1968) selection task for deductive reasoning and Tversky and Kahneman's (1982) representative description task for probabilistic reasoning indicate that otherwise rational human reasoners seemingly fail to reason in logically correct ways. More generally, Godden (2012b) notes that we seem not to be reason trackers in the way required by the standard view. And, if we cannot abide by the rational norm of tracking, it is difficult to understand how we should be expected to adhere to the norms of consistency or closure either. Results such as these seem to indicate that the factual, or descriptive, element of the ambitious claim is false (Perkins 2002).

2.2. The Contrapositive of Ought Implies Can

The normative consequences of these empirical findings are not insignificant. After all, the contrapositive of the *ought implies can* principle is: *cannot implies ought not*. Thus, evidence that ordinary reasoners cannot do something is evidence that they ought not to do it—i.e., that they are under no obligation to do it, and that they cannot be faulted for not doing it. And, the line of reasoning continues, what better evidence is there that we cannot do something than that, ordinarily, we don't?

Accepting the contrapositive of *ought implies can* involves at least some commitment to the general idea of *bounded rationality* (Simon, 1957), that the prescriptive force of rational norms derives, at least in part, from our ability, whether ordinarily or in principle, to adhere to those norms. Thus, Baron (1985, p. 11) distinguishes between *normative* and *prescriptive* models according to whether they incorporate a boundedly rational element. "A good prescriptive model takes into account the very [cognitive and situational] constraints on time [and resources], etc., that a[n idealized] normative model is free to ignore." This distinction allows us to ask whether a norm should have prescriptive force over an agent (in some situation).

Minimally, the position of bounded rationality commits us to a psychologism whereby the results of psychology are relevant to setting prescriptive rational norms, or at least to explaining the prescriptivity of rational norms. Yet, the extent of any resulting psychologism can still vary greatly. At one extreme, one might insist that the standards of logic remain the proper standards of rationality against which our reasoning ought to be judged, such that our reasonings may rightly be said to be faulty to the extent that we fail to reason in accordance with logical norms. Yet, it might also be said that, to the extent that we are genuinely incapable of reasoning in accordance with logical norms, we should not be held under any obligation to do so, and nor should we be held blameworthy for failing to do so—even though we would be in a better position, rationally speaking, were we better able to do so. Here, the standards of rationality remain unaffected by psychological considerations, though their binding or prescriptive force over us is mitigated by our cognitive abilities. At the other extreme, by contrast, some (e.g., Pelletier & Elio 2005, p. 20) opt for a complete psychologizing of rational norms for non-deductive, default reasoning. They argue: "deductive reasoning has a 'normative standard' that is 'external' to people whereas default reasoning has no such external normative standard … Here there is no external standard of correctness other than what people actually infer." On this view, the very nature and content of rational standards is determined, at least partly, by psychological considerations. Whatever the extent of the psychologism we adopt in the end, the effect of the *cannot implies ought not* maxim on the prescriptive force of logical principles and the content of rational norms is significant. To appreciate this point, it is useful to consider an account of argumentative rationality and some of the different assumptions embedded within it.

3. Argumentative Rationality

3.1. The Basic Idea: Rightly Responding to Reasons

The basic idea of rationality, as it is found in argumentation, is that of rightly responding to reasons (Godden, 2015). Rational belief and action are based on reasons, such that one's beliefs, actions, and changes in view are properly justified and explained in terms of the reasons one has for them (Brown, 1988). Reasoning and deliberation are cognitive processes or activities in which reasons are considered and acted upon. Rightly responding to reasons involves appraising the strength of one's reasons, and adhering to a standard of *evidence proportionalism* whereby one accords the degrees of commitment to one's views with the strength of the evidence one has for them (Pinto, 2006, p. 287). As Siegel (1997, p. 2) claims: "to say that one is *appropriately* moved by reasons is to say that one believes, judges, and acts in accordance with the probative force with which one's reasons support one's beliefs, judgments and actions." Argumentation, in turn, can be understood as an interpersonal, communicative activity of "reasoning together" (Campolo, 2005, pp. 38ff.).

3.2. Aspects of Argumentative Rationality

Godden (2015) analyses this picture of argumentative rationality, articulating five rough kinds of assumptions contained within it: (i) the normativity assumption, (ii) deontological assumptions, (iii) structuralist assumptions, (iv) internalist assumptions, and lastly (v) the assumption of reflective stability. It is worthwhile to briefly review these assumptions, as it will be shown that several of them are directly challenged by the picture of bounded rationality discussed in the previous section.

First, the normativity assumption highlights the idea that attributions of rationality are honorific value judgements, rather than descriptive statements, and are made by applying behavioral or observational criteria for a prescriptive (e.g., epistemic) standard that we *ought* to meet even in cases where we *do* not.

The deontological assumptions of accountability, obligation, entitlement, and voluntarism identify a cluster of presuppositions latent in our practices of praising and blaming rational agents for their rational behavior, including their cognitive behavior, and in taking ourselves to be right in doing so (Godden, 2010). By holding rational agents accountable for their rational behavior, we not only presuppose that those agents can act rationally—i.e., that they can accord the credence they place in their views to the strength of the reasons they have for them (the assumption of voluntarism), but we also ascribe rational obligations to those agents, such as the obligation to successfully support their views with reasons when called upon to do so and to surrender those views when they cannot meet this obligation, such that they continue to adhere to a standard of evidence proportionalism (Godden, 2014). Rational entitlements are the permissive counterpart to rational obligations: by satisfying our rational obligations we take ourselves to have demonstrated our entitlement to our views, and thereby to be rationally permitted to hold them and to use them in certain ways (e.g., as premises in further inference).

The structuralist assumptions of basing, causal, rule, and tracking unpack some of the ideas implicit in the claim that rationally held views are based on reasons. The basing assumption reiterates the idea that, when a view is based on reasons, there is an explanatory, as well as a justificatory, relationship between the view and the reasons on which it is based, such that changes in the reasons ought to occasion changes in the view. One rationally holds the views one has *because of* the reasons one has for them, and any decoupling here is an indication that one's espoused reasons are not their *actual* reasons. As such, often the force of reasons is understood causally as well as normatively, at least to the extent that the force of reasons can outweigh any non-rational, psychological forces that might affect one's views. The normative force of reasons is typically expressed in terms of warranting or basing rules which explain not only one's *acceptance* of a claim, but the *acceptability* of the claim itself. Understood this way, reasoning is a rule-governed activity, as distinct from mere psychological processes affecting our mental states and attitudes. The tracking assumption articulates the idea that basing one's views on reasons requires monitoring and keeping some account of the acceptability of reasons and the claims one bases on them.

Both the deontological and structural dimensions of argumentative rationality reveal the internalist assumptions of accessibility and articulability. In order that we can rightly be said to have obligations to base our views on reasons which we keep some track of, we must take some cognizance of those reasons (the accessibility assumption), at least to the extent that we are able to produce or articulate them on demand—e.g., when challenged to demonstrate our rational entitlement to our views.

Finally, in order to rightly respond to reasons, our rationally adopted views must be reflectively stable: once adopted on the basis of reasons, our views should remain settled, and not change due to irrational or non-rational forces, unless and until they are displaced by the force of some stronger reason.

4. Bounded Argumentative Rationality

Having set out a picture of argumentative rationality, let us consider the effects of bounded rationality on it. To do this, it is worthwhile to first recognize some of the similarities and connections between rational principles on the standard view and some aspects of the picture of argumentative rationality just presented.

4.1. Logic and Rationality Revisited: Paradigms of Irrationality

Despite the prescriptivity gap, logic still seems to make some essential contribution to the bases of rational norms. Consider, for example, the following paradigms of irrationality:

Manifest inconsistency: maintaining recognizably inconsistent 'local' beliefs (i.e., some limited set of beliefs, among which is not the belief that one is mistaken about at least one of *them*);

Manifest denial of closure: refusing to accept the recognizable logical consequences of one's beliefs;

Manifest intransigence: refusing to update one's beliefs when confronted with a recognizable failure of one's reasons.

I take each of these cases to be paradigmatically, and hence uncontroversially, manifestly irrational. Yet, these paradigms of irrationality each seem to be explained, at least partly, by their patent illogicality—i.e., by their manifest violation of, or inconsistency with, logical principles.

Each of these paradigms of irrationality has an important connection with some aspect of the picture of argumentative rationality just described. Most generally and importantly, each case seems to violate the principle of evidence proportionalism, and thereby the very idea of rationality: that of responding rightly to reasons. More specifically, each involves the failure of tracking some relation, either logical or evidentiary, among one's reasons and claims, and so contravenes the tracking assumption. The obligation assumption, that we have an obligation to surrender a view whenever we are unable to successfully support it with reasons, is blatantly violated by manifest intransigence. And, insofar as logical inconsistency is the strongest possible reason against some view and logical consequence is the strongest possible support for some view, manifest inconsistency and manifest denial of closure also violate that aspect of the obligation assumption. The manifest-ness of each of these failures implicates that each case also contravenes the other dimension of the obligation assumption, that we have an obligation to successfully support our views with reasons. Unless the failure to meet this obligation is taken as a sign that one's espoused reasons are not one's actual reasons, each case also contravenes the basing, rule and causal assumptions. Manifestly and paradigmatically irrational views are not held on the basis of reasons, which are connected to claims by rules, and so our holding some irrational view is neither caused by, nor explained in virtue of, any reasons we might have or offer for it. Nor, indeed, is it licensed by any rule or reason. In contravening the obligation assumption, each case also contravenes its counterpart, the entitlement assumption, so long as one is never entitled to manifestly, paradigmatically irrational beliefs of the sorts just listed.

It would seem, then, not only that paradigmatic cases of irrationality violate virtually every aspect of argumentative rationality (which, incidentally, is a welcome result for the proposed account of argumentative rationality), but that the irrationality of the belief or behavior is, in each case, significantly explained by its illogicality.

4.2. Bounded Rationality and Argumentative Rationality

What is the effect of bounded rationality on this picture? Harman (1986) argues that the logically-based rational norms of consistency, closure and tracking are not suitable as prescriptive rational norms for ordinary human reasoners, on the grounds that we are normally incapable of satisfying any of them. For example, Harman rightly claims that we are incapable of tracking all of the basing relations, and relations of consistency and consequence, that obtain between each of our beliefs. (Indeed, we are incapable of contemplating, in any occurrent sense, an infinite number of beliefs, let alone tracking all of the relations that occur between them.) As such, we are also incapable of adhering to the norms of consistency and closure, since abiding by these requires tracking.

Of paramount importance for our present purposes is the downward pressure that moves of this sort put on rational norms and on argumentative rationality generally. While Harman's position does not directly license any case of paradigmatic irrationality, it comes remarkably close. Consider, for example, the phenomenon of belief perseverance, where beliefs survive "the total destruction of their original evidential basis" (Ross & Anderson, 1982, p. 149), as empirically reported in the debriefing paradigm (Ross, Lepper, & Hubbard, 1975; see Godden, 2012a for a brief survey of this literature). Belief perseverance readily seems to be a case of manifest intransigence, where a subject refuses to update her beliefs despite a recognizable failure of her reasons, and hence seems to be paradigmatically irrational. Yet, Harman (1986, p. 39; 2002; cf. Godden, 2012a, pp. 61ff.) claims that, contrary to any intuition we might have otherwise, since we cannot be expected to track all of the basing relations among our views (being incapable of doing so), belief perseverance is rational.

Now, if Harman is correct that belief perseverance is rational, then the consequences for argumentative rationality are stark indeed. Not only does this view excuse reasoners of one of their basic obligations under argumentative rationality, that of surrendering a view which cannot sufficiently be supported with reasons, but it violates the principle of evidence proportionalism, since, by hypothesis, the subject in the debriefing paradigm has no better reason for her persevering belief than for its contradictory. Thus, and most importantly, by permitting cases of manifest intransigence, such a position undermines any promise that reasoning or argumentation might offer for improving the rationality of one's overall view, since reasoners are excused in cases where they refuse to accept the consequences of evidence that they themselves recognize as defeating their own reasons for some occurrent belief. At this point, the very idea of argumentative rationality as rightly responding to reasons seems to have been lost.

Suffice it to say that, insofar as the preceding diagnosis is correct, my prognosis for rational norms, and argumentative rationality generally, is dire. While it might seem as though lowing our rational norms places rationality better within our grasp, succeeding at this diminished standard might not be an achievement deserving of much praise or even having much epistemic or probative worth in the end. Elsewhere (Godden, 2015), I have sought to address this circumstance by proposing an activity-based account of reasoning that recontextualizes the relationship between reasoning as a justificatory activity and the psychological processes underlying that activity, such that the assumptions of argumentative rationality can better be retained.

5. Extending Cognitive Responsibility

In the remainder of this paper, I explore a different treatment. The proposal I offer here suggests that argumentation and critical thinking, understood as a set of practices, together with the procedures and technologies of argumentation theory, can extend our cognitive abilities such that we are better able to meet logically-based rational norms, and thereby extend our responsibilities to adhere to those norms.

5.1. Extended Cognition

As originally proposed by Clark and Chalmers (1998) and developed by Clark (2008), the extended mind thesis envisions extending the mind artifactually, such that the usual, instrumentalist, account of the role of technologies in the accomplishment of cognitive tasks by human cognitive agents is replaced by an *active externalism* (1998, p. 8) according to which both human agent and technological artifact, when properly connected through a causal coupling that satisfies the 'glue and trust' conditions (Clark, 2010), become constituents of larger cognitive system. The justification for the move to count extra-cranial operations as properly cognitive, and thereby the things effecting those operations as components of cognitive systems, derives from the *parity principle*:

> If, as we confront some task, a part of the world functions as a process which, *were it done in the head*, we would have no hesitation in recognizing as part of the cognitive process, then that part of the world *is* (so we claim) part of the cognitive process. (1998, p. 8)

Clearly, such an account conceives of cognitive processes functionally, and capitalizes on the multiple realizability hypothesis of functionalism, whereby the same cognitive process can be realized by very different causal processes. The classic example here is memory-impaired Otto who, by making assiduous use of a notebook which he always keeps ready-to-hand, extends his cranial memory to include those things he records in his notebook. According to Clark and Chalmers (1998, pp. 12ff.), Otto's notebook, and his use of it in the course of his ordinary activities, is on a par with Inga's neural memory, and her acts of recollecting: "the essential causal dynamics of the two cases mirror each other precisely." (For completely innocuous examples, consider our ordinary use of prescription eyeglasses or hearing aids.)

5.1.1. Socially Extended Cognition

The idea that minds can be artifactually extended so as to include instruments with which cranial cognizers are causally-coupled has been developed to include other 'technologies,' broadly understood to include social practices and institutions. (Indeed, Clark and Chalmers anticipate such developments, speculating in their original article as to the possibility of "socially extended cognition" and a "linguistically-enabled extension of cognition" (1998, pp. 17, 18).) Perhaps the most ambitious version of the extended mind is Gallagher's (2013) *socially extended mind*, which "builds on the enactive idea of social affordances."

> Just as a notebook or a hand-held piece of technology may be viewed as affording a way to enhance or extend our mental possibilities, so our encounters with others, especially in the context of various institutional procedures and social practices may offer structures that support and extend our cognitive abilities. (Gallagher, 2013, p. 4)

Thus, Gallagher and Crisafi (2009) propose the idea of *mental institutions*, arguing that social institutions such as museums or legal systems meet the parity principle just as well as instruments like calculators and notebooks (see also Gallagher, 2011, 2013).

> If we think that cognition supervenes on the vehicle of the notebook, it seems reasonable to say that it supervenes on the vehicle of the museum—an institution designed for just such purposes. (2009, p. 49)

As with *active* externalism, a central aspect of Gallagher's *enactive* externalism is the claim that cognitive processes are distributed across, or realized by, cognitive systems that include both human and technological components. Drawing on De Jaegher and Di Paolo's (2007) idea of *participatory sense making*, Gallagher (2013, p. 8) proposes that social interaction itself has a "certain autonomy ... [that can] transcend the agent's subjective processes" (cf. De Jaegher, Di Paolo & Gallagher, 2010).

> Social interaction and participatory sense making specifically involve patterns of engagement that can acquire their own form of self-organization. [...] Participatory sense making is always shaped by super-individual norms and institutional practices. (Gallagher, 2013, p. 8)

Perhaps the most general and universal mental institution, understood as a kind of participatory sense making, is language itself. Recognizing this, Fusaroli, Gangopadhyay and Tylén (2014, p. 37) propose the *dialogically extended mind*, according to which language extends our individual reasoning capacities when we engage in the intersubjective activity of communicatively reasoning together.

Language enables individuals to coordinate their cognitive processes in evolutionarily unprecedented ways, effectively constituting dialogically extended minds. In the skillful intersubjective engagement of symbolic patterns, human beings rely on each other and on established cultural practices to achieve feats that would otherwise be beyond reach.

5.1.2. *Extended Cognition: Ontological and Responsibilist Elements*

The ontological dimension of the extended mind thesis is one of its most controversial aspects (Adams & Aizawa, 2001, 2008, 2010). According to Adams and Aizawa, for example, active and enactive externalisms are guilty of the "coupling-constitution fallacy" (2010, pp. 67ff.), which mistakes (i) the essential and fully-integrated use of some technology (be it artifactual or procedural) by some agent within some activity for (ii) some new agent comprised of the closely-coupled things. As an alternative, Hubner (2013), for example, argues for a view of cognition as socially *embedded*, such that technologies are conceived of as "contextual factors" and "enabling conditions" for extending cognitive *capacities*, rather than as "constitutive elements" of newly-conceived cognitive *systems* as proposed by active and socially enactive *extended* approaches. To highlight this kind of distinction, I will distinguish between the *extended mind* thesis (which I will take to have the ontological commitments just described) and the *extended cognition* thesis (which I will take to be independent of those ontological commitments).

My interests are in extending our cognition. I seek ways that we can extend our cognitive abilities so that our rational responsibilities can likewise be extended, while still adhering to the *ought implies can* maxim. I suggest that this can be done by employing ideas from the extended cognition literature, without taking on any of its controversial metaphysical baggage. One way to do this might be to explain extended cognition, not as the activity of some awkwardly extended mind, but by considering it as the result of some cognitive agent's use of a *regulative technology*—a technology (be it artifactual or procedural) that is designed or used to aid some cognitive agent in the normative regulation of their, or another's, behavior (including their cognitive behavior).

A second controversial issue arising from unqualified or uncritical versions of extended cognition is "cognitive outsourcing" (Menary, 2012). The problem here is one of assigning credit—i.e., praise or blame—for any cognitive, including rational, accomplishment or failure in cases where cognition is extended. Consider, for example, Gallagher's (2013, p. 5) claim that "I cannot remember where the restaurant is, but I, *plus* my technology, can" (Gallagher, 2013, p. 5). It is difficult to accept that the tourist with a GPS-enabled mapping app on his smartphone has the same knowledge and cognitive abilities as the lifetime London cabbie, certified with 'the knowledge.' Clearly, the credit for their respective navigational abilities (even supposing them to be functionally equivalent) due to the cabbie vastly exceeds that due to the tourist. (Here, one might even imagine the passenger in the cab, whose only role in the cognitive system, which though trivial is essential nevertheless, is to state her destination to the cabbie.)

One solution to this problem has been to incorporate a virtuistic element into the picture of extended cognition such that "cognitive processes that extend outside of the skin of [the] agent can count as part of one's cognitive agency just so long as they are appropriately integrated within one's cognitive character" (Pritchard, 2010, p. 145). Similarly, Roberts (2012, p. 133) argues that "true cognitive extension occurs only when the subject takes responsibility for the contribution made by the non-neural resource."

One account that strikes a promising balance on these points is Menary's (2007, 2012, 2013) *integrationist account* of extended cognition. On an integrationist account cognition is extended through enculturation: "cognitive capacities are extended through socio-cultural practices" called *cognitive practices* (2013, pp. 26, 29ff).

> The practices are patterns of activity spread out across a population. So for example mathematical practices, such as the partial products algorithm, extend the basic biological capacities with which we are endowed. The practice is first learned by manipulating symbols on a page (for example) and becomes a capacity that can be enacted either by bodily manipulation of public symbols, or offline simulations of such manipulations. (2013, p. 26)

A central feature of cognitive practices is that they are essentially normative (2013, p. 29). That is, they can be performed correctly or incorrectly, are situated in other normative activities such as teaching and correction, and are acquired through practice and training. As such, an integrationist account of extended cognition is a good fit with a view of cognition as a rule-governed activity rather than as a psychological process (cf. Godden, 2015). Further, "[m]any of these [cognitive] practices involve artifacts such as tools, writing systems, number systems, and other kinds of representational systems" (2013, p. 29). Yet, the role of artifacts in cognitive practices is not that of a constituent in some larger cognitive system, but that of a regulative technology. In this way, an integrationist account avoids the cognitive outsourcing of strictly causal, artifactual accounts, and instead incorporates the normative and deontological elements demanded not only by credit-based, virtue accounts of knowledge and cognitive achievement, but by the picture of argumentative rationality presented in section 3.

5.2. From Extended Cognition to Extended Responsibility

Importantly, by providing an inherently normative extension of our cognitive abilities, cognitive practices thereby extend our cognitive responsibilities. To see this, consider an example adapted from Menary (2013, pp. 29ff.). Suppose I have an arithmetic test where I am expected to correctly multiply large numbers together. Should I be excused from having to complete the test, or for failing it, on the grounds that the numbers are so impossibly large that I could not conceivably determine the product in my head using only my untutored arithmetical intuitions? Well, no. Instead, I can be expected to learn and apply the partial products algorithm in order to calculate the final product, even if I cannot compute it in my head all at once.

Similarly, suppose it is my job to track a baseball game. That is, I am to make note of, and subsequently report on, every pitch and every play of the game. Such a feat is typically well beyond the normal cognitive abilities of the ordinary baseball spectator. How, then, can some people be obliged to do this as part of their job? Well, quite easily actually. One makes use of a "scorecard": a table on which one records in writing how each pitch and play is called, as a means of fulfilling one's obligation to note and report on the details of the game. Eventually, one might even internalize this practice such that one can accomplish it without having to write it down in pen-on-paper, but instead develop, through practice, the habit of making a mental note of each play such that one can recall, at least for a time, an entire game in one's head.

The initial point here is that there are ordinary cases where one's constitutive yet unaided inability to perform some action does not relieve one from the responsibility of doing so. Oftentimes, there are readily available cognitive practices or regulative technologies (whether instruments or procedures) whose very purpose is to aid one in meeting their obligations in such circumstances. Particularly when significant social value is placed upon either the obligations themselves or the ends to which the obligations contribute, those practices and technologies are frequently instituted as facets of the very activities or social practices in which one is engaged. Thus, rather than be excused from meeting one's obligations in such circumstances, one is instead obliged to avail oneself of the available or instituted cognitive practices or regulative technologies in order that one *can* meet those obligations.

6. Extending Cognition Through Argumentation

Having proposed the idea that cognitive practices and regulative technologies can extend our cognitive responsibilities by extending our cognitive abilities, let me conclude by suggesting some of the ways that the resources of argumentation theory serve to extend our cognitive abilities and thereby extend our rational responsibilities.

6.1. Argumentative Resources for Extending Rational Responsibilities

As noted earlier, the internal and empirical critiques present a significant and direct challenge not only to logically-based rational norms such as consistency, closure and tracking, but also to many of the core elements of argumentative rationality. At least one element of the empirical critique is the claim that we are not sufficiently able reason-trackers that we should be expected to avoid manifest intransigence, as evidenced by the diagnosis of belief perseverance as rational. Happily, argumentation theory provides several resources that can significantly extend our reason-tracking abilities.

6.1.1 Argument Diagramming or Mapping

First among these are the techniques and technologies of argument mapping or diagramming. These come in all varieties, from simple pen-on-paper practices of argument analysis, whereby the premises, conclusions and basic patterns of inference are identified and itemized, to fully automated and scalable mapping software, allowing several agents to collectively construct vast networks of nested claims and reasons where inferential patterns, schemes, required-but-unstated premises and critical questions or potential defeaters are automatically supplied. Basically, argument diagrams work as 'argumentative scorecards,' allowing reasoners to better track the different commitments and claim-reason complexes that are 'on the table' in some given argumentative exchange. Even the technique of Venn diagramming allows some reasoners to recognize cases of valid consequence, invalidity, consistency, and inconsistency in ways that are not wholly apparent to them when a putative syllogism or immediate inference is presented textually or verbally. More robust and versatile mapping technologies facilitate the efficient and effective tracking of a multitude of rationally-significant relations between claims (e.g., consequence, consistency, closure, exclusion, likelihood, evidence, support, relevance, dependency, defeat, coherence, explanation, etc.).

6.1.2 Procedural Norms for Argumentation

A second cluster of resources developed by argumentation theorists stems from procedural approaches to reasoning and argumentation. Again, these come in a wide variety ranging from rough-hewn Waltonian dialogue types, to the Pragma-Dialectical model of a critical discussion, to fully operationalized, program-like rule systems. Procedural approaches seek to provide the 'partial-product algorithms of reasoning.' In their most rigorously articulated versions, they furnish rule-governed, step-by-step procedures that reasoners can follow with relative ease in order to reach a rational resolution to a difference of opinion or to settle the rational acceptability of a standpoint. Here again, the tasks of tracking, and maintaining consistency and closure are internalized into the procedural rules themselves such that merely following the rules at any given point in the process ensures that one meets the rational norms embodied within the model.

6.1.3 Schemes and Fallacies: Commonplace Recipes for Improved Rationality

A less comprehensive but more perhaps more easily acquired and employed set of argumentative devices are the argumentation schemes and their accompanying critical questions (Walton, Reed, & Macagno, 2008), and fallacies. These function more as rational topoi, or commonplaces, and can be seen as the 'on-the-box recipes of reasoning.' By using these devices, one needn't know how to construct or critique an argument from scratch, so long as they can follow directions by assembling and arranging the listed ingredients. Argumentation schemes provide common patterns of cogent yet defeasible presumptive reasoning, each of which can then be evaluated for stereotypical points of defeat by applying the attendant critical questions (each of which is designed to evaluate some aspect of cogency (Godden & Walton, 2007)). The counterparts to schemes are fallacies, which catalogue typical patterns of incogent reasoning that are frequently but mistakenly appraised as cogent. Because these devices are not comprehensive, even their judicious employment cannot guarantee adherence to every rational norm. That said, they are clearly capable of improving one's rational abilities and frequently produce effective, satisfactory results. And, to the extent that they succeed in these respects, they also extend our rational accountability and accomplishments.

6.1.4 Guidance Norms: From Lists of Do's and Don'ts to Proof Systems

A final cluster of resources provided by argumentation theory—drawing upon the results of adjacent fields like critical thinking, epistemology, and logic—is designed for the task of determining, or estimating, and critically evaluating, the probative force of reasons. As with the tools and techniques already discussed, these evaluative technologies exhibit varying degrees of precision, scope, and rigor (or formalization). At one extreme, there are collections of commonplace rules that provide good starting places in our projects of rational evaluation. For example, Feldman's (1999) "Basic Rules of Argument Evaluation" offers a list of Do's and Don'ts for critically appraising premise acceptability and inferential strength, such as: "direct criticisms at individual premises," "don't criticize an argument by denying its conclusion," and "make your criticisms substantial," "don't accept competing arguments." An equally informal, but completely generic technique for appraising inferential strength is provided by the method for constructing counterexamples: describe a possible situation (or, the most plausible situation) in which an argument's premises can all be true while its conclusion false; evaluate the strength of the inference by comparing the relative likelihoods of the counterexample with the truth of the conclusion given the premises. More thorough, though fit for a narrower range of (specifically causal) inferences, are Mill's (1973, pp. 388ff) methods of experimental inquiry (or canons of induction): the methods of agreement, difference, joint method of agreement and difference, residues, and concomitant variation. These methods aid not only hypothesis formation, but also the testing of formulated hypotheses. Methods like these can be operationalized with varying degrees of rigor, completeness, and comprehensiveness that extend well beyond ad-hoc lists of instructions. For example, Flage (2000) provides a set of flowcharts for critical thinking that map out, in step-by-step fashion, a "highly structured decision-procedure" for the evaluation of argument, and include subsidiary decision trees for dealing with things like ambiguity, relevance, presumption, and observation, testimony and surveys. A final technology, of course, is found in the formal proof systems of formal logic and the probability calculus. As with the procedural rules of some dialogue systems discussed above, these logical systems embed the tasks of tracking, and maintaining consistency and closure, within the very structure of the system itself, by fully operationalizing basic norms such as consistency and consequence.

6.2. Closing Remarks: The Virtues of a 'Can Do' Attitude

While these are but four samples of the kinds of wares that argumentation theory has to offer, the more general point should be readily apparent. Argumentation theory has designed, constructed, and sometimes imported a diverse and versatile product line of turnkey cognitive practices and off-the-shelf regulative technologies that are remarkably effective at extending our untutored, unaided, intuitive rational abilities. In view of their ready availability, the "cannot, because does not, therefore ought not" line of reasoning frequently advanced by theorists of bounded rationality, together with its overtly psychologistic counterpart "ought because does," should be received with considerably more skepticism and reservation than has been fashionable of late.

Overall, argumentation theory informs a set of mental institutions into which human reasoners can readily be enculturated. And, this enculturation into the social practices of critical reasoning and argumentation is perhaps the most basic, yet most important, result of the cognitive practices and regulative technologies offered by argumentation theory. That people are willing to hold themselves and those around them rationally accountable, such that they see the value in, and are willing to take on for themselves, the basic rational obligations of giving reasons for their views and changing those views when their reasons don't pan out, is the kernel of rationality understood as rightly responding to reasons. Having this sense of rational accountability makes one a scorekeeper in the game of giving and asking for reasons (to borrow Brandom's (1994) phrase), and provides the impetus to seek out and develop the skills required to rightly respond to reasons in increasing refined and effective ways.

Those of us who feel committed to, or even find something right about, the picture of argumentative rationality sketched above, or to the notion that at least some rational norms derive at least partly from logical ideals, should take heart at the 'can do' attitude embodied by critical thinkers, of every ability, everywhere.

References

Adams, F. & Aizawa, K. (2001). The bounds of cognition. *Philosophical Psychology, 14*, 43-64.
Adams, F. & Aizawa, K. (2008). *The bounds of cognition.* Oxford: Blackwell.
Adams, F. & Aizawa, K. (2010). Defending the bounds of cognition. In R. Menary (Ed.), *The extended mind* (pp. 67-80). Cambridge, MA: MIT Press.
Baron, J. (1985). *Rationality and intelligence.* Cambridge: Cambridge University Press.
Brandom, R. (1994). *Making it explicit: Reasoning, representing, and discursive commitment.* Cambridge, MA: Harvard University Press.
Brown, H. (1988). *Rationality.* London: Routledge.
Campolo, C. (2005). Treacherous ascents: On seeking common ground for conflict resolution. *Informal Logic, 25*, 37-50.
Clark, A. (2008). *Supersizing the mind: Reflections on embodiment, action, and cognitive extension.* Oxford: Oxford University Press.
Clark, A. (2010). Memento's revenge: The extended mind, extended. In R. Menary (Ed.), *The extended mind* (pp. 43-66). Cambridge, MA: MIT Press.
Clark, A. & Chalmers, D. (1998). The extended mind. *Analysis, 58*, 7-19. [Reprinted in Menary, R. (Ed.). (2010). *The extended mind* (pp. 27-42). Cambridge, MA: MIT Press.]
De Jaegher, H., & Di Paolo, E. (2007). Participatory sense making: An enactive approach to social cognition. *Phenomenology and the Cognitive Sciences, 6*, 485-507.
De Jaegher, H., Di Paolo, E., & Gallagher, S. (2010). Does social interaction constitute social cognition? *Trends in the Cognitive Sciences, 14*, 441-447.
Feldman, R. (1999). *Reason and argument* (2nd ed). Upper Saddle River, NJ: Prentice Hall.
Flage, D. (2000). Flowcharts for critical thinking. *Informal Logic: Teaching Supplement #3, 20*, TS57-TS69.
Frege, G. ([1918] 1956). The thought: A logical inquiry. *Mind, 65*, 289-311.

Fusaroli, R., Gangopadhyay, N., & Tylén, K. (2014). The dialogically extended mind: Language as skillful intersubjective engagement. *Cognitive Systems Research, 29-30*, 31-39.

Gallagher, S. (2011). The overextended mind. *Versus: Quaderni di Studi Semiotici*, 55-66.

Gallagher, S. (2013). The socially extended mind. *Cognitive Systems Research, 25-6*, 4-12.

Gallagher, S. & Crisafi, A. (2009). Mental institutions. *Topoi, 28*, 45-51.

Godden, D. (2010). The importance of belief in argumentation: Belief, commitment and the effective resolution of a difference of opinion. *Synthese, 172*, 397-414.

Godden, D. (2012a). Rethinking the debriefing paradigm: The rationality of belief perseverance. *Logos & Episteme, 3*, 51-74.

Godden, D. (2012b). The role of mental states in argumentation: Two problems for rationality from the psychology of belief. In F. Paglieri, L. Tummolini, R. Falcone & M. Miceli (Eds.), *The goals of cognition: Essays in honor of Cristiano Castelfranchi* (pp. 123-143). London: College Publications.

Godden, D. (2014). Teaching rational entitlement and responsibility: A Socratic exercise. *Informal Logic, Teaching Supplement, 34*, 124-151.

Godden, D. (2015). Argumentation, rationality, and psychology of reasoning. *Informal Logic, 35*, 135-166.

Godden, D. & Walton, D. (2007). Advances in the theory of argumentation schemes and critical questions. *Informal Logic, 27*, 267-292

Harman, G. (1984). Logic and reasoning. *Synthese, 60*, 107-127.

Harman, G. (1986). *Change in view: Principles of reasoning.* Cambridge, MA: MIT Press.

Harman, G. (1995). Rationality. In E. Smith, & D. Osherson (Eds.), *Thinking: An invitation to cognitive science, Vol. 3*, 2nd ed, (pp. 175-211). Cambridge, MA: MIT Press.

Harman, G. (2002). Internal critique: Logic is not a theory of reasoning and a theory of reasoning is not logic. In D. Gabbay, R. Johnson, H. Ohlbach, & J. Woods (Eds.), *Handbook of the logic of argument and inference: Turn towards the practical* (pp. 171-186). Amsterdam: Elsevier.

Huebner, B. (2013). Socially embedded cognition. *Cognitive Systems Research, 25-26*, 13-18.

Johnson, R. & Blair, J.A. (2002). Informal logic and the reconfiguration of logic. In D. Gabbay, R. Johnson, H. Ohlbach & J. Woods (Eds.), *Handbook of the logic of argument and inference: Turn towards the practical* (pp. 339-396). Amsterdam: Elsevier.

Menary, R. (2007). *Cognitive integration: Mind and cognition unbounded.* Basingstoke: Palgrave Macmillan.

Menary, R. (2012). Cognitive practices and cognitive character. *Philosophical Explorations, 15*, 147-164.

Menary, R. (2013). Cognitive integration, enculturated cognition and the socially extended mind. *Cognitive Systems Research, 25-26*, 26-34.

Mill, J.S. [1843/1872.] 1973. *A system of logic*, Books I-III: *The collected works of John Stuart Mill*, Vol. 7. J.M. Robinson (ed.). London: Routledge and Kegan Paul.

Pelletier, F., & Elio, R. (2005). The case for psychologism in default and inheritance reasoning. *Synthese, 146*, 7-35.

Perkins, D. (2002). Standard logic as a model of reasoning: The empirical critique. In D. Gabbay, R. Johnson, H. Ohlbach & J. Woods (Eds.), *Handbook of the logic of argument and inference: Turn towards the practical* (pp. 187-223). Amsterdam: Elsevier.

Pinto, R.C. (2006). Evaluating inferences: The nature and role of warrants. *Informal Logic, 26*, 287-317.

Pritchard, D. (2010). Cognitive ability and the extended cognition thesis. *Synthese, 175*, 133-151.

Roberts, T. (2012). You do the maths: Rules, extension, and cognitive responsibility. *Philosophical Explorations, 15*, 133-145.

Ross, L., & Anderson, C. (1982). Shortcomings in the attribution process: On the origins and maintenance of erroneous social assessments. In D. Kahneman, P. Slovic, & A. Tversky (Eds.), *Judgment under uncertainty: Heuristics and biases* (pp. 129-152). Cambridge: Cambridge UP.

Ross, L., Lepper, M., & Hubbard, M. (1975). Perseverance in self-perception and social perception: Biased attributional processes in the debriefing paradigm. *Journal of Personality and Social Psychology, 32*, 880-892.

Siegel, H. (1997). *Rationality redeemed: Further dialogues on an educational ideal*. New York: Routledge.

Simon, H. (1957). A behavioral model of rational choice. In *Models of man, social and rational: Mathematical essays on rational human behavior in a social setting*. New York: Wiley.

Stein, E. (1996). *Without good reason: The rationality debate in philosophy and cognitive science*. Oxford: Oxford UP.

Tversky, A. & Kahneman, D. (1982). Judgements of and by representativeness. In D. Kahneman, P. Slovic, & A. Tversky (Eds.), *Judgement under uncertainty: Heuristics and biases* (pp. 84-98). New York: Cambridge UP.

Walton, D., Reed, C., & Macagno, F. (2008). *Argumentation schemes*. Cambridge: Cambridge UP.

Wason, P.C. (1968). Reasoning about a rule. *Quarterly Journal of Experimental Psychology, 20*, 273-281.

Chapter 6

Don't Blame the Norm. On the Challenge of Ecological Rationality

Maarten Boudry[1], Michael Vlerick[2] & Ryan McKay[3]

[1] Department of Philosophy & Moral Sciences, Ghent University, Ghent, Belgium, maartenboudry@gmail.com;
[2] Department of Philosophy, University of Johannesburg, Johannesburg, South Africa;
[3] ARC Centre of Excellence in Cognition and its Disorders, Department of Psychology, Royal Holloway, University of London Egham, Surrey, United Kingdom.

Abstract. Enlightenment thinkers viewed logic and mathematical probability as the hallmarks of rationality. In psychological research on human (ir)rationality, human subjects are typically held accountable to this arcane ideal of Reason. If people fall short of these traditional standards, as indeed they often do, they are biased or irrational. Recent work in the program of ecological rationality, however, aims to rehabilitate human reason, and to upturn our traditional conception of rationality in the process. Put bluntly, these researchers are turning the tables on the traditionalist, showing that human reasoning often outperforms complex algorithms based on the traditional canons of rationality. If human reason still appears paltry from the vantage point of capital-R Rationality, then so much the worse for Rationality. Maybe the norms themselves are in need of revision. Perhaps human reasoning is *better than rational*. Though we welcome the naturalization of human reason, we argue that this backlash against the classical norms of rationality is uncalled for. Ecological rationality presents two apparent challenges to the traditional canons of rationality. In both cases, we contend, the norms emerge unscathed. In the first category, norms of rationality that appear violated by individual reasoners re-emerge at the level of evolutionary adaptation. In the second category, the norms under challenge simply turn out to be not applicable to the case at hand. Moreover, we should keep in mind that, when they are assessing the efficiency of human reasoning, advocates of ecological rationality still use the traditional norms of rationality as a benchmark. We conclude that, even if we accept all the fascinating findings garnered by the advocates of ecological rationality (and there is ample reason to do so), we need not be taken in by the rhetoric against classical rationality, or the false opposition between logical and ecological rationality. When the dust has settled, the norms are still standing.

1. Introduction

The rationalist tradition of the Enlightenment conceived of human reason as a unique and defining faculty lifting us above the realm of nature and radically separating humans from animals (Talmont-Kaminski, 2007). The canons of rationality, according to Enlightenment thinkers, were logic and mathematical probability. Of course, Enlightenment thinkers were well aware that the average human falls short of this ideal. In practice, they admitted, reason is often clouded by prejudice, emotion and magical thinking. In one of the dualities typical of Enlightenment thinking, rationality was set in stark contrast to the dark and irrational forces of human nature. Irrationality, indeed, was characterized as a different mode of thinking altogether, a *sui generis* form of deficient reasoning.

In the wake of Darwin's theory of evolution by natural selection and the maturation of psychology as a scientific discipline, human reason has been brought down to earth. Human rationality is a natural faculty acquired over a long process of undirected evolution, not a spark of divinity separating us from the lower animals. Still, the traditional canons of rationality were upheld as a yardstick to measure the performance of human reason. Not that human reasoning lived up to the ideal. As cognitive psychologists began to investigate actual human reasoning, they amassed a wealth of evidence that is quite an embarrassment to our self-image as Rational Animals: we suffer from all sorts of incorrigible biases and commit elementary fallacies of logic and probability. Popular summaries of research on irrationality are fond of putting humans down: we are characterised as stupid, irrational, mindless, biased and stubborn (Sutherland, 2007; Singer & Benassi, 1981; Shermer, 2011; Ariely, 2009; Piattelli-Palmarini, 1996; Polonioli, 2013). De-biasing efforts seem undertaken to little avail, or even backfire: we often fail to learn from our mistakes. Many share the sentiment expressed by Bertrand Russell: "Man is a rational animal—so at least I have been told. Throughout a long life, I have looked diligently for evidence in favor of this statement, but so far I have not had the good fortune to come across it." (Russell 2009, p. 69)

Luckily, *Homo sapiens* is not completely beyond the pale. The psychologists who have documented and corrected our mistakes are members of the human species, after all. If they can spot the errors of their fellow human beings, then perhaps we can correct each other's blind spots, and all is not lost. This line of reasoning assumes, however, that those who chastise human reason are not deluded themselves. Psychologists, when accusing their subjects of cognitive error, usually use some established rule of logic and probability as a benchmark. (Nisbett & Ross 1980; Kahneman, Slovic *et al.* 1982). But what if the norms themselves are in need of revision? Perhaps human reason looks irrational only because we are applying bad or inappropriate norms.

This is exactly the promise held out by the program of *ecological rationality*. Launched by Gerd Gigerenzer & Peter Todd in the late 1990s, and mostly centered around the Adaptive Behavior and Cognition (ABC) centre in Berlin, this research program aims to unseat our traditional view of rationality and thereby rehabilitate human reason. Starting from the finding that our simple heuristics perform very well in ecologically valid contexts, even outperforming algorithms based on classical rationality (on which more below), these researchers have turned the tables on Classical Rationality: if human reason appears paltry seen from the vantage point of Rationality, then so much the worse for our capital-R ideal of Rationality. In other words: down with the norms.

Is this backlash against traditional normative criteria of rationality warranted? We explore two ways in which the program of ecological rationality might seem to pose a threat to the classical norms. In both cases, we will see that the norms are actually left intact. First, by presenting the prowess of human reasoning in an "ecologically valid" context, defenders of ecological rationality have not so much jettisoned the traditional norms as relegated them to the locus of evolutionary adaptation. The "rationality" of human cognition simply re-emerges at the evolutionary level. Sensible "decisions" made by evolution (i.e. solutions to adaptive problems) may translate into mindless or even norm-violating behaviour at the level of individual reasoners. This point is based on the concept of *locus shift*, which we developed in an earlier paper, in the context of human irrationality (Boudry, Vlerick, & McKay, 2015).

Second, we discuss cases where human reasoning *appears* to violate some norm of rationality, but on closer inspection, these instances simply fall outside the domain of application of those norms, through no fault of the latter. Again, the norms are left standing.

Once the dust has settled, the backlash against traditional canons of rationality seems unwarranted. Traditional conceptions of rationality remain valid after the human mind has been thoroughly naturalized.

2. Ecological Rationality

The debate over human rationality has been waged for decades (Stein, 1996; Evans & Over, 1996; Krueger & Funder, 2004; Stanovich & West, 2000; Mercier & Sperber, 2011). Ever since its emergence in the 1970s, the heuristics and biases program spawned by Kahneman and Tversky has had its detractors. For example, psychologist David Funder lamented the negative slant in much cognitive and social psychology (Funder, 1987), particularly the bad habit of interpreting any deviation from content-blind norms of reasoning as *ipso facto* demonstrations of "irrationality". This "one-sided emphasis on what people do wrong" (Funder, 1987, p. 83) results in an endless compendium of fallacies and biases, with little understanding of underlying cognitive processes. In a paper co-written with Dianne Krueger, he writes that many of his colleagues still have an "inordinate fondness for errors" (Krueger & Funder, 2004, p. 317). One could take Funder's lament one step further: what if there were no errors in the first place? What looks like an error, viewed from an idealized angle, may turn out to be a successful judgment in real life. Proponents of *ecological rationality* have argued precisely this, challenging the bleak view of human reason. Many of the classical demonstrations of irrationality, fallacies and illusions (Kahneman, 2011; Kahneman *et al.*, 1982), they claim, evaporate once we view human cognition in its proper ecological framework.

Gigerenzer sees a clash between two competing conceptions of rationality. The traditional view is based on the content-free canons of logic and probability theory, whereas ecological rationality consists of an "adaptive toolbox" of heuristics (Todd & Gigerenzer, 2012; Gigerenzer & Todd, 1999; Gigerenzer, 2008), each suited to a particular set of challenges endemic to a particular environment. In contrast with unbounded models of rationality, which typically assume unlimited resources both with regard to information gathering and computational processing, ecological rationality is "fast and frugal".

In order to appreciate the ecological rationality of a heuristic, one needs to look at the way in which it exploits the structure of the environment (Gigerenzer & Todd, 1999, p. 13). Environment and cognition work hand in hand to produce good judgments and smart decisions. In Herbert Simon's famous metaphor, mind and environment act like two blades of a pair of scissors (Simon, 1955). As Gigerenzer argues, rationality does not reside in the mind alone, but is a feature of the mind plus the environment.

Fast and frugal heuristics are quick and computationally cheap, requiring few and simple computational steps, and working on a limited input. Based on the traditional conception of rationality, one would expect that such quick-and-dirty rules sacrifice accuracy for frugality and speed. More information and computation, after all, can only improve judgment. Gigerenzer and colleagues have challenged this common wisdom, documenting many cases in which ignoring information leads to better decision-making. "Humans and other animals rely on heuristics in situations where these are ecologically rational, including situations where less information and computation lead to more accurate judgments" (Gigerenzer *et al.*, 2011, p. 261). One striking instance of such less-is-more effects is the recognition heuristic, whereby people take advantage of their own ignorance to derive inferences about the unknown. In judging which out of two German cities has a larger population, for example, people use name recognition as a cue for population level. If they recognize only one city, they assume that must be the most populous one. A certain level of ignorance, it turns out, is the secret to excellence in this task (Goldstein & Gigerenzer, 2002).[1] The recognition heuristic has been shown to be remarkably successful in a wide range of domains. For example, a stock portfolio consisting of companies recognized by naive passers-by has been demonstrated to beat market analysts and financial experts (Borges *et al.*, 1999).

Gigerenzer has also demonstrated cases where ignoring information leads to better decision making. There is a family of heuristics for making decisions based on one reason only. The take-the-best heuristic, for example, chooses between two items by searching for the first cue that discriminates between them, and ignoring all the other cues. Again, one would expect that such simple-minded heuristics would perform worse than more complex decision algorithms that integrate the information of different cues. But this turns out to be false: one-reason decision is highly successful in a range of circumstances.

Fast and frugal heuristics seem to violate one of the cornerstones of our traditional conception of rationality: the assumption that more information and computation leads to better decisions. How can ignorance possibly lead to better decisions than knowledge? How can someone who knows less about German cities be at an advantage on this topic compared to a more knowledgeable person? Less-is-more effects in ecological rationality, where using less information leads to better decisions, seem to violate what Rudolf Carnap has called the *principle of total evidence* (Gigerenzer & Sturm, 2012), according to which the rational course of action always takes into account all available information.

Nothing beats success, says the proponent of ecological rationality. If decision rules based on classical norms of rationality fail, then we should reject those norms. In the real world, the content-blind norms of classical rationality are not "reasonable norms" (Gigerenzer, 2008, p. 12), because the real world is too messy and complicated to be captured by formal norms of rationality, and because human reason has "other goals than logical truth or consistency" (p. 12). If you use the classical norms as a benchmark for human reasoning, all you end up with are spurious demonstrations of "irrationality" and "simple-mindedness" that fail to do justice to human reasoning: "Often what looks like a reasoning error from a purely logical perspective turns out to be a highly intelligent social judgment in the real world" (Gigerenzer, 2007, p. 103). In a recent volume on adaptive rationality, Fiedler and Wänke (2013) wrote: "At the normative level, we have seen that absolute, unique standards of rationality can hardly be upheld." The traditional canons of rationality must be supplanted by a new set of ecological canons – enter ecological rationality.

In the next section we show how the tension between classical and ecological rationality rests on a false opposition, based on a slippage between different *loci* of rationality. The program of ecological rationality need not lead to a backlash against classical rationality. It is possible to defend human reason without relinquishing classical canons of logic and probability.

---

[1] Support for the ecological conception of human reason has also come from evolutionary psychology, in particular adaptationist analyses of human reason in the Santa Barbara school. Evolutionary psychologists are not surprised that the human mind is better designed than the pessimists would have us believe. The human brain is the craftwork of natural selection, after all, and evolution does not tolerate slapdash engineering (Pinker, 1997, but see Boudry & Pigliucci, 2013). Evolutionary psychologists Haselton & Nettle have urged us to dispel the "unnecessarily dreary outlook on human cognition" (Haselton & Nettle, 2006, p. 62).

3. Evolutionary Rationale

Ecological rationality resides not so much in either blade of Simon's pair of scissors (mind/environment), but in the alignment of the two blades. But what process ensures that the blades align? In principle, research on fast and frugal heuristics is neutral with respect to how heuristics originate. Some may be constructed on the basis of learning or experience, some may be innate. And there are intermediate possibilities: for example, it may be the case that we have an innate toolbox of heuristics, but still need to learn under what conditions to use them (Hutchinson & Gigerenzer, 2005). In the end, however, the fundamental building blocks of cognition are the result of evolutionary processes. The work of Gigerenzer & Todd contains little explicit evolutionary theorizing, but they do acknowledge the ultimate evolutionary origins of heuristic reasoning: "evolution would seize upon informative environmental dependencies such as this one and exploit them with specific heuristics if they would give a decision-making organism an adaptive edge." (Gigerenzer & Todd, 1999, p. 19)

In other words, natural selection has aligned our minds with the environment, equipping us with an adaptive toolbox for navigating the world. But evolution often leaves animals perfectly clueless about the rationale for their behavior (Hutchinson & Gigerenzer, 2005; Sterelny, 2006). Just as the spider has no idea of the intricacy of her web, and the cicada has no clue about prime factoring, human reasoners can be blissfully unaware of the rationale for their heuristics. In a previous paper, we have argued that, though it is important to appreciate the evolutionary roots of human reason to understand its foibles, we should be wary of a certain *locus shift* in attributions of rationality (Boudry *et al.*, 2015). In some of their arguments, advocates of ecological rationality have tried to get human reasoners off the hook by providing an adaptive rationale for their weird beliefs (e.g. superstitions). But human folly is perfectly compatible with adaptive behaviour. Sometimes, given costs and energy constraints, the most fitness-enhancing strategy is the mindless or stupid one. This distinction between the evolutionary and individual locus of rationality, introduced in our earlier paper, is also helpful to clarify the confusion about norms of rationality.

Heuristics are the proximal implementation of evolutionary adaptations. If we argue that the traditional norms of rationality stand in need of revision because certain evolved heuristics, while demonstrably successful, seem to violate them, we are oblivious of the evolutionary process which shaped these heuristics. Much of the "credit" for the R&D of fast and frugal heuristics, particularly of course in the lower animals, does not pertain to the executive control of the organism itself, but to the evolutionary process responsible for its cognitive make-up. This means that, if the decision rules and heuristics enacted by human reasoners flout the traditional canons of rationality, while being 'successful' nonetheless, we should not jump to the conclusion that those standards of rationality are otiose. "Seizing" upon informational dependencies in the environment, as evolution did, by implementing simple heuristics to exploit them, is hardly a violation of the traditional standards of rationality.

Evolution, to be sure, is a mindless process, but that is irrelevant to the current point about the locus of rationality. If heuristics are successful, that is because evolution has been able to track recurrent statistical properties in the environment, and has equipped us with the appropriate cognitive mechanisms to successfully navigate that environment. *These* statistical principles, it will turn out, do not violate the traditional canons of rationality. If we can reach better decisions by ignoring information, this is because evolution by natural selection has first "learnt" when it is useful for an organism to do so. This is not magic or cognitive luck. Without the preceding R&D work carried out by evolution, stumbling in the dark would be a very bad idea indeed.

In contrast to other animal species mindlessly using their evolved heuristics, humans have unprecedented higher reasoning faculties with which they can reconstruct and spell out their own intuitive heuristics. Thanks to the human invention called 'science', we can identify the underlying statistical principles *and* explain why they work so well in the real-life contexts in which they are being applied. However, in order to appreciate the effectiveness of our heuristics in their proper environment, we need the tools of classical rationality. We use Bayesian probability theory to evaluate the performance of different heuristics against classical approaches (Martignon & Laskey, 1999). Moreover, we need these same tools of classical rationality to *understand* the statistical principles underlying the success of fast and frugal heuristics.

The success of simple heuristics is not miraculous. It is based on solid statistics. By using the tools of classical rationality, we also understand why simple rules of thumb are not *always* better. Statisticians can now identify the conditions under which simple heuristics outperform or are beaten by complex decision-making strategies. In complex, noisy environments with a small sample size, simple heuristics outperform more complex and information-hungry algorithms. That is because the latter have the tendency toward overfitting, translating noise into complex patterns. With larger sample size and less noise, by contrast, complex weighting strategies have an edge. Philosophers of science have begun to appreciate the rationale behind our intuitive preference for simple explanations, and now see that the value attached to parsimony in science is not a purely aesthetic one (Forster & Sober, 1994; Hitchcock & Sober, 2004). In a way, they are rediscovering what natural selection "knew" all along.

For example, when we employ the recognition heuristic, we intuitively rely on statistical correlations holding between the probability that an item is recognized and the value of interest. If the heuristic is to work, this "recognition validity" needs to outweigh the "knowledge validity", which is the probability of giving a correct answer when both items are recognized. In other words, your ignorance must be more valuable than your knowledge. Evolution must have hit upon this rationale, even though it is not inscribed anywhere in our brains or in our genes. What this means is that classical rationality *itself* is not threatened by the victory of simple and frugal heuristics over Bayesian networks, multiple regression or other standard approaches. In fact, the traditional approaches are needed to show why in certain contexts the particular heuristics with which evolution endowed us outperform traditional algorithms.

Apart from the fact that you actually need the traditional canons of rationality to appreciate the excellent performance of heuristics, there is another problem with the view that the success of heuristic-based reasoning casts a bleak light on classical rationality. Suppose that an engineer is asked to design robotic life forms that are capable of surviving on a distant planet. Such an engineer may equip her creatures with simple decision rules if she knows that those will prove useful in the particular environment in which she intends the robots to live. The creatures may blindly use these heuristics, because there is no need for them to know anything about the structure of the environment. If the designer has done her job well and the heuristics are successful, however, we would be mistaken to conclude that the "simple-mindedness" of these robots violates the traditional conception of rationality. The ultimate rationality of these heuristics does not reside in their decision rules, after all, but in the foresighted work of their designer. The robots were just programmed to carry out the designer's instructions. The programmer is calling the shots, and for her the benchmark will still be the familiar framework of statistics, Bayesian updating, logic, etc. The same goes for that blind engineer we call natural selection. What underlies the success of heuristics is precisely the statistical and logical relations that natural selection has exploited. Indeed, if the ancestral environments had been such that, say, the average validity of the recognition cue were low (weighted over fitness costs and benefits), then evolution would not have endowed us with the recognition heuristic in the first place. In a similar way, an engineer who develops a computational toolkit for a robot to survive in a novel environment will only design and implement heuristic principles if they promote success in the kind of environment that the robot will encounter.

It is tempting to think that the remarkable success of fast and frugal heuristics flouts the standards of rationality, if you focus only on the proximal implementation of these heuristics. For example, as a violation of the rationality of coherence norms, Gigerenzer mentions that prey animals often display "inconsistent behavior" when interacting with their predators, to reduce the predictability of their behavior (Gigerenzer & Todd, 1999, p. 22). But this is hardly a violation of coherence norms. When taking penalty shots, soccer players are trying to outsmart the goalkeeper: the goalkeeper tries to predict which corner the shooter is going to aim at, while the shooter tries to behave in an unpredictable manner, possibly also tricking his opponent into *thinking* that he (the shooter) is predictable (and then aiming for the other corner). Nothing in this strategic game violates coherence criteria of rationality: a rational agent can decide to engage in unpredictable behavior for strategic reasons. Whether or not hunted animals engage in any *conscious* deliberation as to what their next move will be, it is clear that the logic of deception and counter-deception is undergirding their behavior (von Hippel & Trivers, 2011). If the animal hasn't figured this out, then surely evolution must have. As Dennett wrote, if we discover that an animal is too simple-minded to harbor an adaptive rationale, we do not discard the rationale, but are simply forced to "pass the rationale from the individual to the evolving genotype" (Dennett, 1983, p. 351).

3.1 *General and Content-Free Rationality*

At the level of the heuristic-wielding organism, reasoning looks like a jumble of simple tools and tricks. But the rationality that arises at the evolutionary level is general and context-free in the sense envisaged by proponents of classical rationality. Natural selection is a content-free and general process, an abstract algorithm that works whenever certain minimal conditions are satisfied (variation, heritability, differential reproduction) (Dawkins, 1983; Dennett, 1995). It "learns" by aggregating statistical information about the success of various genotypes in various environments. It does not learn by using a bag of tricks. This mindless process produces only a simulacrum of rationality (de Sousa, 2007), but it is a simulacrum that respects the norms of classical rationality, and that we can understand using classical tools. For example, the theory of error management in evolutionary biology (Galperin & Haselton, 2012; Haselton & Buss, 2000) is essentially an application of expected utility theory and signal detection theory to adaptive problems encountered by evolution.

There is thus something ironic about the claim of evolutionary psychologists Tooby and Cosmides that human reason is "better than rational", in the sense that it beats traditional methods:

> For the problem domains they are designed to operate on, specialized problem-solving methods perform in a manner that is better than rational; that is, they can arrive at successful outcomes that canonical general-purpose rational methods can at best not arrive at as efficiently, and more commonly cannot arrive at at all. Such evolutionary considerations suggest that traditional normative and descriptive approaches to rationality need to be reexamined. (Cosmides & Tooby, 1994, p. 329)

In a trivial sense, this cannot be true. If simple heuristics breed more success than sophisticated methods in certain environments, given some agreed goal or benchmark, then *ipso facto* it is (instrumentally) rational to prefer these heuristics. It is rational to prefer methods with demonstrable success. More importantly, the only reason why our fast and frugal heuristics outperform general-purpose methods of rationality is that they were designed by a general-purpose information-processing algorithm in the first place: evolution by natural selection. If evolution had not done the hard work for us, we would be stumbling in the dark. Or we would not be here at all. Tooby and Cosmides, of all people, should be sympathetic to this point.

If one loses sight of the crucial role of evolution in aligning the two blades of Simon's pair of scissors, the proficiency of our heuristics and intuitions seems almost miraculous, or a matter of sheer luck. Matheson (2006, p. 142) worries that the program of ecological rationality situates rationality "partly outside of the mind", as heuristics are only successful relative to a certain environment. This would amount to a form of "cognitive luck", according to Matheson, which abandons one of the central tenets of the Enlightenment view of rationality, according to which rationality inhabits the mind alone. But how could we be the source and imprimatur of our own rationality? There was a time when there were no humans around. If the source of our rationality were "wholly within our minds", as Matheson put it, we would have pulled ourselves up by our own hair, a godlike feat. It would be more accurate to say that Gigerenzer's program shifts part of the credit for the "rationality" of human behavior to the evolutionary process giving rise to it. It is evolution by natural selection that ensures that there is a *match* between our mind and the environment. This may be a fortuitous arrangement as far as human beings are concerned (at least most of the time), but it is certainly not a matter of *accident*.

4. Individual-level rationality

The traditional canons of rationality have not been vanquished. They have merely been relegated to the locus of evolutionary adaptation. As Orgel's second rule has it, Evolution is cleverer than we are. In our previous paper on the program of ecological rationality (Boudry *et al.*, 2015), however, we discerned a second strand in the research spearheaded by Gigerenzer, which does not involve any adaptive locus shifts. This second strand argues that experimental demonstrations of "irrationality" are often the result of artificial set-ups, which truncate the nuances and complexities of real-life contexts. Psychologists hold subjects accountable to a norm of reasoning that simply fails to capture the ecological complexity of real life.

Take the phenomenon of preference reversals, which have (previously) been interpreted as a form of inconsistent behavior, violating the norms of transitivity (if A < B, B < C; then it follows that A < C). Recent research, however, shows that such behavior can be seen as adaptive, provided we take into account the changing context under which decisions are being made (Schuck-Paim *et al.*, 2004). For instance, the organism may be in (slightly) different states when making choices, which affects the fitness value assigned to A, B and C. Does this show that there is anything wrong with coherence criteria of rationality? No. If choice A turns out to be valued differently by the organism, depending on the context, there is no such thing as 'the' fitness value of A, and the alleged violation of transitivity disappears. Houston *et al.* (2007) have also shown that 'intransitive' behavior may be fitness-maximizing even if the organism is in the same state when making different choices. In their model, the organism is presented with two out of three alternatives each time (A or B; A or C; B or C), while having varying internal energy levels. Within a certain range of energy levels, choices across settings may appear intransitive, for instance ranking A over B, B over C, and C over A. This phenomenon occurs because the availability of an option, even when not chosen, changes the fitness value of the preferred option, because its presence "may act as an insurance against a run of bad luck in the future" (Houston *et al.* 2007, p. 365). Humans and other organisms often (reasonably) assume that the same options will persist in the future (i.e. that the environment will stay roughly the same). Because organisms make different predictions about future options, sometimes intransitive ranking may maximize fitness. But of course that is no reason to jettison the transitivity norm as an axiom of rationality. As Houston *et al.* explain:

> Decisions appear to violate transitivity if an observer interprets a single choice by an animal with given reserves as indicating a straightforward preference for one option over another, instead of viewing the choice as a consequence of following the optimal reserve-dependent strategy. (Houston *et al.*, 2007, p. 367)

In other words, the violation of transitivity is in the eye of the beholder. To use a well-known analogy from Karl Popper (1963, 2002), when you add two drops of water, they join and form a single drop. But that does not mean that arithmetic (1+1=2) has been falsified.

As another example, let us consider the conjunction rule of probability theory: the principle that the conjunction of two events can never have a higher probability than that of either event happening alone. This rule was allegedly violated in the famous Linda problem:

> Linda is 31 years old, single, outspoken, and very bright. She majored in philosophy. As a student, she was deeply concerned with issues of discrimination and social justice, and also participated in anti-nuclear demonstrations.
> Which is more probable?
> (A) Linda is a bank teller.
> (B) Linda is a bank teller and is active in the feminist movement.

In their original research on what became known as the "Linda problem", Tversky and Kahneman (1983) held human reasoners accountable to the conjunction rule, which entails that B cannot be more probable than A. Still, the majority of subjects answered B.

Gigerenzer, however, pointed out that "probable" can also mean plausible, sensible, or supported by evidence (Gigerenzer, 1996; Hertwig & Gigerenzer, 1999). Kahneman and Tversky expect subjects to interpret "probable" in the sense of mathematical probability, but as a number of researchers have pointed out (e.g. Dulany & Hilton, 1991), this construal violates pragmatic rules of conversational inference, in particular the maxim of relevance (Grice, 1989).[2] If subjects interpret the story as Kahneman wanted them to, it becomes a trivial logical exercise, in which the whole description of Linda becomes irrelevant.

We think Gigerenzer is on solid ground here. But does this mean there is anything wrong with the conjunction rule, or that subjects have violated it? Gigerenzer strongly implies that the answers to these questions is yes: "[T]he Linda problem creates a context […] that makes it perfectly valid not to conform to the conjunction rule" (Gigerenzer, 1996, p. 593); "[A]dhering to social norms, here conversational maxims, is rational, although it conflicts with classical rationality" (Hertwig & Gigerenzer, 1999, p. 300). Whereas Kahneman & Tversky choose to retain the norms, after having demonstrated that humans routinely violate them, Gigerenzer urges us to "rethink the norms" themselves (Gigerenzer, 2008, p. 7). Similarly, Polonioli's (2013, p. 6) conclusion about the work of the ABC

---

[2] For a relevance explanation of human performance on the Wason Selection Task, see Sperber, Cara, and Girotto (1995).

research group on the Linda problem is that "violating the norms of coherence might be a key condition for successful communication".

But the conjunction rule of classical probability theory emerges unscathed after the ecological reframing of Linda, because if subjects construe the problem as Gigerenzer thinks they do, the rule is never "violated" in the first place. The conversational maxim of relevance can hardly clash with the conjunction rule if the former simply dictates a construal of the problem that falls outside the latter's domain of application.

Part of Gigerenzer's demonstration that Kahneman has underestimated his subjects is that, when presented in a frequency format, the so-called "cognitive illusion" disappears. In this version of the problem, subjects are told that

> There are 100 people who fit the description above (i.e. Linda's). How many of them are
> (A) Bank tellers?
> (B) Bank tellers and active in the feminist movement?
> (Gigerenzer, 2000, p. 250)

This presentation of the problem avoids the ambiguity of the term "probable" and narrows down on the construal of mathematical probability. In this case, subjects give the appropriate response, as indeed they *should* do. Even Gigerenzer implicitly admits that the conjunction rule is normatively binding *to the extent that it is applicable*. In other words, the problem lies not with the norm, but simply with certain misapplications of the norm.[3]

5. Discussion

There are two ways in which the program of ecological rationality can be construed as a challenge to the canons of classical rationality. First, advocates of ecological rationality have pointed to forms of heuristic intelligence that, though apparently at odds with classical rationality, fare quite well in the real world. What looks like an error, from the myopic view of classical rationality, turns out to be an effective judgment in proper ecological context.

If one loses sight of the 'hidden engineer' – i.e. natural selection – invisibly shaping these successful heuristics, it is tempting to conclude that the benchmark of traditional rationality against which human behavior had been judged is somehow deficient. It is not. The norms simply re-emerge at the evolutionary level, where the adaptive rationale of human cognition resides. The cognitive heuristics of the gene-carrying vehicles – us – are merely proximal implementations of evolution's acquired wisdom about ancestral environments – their success is not miraculous, and they would not work in all environments. Fast and frugal heuristics can beat sophisticated algorithms only because the R&D has already been carried out by evolution. And that R&D exploits the traditional canons of rationality. Moreover, heuristics can and do err, for instance when there is a

---

[3] In this context, Gigerenzer has also made the additional argument that Kahneman and his colleagues stick with one conception of probability, ignoring alternative conceptions in the field of statistics (e.g. Gigerenzer, 2000, pp. 241-266). In particular, Kahneman adopts a Bayesian conception of probability and neglects the influential frequentist school, according to which single-event probabilities such as the ones used in the Linda problem are meaningless in any case. If the human mind is frequentist, as Gigerenzer claims it is, people cannot make the mistakes that Kahneman attributes to them. However, we think this is a red herring that detracts from the main problem with Kahneman's demonstration of the conjunction fallacy, which is the ambiguity of the word "probable" and the conversational implicatures. For a good criticism of Gigerenzer's resort to pluralism about statistical norms, see Bishop and Trout (2005, pp. 118-137).

mismatch between the environments to which they were adapted and the one in which they are now being applied. In such cases, evolution cannot get us off the hook: adaptive explanations cannot exculpate blatant forms of human irrationality (Boudry et al., 2015).

As for the second challenge, when we consider how content-free logical norms fail to capture the intricacies and nuances of human judgment, it is tempting to conclude that there must be something wrong with those norms. But again, this is mistaken. Effective judgment can hardly violate the traditional norms of rationality when these norms do not apply. Bayesian theory, the conjunction rule and the transitivity rule (and other coherence norms) emerge intact after the ecological gestalt switch. The problem here is the rigid (mis)application of content-free logical norms, not the norms as such. Indeed, we need those tools to understand why fast and frugal heuristics are successful in the first place.

While we welcome a naturalized and evolutionary take on human reason, and we agree with Gigerenzer that it is time to correct the bleak picture of human reason, we have argued that the rhetoric against the traditional canons of rationality is unwarranted. Ecological rationality does not challenge, let alone refute, the classical norms of rationality.

Acknowledgments

We would like to thank the anonymous reviewer as well as Fabio Paglieri (unless those are co-referential) for their valuable suggestions. The research of the first author was supported by the Fund for Scientific Research Flanders (FWO).

References

Ariely, D. (2009). *Predictably irrational, revised and expanded edition: The hidden forces that shape our decisions*: HarperCollins.
Bishop, M. A., & Trout, J. D. (2005). *Epistemology and the psychology of human judgment*. Oxford: Oxford University Press.
Borges, B., Goldstein, D. G., Ortmann, A., & Gigerenzer, G. (1999). Can ignorance beat the stock market? In G. Gigerenzer and P. M. Todd (Eds.), *Simple heuristics that make us smart* (pp. 59-72). New York: Oxford University Press.
Boudry, M. & Pigliucci, M. (2013). The mismeasure of machine: Synthetic biology and the trouble with engineering metaphors. *Studies in History and Philosophy of Biological and Biomedical Sciences, 44*, 660–668
Boudry, M., Vlerick, M., & McKay, R. T. (2015). Can evolution get us off the hook? Evaluating the ecological defence of human rationality. *Consciousness and Cognition, 33*, 524–535.
Cosmides, L. and J. Tooby (1994). Better than rational: Evolutionary psychology and the invisible hand. *The American Economic Review, 84*(2), 327-332.
Dawkins, R. (1983). Universal Darwinism. In D. S. Bendall (Ed.), *Evolution from molecules to man* (pp. 403-425). Cambridge: Cambridge University Press.
de Sousa, R. (2007). *Why Think? Evolution and the Rational Mind: Evolution and the Rational Mind*. New York: Oxford University Press, USA.
Dennett, D. C. (1983). Intentional systems in cognitive ethology: The "Panglossian paradigm" defended. *Behavioral and Brain Sciences, 6*(03), 343-355.
Dennett, D. C. (1995). *Darwin's dangerous idea: evolution and the meanings of life*. New York: Simon & Schuster.

Dulany, D.E. and D. J. Hilton (1991). Conversational Implicature, Conscious Representation, and the Conjunction Fallacy. *Social Cognition*, 9(1), 85-110.

Evans, J. S. B. T. and D. E. Over (1996). *Rationality and reasoning*: Taylor & Francis.

Fiedler, K. and M. Wänke (2013). Why Simple Heuristics Make Life Both Easier and Harder: A Social-Psychological Perspective. In R. Hertwig, & U. Hoffrage (Eds.), *Simple heuristics in a social world* (pp. 487-515). New York: Oxford University Press.

Forster, M. R., & Sober, E. (1994). How to Tell When Simpler, More Unified, or Less Ad-Hoc Theories Will Provide More Accurate Predictions. *British Journal for the Philosophy of Science*, 45(1), 1-35.

Funder, D. C. (1987). Errors and mistakes: evaluating the accuracy of social judgment. *Psychological Bulletin*, 101(1), 75-90.

Galperin, A., & Haselton, M. G. (2012). Error management and the evolution of cognitive bias. In J. P. Forgas, Fiedler, & C. Sedikedes (Eds.), *Social Thinking and Interpersonal Behavior* (pp. 45-64). New York: Psychology Press.

Gigerenzer, G. (1996). On narrow norms and vague heuristics: a reply to Kahneman and Tversky. *Psychological Review*, 103(3), 592-596.

Gigerenzer, G. (2000). *Adaptive thinking: Rationality in the real world*. New York: Oxford University Press.

Gigerenzer, G. (2007). *Gut feelings: The intelligence of the unconscious*. New York: Viking Press.

Gigerenzer, G. (2008). *Rationality for mortals: How people cope with uncertainty*. Oxford: Oxford University Press.

Gigerenzer, G., Hertwig, R., & Pachur, T. (2011). *Heuristics: The foundations of adaptive behavior*. New York: Oxford University Press.

Gigerenzer, G., & Sturm, T. (2012). How (far) can rationality be naturalized? *Synthese*, 187(1), 243-268.

Gigerenzer, G., & Todd, P. M. (1999). *Simple heuristics that make us smart*. New York: Oxford University Press.

Goldstein, D. G., & Gigerenzer, G. (2002). Models of ecological rationality: the recognition heuristic. *Psychological Review*, 109(1), 75-90.

Grice, P. (1989). *Studies in the Way of Words*. Harvard University Press.

Haselton, M. G., & Buss, D. M. (2000). Error management theory: a new perspective on biases in cross-sex mind reading. *Journal of Personality and Social Psychology*, 78(1), 81-91.

Haselton, M. G., & Nettle, D. (2006). The paranoid optimist: An integrative evolutionary model of cognitive biases. *Personality and Social Psychology Review*, 10(1), 47–66.

Hertwig, R., & Gigerenzer, G. (1999). The Conjunction Fallacy Revisited: How Intelligent Inferences Look Like Reasoning Errors. *Journal of Behavioral Decision Making*, 12, 275-306.

Hitchcock, C., & Sober, E. (2004). Prediction versus accommodation and the risk of overfitting. *The British Journal for the Philosophy of Science*, 55(1), 1-34.

Houston, A. I., McNamara, J. M., & Steer, M. D. (2007). Do we expect natural selection to produce rational behaviour? *Philosophical Transactions of the Royal Society B: Biological Sciences*, 362(1485), 1531-1543.

Hutchinson, J., & Gigerenzer, G. (2005). Simple heuristics and rules of thumb: where psychologists and behavioural biologists might meet. *Behavioural Processes*, 69(2), 97-124.

Kahneman, D. (2011). *Thinking, fast and slow*. London: Penguin Books.

Kahneman, D., Slovic, P., & Tversky, A. (1982). *Judgment under uncertainty: Heuristics and biases*. Cambridge: Cambridge University Press.

Krueger, J. I., & Funder, D. C. (2004). Towards a balanced social psychology: Causes, consequences, and cures for the problem-seeking approach to social behavior and cognition. *Behavioral and Brain Sciences, 27*(3), 313-327.

Martignon, L., & Laskey, K. B. (1999). Bayesian benchmarks for fast and frugal heuristics. In G. Gigerenzer and P. M. Todd (Eds.), *Simple Heuristics That Make Us Smart* (pp. 169–188). New York: Oxford University Press.

Matheson, D. (2006). Bounded rationality, epistemic externalism, and the Enlightenment picture of cognitive virtue. In R. J. Stainton (Ed.), *Contemporary debates in cognitive science* (pp. 134-144). Oxford, UK: Blackwell.

Mercier, H., & Sperber, D. (2011). Why do humans reason? Arguments for an argumentative theory. *Behavioral and Brain Sciences, 34*(2), 57-74.

Nisbett, R., & Ross, L. (1980). *Human inference: Strategies and shortcomings of social judgment*. Englewood Cliffs, NJ: Prentice-Hall.

Piattelli-Palmarini, M. (1996). *Inevitable Illusions: How Mistakes of Reason Rule Our Minds*: Wiley.

Polonioli, A. (2014). Blame It on the Norm The Challenge from "Adaptive Rationality". *Philosophy of the Social Sciences, 44*(2), 131-150.

Pinker (1997). *How the Mind Works*. New York: Norton

Popper, K. R. (1963/2002). *Conjectures and refutations: The growth of scientific knowledge*. London: Routledge.

Russell, B. (2009). *Unpopular Essays*: Taylor & Francis.

Schuck-Paim, C., Pompilio, L., & Kacelnik A. (2004) State-Dependent Decisions Cause Apparent Violations of Rationality in Animal Choice. *PLoS Biol, 2*(12), e402

Shermer, M. (2011). *The Believing Brain: From Ghosts and Gods to Politics and Conspiracies. How We Construct Beliefs and Reinforce Them as Truths*. Macmillan.

Simon, H. A. (1955). A behavioral model of rational choice. *The quarterly journal of economics, 69*(1), 99-118.

Singer, B., & Benassi, V. A. (1981). Occult beliefs: Media distortions, social uncertainty, and deficiencies of human reasoning seem to be at the basis of occult beliefs. *American Scientist, 69*(1), 49-55.

Sperber, D., Cara, F., & Girotto, V. (1995). Relevance theory explains the selection task. *Cognition, 57*(1), 31-95.

Stanovich, K. E., & West, R. F. (2000). Individual differences in reasoning: Implications for the rationality debate? *Behavioral and Brain Sciences, 23*(5), 645-665.

Stein, E. (1996). *Without good reason the rationality debate in philosophy and cognitive science*. Oxford: Clarendon press.

Sterelny, K. (2006). Folk logic and animal rationality. In H. Susan, & M. Nudds (Eds.), *Rational animals* (pp. 293-312). Oxford: Oxford University Press.

Sutherland, S. (2007). *Irrationality: The Enemy Within* (2nd edition). Pinter & Martin.

Talmont-Kaminski, K. (2007). Reason, Red in Tooth and Claw: Naturalising Enlightenment Thinking. In G. Gasser (Ed.), *How successful is naturalism?* (pp. 183-199). Frankfurt: Ontos Verlag.

Todd, P. M., & Gigerenzer, G. (2012). *Ecological rationality: Intelligence in the world*. New York: Oxford University Press.

Tversky, A. & Kahneman, D. (1983). Extensional versus intuitive reasoning: The conjunction fallacy in probability judgment. *Psychological Review, 90*(4), 293-315.

von Hippel, W., & Trivers, R. (2011). The evolution and psychology of self-deception. *Behavioral and Brain Sciences, 34*(1), 1-16

Chapter 7

The Fragility of Argument

John Woods

Director, Abductive Systems Group, Department of Philosophy, University of British Columbia, 1866 Main Mall, Vancouver, BC, Canada V6T 1Z1; john.woods@ubc.ca; www.johnwoods.ca

> "But the old connection [of logic] with *philosophy* is closest to my heart right now [...]. I hope that logic will have another chance in its mother area."
> Johan van Benthem

> "There is no substitute for philosophy."
> Saul Kripke

> "There is no statement so absurd that no philosopher will make it".
> Cicero

> "We know that people can maintain an unshakable faith in any proposition, no matter how absurd, when they are sustained by a community of like-minded believers."
> Daniel Kahneman

Abstract. Here is a brief tour of the paper to follow. In section 1, I invoke a three-part distinction between a proposition's having consequences, a person's spotting or recognizing them, and a reasoner's drawing them. I argue for the centrality of this trichotomy to the logics of argument. In section 2, I find the first intimation of the trichotomy in Aristotle's distinction between syllogisms as-such and syllogisms in-use, and in his perfectability thesis as well. The trichotomy is also loosely discernible in many present-day writings, but in a less controlled way than in Aristotle. Section 3 turns to the distinction between implication and inference, and to an allied distinction between inference and argument, in which I argue that, all things considered, a good theory of argument cannot be a good model of human inference. Section 4 discusses the normativity problem, which arises from the presumption that idealized models of human cognitive behavior – for example, the probability calculus for belief intensities – are normatively authoritative for real life reasoning. I argue that the normativity presumptions of idealized rationality have yet to be justified satisfactorily. Section 5 calls upon differences, lightly touched upon in section 3 between conscious and unconscious cognitive processes or "cognition up above" and "cognition down below." Section 6 turns to some recent developments in the more formal precincts of argument theory, whose purpose is to bring mathematical models of intelligent interaction into closer harmony with what actually goes on in real-life contexts. I argue that the extent to which that objective is fulfilled those models fall prey to the normativity problem. Section 7 investigates the vulnerability of nonmonotonic premiss-conclusion links to the openness of the world. Section 8 follows up with the suggestion that nonmonotonic consequence relations aren't consequence relations at all, but rather epistemic relations; which calls attention in turn to the depth of the implication-inference divide. Section 9 emphasizes the importance of facts on the ground for the logic of argument, citing the behaviorally discernible trichotomy between arguing *for,* arguing *against* and arguing *with.* It is tied to our earlier consequence trichotomy of having, spotting and drawing, and also to the difference between case-making and face-to-face engagement. I argue that face-to-face combatative case-makings are comparatively rare and in general strikingly difficult to do well. Section 10 offers a tentative (and limited) solution of the normativity problem. It proposes the default condition, that in the case of premiss-conclusion reasoning, how reasoning normally plays out in the conditions of real life is the way it should play out in those circumstances. Section 11 invokes the distinction between arguing and meta-arguing, and cites the difficulties inherent in the latter as one of the reasons for the relative infrequency of face-to-face combat arguments. Section 12 discusses the distortive influences of paradigmatic theories such as Bayesianism, especially as a constraint on what is considered good research practice. It concludes that an uncritical deference to dogmatic paradigm can be intellectually pathological. A steady-handed naturalism is the recommended antidote.

1. Data, Consequence and Inference

Universities are large complexes of disciplines whose root-and-branch purpose is the transmission and advancement of learning by various methods of enquiry. One of these is the collection, storage and communication of information accrued to date. Another searches for data not yet known. The first of this pair *archives* our epistemic inheritance. The second *prospects* for new veins of epistemic gold. Quite often the things we want to know – our epistemic targets, so to speak – can be catered for in these very ways; but not by any means always, or even oftener than not. Of at least equal importance, therefore, are mechanisms for mining both sources of information in ways that release their stored potential for new knowledge – not, mind you, new facts, but rather the investigator's newly acquired grasp of them. Even if the archiving and prospecting aspects of our agendas for hitting epistemic targets were fairly straightforward (they aren't alas!), the *mining* component would be several orders less so. Here is why. At the center of it all is the plain fact of logic and life that the information stored in data bases stands in various kinds of *following-from* relations to other data not in the express purview of the data on hand. Philosophers, and nearly everyone else, have adopted a standardized way of making this point. If we think of data on hand as propositions available for premissory use and a following-from relation as a relation of premiss-conclusion *consequence* or *implication,* it is easy to see that the data-mining[1] that I have in mind here pivots on a threefold distinction between and among

(1) consequences that premisses *have,*

(2) *spotting* consequences that premisses have,

and

(3) *drawing* spotted conclusions.

---

[1] I borrow the term "data mining" from an interdisciplinary subfield of computer science. It is a computational procedure for detecting patterns in large data-bases, drawing upon methods at the intersection of AI, machine learning, statistics and database systems.

I want for now to refer to these premiss-conclusion patterns neutrally, as "premiss-conclusion *structures*". In short order, I'll explain why.

Notwithstanding their common attachment to the notion of consequence, there are key differences among these three. Let B be some remote and arcane consequence of the axioms of Robinson Arithmetic[2]. It is hardly possible that more than a comparative handful of people would know of this, if any at all. Suppose that no one does know or ever will. It is taken as given by working mathematicians, and any philosopher free of epistemological neuroses, that this not-knowing in no way molests that fact that B really is a consequence of Robinson Arithmetic. Their view, though not in these words, is that since consequence-having obtains in logical space, what people know or don't is irrelevant to what holds there. The notion of logical space is a metaphor. Its principal intent is to emphasize that propositions have consequences, or not, independently of cognitive engagement.

If consequence-having obtains in logical space, consequence-spotting occurs elsewhere. It occurs in psychological space – in the spotter's head, indeed in his "recognition subspace", as we might say. If so, consequence-drawing likewise occurs in psychological space, in a sub-region of psychological space which I'll call his "inference subspace", within which consequences are believed for a reason, and the reason is supplied by the premisses from which that conclusion follows. It is easy to see that consequence-spotting and consequence-drawing are natural processes. When they occur, they do so on the four-dimensional wordline of some or other individual. It strains credulity to think that anything like this is what happens when B is merely a consequence of A. We have here two more spatial metaphors, each of which ties consequences to minds. In psychological space, consequences can be recognized without being drawn. In inferential space, recognized consequences are drawn.

I said three paragraphs ago that I wanted to be able to speak of premiss-conclusion arrangements as structures, and to use the word "structure" in a neutral way. Readers might have wondered why. Where is the need for all this mystery and caution when the very words "premiss" and "conclusion" blow the mystery away? As every logician will know, surely premiss-conclusion structures are *arguments*. Why would I be so coy about *them*? Here is why. Logicians have also learned to expect that whenever C is a consequence, say, of A and B there is a derivationally cogent argument ⟨A, B, C⟩ whose conclusion is the consequence C and the other two its premisses. If C is a truth-preserving consequence of A and B, then the cogency possessed by the corresponding argument is conferred by the argument's validity.

Since consequence-havings obtain in logical space, it would seem that their corresponding arguments also do. This alone lends to the words "premiss" and "conclusion" a somewhat strained and artificial feeling. An argument in logical space is nothing but a sequence of formulas, whose "conclusion" is just its last member, and whose "premisses" are the ones left over. "Conclusion" is especially suspect. Conclusions are the result of concludings, but there are no concludings going on in logical space. The reason why is that there are no people there.

2. Arguments, One and Many

Logicians have known since the very founding of systematic logic that two notions of argument have been in play, each under the name of *syllogism*. A syllogism is a valid

---

[2] Robinson Arithmetic, also known as Q, is a finitely axiomatized fragment of Peano Arithmetic, also known as PA. In all essentials Q just is PA minus the axiom schema for mathematical induction. See Robinson (1950).

argument subject to tough conditions on premiss-selection. The core notion of syllogism is simply an ordered sequence of propositions fulfilling the validity condition and the others purpose-built for syllogisity. These, we might say, are syllogisms-*as-such*, and they too reside in logical space.

In contrast to these are syllogisms-*in-use*. They must meet the core as-such conditions, plus such further requirements as serve the particular objectives at hand. Some syllogisms-as-such occur in what we might think of as *public space*, in which argument is conversationally expressed, face-to-face and in real time. Along with this division of syllogisms as to general type, correspondingly different names arose for their respective theoretic treatments. Syllogism-as-such would be handled by analytics (or, as we now say, by logic) and in-use syllogisms would be the business of dialectic. At least this would be so when syllogisms-in-use are refutations. There are two things wrong with these baptisms. One is that the name "analytics" didn't stick. The other is that the word intended to replace it came to be used with a permissiveness that dishonored the dichotomy it was originally intended to draw. In time, "logic" would take on an ambiguity enabling its employers to speak openly of dialectical and dialogue logics, as well as logics that are neither. In the *Organon,* Aristotle selects four types of in-use argument for particular attention. In addition to refutation arguments, they are instruction arguments, examination arguments, and demonstrative arguments from the first principles of science[3]. Aristotle treats refutation arguments in *On Sophistical Refutations,* and demonstration arguments in the *Prior Analytics* and, to some extent, the *Posterior Analytics* too. Instruction and examination receive no like treatment. They are noted without much in the way of follow-up development. More's the pity, too. Dialectical considerations arise for refutations, but are neither intrinsic to nor frequently present in instruction and examination arguments, nor indeed in demonstration arguments either.[4] Public space is another metaphor. It is inhabited by entities with minds, public languages and interpersonal goals, hence is a subspace of psychological space.[5]

A further example of syllogisms-in-use is examined in the *Rhetoric*, which centers on arguments whose aim is persuasion and whose core underpinnings are enthymematic syllogisms, some of whose premisses, while missing, nevertheless remain operational.[6] All these types of syllogisms-in-use, in one way or another engage with their advancers' heads, and usually with their voice-boxes too, and therefore inhabit the public domain, which is in turn is part of the natural world.

My remarks about Aristotle are not an idle historical diversion. I recur to Aristotle because of his foundational importance for the theory of argument. The having-spotting-drawing trichotomy is clearly discernible in his distinction between as-such and in-use syllogisms and the perfectability proof of *Prior Analytics*. Distinctions such as these are crucial for the theory of argument. We ignore them at our peril.

For any syllogism-in-use, it is a necessary condition of adequacy that at its core there be a syllogism-as-such. In which case, the rightness of arguing in public space partly depends on what obtains in logical space. In so saying, philosophical anxieties start to stir all over the place. How, it is asked, can what happens in logical space exercise any dominion over what happens on the four-dimensional worldlines of natural organisms, at the intersection of which is precisely where we find the public domains of mankind? Whether under the name of Benacerraf's Dilemma or none, this is a question as old as the hills (Benacerraf,

---

[3] Strictly speaking, these need not be conversationally expressed.
[4] For more of the logic of syllogisms, readers might wish to consult my *Aristotle's Earlier Logic* (2014).
[5] It is interesting to speculate on whether an argument that occurs in psychological space might fail to occur in public space. My inclination is to think not. It depends on what we think of the relation between argument and inference. I return to this in section 2.
[6] Fabio Paglieri and I (2011a, 2011b) have examined these arguments in greater detail.

1973). I propose to give it no further mind here, beyond noting a recent occasion for ridding ourselves of the problem by the systematic reinterpretation of logical-space properties as public-space properties. In which case, successful reinterpretation might be said to reduce all logical space argument to in-use argument, to face-to-face goings-on in the agora, the senate, the talk show, the kitchen table and Flanagan's Bar and Grill. Let's briefly turn to this now.

In its exuberant expansion from its founding by von Neumann and Morgenstern in 1944 and the refinements of Nash[7] in the early 1950s, the game theoretic orientation adopted by logicians began innocently enough with the dialogical work of Lorenzen and Lorenz in the 1960s (Lorenz, 1961; Lorenzen & Lorenz, 1978). We don't get to game-theoretic logic proper until the latter 1960s (see, e.g., Hintikka, 1968). Game theoretic logic is adapted from the mathematical theory of logic games. It owes its peculiarities to a distinction between two kinds of rules for logic. Some are rules of procedure that regulate the exchanges between rival payers. These are called *structural* rules (although not, to be sure, in Gentzen's sense). The others are the *logical* rules, whose function is to define the logical particles of the players' language – its connectives, quantifiers and the like – and its key metalogical properties, such as deducibility, semantic consequence and logical truth. What makes this logic a game theoretic logic is that all these logical items are wholly defined by rules providing selective constraints on combatants' respective moves and countermoves. Take the universal quantifier as an example. Let $\ulcorner A[x] \urcorner$ be a formula, with x's occurrence possibly free. When one party advances $\ulcorner \forall x \, A[x] \urcorner$, his opposite number must select a constant a for x and challenge the first party to defend $\ulcorner A[x]a \urcorner$. This defines the meaning of the universal quantifier "$\forall$". Accordingly, logics of games are not open to truth conditional semantic interpretation[8].

A game theoretic logic is intrinsically dialectical, both at its core and through-and-through. One *might* argue that since games and gamesters are occurrences in the public spaces of the natural world, *all* of logic – not only the parts governed by the structural or procedural rules but also the part covered by the logical rules – obtains, not in logical space, but rather right down here *in terra firma*. If that worked out properly, perhaps we would have rid ourselves of the dreaded Benaceraff's Dilemma by the simple expedient of denying to logical space anything of value for it to offer to logic. Strange to say, as far as I can determine no game theoretic logician to date has availed himself of this relief. It is quite the other way. Logical space is busier than ever, now liberally stocked with *idealized representations* of agents and actions and of flurries of forms of intelligent dynamic interaction. It is the very place to which the modern game theoretic logician redirects talk about what happens on the ground.

3. Inference and Logic

In a justly famous paper from 1970, Gilbert Harman mounted an argument for freeing consequence-drawing from the over-strong conditions on consequence-having[9]. When consequence is truth-preserving, it is Harman's view that the rules of logic apply without exception to consequence-having, but break down convincingly in their application to

---

[7] John Nash, 1950a, 1950b, 1950c. For the classically foundational papers, see also Kuhn, 1997.
[8] For a good state of the art source, see van Benthem, Gupta, and Pacuit, 2011; and for a first-rate senior textbook, van Benthem, 2014. It is necessary to note, before moving on, that not all logics of dialogue games – for example, Walton's *Logical Dialogue Games and Fallacies* (1984) – are game theoretic logics. They can't be unless they define logical particles and metalogical relations by the logical rules of two-party warfare.
[9] See also Harman, 1986; especially chapters 1 and 2.

drawing or inferring. He cites *modus ponens* as an example. MP, he says, is a valid condition on having, but when re-expressed as a rule of inference, it is easy to see that it fails in lots of quite commonplace situations. If, for example, we believe at t that A and also that if A then B, we are presented with options. One is to do MP's bidding and draw B. Another is to refuse B and drop A. Yet another is to retain A, refuse B, and, then to drop "if A then B". Put Harman's way, truth conditions on implication can't, just so, function as valid rules of inference. Put my way, it comes to the same thing. The conditions on having can't, just so, function as valid rules for drawing.[10] By now this is old news for argument theorists, certainly anyone who has some acquaintance with Harman's 1970 paper or his 1986 book. Logicians at large are less aware of Harman's point, or perhaps not as much impressed with it as they should be. One reason for this is that some of them actually think that the ideally rational reasoner will draw every logical consequence of anything he believes.

Harman also finds fault with the idea that inductive reasoning is regulated by the axioms and rules of the probability calculus. He takes belief-change as an example. It is a commonplace of daily life that, second-by-second, new information hits the processing sensors of the human organism, giving both occasion and necessity for belief-update and/or belief-revision. If our belief-change inductive procedures were indeed determined by the laws and calculation rules of the probability calculus then, upon the arrival of twenty items of new evidence, the human inferer would have to perform one million operations, and a billion were those twenty new bits of information to expand to thirty. This is computational explosion beyond any remotely credible reach of even a really terrific human belief-changer. So Harman concludes that, while the rules of the probability calculus *might* be all right as conditions on inductive consequence-having, they *cannot* be right for inductive consequence-drawing.

If Harman is right, it raises an important question for argument. We would seem to have it that since conditions on consequence-having don't direct all the traffic for consequence-drawing, the same might well be said of argument. Using Aristotle's example again, the conditions on arguments in the form of syllogisms-as-such would not call all the shots for arguments having the form of syllogisms-in-use. This is manifestly the case for Aristotle. The question is whether it should also be the case for us. My own answer is manifestly yes. If Harman is right about the implication-inference divide and I about the logical space-public space distinction as to argument type, a further question arises. Do these two distinctions coincide in any metaphysically principled way? Is inferring consequences in psychological space the same as arguing for them in public space?

I have a brief answer to this question, here more promissory than definitive[11]. If we allow, what virtually every cognitive scientist avers, that psychological space operationally divides into a subspace that falls within the eye of the conscious mind, the conscious reach of the heart's command and the human jaw-bone and voice-box, and other regions in which none of this is true, then we have a rough working distinction between cognition "up above" and cognition "down below". It is an operationally significant division, made so by the abundantly attested-to fact that nothing goes well up above unless requisite things go well further down. Should these claims hold true, there is a large likelihood that cogent arguments transacted up above would be impossible in the absence of smooth sailing further down. Which leaves the further suggestion that, while public space arguing is not,

---

[10] Let me quickly add that Harman is someone of a one-true-logic mind about logic. He thinks that the *only* thing that logic is classical first order logic. In this respect, he is more conservative than Quine even. But this idiosyncrasy doesn't impair the accuracy of the point at hand.

[11] A fuller answer is developed in *Errors of Reasoning*, chapter 4, sections 4.4 to 4.7, chapter 6, sections 6.9 to 6.10, and chapter 10, sections 10.9 to 10.14.

just so, *inference*, it is nevertheless causally supported by inferential traffic below. But now the question is whether inference down below can be *modelled* by arguing up above. My own indication is that in the absence of a representability proof all modelling talk is *flatus vocis*. Think here of the formal representability proof for primitive recursive functions in Gödel's incompleteness theorem for formal arithmetic (1967), as well as the statistical representation theorems for Field's nominalistic recasting of thermodynamics (1980).[12]

There is these days more modelling going on than you can shake a stick at. We all do it, certainly all the authors discussed in this essay. For economists it is as natural as breathing, just as it is mother's milk for physics. There is an amusing story making the rounds. Physicists are said to have two reservations about biologists. They aren't as good as they should be at *data-analysis*, and they aren't good enough at *modelling*. Perhaps this is more of a joke than strictly true, but it makes a serious methodological point even so. It tells us that models aren't free for the taking and that *saying* that this models that is not enough to make it true that it does.

As we cast our eye over the startlingly numerous claims of modellability, whether in informal logic, pragma-dialectics, most by of dialogue logic, public announcement logics, and on and on, there is scant trace of such theorems, or any notice of their importance. So the answer to which I default is that

- Logics of arguing do not model logics of human inference.

4. The Normativity Problem

A growingly influential development in contemporary philosophies of knowledge carries a warning label, albeit wholly free of admonitory intent, calling itself by the name of *formal epistemology*. It over-simplifies things to say that formal epistemology is the probability calculus in application to procedures for drawing inferences, refreshing belief, and making decisions. It is nearer the mark to say this of Bayesianism, and even closer still to say that Bayesianism is a dominating presence in formal epistemology. For present purposes this is domination enough to let Bayesianism stand in for them all. Harman's point is that virtually all of formal epistemology is wrong for inference, made so by its subscription to the idea that good inference requires comportment with the theorems and rules of the probability calculus. If, as it certainly must be, the probability calculus is empirically false for human belief change and inference, why wouldn't we just jettison probability theory as the way to model it?

There is a widely respected answer to this challenge. It proposes that Bayesian probability theorists never intended empirically accurate descriptions of how inference actually plays out on the ground. Its purpose rather was to construct a set of idealized and normatively binding procedural rules for drawing inferences in the right way, in the way that inferences *should* be drawn. In the interests of space, I'll limit myself to just a few of the traditional examples of these idealized norms. One is that the human consequence-drawer has perfect information with respect to premiss-selection. Another, as mentioned, is that he closes his beliefs under truth-preserving consequence. A third is that the ideally rational reasoner knows every logical truth. None of these assumptions stands to any finite degree a chance of being even approximately right empirically. It would be a mistake to over-rely on these extreme absurdities. Various remedies for reducing their offensiveness

---

[12] Krantz, Luce, Suppes, and Tversky (1971) present a number of theorems providing that, under requisite assumptions, qualitative relations imply the existence – and uniqueness relative to stated changes of scale – of quantitative functions "appropriately" representing them.

float about in the literature. But the basic point still holds. It is that the question of what *grounds* the normative authority of these transfinite falsehoods and various others of only finite awfulness is not only rarely posed by the very people who invoke them but still less attracts a satisfactory answer.[13] To the best of my knowledge, there are only two veins of thought about this. One is that the site of the normative authority of empirically false idealized norms of inference lies in the meaning of the word "rational". The other locates their authoritative source in social harmony. Let's take these in order.

Concerning the first, it suffices to ask whether any known theory of the meanings of English by an empirical linguist supports the proposition that it follows from the meaning of the English word "rational" that the rational agent closes his beliefs under consequence. The social harmony answer is a variation of what later would be called reflective equilibrium. The source of it all was Nelson Goodman's 1954 book *Fact, Fiction and Forecast*. Goodman writes:

> Principles of deductive inference are justified by their conformity with accepted deductive practice. Their validity depends upon accordance with particular deductive inferences we actually make and sanction. If the rule yields unacceptable inferences, we drop it as invalid. Justification of general rules then derives from judgments rejecting or accepting particular deductive inferences. (pp. 63-64)

Goodman provides the same test for the correctness of the rules for inductive inference, in *Ways of Worldmaking* (1978). When those principles are taken to be, or to ineliminably include, the principles and rules of the probability calculus, the Goodman Test provides as follows:

- The rules of the calculus are justified by their fit with confident practice, and practice is something to be confident about by its fit with the probability rules.

Suppose that the Goodman Test is the right test[14]. To see whether the probability rules fit with it, it is necessary to identify the "we" of the quoted passage. If *we* are the we of that passage – that is, if we is all of us – then the probability rules fail the Goodman Test hands down. On the other hand if "we" denotes the considered judgements of Bayesian epistemologists, the probability rules pass Goodman's test, leaving a further question to ask and an embarrassing fact to notice. The question is "What justifies this deference to the very folk who condemn all the rest of us for our systematic inferential irrationality?" And the fact to notice is that, in their own inferential lives, these same experts massively discomfort with their own professional judgements of inferential rightness when, after a hard day at the office, they head off to Flanagan's Bar and Grill. So we are only left to ask, *tu quoque,* why their own heedlessness hasn't disturbed their confidence in their professional judgement[15]? See the story about Raiffa and Nagel forthcoming in section 6.

5. A Neuroscientific Response

These days there is simply no shaking Bayesian confidence. This should give us pause.

---

[13] "Transfinitely false"? Every proposition has a denumerable infinity of deductive consequences. A logically omniscient agent knows a denumerable infinity of logical truths.
[14] For reasons to think not, see Stitch & Nisbett, 1980; Thagard, 1982; Woods, 2003, chapter 8; and Gabbay & Woods, 2003, chapter 10.
[15] John Pollock was a refreshing exception. See his "Defeasible reasoning" (2008).

Mightn't there be some reason in this steadfastness to look for *something* about which Bayesianism could be plausibly held to be true? To explore this further, it will be helpful to re-invoke our distinction between cognition up above and cognition down below. In *Agenda Relevance,* Gabbay and I noted the vastness of the difference between the two regions. Cognitive processing up above is in varying degrees, conscious, agent-centred, controlled, attentive, voluntary, linguistically embodied, semantically loaded, surface-processing, linear and computationally weak. Processing down below has none of these attributes and in varying degrees at least most of their opposites. It is unconscious, agentless, automatic, inattentive, involuntary, nonlinguistic, semantically inert, deep, parallel, and computationally luxuriant.

Perhaps this has something to do with the thermodynamic fragility of conscious states. In the sensorium which stores information from the five senses, the information at any instant is reckoned to be 11,000 bits. When admitted to consciousness, there is an informational collapse from 11,000 bits to 40. If the information is given linguistic expression, there is more attrition still, from 40 bits to 16.[16] We already have reason to think that up above processing won't go well unless things also go well in a supporting way down below. The thermodynamic costliness of consciousness suggests a reason why. It suggests that the up above is much too resource-poor to handle the load and complexity of our cognitive agendas. This leaves as a further suggestion that the thing that Bayesians think that they are right about they aren't and what they might be right about isn't the thing they originally had in mind. It is wrong for up above, but it *might* be right for down below.

This is where neuroscience might be of some help. In recent studies the thesis has been advanced that Bayesian frameworks are not just a plausible way of *interpreting* the brain's behavior but are a way of achieving *descriptive accuracy* about the brain's mechanisms and operations.[17] Of course, one of the discouraging things about the brain sciences is that almost nothing of the brain lies exposed to conscious inspection. The brain is needed for up above processing but it is not itself an up above resident. We experience ourselves as thinking but we do not experience *ourselves* as doing what the brain does.[18] Most of what scientists know (or think they do) about the brain is conjectural, using the devices of Peircean abduction. And, as Peirce himself would assuredly have supposed, the routines of abductive reasoning lie mainly in the down below. The critical question for us is whether supposing the brain to implement the *regulae* of the probability calculus would overcome Harman's computational explosion problem. It is true that, when compared to the up above, the down below is a computationally robust parallel processor. But does it have thrust enough to solve an intractability problem as hard as Harman's? The short answer is that we don't know (yet). It wouldn't hurt to try to find out.

Meanwhile, there are two further things to note about the descriptive accuracy thesis for Bayesianism. One is that it has its critics, Stephen Grossberg for one:

> The world is filled with uncertainty, so probability concepts seem relevant to understanding how brains learn about uncertain data. This fact has led some machine learning practitioners to assume that brains obey Bayesian laws (e.g. Knell and Poujet (2014) and Doya *et al.* (2007)). However, the Bayes rule is so general that it can accommodate any system in Nature. This generally makes Bayes a very useful statistical method. However, in order for Bayes concepts to be part of a physical theory, additional computational principles and mechanisms are needed to augment the Bayes rule to distinguish a brain from, say, a hydrogen atom or a hurricane.

---

[16] Agenda Relevance, chapter 2, section 2.6.1.
[17] See, for example, Doya *et al.*, 2007; Hohwy, 2013; Dayan *et al.*, 1995; Friston, 2009, 2010; and Clark, 2013.
[18] For more on this readers could consult my "Reorienting the logic of abduction" (2016a). A pre-publication version is available from my website at www.johnwoods.ca

> Because of the generality of the Bayes rule, it does not, in itself, provide heuristics for discovering what these distinguishing physical principles might be. (Grossberg, 2013)[19]

The second thing to heed, should the descriptive accuracy thesis turn out to be true, is that this alone removes the need for normative presumptiveness as a means of compensating Bayesianism for its descriptive *in*accuracy. If it is not descriptively inaccurate *in re* the brain, there is nothing in it that requires compensation; and as long as they stick to the brain there is no reason why Bayesians can't leave the normativity question open. Indeed it is not clear whether it even arises for brains.

Where does this leave us, then? Our question in this section has been whether human consequence-drawing falls under the normative control of the calculus of probability. The question that preceded that question was whether righteous consequence-drawing can be modelled to advantage in a theory of social argument. I have already put in my two cents about that matter. I have said that I default to the negative answer. What I now want to say is that if the rules of social argument did faithfully reflect the rules of the probability calculus, the answer to the prior question would now become *definitely* no. Of course, the extent to which "no" is the right answer may well depend on how much of social argument is transacted in the public spaces of up above.[20]

6. Heavy Equipment Technologies

What is particularly striking about the modern logician's affection for logical space, is the sheer lavishness of his provisions for it. In earlier times its population, though transfinitely large, was not an especially varied one. Its inhabitants were abstractions from utterances, a consequence relation defined over them and sequences of these, and additional properties of interest defined for them. But there were no people in logical space, no places or times, no actions or interactions, no change, no contexts; and needless to say, no properties of them and no rules or norms governing performance. These days, however, all these omissions have been repaired, not literally mind you, but figuratively. Logical space is fairly chock-a-block with idealizations and mathematical representations of these real-life entities. Why, logical space is now a flourishing cosmopolis that thrills to the discoveries that heavy equipment technologies make possible. Since our topic is argument, I'll confine myself to a brief description of two examples of the work of the heavy equipment logic of argument.

In his dynamic epistemic logic, Johan van Benthem and like-minded colleagues have constructed a complex technology for the execution of what they call the "dynamic turn". It is an impressive instrument of many moving parts. Here is partial list: categorical grammar, relational algebras, cognitive programming languages for information transfer, modal logic, the dynamic logic of programs, whereby insights are achieved (or purported) for process invariances and definability, dynamic inference and computational complexity logics. Synthesis give rise to a unified theoretical framework for the investigation of every variety of intelligent interactions within human societies. And when it comes to argument van Benthem sees interactive argument "with different players as a key notion of logic, with proof just a single-agent projection" (van Benthem, 2011, p. ix).

---

[19] While I like this paper, I dislike its subtitle. It is too anthropomorphic for serious belief, if literally intended.
[20] For other imputations of the descriptive adequacy of the Bayesian inference rules, a good place to start is with a special issue of *Trends in Cognitive Sciences* on probabilistic models of cognition. There is a very useful brief introduction by the issue's guest editors Nick Chater, Joshua B. Tenenbaum and Alan Yuille (2006).

The second example is the heavy equipment technology of the logic of attack-and-defend networks (ADN), developed by Howard Barringer, Dov Gabbay, and your obedient servant in a number of recent papers (Barringer *et al.*, 2005; Barringer *et al.*, 2008; Gabbay, 2012; Barringer *et al.*, 2012a, 2012b). Here, too, we find a good many moving parts – from unconscious neural nets to adjustments for various kinds of conscious reasoning. The ADN paradigm unifies across several fields, from logic programs to dynamic systems. ADN systems pick up interesting properties along the way – some pertaining to equational algebraic analyses of connection strength, where stability can be achieved by way of Brouwer's fixed point theorem. When network processes are made sensitive to time, logic re-enters the *tableau*, involving quite novel modal and temporal languages.

I have two things to say against these heavy-equipment argument technologies. One is the myriad ways in which they advance and sanction empirical falsehoods. The other is the failure to convincingly ground the presumption that these empirical falsehoods are redeemed by their normative authority.

Lots of successful theories depend irreducibly on empirically false axioms or assumptions. Population genetics is like this. It embeds the falsehood that populations are infinitely large, which makes it the case that no actual or physically possible real-world population approximates to this size in any finite degree. A ten-membered population and a trillion-membered population lie *equally* far removed for an aleph null-membered one. The assumption, we might say, is not only false; it is, as we said, transfinitely false. However, notwithstanding this transfinite falsehood, population genetics is a considerable empirical success. It gets natural selection right as it unfolds on the ground. Its predictions are well-confirmed at the empirical checkout counter.

There exists a substantial majority who think that empirical falsehoods required for a theory's success at the empirical checkout counter are vindicated by this nice come-uppance. But that won't do our DEL and ADN technologies a lick of good. Not only do they embed transfinite falsehoods – e.g. that agents have perfect information or that they close their beliefs under consequence – but they are also a train wreck at the empirical checkout counter.

In the heavy equipment world, all this is well known and cheerfully acquiesced to. As previously noted, empirical adequacy is not what their theories strive for, not anyhow when they are theories of empirically instantiated and normatively assessable kinds of human behavior. On the contrary, their theories succeed when their empirically false laws and theorems correctly describe idealized rational agency under idealized performance conditions, and do so in ways that make these empirical falsities normatively binding on *us*. Despite my own complicity in the ADN technology, I don't for a moment suppose it to have rightful normative sway over what we do open-mindedly on the ground when we argue for or against something, or with whom.[21] I believe that this is also true of my co-conspirators Barringer and Gabbay. Things are different with DEL. Van Benthem's purpose in putting it together in the first place was to make logic more faithful to what actually happens. The trouble is that the more complex and mathematically virtuosic it gets, the less the DEL machinery connects with actual happenings, leaving its engineer with two options. One is to scrap the equipment. The other is to play the normative authority card, which indeed is the option van Benthem has exercised.

The normative authority card lacks for a convincing justification, as I've already said. One of its failures is the meaning-of-"rational" justification. The other is the social harmony justification. The first is false on its face. The second calls to mind a story about

---

[21] My attraction to ADNs, aside from the fun of thinking them up, is motivated by the fact that, at times, made-up things can turn out to have instrumental and empirical value. Think of Riemannian geometry in relation to relativity theory, especially after experimental confirmation of the latter.

Howard Raiffa too good to be true, but told as true by Paul Thagard at the Conference on Model-Based Reasoning in Sestri Levante, in June 2012. Raiffa is one of the pioneers of rational decision theory who's played a key role in formulating the decision-tree model of rational decision. Decision-trees embed empirically false assumptions and are governed by empirically unperformable rules. But they are said even so to be normatively binding on beings like us. One day, the story goes, Raiffa, then at Columbia, received an offer from Berkeley. He found himself in a quandary, both wanting to go and yet not wanting to leave. So Raiffa sought the advice of his colleague and friend Ernest Nagel. Nagel advised him to submit the facts of the case to a decision-tree, and wait to see what popped out. "Come on Ernie", cried Raiffa, "this is *serious!*"

7. Closing the World

There is not a single logic student in the wide world who, having been formally introduced to the concept of consequence hasn't been admonished, in the first instance at least, to think of consequence as truth-preserving at its very best – which is what most relations of deductive consequence assuredly are. Non-truth-preserving consequence would be an acceptable but still lesser thing. In the more relaxed milieux of a really good first-year textbook, the student will be subsequently allowed exposure to a further two facts about these relations. One is that truth-preserving consequence closes the world, whereas non-truth-preserving consequence relations do not. The other is that non-truth-preserving consequence relations are the stock-in-trade of a whole host of our most intellectually and practically valuable modes of enquiry, data-mining included.

The metaphor of world-closure comes from computer science, with a provenance now converging on forty years (Reiter, 1978). Under whatever name or none, any neurotypical human being who has drawn breath in the last fifty thousand years has known as a matter of course that the world is open. He knows that today's fact can be tomorrow's toast, and that consequences drawn today are often tomorrow's fallacies. When computer scientists introduced the metaphor they did so with no literal intent. When they spoke of investigations that closed the world, they didn't mean the world; they meant our present knowledge of it. They knew full well that when it comes to our knowledge of the world, it is not we who call the shots. It is the world itself that does. Because of its control of truth, the world that has veto on knowledge.

It has been long known that premiss-conclusion inferences are vulnerable to the openness of the world in two places. As the world turns, premisses could turn out to have been false or to be so now, and conclusions could likewise be so fated. It is not entirely clear at which point in its adaptive development the human organism became aware of a quite striking point of invincibility to the world's mischievous flux. No doubt this awareness arose on the wings of the Attic revolution of *logos*, but it may well have had an earlier start. The point at which premiss-conclusion data-mining closes the world is precisely when the underlying consequence relation is *monotonic*.

Let ⊨ be an arbitrarily selected consequence relation. Then

- ⊨ is monotonic just in case for any $A_1, \ldots, A_n$ and B, for which $\{A_1, \ldots, A_n\} \models B$, we also have it that $\{C, A_1, \ldots, A_n\} \models B$, for any C.

In premiss-conclusion terms, if

$$A_1$$
$$\vdots$$
$$\underline{A_n}$$
$$B$$

is a truth-preserving sequence, so is

$$C$$
$$A_1$$
$$\vdots$$
$$\underline{A_n}$$
$$B$$

for any C, as often as you like; even when $C = \ulcorner \sim A_i \urcorner$. The definition of *non*monotonicity follows by straightforward negation of monotonicity.

Nonmonotonic consequence relations are a premiss-conclusion argument's third point of vulnerability. When the openness of the world is uncurtailed by a consequence relation that binds premisses to conclusions, consequence itself is liable to rupture as the world turns. Bearing in mind that it is ultimately the world that calls the shots, it is precisely here that a question of foundational importance for logic arises.

- If monotonic consequence is a strictly *logical* matter, how could nonmonotonic consequence not be a strictly *epistemic* one?

It may be true that the world holds the trump card for knowledge, but knowledge is also trumped by what goes on in the knower's head. If his head contains the requisite belief that A, the world can make it *true* that A; but without the interplay of a head that would not be *known*. Justification might be like this too. If we think that justification also goes on in the head, the up-above head is doubly-trumpable for knowledge. Mind you, the literature on epistemic justification fairly staggers under its own Gargantuan weight and lawyerly cleverness.[22] So who really knows?

Still, this matters in a rather crucial way. It is easy to see that epistemic consequence-spottings and drawings are things that go on the head. What is not so clear is whether these facts influence where *epistemic* consequence-*having* resides. My own view is that they defeat the idea. There are no heads in logical space. If being evidence for is an epistemic or inferential property, that is, a property that's tied to *believing,* how could it not be in the head? So why wouldn't we have it that epistemic consequence is also in the head? Why wouldn't we have it that epistemic consequence-having supervenes on epistemic spotting and drawing? In which case, why would we give the time of day to the idea that evidential relations are *any* kind of consequence-having, or spotting or drawing either? Why not call them what they are? They are premiss-conclusion relations alright. They are inferential relations and they arise and die in the heads of human reasoners. This is not to say that when A bears an inferentially evidential relation to B, it isn't possible to abstract from this a logical space entity instantiated by the link to logical space proposition A to its logical space relatum B. Call this link R. The trouble is that there is no natural and unforced way of interpreting R as implication. Perhaps R does exist in logical space, but if so, not as a

---

[22] Concerning which, see my lamentations in *Errors of Reasoning* (2014), chapter three.

consequence relation. Recall Harman's insistence that implication and inference differ in kind.

In the section to follow, I want to give some notice of attempts to save the consequencehood of inferential relations. After that we'll return full-steam to argument.

8. D-Logics

By a D-logic I mean a logic whose consequence relations are defeasible by virtue of their rupturability upon promissory expansion. These, of course, are generally speaking non-deductive relations of the nonmonotonic kind.[23] Of the vast literature on nonmonotonic consequence relations, the vast bulk of the logics for dealing with them arise from classical logic – which is monotonic – by attaching to classical consequence various context-sensitive clusters of constraints in which logical particles retain their classicality.[24] In his 2005 book, David Makinson shows that, these exceptions aside, nonmonotonic consequence is a relation at two removes from classical consequence. At first remove, classical consequence becomes paraclassical consequence which, at second remove becomes nonmonotonic consequence. Motivating this bridge between classical and nonmonotonic logic is the theorist's desire to obtain more from premises on hand than can be delivered by classical consequence.

A further and I think deeper, motivation is the desire to preserve this now teeming prosperity of nonmonotonic consequence relations against the rupture threatened by the open world, against which their nonmonotonicity offers no protection. In lots of cases, the world is made to close on nonmonotonic consequence-drawing by theoretical stipulation. The logic simply embodies a closed world *assumption*.

In other cases, protections against rupture are differently wrought. One of these seeks relief in adverbial hedging: "Presumably, B is a consequence of A". Another imposes a qualification on the consequence relation itself: "B is a presumptive-consequence of A", or "A presumatively-implies B". From evidence on hand, Spike is now pretty close to chargeable for last night's homicide. Suppose that tomorrow brings new evidence that significantly weakens the case against him. Suppose, even so, that this new evidence is consistent with yesterday's and also with the judgement that Spike is indeed the guilty party. The weight of the new evidence falls squarely on the link between today's premises and the consequence that Spike is their guy. Accordingly, in this new light of day it is no longer the case that Spike's guilt is a presumptive-consequence of it. But that doesn't change the fact that on yesterday' evidence, this is precisely what does presumptively follow from it. So yesterday's reasoning was correct yesterday. It would also be correct today if restricted to yesterday's evidence, but isn't correct on today's evidence, and would never have been advanced even presumptively on this updated basis.

---

[23] Not all deductive consequence relations are monotonic. The consequence relations of linear logics are deductive, but open to rupture. This is just to say that new premises can deny a truth-preserving deduction its linearity without laying a glove on its deductive validity or its truth-preservation property. See here Girard, 1987. Another example is syllogistic consequence. If ⟨A, B C⟩ is a syllogistically valid argument, it is valid and truth-preserving no matter what. But add one new premiss D to A and B, and the result ⟨A, B, D, C⟩ remains valid and truth-preserving, but loses its syllogisity.

[24] Autoepistemic logics are an exception, what with their employment of the so-called introspective modal operator. Other purported exceptions are the logics of defeasible inheritance-sets and the abstract logic of argument defect. Exceptive status is also claimed for logic programming systems in which "not" expresses negation-as-failure. But it might be possible – I am one who thinks so – to restore these latter rules to classical dignity. See Makinson, 2005, chapter 4.

I have two reservations about this remedy (I was going to say "ploy"). One is that no one has much of a clue about the semantics of "presumptive" in apposition to a consequence relation.[25] The other is that it offers a weak defence against open-world consequence rupture *yesterday*, when yesterday's presumptive consequence is simply now out of date. Which means that yesterday's inference could have been apple-pie, but today it's not worth the paper it was written on in that downtown station of the Vancouver Police Service. So I recur to my recently expressed doubt:

- The nonmonotonic link between premises and conclusions isn't an implication relation after all.

If this is right, how could it matter for social argument? I think it would matter fundamentally. If arguing is public-space consequence-drawing, then there is hardly any social arguing. Even if argument is agent-centred logical space argument, there is hardly any of that either. The dogma that argument is consequence-drawing ethnically cleanses and radically depopulates both places.

I also see another encouraging discomportment with the D-logical approach to multiplicities of nonmonotonic consequence relations in the Gricean-inspired precincts of psycholinguistics, in which, for example, presumptiveness is a property of meanings, and presumptive meaning is generalized conversational implicature. That's *implicature*, not implication.[26] There is a growing literature about presumptive reasoning, especially in logical and computer models of legal reasoning. In much of this work presumability is a property of defeasible rules of reasoning. Douglas Walton's recent book *Burden of Proof, Presumption and Argumentation* (2004) provides an accessible over-view.[27] But there is no mention there of presumptive-consequence as a species of the genus consequence.

Many of the people who write these works are certainly not unfamiliar with the D-logic literature, but in their approach to the presumptive links of premises to conclusions is an absence of what I'll call Makinson's Ascent. The view that nonmonotonic premiss-conclusion links are at two adaptative moves from consequence-having might well be correct, but in Makinson's approach these links retain their membership in the family of consequence relations. It can't be that the computer science modellers of law, especially they, wouldn't know of Makinson's Ascent. Even so, it is notable that in their writings on presumptive reasoning writings, there is virtually no mention of the idea that when premiss-conclusion reasoning is presumptively good, it is underlain by a relation of presumptive-consequence. Speaking again for myself, I see in such circumspection nothing but theoretical and methodological encouragement. Indeed, the general tone of that literature suggests an unannounced retreat from the still dug-in idea that nonmonotonic premiss-conclusion links are a species of implication.

The computer modelling people, also Walton in his non-computability writings, and pragma-dialecticians and virtually all informal logicians transact their business in normative models of empirically instantiated, normatively assessable human behavior. I would make of them the same two pleas I made to the heavy equipment crowd:

- May we have your representability theorems?

- May we also have your solution to the normativity problem?

---

[25] See again *Errors of Reasoning* (Woods, 2014), especially chapter 7.
[26] See Levinson, 2000.
[27] See also Bench-Capon, 1997; Prakken & Vreeswijk, 2001; Prakken & Sartor, 2006; Sartor, 1993; and Walton, 1997.

Perhaps these same pleas would better be self-directed. Here is how I would respond to them. In my own heavy-equipment involvements I have abandoned my former allegiance to the normative models assumption. Whatever Barringer and Gabbay may think, I've already said that I have no inclination to think of ADNetworks as anything but a device for stipulating *new* concepts and the conditions that regulate their instantiations in logical space. Should those concepts thus conditioned prove in due course to be usable in various ways, that would be a bonus. But the last thing that I see ADNetworks as doing is modelling fight-to-the death combat in public space. That being so, the call for a formal representability theorem doesn't arise, and the normative authority question has no occasion to. As for my informal writings, the same answers apply. I have no idea of what are the right rules of public-space/inference-space premiss-conclusion reasoning, if indeed rules there be. Most of what goes on there is regulated down below, where the "rules" are at best a figure of speech for causal (and truth) conditions. Why would I set out to model what I have so little a grasp of what I would want the model to be a model *of*? Since I don't seek such models, once again the question of their normative authority gets no purchase.

It is now time to get off this high horse of heavy equipment technologies and to direct our feet and our attention on the ground, which is where arguing actually goes on.

9. Arguing

When it comes to arguing, logicians and epistemologists have routinely deferred to the noun and given less (or no) attention to the verb. They have also insufficiently heeded the differences among arguing's three modifiers: we can argue *for,* we can argue *against*, and we can argue *with*. Speech communication theorists have been more attentive, advancing a sensible distinction between argument as product and argument as process. My view of the matter is that, in this and all allied matters, priority extends to process. It is a priority that confers some methodological advantage. It tells us to reserve our theoretical judgements until an open-minded inspection of argument-*making* is brought to completion, as nearly as possible without empirically unmotivated philosophical preconception.

In this spirit, consider the following case of a conversationally expressed difference of opinion:

> *Harry*: Gramercy Grill is at the corner of Arbutus and 10th.
>
> *Sarah*: No it isn't, it's at the corner of 11th.
>
> *Harry*: No kidding, I thought it was two blocks from 12th.
>
> *Sarah*: Nope.
>
> *Harry*: Well, I'll be darned. See you on Tuesday.

There is something to be learned here, indeed two if we pay attention:

- Conversationally expressed differences of opinion are not in any sense inherently *arguings-with*.

- The frequency of these *correction-by-contradiction* exchanges considerably outpaces the frequency of arguings-with.

Of course, this is not to deny or play down the fact that in the ordinary sense of the word a great deal of conversationally expressed differences of opinion are indeed arguments. Quarrels are a common enough example:

Sarah: By God, Harry, look at the job your mother did on you!

Harry: Shut your mouth about my mother!

A lot of the time, such exchanges of differing opinions is a good deal more equable. Michael Oakshott is good on this point. According to his former student and colleague,

> … he remarked once that, in conversation, one makes clear the reasons why one sees things the way one does, and conversational partners respond why they see things as they do. Conversation is not, he thought, about winning a debate or besting someone in argument – it is rather mutual self-disclosure in order to understand better what one already understands in part, while accepting that such exchanges can be enjoyed for itself, and need not point to some extrinsic purpose. Refutations and victories are not the goal of conversation (Fuller, 2014).

Oakshott's point is that conversationally expressed differences of opinion are not inherently – not even close – contests of rebuttal and counterargument.[28] This inclines me to think that

- *Explaining is more effective than besting*: Giving reasons for one's side of a disputed opinion in the spirit of making oneself better understood has a better record of dispute resolution than mounting crushing refutations of one anothers' positions.

10. Respecting Data

A perfectly natural reaction to all this downplaying of argumentative clashing is that it is intellectually more honorable to take problems on than to shirk them. My answer to this is that I am not so much a shirker as a prioritizer. For me, one of the guiding rules of procedure is what I call the Respect for Data Principle, which bids us to hold theory and modelling in abeyance until we've achieved some principled command of facts of interest on the ground. Recall the importance of data analysis in the physicists' remonstrations with

---

[28] There appears in the March 6th issue of the 2015 *Times Literary Supplement* are a heartening splash of *pure laine* Oakeshottism. On p. 30 we find a photo of a 1929 poster by Willard Frederic Elmes, entitled "Why Bow Your Back?" This line appears at the top of the poster on a background of what used to be known as hospital green. Below are a large black cat whose back is arched heavenwards, overlooked menacingly by an evil-looking bull terrier in shades of brown and ochre. In the space between the upper line and the two unhappy critters appear three smaller lines:
**Arguing wastes time –
spoils tempers – kills teamwork
– stalls progress**
followed by a fourth in larger type.
Let's Agree to Agree
There is a lot of wisdom in Mr. Elmes' poster. It should be necessary viewing for theorists of argument

biologists. Fair criticism or foul, it is undoubtedly more difficult to analyse biological data than the comparative paucity of those that fire the engines of physics. How less so could this be for the data that drive the engines of theories of human cognition? The physicist example also help in seizing upon a further fact of sound data analysis. It is that *data collection* precedes data analysis, which in turn often loopingly activates still more data collection. There now arise three questions of methodological substance for theories whose subject matters are empirically instantiated normatively assessable modes of human behaviour:

- What are the data that *motivate* your theory?

- Where are those data to be *found*?

- How should these data be *analyzed* prior to their engagement with the apparatus of your theory?

When it comes to facts on the grounds, it is striking how difficult it is to come upon behavioral indications of the cognitive processings of everyday life. Where, for example, would we expect to find behavioral indications of arrivals at knowledge? The answer is that, relative to the high frequency of such arrivals, the frequency of their *express* behavioral reflection is strikingly low, still less, beyond *assertion*, reflected in everyday speech. One search-device which is mainly a failure is to ask a colingual on the ground, having asserted that A, for a report of the state that he was in when the assertion arose and in virtue of which the assertion is accurate. Mind you, some questions are more easily answered than that one. "What makes you think that?" is easy in comparison to "What justifies your thinking it?", which – never mind the shenanigans of internalist-externalist wrangles – is comparatively rarely handled successfully. "Sarah told me so" has a better success-rate in the first instance than it does in the second.

Epistemologists tend to default to the view that to know what human knowledge is it is necessary to unpack "our" concept of it. This inclines me again towards the physicist's reservation about biology's data analysis shortcomings. Much the same can be said of behavioural indications of premiss-conclusion reasoning on the ground. Premiss-conclusion reasoning is transacted in good part down below, which puts first person reportage of its doings beyond reach, leaving distinctly less of a distinct behavioural footprint than we might think or wish. Of course, here too, some signs are more accessible to observation. At the linguistic level, "therefore", "so", "hence" are useful indicators. But let's not overlook that therefoeing, so-ing and hencing occur to a significant extent down below. One source of comparative richness are patterns of on-the-ground assertion-*challenge*, in which we routinely find helpful linguistic indicators, "I don't think you know what you're talking about"; "How could you possibly know that!"; "I think your reasoning is defective."; "Surely that doesn't in the least follow."; and on and on. Where does this leave us now?

Suppose that we granted that all good nonmonotonic reasoning is underlain by a nonmonotonic premiss-conclusion link which, by virtue of its epistemic nature, is instantiated in psychological space and has no real appearance in logical space. (Of course, in agent-centred time-and-change formalisms, it has a *formal representation* in logic space; to what good end is part of our question here.) A central question remains unanswered. What are the rightness conditions for nonmonotonic conclusionality? If our earlier speculations held true, the Bayesian option would no longer be available, not at least for premiss-conclusion traffic up above. This is what I say is so. Many others think otherwise:

> We humans live in a tiny range of the total scale of magnitude, where our body movements bring new objects of the right size under deliberate control. 'Below' us is the statistical molecular and atomic reality over which we have no control, 'above' us is the large-scale structure of the universe with the same lack of control, cognitively, we live in a tiny personal zone.

So far, so good. Invoked here is our distinction between cognition up above and cognition down below. But some very smart people go from good to not so good:

> Likewise, cognitively, we live in a tiny personal zone of deliberation and decision described by logical and game-theoretic models, with below us the statistical physics of brain processes, and above us the statistical realities of long-term social group behavior (van Benthem, 2012).[29]

Bayesianism redux? Or something even more remote? If so there is work to do, as I keep saying. Kindly do one or more of the following:

- Solve the normativity problem.

- Invoke the Bayesian Brain hypothesis if you must, but only after overcoming the computational explosion problem.

- Certify the formal modelling by way of convincing representability proofs.

We've now arrived at a point at which yet another distinction might be of some use. Pragma-dialecticians build their approach around what they call "critical discussions".[30] The discussion-part sounds Oakeshottian, but the critical-part sounds rather otherwise. There are many discussions that arrive at an agreement – e.g. about which movie to see – without the slightest whiff of differing opinions, still less of the need or occasion to resolve them. But PD discussions, from early on until now, center on critical discussions aimed at the resolution of conflict. Except in the adventitious ways indicated in the bulleted passage just above, there is nothing Oakeshottian about critical discussions, beyond the assumption that they be transacted civilly and forthrightly.

Keeping in mind that most of what we could call premiss-conclusion transition occurs down below, it bears repeating that the frequency of its public announcement via assertion, in relation to the frequency of its occurrence is very low. Whatever behavioral manifestations there may be will reside in non-linguistic, highly contextualized action and reaction, with concomitant difficulties for specification. My working hypothesis is that, owing to the difficulties in collecting them, good data analyses for premiss-conclusion reasoning are difficult to get hold of, and can't be pulled off at all simply (or even mainly) by examining "our" *concept* of inference. If that were the right inclination, then it would also be fair to say that argument theorists have the same data analysis problem that physicists attribute to biologists, only more so. The Respect for Data Principle bids us to respect the central importance of good data collection and analysis, and that is a good reason to respect the principle.

We are now in a position to advance what I take to be a plausible conjecture about normativity. To the extent that they are behaviorally discernible, indications of premiss-conclusion adjustments, and also of dissatisfaction with such, disclose a strikingly low

---

[29] But see Xie & Xiong, 2012.
[30] See van Eemeren & Grootendorst, 1984, 2004.

frequency of the latter in relation to the frequency of the former. From this there flows a suggestion which I formulate with a tentativeness that is due it:

- The NN-convergence thesis: In matters of premiss-conclusion transitions, a principal datum for a theory of right reasoning is that the way that premiss-conclusion reasoning normally happens in real life is defeasibly the way it *should* happen there.

It is necessary to emphasize that NN-convergence is an investigative default. It is open to telling exceptions, not least on traditional tellings of the reasoning errors known as fallacies.[31] No such default principle will work for premiss-selection. Indeed most of the errors we make flow from our vulnerability to misinformation, hence bad premiss-selection. What I mean by this is that even when our belief-forming devices are in apple-pie order, we end up with lots of false beliefs. However when our conclusion-forming devices are in equally good order the success rate is markedly higher. The question here is not whether believed premisses are true, but whether the conclusions drawn from them are rightly drawn.

This goes a long way in accounting for the behaviorally discernible fact that the human animal makes errors, lots of them. But when applied to premiss-conclusion transitions, a compensating abundance arises. All told, the human organism is *very* much better at rightly arriving at conclusions than he is at selecting premisses. Whereupon we would have it that, in addition to making lots of errors, beings like us know things, lots and lots of things. Negotiating this tension between getting things wrong and getting things right reflects a further empirically indicated pattern:

- *The enough already thesis*: Beings like us get enough of the right things right enough enough of the time to survive, prosper and, from time to time, build great civilizations.

11. Metadialogues

Let's come back now to arguing. To the extent to premiss premiss-conclusion arguing is facilitated by the mechanisms of premiss-conclusion transition down below, it really should be the case that the arguer's principal point of vulnerability is premiss-selection, and much less so his conclusion-reaching prowess. If arguing actually did model premiss-conclusion reasoning – both the premiss-selection part and the conclusion-drawing part – we would expect much higher rates of premiss-selection challenge than conclusion-drawing challenge. Against this is the empirically evident fact that both sorts of challenge have robust frequencies. The standard complaints against premiss-selection run from begging the question to abusive *ad hominem,* and to babbling (in Aristotle's sense). There are at least equally high rates of challenges *non sequitur*, not just in the consequence-drawing sense, but in any sense in which a conclusion is alleged to be inadequately supported by its premisses.

Especially interesting is a related feature of argumental dissatisfactions, for which Erik Krabbe has coined the term "metadialogues" (Krabbe, 2003).[32] When dialectical slugging matches aren't going well, it frequently happens that the discussion transitions from the

---

[31] For a discouragement of this view, readers may wish to consult *Errors of Reasoning* (Woods, 2014).
[32] See also Finocchiaro, 2013, and Woods, 2007.

point the argument is presently about to one or other or both of the parties' *arguments*. These transitions therefore change the subject. This is what I call "dialogic ascent" from the original argument to what Krabbe calls a metadialogue, which is a dialogue about the adequacy of one another's arguments, or what Finochiaro calls a meta-argument. Meta-arguments carry high risk of failure. A good part of why is that criteria of good argument-making aren't well known. Theorists of argument may think that they are well-known to *them*. Perhaps that is true (though I am a doubter), but there is no doubt that they aren't well known to people at large. How, for example, would the man down the lane handle the following challenges?

Man: "Since A, B".

Challenger: "A is of no relevance to B!"

Man: "Sure it is."

Challenger: "Then tell me your definition of relevance!"

Man: [Fizzles]

Or

Man: "My argument for B is $A_1, \ldots, A_n$"

Challenger: "That's a simply awful argument."

Man: "What's wrong with it?"

Challenger: "The conclusion doesn't follow."

Man: "Why not?"

Challenger: "Because B is false and the A_i are true."

Man: "You're just begging the question!"

Challenger: "I bloody well am not!"

Man: "Oh yeah? Tell me why?"

Challenger: [Punches man in the head.]

The moral is that there is plenty of occasion for arguments to ascend to meta-arguments.
When that happens they rarely turn out well. And that is a good reason to moderate one's enthusiasm for duking it out in the first place. This gives further support to the thesis that human inference is inadequately modelled even by adequate theories of human arguing (if there were any). We're better at inference than we are successful at arguing. What's an inferer to learn from some other thing he's not really good at?
As previously noted, it is perfectly possible to give conversational expression to arguments in the reason-giving sense in the hope that agreement might be achieved as a

byproduct. But reason-giving arguments for and against don't require conversational expression, hence any point of contact with arguings with. I can argue against the normative presumptions of pragma-dialecticians without arguing with them. For years I have argued against this presumption, but I don't readily recall any episode of arguing with its proponents, still less in fight-to-the-death mode. Perhaps I am too polite for combat. Besides, who has ever defeated a Dutchman in a face-to-face fight?

Journal articles are not conversations even when they soundly reason for and against various positions. Neither is it required, or anything close to frequent, that journal articles are conversationally rendered defeats of opponents. You can defeat a position on your i-pad, and in so doing perhaps embarrass its proponents. But there is no need to do it face to face and no need and little occasion to implicate them in the proposition's defeat.

Still, the fact remains that in this foundational PD book, the role of discussions is conflict resolution – "solving conflicts of opinion". This overlooks a large class of discussions in which the objective is not to dissolve conflict but rather to reach agreement in such a way that leaves the differences of opinion intact. Think here of collective agreement negotiations which in virtually all jurisdictions as a requirement of labour law, leaves at least part of what divides the parties untouched. The immediate moral here is that the PD model of critical discussions doesn't generalize very well. Collective bargaining negotiations exchange arguments for positions that aren't in the least interrogational, hence not dialogical either, or dialectical in the manner of Greek refutation or, for that matter, of cross-examinations in common law trials. Concerning these latter, it is true that a key phase of a trial are the closing arguments of counsel. But at no stage does the lawyer who makes one address anyone but the judge, certainly never his opposite number and never a witness, the examination of whom having now being concluded. Not dialogical, and hence not dialectical either, these arguments do have destructive intent – to defeat the presumption of innocence if it's the Crown's argument, or disarm the state's proof if it's the defence's argument.

Arguing with someone presupposes an intimacy that is hardly ever present. Or it requires fulfillment of a whole ballet of socio-professional enabling conventions – e.g. college debating tournaments or meet the leaders go-rounds in runups to general elections. Contentious argument is nearly always rude or out of place, and argumental attack is nearly always for boors.

It is a hallmark of PD critical discussions that they be courteously wrought. Some argument theorists make careful provisions for exceptions to this rule. Rudeness is allowed in philosophy but only when it is cleverly disguised, or wittily or ironically rendered in ways that tend to de-personalize it; euphemism helps here.[33] There are however, exceptions. Question Period in the Canadian House of Commons is one. Prime Minister's Questions in Westminster fares a little better, but not by much. However, not even in Ottawa or London does this rudeness have the slightest thing to do with face-to-face dialogue. They are never arguments *with*. In parliamentary democracies the only addressee of a member's remarks is the Speaker of the House, and MPs are not permitted to argue with him (or, upon occasion, her.).

---

[33] There appeared in the *Philosophical Review* of 2007 a review by Colin McGinn of a book entitled *On Consciousness*. It begins as follows: "This book runs the full gamut from the mediocre to the ludicrous to the merely bad. It is painful to read, poorly thought out, and uninformed. It is also radically inconsistent …. Throughout, the book is woefully uninformed about the work of others and at best amateurish. [The author's] understanding of positions he criticizes is often weak to nonexistent, though not lacking in chutzpah. And the view he ends up defending is preposterous in the extreme and easily refuted." Now this is rude (even if non-conversational), not in the least clever or witty or ironic, and certainly not euphemistic. Mind you, the book's author, whom I have known since 1954, is himself no pussycat. But in this instance, at least, it is McGinn who's the boor.

Arguing contentiously is an acquired skill, all the more so when it is fight-to-the-death arguing. It is often much less successful than not, what with the near-constant threat of dialogic ascent to meta-argument. The truth is that beings like us aren't very good at this sort of thing. It is too much inclined to bring out the McGinn in us. One thing is as clear as glass, though I repeat myself in saying so. We are vastly better at inference than we are at argumentative combat. Why then would we want to model something that we're so good at in something we're so bad at? This sort of carrying-on is expensive, engendering costs too heavy for intelligent indulgence. The costs are both direct and lost. Direct costs include time, effort, wear and tear, uncompensated by good levels of success. Opportunity costs speak for themselves, among them going to the movies, getting back to business, having a nice lunch with someone less bellicose, or frolicking in the park with the kiddies.[34] They are also dangerous pastimes, rupturing friendships and infuriating heads of state (I wonder who?). They occasion embarrassment, which induces in turn fallacy-making and other forms of deception and trickery. Arguments of this sort are mainly for losers. That's why, like Hertford, Hereford and Hampshire where hurricanes hardly ever happen, argumental combat hardly ever happens either. Inferring, on the other hand, is simply everywhere and always.

12. Paradigm Creep

In Thomas Kuhn's *The Structure of Scientific Revolutions*, paradigms are scientific orthodoxies. In a slight extension of the term, a paradigm is any field or form of enquiry centered in a structured organization of received considered opinion and methods of operation. Paradigms are where a discipline's establishment resides, the home of the subject's leading lights. Along with the experts are the established ways of thinking, encoded in the subject's operating manual. Paradigms thus are begettors of doctrines imposing impressive levels of world-closure. A world-closing doctrine is one that closes questions in corresponding degrees. Closed questions are like cold cases; neither is active any more. A closed-minded doctrine is in turn the end result, a promoter of *dogmatism*. Which is only to say that their doctrines remain authoritatively in control even without reconfirmation, or indeed any kind of systematic reconsideration. Dogmas at their most deeply dug-in are nonnegotiable. Consider for example,

> That correct reasoning accords with Bayesian principles is now so widely held in philosophy, psychology, computer science and elsewhere that the contrary is beginning to seem obtuse or at least quaint (Glymour & Danks, 2007, p. 464).

Dogmas are valuable things to have. They are richly value-adding in the cognitive economies of humanity. The savings in time, effort, money and wear-and-tear effected by the closing of our minds to questions speak for themselves. Their greatest value is in not having to re-invent the wheel every other Thursday.

Another virtue of paradigms is their capacity for constructive expansion. The way to go in given areas of enquiry may, with suitable adjustment, turn out also to be the way to go in adjacent areas, and not infrequently even in those that are intuitively far removed, as when it was decided to submit the processes of ampliative reasoning to the regulatory control of

---

[34] Concerning the cost-benefit fragilities of combat, see two important papers, one by Karunatillake and Jennings (2004), and the other by Paglieri and Castelfranchi (2010).

the probability calculus.[35] For logicians the dominant paradigm is the classical treatment of truth-preserving consequence, with high levels of adaptive expansion into nonmonotonic environments. The old paradigm is still present there, what with whole families of nonmonotonic consequence relations at two adaptative removes from monotonic home-base. One could even say that established opinions and methods have an appetite for expansion. The good of it is the wealth that they spread. The downside is imperialistic excess using doctrines and methods that do extremely well at home but turn out to be wrong for the territories they seek to colonize.

A paradigm colonizes successfully to the degree that doing things its way enable us to advance our knowledge of the colony's original subject matter, unattainable by its own old-way methods. Successful colonizers have good track records in Mergers & Acquisitions. But, as is widely known in financial and commercial circles, M & A successes are often vehicles for creative destruction. In scientific circles, creative destruction, whatever its merits, nearly always changes the subject.[36]

Paradigms that overreach themselves usually end up doing more harm than good. When this happens, the paradigm falls prey to what I call "paradigm creep." Paradigm creep resembles mission creep, as when limited military missions spin out of control. When a mission creeps it goes astray, and often enough severs the mission's enabling authority and overwhelms its original rationale. Paradigm creep is different, as different as establishment science and logic are from military engagements. But telling similarities remain, undisturbed by their sundry differences. Methods and doctrines that do wonderfully both at home and in some fruitful outward expansions, start to do badly when creep sets in. The overstretched paradigm's theorems start to strike us as forced. However, since paradigms are dogmas and dogmas are ineligible for review, creep tends not to signal the need for reconsideration. What this means in plain words is that a paradigmer's mind is not going to change, even when creep sets in. See again the quotation from Glyman and Danks. I have high regard for this talented pair, but this time, I fear, they are just being silly. (Of course they might have just been kidding.)

The mathematical turn in logic engineered some powerful reductive successes in the foundations of mathematics and gave rise to a maturely configured model theory and reassuring developments in how to reconcile high points of formal syntax with high points in formal semantics, by way of soundness and completeness theorems. The great achievements of mathematical logic are contributions to mathematics. But they are in the main paradigm creep for logic. By this I mean for the logic of human reasoning.

In the course of this essay I have marked, without announcement, various instances of paradigm creep, or anyhow of what I regard as such. Bayesianism may be wonderful for the reckoning of chances in dice-games, but it is paradigm creep for belief revision, decision making, and the other forms of human inference and reasoning. The logics of truth-preserving consequence relations do marvelously well for entailment, but for spotting and drawing they are paradigm creep. The paradigmatic dominance of consequence relations pays good dividends in theories of deductive reasoning, but it is paradigm creep in theories of nonmonotonic premiss-conclusion reasoning. Makinson's Ascent isn't much of an ascent, after all, but pretty soon it gets to be creep. Heavy equipment technologies of argument are wonderful for proving new theorems, but in application to intelligent-interaction warfare on the ground they are paradigm creep. Formal modelling flourishes in many areas of mathematically expressible enquiry, but in the absence of representation

---

[35] We really shouldn't lose sight of the initial culture shock embodied in the bold decision to model belief intensities in the real numbers.
[36] See here Woods, 2016a. See also Woods & Rosales, 2010; Woods, 2016b and Woods, 2016c.

proofs it is questionable in general, and in application to most of what goes on in premiss-conclusion environments it is paradigm creep.

Aristotle's refutational paradigm, though somewhat artificialized and unrealistic, captures just about perfectly a part of arguing whose bandwidth is small and whose success-rate equally small. Refutations are dialogical attack-arguments whose object is the production of honest answers to cherry-picked questions, whose answers contradict the answerer's own thesis. Most attack-arguments end up begging the question, by importing into one's argument premisses the opponent hasn't conceded, and wouldn't if asked. But since Aristotle's refutations convict the loser of inconsistency with premisses he himself has supplied, there is no question-begging here. The loser's conviction proceeds from his own mouth. This phenomenon is interesting enough to have a name. Let's call it "self-conviction".

In my view, Aristotle's handling of this narrow class of refutation arguments is virtuously paradigmatic for self-conviction, and no modern variation of it is markedly better, whether in formal or informal dialectic or game theoretic logic. But, as mentioned earlier, for the other four classes of arguments he mentions – instruction arguments, examination arguments, demonstration arguments, and enthymematic persuasion arguments – Aristotle's paradigm is paradigm creep. It is helpful to keep in mind that in Aristotle's dialectic, losses are not occasioned by giving up. Giving up must be rooted in self-inflicted contradiction. So let's ask ourselves the obvious two questions:

1. In our own contestational arguings what is the relative frequency of self-convictional intent?

2. What in our self-convictionally intended arguments is the relative frequency of success?

These questions are meant to answer themselves. Negatively.

I conclude from all this that employment of the self-conviction paradigm for the modelling of contestation arguments of real life is already a bit shifty, and that extending it to what PD theorists call critical discussions is paradigm creep through and through. Making it canonical for conversationally voiced differences of opinion is entirely off the radar of credible theoretical pretence, and doubly so when any variant of it is made canonical for inference.

Acknowledgements

This essay is a sequel to "Advice on the logic of argument", which I was my response to a welcome invitation to write for the inaugural issue of *Revista del Instituto de Filosophia de Valparaiso*, 1 (2013), 7-34. Earlier versions of that one were given as talks to the Symposium on New Developments in Dialogue Logics at the Congress on *Logic, Methodology and Philosophy of Science*, Université de Nancy, in July 2011, and to the Stanford CSLI Conference on Logic, Rationality and Intelligent Interaction in May 2013. Earlier versions of (parts) of this follow-up essay were given as talks to a university-wide audience at Sun Yat-sen University in November 2012, and to the day long Philosophy Colloquium at the Okanagan campus of UBC in March 2015. For stimulating conversation, I warmly thank Shahid Rahman, Matthieu Fontaine, Gerhard Heintzman, Ulrike Hahn and Hartley Slater in Nancy, Minghui Xiong, Yun Xie, Zhicong Liao, Xiajing Wu and Bin Wei in Guangzhou, Solomon Fefferman, Thomas Icard, Anna-Sara Malmgren and Carl Hewitt in Palo Alto, and Johan van Benthem for post-conference correspondence, as well as Otávio Bueno, Andrew Irvine, Dan Ryder, Holger Andreas, and Manuela Ungureanu in Kelowna. I also want to thank Henry Prakken, Giovanni Sartor and Sina Fazelpour for some very helpful discussions. Fabio Paglieri and an anonymous referee have also favoured me with constructive suggestions. For technical support, I am blissfully indentured to Carol Woods for life.

I dedicate this essay to our dearly departed friend Jaakko Hintikka.

References

Barringer, H., Gabbay, D. M., & Woods, J. (2005). Temporal dynamics of support and attack networks: From argumentation to zoology. In D. Hutler & W. Stephan, (Eds.), *Mechanizing Mathematical Reasoning: Essays in Honor of Jörg Siekmann on the Occasion of His 60th Birthday* (pp. 59-98). Berlin and Heidelburg: Springer-Verlag.

Barringer, H., Gabbay, D. M., & Woods, J. (2008). Network modalities. In G. Gross & K. U. Schulz (Eds.), *Linguistics, Computer Science and Language Processing: Festschrift for Franz Guenthner On the Occasion of his 60th Birthday* (pp. 70-120). London: College Publications.

Barringer, H., Gabbay, D. M., & Woods, J. (2012a). Temporal argumentation networks, *Argument and Computation, 2-3*, 143-202,

Barringer, H., Gabbay, D. M., & Woods, J. (2012b). Modal argumentation arguments, *Argument and Computation, 2-3*, 203-227.

Benaceraff, P. (1973). Mathematical truth, *Journal of Philosophy,* 70, 661-679.

Bench-Capon, T. J. M. (1997). Arguments in Artificial Intelligence and Law, *Artificial Intelligence and Law,* 5, 249-261;

Chater, N., Tenenbaum, J. B., & Yuille, A. (2006). Probabilistic models of cognition: Conceptual foundations, *Trends in Cognitive Science, 19*, 287-291.

Clark, A. (2013). Whatever next? Predictive brains, situated agents and the future of cognitive science, *Behavioral and Brain Sciences,* 36, 181-204.

Dayan, P., Hinton, G. E., Neal, R. M., & Zemel, R. S. (1995). The Helmholtz machine, *Neural Computation, 7*, 889-904;

Doya, K., Ishii, S., Pouget, A., & Rao , R. P. N. (Eds.) (2007). *Bayesian Brain: Probabilistic Approaches to Neural Coding,* Cambridge, MA: MIT Press.

Field, H. (1980). *Science Without Numbers: A Defense of Nominalism.* Princeton: Princeton University Press.

Finocchiaro, M. (2005). Arguments About Arguments: Systematic, Critical and Historical Essays in Logical Theory, New York: Cambridge University Press.

Finocchiaro, M. (2013). *Meta-argumentation: An Approach to Logic and Argumentation.* Studies in Logic (Vol. 42). London: College Publications.

Friston, K. (2009). The free-energy principle: A rough guide to the brain. *Trends in Cognitive Sciences, 13*, 293-301.

Friston, K. (2010). The free-energy principle: A unified brain theory. *Nature Reviews Neuroscience, 11*, 127-138.

Fuller, T. (2014). The compensations of Michael Oakeshott. *The New Criterion,* November.

Gabbay, D. M. (2012). Equational approach to argumentation networks, *Argument and Computation, 2-3*, 87-142.

Gabbay, D. M. (2013). *Meta-logical Investigations in Argumentation Networks.* Studies in Logic (Vol. 44). London: College Publications.

Gabbay, D. M. & Woods, J. (2003). *Agenda Relevance: A Study in Formal Pragmatics.* Amsterdam: North-Holland.
Girard, J-Y (1987). Linear logic. *Theoretical Computer Science, 50,* 1-102.
Glymour, C. & Danks, D. (2007). Reasons as causes in Bayesian epistemology, *Journal of Philosophy, 104,* 464-474.
Gödel, K. (1967). On formally undecidable propositions of *Principia Mathematica and Related Systems I.* In J. van Heijenoort (Ed.), *From Frege to Gödel: A Source Book in Mathematical Logic, 1879-1931.* Cambridge, MA: Harvard University Press.
Goodman, N. (1978). *Ways of Worldmaking,* Indianapolis, IN: Hackett.
Goodman, N. (1983). *Fact, Fiction and Forecast,* Fourth edition, Cambridge, MA: Harvard University Press.
Grossberg, S. (2013). Adaptive resonance theory: How a brain learns to consciously attend, learn, and recognize a changing world. *Neural Networks,* 1-48.
Harman, G. (1970). Induction: A discussion of the relevance of the theory of knowledge to the theory of induction. In M. Swain (Ed.), *Induction, Acceptance, and Rational Belief,* Dordrecht: Reidel.
Harman, G. (1986). *Change in View: Principles of Reasoning,* Cambridge, MA: MIT Press.
Hintikka, J. (1968). Language games for quantifiers. In N. Rescher (Ed.), *Studies in Logical Theory,* (pp. 46-72). Oxford: Blackwell.
Hohwy, J. (2013). *The Predictive Mind,* New York: Oxford University Press.
Karunatillake, N. & Jennings, N. (2004). Is it worth arguing?. In I. Rahwan, P. Moralïs, & C. Read (Eds.), *Argumentation in Multi-Agent Systems,* (pp. 234-250). Berlin: Springer-Verlag.
Krabbe, E. C. W. (2003). Metadialogues. In F. H. van Eemeren, J. A. Blair, C. A Willard & F. Snoeck Henkemans, (Eds.), *Anyone Who Has a View: Theoretical Contributions to the Study of Argumentation,* (pp. 83-90). Dordrecht: Kluwer.
Krantz, D. H., Luce, R. D., Suppes, P. & Tversky, A. (1971). *Foundations of Measurement Theory.* New York: Academic Press.
Kuhn, H. (1997). *Classics in Game Theory.* Princeton: Princeton University Press.
Levinson, S. C. (2000). *Presumptive Meanings: The Theory of Generalized Conversational Implicature.* Cambridge, MA: MIT Press.
Lorenz, K. (1961). Arithmetik und Logik als Spiele, PhD dissertation, Christian-Albrechts-Universität, Kiel.
Lorenzen, P., & Lorenz, K. (1978). *Dialogische Logik.* Darmstadt: Wissenschaftkiche Buchgessllschaft,.
Makinson, D. (2005). *Bridges From Classical to Nonmonotonic Logic.* Texts in Computing (Vol. 5). London: College Publications.
Nash, J. (1950a). The bargaining problem, *Econometrics, 18,* 155-162.
Nash, J. (1950b). Equilibrium points in n-person games, *Proceedings of the National Academy of Science, 36,* 48-49.
Nash, J. (1950c). Non-cooperative games, *Annals of Mathematics Journal,* 54, 286-295.
Paglieri, F. & Castelfranchi, C. (2010). Why argue? Towards a cost-benefit analysis of argumentation, *Argument and Computation, 1,* 71-91.
Paglieri, F. & Woods, J. (2011a). Enthymematic parsimony, *Synthese, 178,* 461-501.
Paglieri, F. & Woods, J. (2011b). Enthymemes: From resolution to reconstruction, *Argumentation, 25,* 127-139.
Pollock, J. (2008). Defeasible reasoning. In J. Adler, & L. Rips (Eds.), *Reasoning: Studies in Human Inference and its Foundations,* (pp. 451-471). Cambridge and New York: Cambridge University Press.

Prakken, H. & Sartor, G. (2006). Presumptions and burdens of proof. In T. M. van Engers (Ed.) *Legal Knowledge and Information Systems: JURIX 2006: The Nineteenth Annual Congress,* (pp. 21-30). Amsterdam: IOS.

Prakken, H. & Vreeswijk, G. (2001). Logics for defeasible argumentation. In F. Guenthner, & D. M. Gabbay (Eds.) *Handbook of Philosophical Logic* (2nd ed., vol. 4, pp. 219-338). Dordrecht: Kluwer.

Reiter, R. (1978). On closed world data bases, *Journal of Logic and Data Bases, 1,* 55-76.

Robinson, R. M. (1950). An essentially undecidable axiom system, *Proceedings of the International Congress of Mathematics,* 729-730.

Sartor, G. (1993). Defeasibility in legal reasoning, *Rechstheorie, 24,* 281-316.

Siegel, H. (1992). Justification by balance, *Philosophy and Phenomenological Research, 52,* 27-46;

Stitch, S. & Nisbett, R. E. (1980). Justification and the psychology of human reasoning, *Philosophy of Science, 47,* 188-202.

Thagard, P. (1982). From the descriptive to the normative in psychology and logic, *Philosophy of Science, 47,* 24-42.

van Benthem, J. (2011). *Logical Dynamics of Information and Interaction,* New York: Cambridge University Press.

van Benthem, J. (2012). The nets of reason, *Argumental Computation, 3,* 83-86.

van Benthem, J. (2014). *Logic in Games.* Cambridge, MA: MIT Press.

van Benthem, J., Gupta, A. & Pacuit, E. (Eds.) (2011). *Games, Norms and Reasons: Logic at the Crossroads.* Dordrecht: Springer.

van Eemeren, F. H. & Grootendorst, R. (1984). *Speech Acts in Argumentative Discussions: A Theoretical Model for the Analysis of Discussions Directed Towards Solving Conflicts of Opinion.* Dordrecht: Foris.

Van Eemeren, F. H. & Grootendorst, R. (2004). *A Systematic Theory of Argumentation.* New York: Cambridge University Press.

Von Neumann, J. & Morgenstern, O. (1944). *The Theory of Games and Economic Behavior.* Princeton: University of Princeton Press.

Walton, D. (1984). *Logical Dialogue Games and Fallacies.* Lanham, MD: University Press of America.

Walton, D. (1997). *Argumentation Schemes for Presumptive Reasoning.* Mahwah, NJ: Erlbaum.

Walton, D. (2004). *Burden of Proof, Presumption and Argumentation.* New York: Cambridge University Press.

Woods, J. (2003). *Paradox and Paraconsistency: Conflict Resolution in the Abstract Sciences.* Cambridge: Cambridge University Press.

Woods, J. (2004). *The Death of Argument: Fallacies in Agent-Based Reasoning.* Dordrecht: Kluwer.

Woods, J. (2007). Agendas, relevance and dialogic ascent, *Argumentation, 3,* 209-221.

Woods, J. (2014). *Errors of Reasoning: Naturalizing the Logic of Inference.* Studies in Logic (Vol. 45). London: College Publications.

Woods, J. (2016a). Reorienting the logic of abduction. In L. Magnani & T. Bertolotti (Eds.), *Springer Handbook of Model-Based Science.* Dordrecht: Springer.

Woods, J. (2016b). Does changing the subject from A to B really provide an enlarged understanding of A? A puzzle and a muddle. In L. Magnani, & T. Bertolliti (Eds.), *Conference on Model-Based Reasoning.*

Woods, J. (2016c). *How globalization makes inconsistency unrecognizable,* forthcoming.

Woods, J. & Rosales, A. (2010). Virtuous distortion in model-based science. In L. Magnani, W. Carnielli, & C. Pizzi (Eds.), *Model-Based Reasoning in Science and*

Technology: Abduction, Logic and Computational Discovery, pp. 345-388. Berlin: Springer.

Woods, J. (2014). *Aristotle's Earlier Logic* (2014), Studies in Logic (2nd expanded ed., Vol. 53). London: College Publications. [First appeared as Woods, J. (2001). Aristotle's earlier logic. Hermes Science Publishing, Paris].

Xie, Y. & Xiong, M. (2012). Whose Toulmin, and which logic? A response to van Benthem, *Cogency, 4*, 115-134.

Chapter 8

Arguments and Their Sources

Peter Collins & Ulrike Hahn[1]

Department of Psychological Sciences, Birkbeck, University of London; [1] u.hahn@bbk.ac.uk

Abstract. As argumentation theory has moved away from classical logic as a standard, sources have played an increasingly important role in the psychology of argumentation. Considering the connections between arguments and their sources is important for both descriptive and normative projects. This chapter draws together different strands of research in the psychology of argumentation and their differing views on source characteristics: namely, procedural rules, pragmatics, argumentation schemes and Bayesian Argumentation. We argue for a reconciliation of these different approaches around a probabilistic notion of relevance.

1. Introduction

To have a theory of argumentation requires us to choose from, or weave together, different senses of the term 'argument'. An argument can be a reason given in support of an action or idea. Relatedly, an argument can be the combination of the reason and the claim it is taken to support. In other words, an argument, as an inferential argument, will be a set of premises (evidence) and a conclusion (claim) bound together in some way. Lastly, an argument can be a dialogue in context: a dialectical process in which discussants propose and oppose claims. These argumentative dialogues vary in character, including, for instance, both reasonable attempts to resolve or clarify an issue and personal conflicts in which participants verbally hit out at each other (for a typology, see Walton, 2008). Different theories of argumentation have focused on different senses of the word 'argument' and, hence, on different aspects of argumentation (on different senses of 'argument' see also O'Keefe, 1977, and Hornikx & Hahn, 2012).

Classical logic focuses on the inferential object sense. Thus, arguments are sets of propositions linked through a set of logical rules (see, for instance, Arthur, 2011), and arguments are sound solely in virtue of their structure and the truth of their premises. But since at least Toulmin (1958), argumentation theory has taken a dialectical turn: theories emphasize an argument's use in context. This dialectical turn brings the different senses of argumentation tightly together. Arguments are reasons for actions or ideas, but they are reasons for *someone*: that is, they are relative to the arguer and audience. And while arguments may have a complex structure, even as inferential objects, they are, on this view properly understood only as part of the wider argumentative process.

In the same way that a dialectical perspective brings together argument and recipient, it also brings into view the argument's proponent. Historically sources (to wit, proponents of arguments) have been treated as irrelevant to argument quality. On this view, a source's standing should neither improve nor worsen an argument. Hence, numbering among the traditional fallacies are appeals to a source's expertise (*argumentum ad verecundiam*) or arguments attacking a source (*argumentum ad hominem*) (Hahn, Oaksford, & Harris, 2012). However, while such a view seems natural from a logical perspective, the dialectical turn means that arguments and sources can no longer be treated as wholly separable. Arguments always come from a source, and properties of the source bear on the strength of the argument. For instance, as we will see below, sources seem subject to procedural rules that govern dialogue; if they violate these rules, they risk rendering their contributions inappropriate or unreasonable.

In this chapter we will discuss the multiple, intrinsic connections between arguments and their sources. We will outline the different approaches taken to accounting for the role of sources, and argue for a division of labour between the approaches.

2. Procedural Rules

In argumentation theory, a large and influential body of work situates arguments firmly in the context of discourse. This work focuses on deriving procedural rules for arguments, which is to say principles for how discussants should reasonably behave. Procedural rules bear the imprint both of Toulmin's seminal work on the use of arguments and of natural-language pragmatics.

Pragmatics resists straightforward definitions (Huang, 2007; Levinson, 1983). But popular definitions set pragmatics alongside semantics: both are the study of (or mental faculties for) meaning, distinguishable (crudely) in the following ways. Semantics treats the meanings of words and sentences in abstract. Pragmatics treats the meanings of words and sentences as used by a speaker in a specific context. Hence, we can speak of sentence meaning (semantics) and utterance meaning (pragmatics) (Levinson, 1983). Pragmatics includes a wealth of information that speakers intentionally communicate but leave implicit. Hence, we can also speak of a code and inference model of language: semantics is the code, associating words, concepts and the world; pragmatics is the inference, linking the code to the speaker's intended meaning (Clark, 2013).

One account of procedural rules dominates the literature: namely, pragma-dialectics (see, e.g., van Eemeren & Grootendorst, 1984, 1992, 2004). Pragma-dialectics views argumentation as "verbal moves ideally intended to resolve a difference of opinion" (van Eemeren & Grootendorst, 1995). On this view, argumentation theory becomes a sort of complex pragmatics, the aim of which is to analyze speech acts that are chained together in argument. Pragma-dialectics breaks argumentative discourse down into various stages. During the confrontation stage, discussants encounter a difference of opinion. During the opening stage, they identify themselves and establish their initial standpoints. During the argumentation stage, they begin the discussion proper, outlining their standpoints and defending them against critical questioning. During the conclusion stage, the discussants reach their final standpoints: ideally they reach agreement. Complementing these stages is a set of procedural rules, which draw on Grice's (1975) Cooperative Principle and Maxims of Conversation. The rules derive from the qualitative study of real arguments. Table 1 below lists these rules alongside a competing set.

Standards of Fairness	Pragma-dialectic Rules
I. Faulty Arguments	
(1) Violation of Stringency *Arguments presented non-stringently*	(1) Freedom Rule *Don't prevent advancing standpoints or questioning them*
(2) Refusal of Justification *No, or insufficient reason, given for assertion*	(2) Burden of Proof Rule *Must defend standpoint when requested*
II. Insincere Contributions	
(3) Pretence of Truth *Assertion known to be false, or subjectively true, presented as objectively true*	(3) Standpoint Rule *Only attack standpoints that have really been introduced*
(4) Shifting of Responsibility *Unwarranted denial, claim or transfer of responsibility*	(4) Relevance Rule *Real arguments, relevant to standpoint*
(5) Pretence of Consistency *Arguments presented that are incongruent with actions/other arguments*	(5) Unexpressed Premise Rule *Don't falsely attribute implicit premises or deny your own*
III. Unjust arguments	
(6) Distortion of Meaning *Intentional distortion of any contributions or facts*	(6) Starting Point Rule *Don't falsely present or deny an accepted starting point*
(7) Impossibility of Compliance *Don't demand anything of others that you know they won't be able to do.*	(7) Validity Rule *Supposedly conclusive arguments must be logically valid*
(8) Discrediting of Others *Intentional or negligent discrediting of other participants*	(8) Argument Scheme Rule *Non-conclusive arguments must respect argument schemes*
IV Unjust interactions	
(9) Expressions of Hostility *Don't treat your adversary as your enemy*	(9) Concluding Rule *Respect inconclusive and conclusive defences in settling argument*
(10) Hindrance of Participation *Don't impede others' participation*	(10) Usage Rule *Don't use confusing/ambiguous language; don't intentionally misinterpret opponent's language*
(11) Breaking Off *Don't break off argumentation without justification.*	

Table 1. Fairness Rules versus Pragma-dialectic Procedural Rules; wordings adapted from Schreier et al (1995) and van Eemeren, Garssen and Meuffels (2009)

To illustrate, consider the first two rules:

> Rule 1 (Freedom Rule): Discussants may not prevent each other from advancing standpoints or from calling standpoints into question.
>
> Rule 2 (Burden of Proof Rule): Discussants who advance a standpoint may not refuse to defend this standpoint when requested to do so.
> (van Eemeren, Garssen, & Meuffels, 2009, pp. 21-22)

These rules have good face validity: we would intuitively consider someone who followed the rules reasonable, and someone who violated them unreasonable. The rules also potentially account for certain fallacies. A discussant, for instance, might violate the Freedom Rule by threats of force (*argumentum ad baculum*) or appeal to compassion (*argumentum ad misericordiam*) or personal attack (*argumentum ad hominem*) (van Eemeren & Grootendorst, 1995). A discussant might violate the Burden of Proof Rule by refusing to justify his/her own standpoint, instead challenging his/her opponent to disprove it (van Eemeren & Grootendorst, 1995).

Pragma-dialectics has accumulated considerable evidence to support its procedural rules. Much of this evidence is naturalistic (e.g. van Eemeren & Houtlosser, 1999), but lately there has been an increasing interest in experimental investigations (van Eemeren, Garssen, & Meuffels, 2012; van Eemeren *et al.*, 2009). These experiments typically present short arguments that respect or violate procedural rules. Participants tend to judge arguments that respect the rules as reasonable and those that violate the rules as unreasonable. Pragma-dialectic rules also fare well with extended arguments: witness van Eemeren and Houtlosser's (1999) detailed discussion of a sophisticated 63-line argument.

These data suggest that insight can be gleaned from generalized rules that treat procedure rather than specific content. But in the rules' procedural nature lies an important limitation, for the rules do not ultimately measure the strength of a specific argument. Imagine, for instance, an argument between two perfectly reasonable discussants, in the sense that they obey all the procedural rules. Procedural rules do not give us a way to decide between these discussants' positions. Who should we take to have provided the stronger argument?

There is also the question of how binding these pragma-dialectic rules are. As we have seen, these rules derive from the study of real arguments, much in the same way that Grice's maxims derived from the study of real language use. But it is not clear how and why pragma-dialectic rules have normative force. It might be desirable to follow the rules: our arguments might be more civilized if we did. It might also be practical to do so: our arguments might be more persuasive if we did. But it is not obviously irrational to disobey the rules, in the way that it is irrational to disobey the rules of logic.

An alternative set of procedural rules comes in the form of fairness rules for argumentation (Christmann, Mischo, & Flender, 2000; Christmann, Mischo, & Groeben, 2000; Mischo, 2003; Schreier, Groeben, & Christmann, 1995). This theory takes as its goal an account, not of reasonableness, but of fairness. That is, the theory is intended as an ethical theory, not an epistemological one (Schreier et al, 1995). According to Christmann, Mischo and Flender (2000), fair argument is governed by rationality and argumentation. There are, on this account, four general aspects to fair argumentation: formal validity (covering, in some unspecified way, abduction, deduction and induction); sincerity or truth; justice; and procedural justice. Arguments are judged unfair if there is rule violation and subjective awareness, that is, if the violator was aware of or intended the violation. Fairness is also subject to contextual factors. The following factors, due to Christmann, Mischo and Groeben (2000), count as mitigating circumstances. (1) Variables in the exchange: the rule violator may, for example, notice the violation, and make amends for it. (2) Person variables: the rule violator may prove incompetent rather than intentionally unfair. (3) Situation factors: the fairness of moves seems to depend on whether the argument is planned or unplanned, public or private, prepared or unprepared.

This fairness account has been subjected to empirical testing in the following ways. To compile a detailed list of fairness rules, Schreier *et al.* (1995) had participants rate the fairness of 35 commonly discussed rhetorical strategies and classify them into the four general aspects of fair argumentation mentioned above. A cluster analysis then revealed 11 clusters, which are presented in the left-hand side of Table 1 above. Christmann, Mischo and Groeben (2000) then tested these rules by presenting participants with scenarios that varied in their degree of conformity to fairness rules, in their contexts, and in the degree of subjective awareness shown. Participants indicated whether they would personally reproach a transgressor (in Christmann *et al.*'s terms, the 'objective data'), and gave a free response to an open-ended question asking them to indicate which aspects of the scenarios most influenced their decision (in Christmann *et al.*'s terms, the 'subjective data'). The resulting data suggested that participants were sensitive to violations of fairness rules, and that these violations could be mitigated by contextual factors. Rule violations and contextual factors significantly affected fairness judgments for both objective and subjective data. However, subjective awareness (whether a violation was intentional) significantly affected judgments only in the subjective data, thus offering somewhat weaker evidence of its importance.

Fairness rules offer a similar picture of argumentation to procedural rules, as witnessed by the overlap between the sets of rules in Table 1. This similarity raises the question of whether the theories are meaningfully distinct. Nevertheless, that these theories are so similar, and can both call on supporting data, suggests that they are tapping into a real phenomenon. Fairness rules do not, however, add an ultimate measure of argument strength.

Thus far, the source of an argument has been somewhat spectral: procedural rules invoke the arguer, but only in a highly generalized way. Pragma-dialectics makes more concrete reference to discussants as people through the notion of strategic maneuvering. On this view, argumentation has a dual purpose, discursive and rhetorical: that is, argumentation is aimed at resolving a difference but also at doing so in one's own favor (van Eemeren & Houtlosser, 1999). In strategic maneuvering, discussants take advantage of 'topical potential'. Protagonists have available to them, in any given context, a large set of potential topics and argumentative moves, from which they can select the best candidates to achieve their goals. Thus, protagonists can strategize throughout the dialogue. The following points are due to van Eemeren and Houtlosser (1999). At the confrontation stage, a protagonist can choose a standpoint to minimize the 'disagreement space'. To wit, if a protagonist chooses a standpoint fairly close to their antagonist's, there will be less persuading to do. At the opening stage, she can target the most helpful concessions from her antagonist. At the argumentation stage, she can choose the most strategic line of defense. And at the conclusion stage, she can invoke the happy consequences of accepting her standpoint, or the unhappy consequences of not accepting it. All this must be achieved without flouting the antagonist's expectations and preferences. It should be clear that this strategizing places a heavy demand on protagonists. The protagonist, for instance, must formulate general goals and specific strategies to achieve them, and must anticipate and compare the antagonist's responses to a range of hypothetical strategies.

There is a body of evidence to support strategic maneuvering, both within pragma-dialectics and in neighboring social psychology. Strategic maneuvering has been tested in recent experimental work on ad hominem arguments (van Eemeren *et al.*, 2012). An ad hominem argument is an attack against the person, typically occurring in response to another argument. The protagonist of an ad hominem argument does not attack the content of the original argument but rather its protagonist, dismissing them as a source. Ad hominem arguments are sometimes legitimate, as, for example, when levelled against an argument from expertise, in which the source's expertise is the crucial factor. But ad hominem arguments are very often considered unreasonable or abusive (Walton, 1998).

Van Eemeren et al. (2012) presented participants with ad hominem arguments in various forms: as straightforwardly abusive arguments, as disguised abusive arguments, and as legitimate personal attacks. For present purposes, the crucial distinction is between straightforward and disguised abusive arguments. The disguised argument was an abusive ad hominem attack made in the guise of a legitimate personal attack, as in the following item from their materials:

> Context: The art museum is renovated and that is the reason why it has been inaccessible to the public for some time. The museum curator discusses this with a journalist.
> Curator: I think the museum can be open again for the public. The building is in excellent shape now and it is perfectly safe.
> Journalist: As a curator you may know about art but you are not knowledgeable about the safety of the building.

The point, here, is that it is quite reasonable for a curator to know about the health and safety aspects of his building, and to be able to testify about these matters. Hence, the attack is only superficially legitimate. Van Eemeren et al. found that participants considered straightforwardly abusive *ad hominem* arguments unreasonable ($M = 2.44$ on a scale from 1-7 where 1 is 'very unreasonable' and 7 is 'very reasonable') and disguised abusive *ad hominem* arguments significantly less so ($M = 4.09$).

Elsewhere, O'Keefe (2009) has linked strategic maneuvering to a large body of research in persuasion theory. He interprets strategic maneuvering somewhat broadly to include various presentational devices, which is to say ways of portraying the same information. Presentational devices do seem to affect persuasion. Arguments are more persuasive when they contain explicit conclusions, identify their sources, are more complete, or use figurative language (O'Keefe, 2009). Arguments are also more persuasive when they are culturally adapted to their audience (O'Keefe, 2009). But all these effects are rather small (effect sizes up to $d=0.28/r=0.14$[1]).

3. Pragmatics

Pragmatics has featured so far as the inspiration for procedural rules. As we have seen, pragma-dialectics, for instance, draws on Grice's (e.g. 1975) work to cast argumentation as a complex series of speech acts. However, pragmatics bridges content and source in its own right.

Pragmatic theories generally hold that comprehension is based in large part on making assumptions about the source. Indeed, comprehension is often thought to require recognition of (or at least making assumptions about) speakers' intentions. Take the following stock examples, presented here as in Clark (2013):

---

[1] Cohen (1988, 1992) makes the following suggestions for interpreting effect size. $d = 0.2$ is small; $d = 0.5$ is medium; $d = 0.8$ is large. $r = 0.1$ is small; $r = 0.3$ is medium; and $r = 0.5$ is large. Note that, unlike d, r is capped at 1, a perfect correlation.

(1) Those spots mean she's got measles.
(2) Those three rings on the bell (of the bus) mean that the bus is full.

Sentence (1) is an example of natural meaning: we can infer something from the state of the world (Grice, 1957/1989; Clark, 2013). Sentence (2) is an example of non-natural meaning: the hearers of the bell can infer that the bus is full because that is what the bus driver intended them to infer (Grice, 1957/1989; Clark, 2013). More formally, non-natural meaning is defined as follows:

"U meant something by uttering x" is true iff, for some audience A, U uttered x intending:
(i) A to produce a particular response r.
(ii) A to think (recognize) that U intends (i).
(iii) A to fulfil (i) on the basis of his fulfilment of (ii).
(Grice, 1969/1989, p. 92)

Subsequent work has considered this definition too broad, and added that the speaker's intentions must be part of mutual knowledge – mutual, that is, between the speaker and hearer (e.g. Clark, 1996; Schiffer, 1972; Strawson, 1964; though see Sperber & Wilson, 1995, for further qualifications). It follows from this definition that comprehension requires extensive assumptions about the source of an utterance and, by extension, that comprehension of an argument requires extensive assumptions about the source of utterances chained together in argumentation (Breheny, 2006). Equally, when crafting their utterances, or arguments, sources must make assumptions about how their audience will receive them. This does not, of course, mean that all aspects of comprehension require such assumptions (see the definition of semantics and pragmatics above), but it does mean that full comprehension requires them.

Alongside this definition of meaning we need an account of how people go about making reasonable assumptions about a source's intended meanings. Grice (1975) offered the following account in the form of the Cooperative Principle and complementary Maxims of Conversation:

Cooperative Principle:
"Make your conversational contribution such as is required, at the stage at which it occurs, by the accepted purpose or direction of the talk exchange in which you are engaged." (Grice, 1975, p. 45)

Maxims:
Quantity:
   1. Make your contribution as informative as required (for the current purposes of the exchange)
   2. Do not make your contribution more informative than required.
Quality: Super-maxim, "Try to make your contribution one that is true"
   1. Do not say what you believe to be false
   2. Do not say that for which you lack adequate evidence
Relation: Be relevant
Manner: Super-maxim, "Be perspicuous"
   1. Avoid obscurity of expression
   2. Avoid ambiguity
   3. Be brief (avoid unnecessary prolixity)

4. Be orderly
  (Grice, 1975, pp. 45-46)

For pragmatics to proceed, speakers observe or ostentatiously flout maxims. Hearers then calculate the intended implication (implicature). For example:

Maxim Observed
A: I've just run out of petrol.
B: Oh, there's a garage just around the corner.
Implicature: A may obtain petrol there.

Maxim Flouted
A: Let's get the kids something to eat.
B: Okay, but I veto I-C-E-C-R-E-A-M.
Implicature: B would rather not mention ice cream directly in case the children then demand some.
(Levinson, 1983, pp. 104).

Thus the maxims help to explain intuitions about qualitative linguistic data (for discussion, see, for example, Huang, 2007, and Clark, 2013), and have also proved a springboard for rich explorations of pragmatics. But, as Grice himself seems to have acknowledged, there is both vagueness and redundancy in the account (Clark, 2013). When it comes to discussion of natural conversations, most analyses end up citing the Maxim of Quantity and the Maxim of Relation, and it is difficult to tease the two maxims apart (Clark, 2013). This closeness raises the question of whether the maxims are genuinely distinct. Indeed, leading successors to Grice have re-worked the maxims, preferring a fleshed-out notion of quantity[2] (e.g. Horn, 1984, 1989, 2004) or of relevance (Sperber & Wilson, 1995). This fleshed-out relevance, embodied in Relevance Theory (Sperber & Wilson, 1995), has achieved greatest popularity.

Relevance Theory takes cognition in general, and communication (pragmatics) in particular, to be geared towards the maximization of relevance. In other words, cognition and communication focus on deriving the greatest, positive cognitive effects (e.g. true conclusions, strengthened or contradicted assumptions) for the least processing effort (Clark, 2013). Speakers produce utterances that are worthwhile for the hearer to process, and as relevant as possible given speakers' abilities and preferences (Clark, 2013; Sperber & Wilson, 1995). With this notion of relevance, the other maxims are arguably superfluous (for a discussion, see Sperber & Wilson, 1995, or Clark, 2013).

There are, of course, alternative theories in both pragmatics (e.g. Levinson, 2000) and formal semantics (e.g. Jaszczolt, 2007; Kempson, Meyer-Viol, & Gabbay, 2000) that all aim at more minimal accounts than the Cooperative Principle and Maxims of Conversation. It suffices, however, to note that, while the theoretical framework of pragmatics has been reduced substantially, the purview of pragmatics has expanded. For Grice, semantics seems to have contributed what is said - roughly, the propositions expressed - and pragmatics what is meant (but not said) (Clark, 2013). On the whole, Grice's successors - and, certainly, relevance theorists - have argued that pragmatics contributes at even lower levels (Clark, 2013). That is, we even need pragmatics to identify the propositions that we would intuit as being literally expressed.

It is easier to appreciate how these aspects of pragmatics bear on argumentation by

---

[2] As Clark (2013) argues, Horn's Q and R principles (see references above) can be seen as an expanded version of the Maxim of Quantity: say enough, but not too much.

referring to an example. Take, then, the following toy conversation and suggested pragmatic contributions, which broadly conform to the style of analysis used in Relevance Theory (see, for example, Clark, 2013).

> A: John looks grumpy.
> B: He hasn't had breakfast, so he is starving.
> (3) [John] hasn't had breakfast, so [John] is starving.
> (4) [John] hasn't had breakfast [today], so [John] is starving.
> (5) [John] hasn't had breakfast [today], so [John] is [very hungry].
> (6) [B believes that][John] hasn't had breakfast [today], so [John] is [very hungry].
> (7) [B believes that John looks grumpy because (or so B believes)[John] hasn't had breakfast [today], so [John] is [very hungry].

Sentences (3) to (4) are low-level pragmatic intrusion into what is said: reference must be assigned to both instances of the deictic pronoun 'he'; temporal reference must be assigned to the sentence, as "John hasn't had breakfast" presumably does not mean that John hasn't *ever* had breakfast. Sentence (5) is needed to fix the sense of 'starving': the appropriate sense is 'very hungry', not 'in danger of death through malnutrition'. Sentence (6) makes clear that B is expressing his/her belief, a belief that corresponds to reality to a greater or lesser degree. Sentence (7), finally, identifies the real, but implicit, point of the utterance: a possible explanation for John's apparent grumpiness. There are two observations to make, here, about pragmatics and argumentation. Firstly, the pragmatic contributions above resemble an argument; together they seem to constitute a complex inferential object. Secondly, if pragmatic contributions like these are typical of comprehension, then they will also be typical of the comprehension of arguments. In other words, it is pragmatics that offers up the actual content of the argument; it is pragmatics that lets us understand what the argument actually is.

How has pragmatics, in this sense, contributed to the study of argumentation? Certainly, pragmatics has contributed to the study of rationality more generally. Hilton (1995) considered the influence of pragmatics in developmental psychology, judgment and decision-making, and argued that, whenever we assess the rationality of participants' behavior, we must also consider pragmatic cues in tasks. The provision of instructions and materials to participants in psychology experiments is a communicative situation, and participants will naturally draw on natural language pragmatics for their interpretation. Hilton (1995) provides examples of experimental findings on human (ir)rationality, such as the use of base-rate information, the presumed relevance of non-diagnostic information, false-memory effects, and the conjunction fallacy that may require some degree of questioning once the pragmatic context of the experimental situation, as viewed by the participant, is properly taken into account[3].

Hilton (1995) used a definition of pragmatics that is rather similar to that presented above, but included it in his broader Attributional Model. In this model, pragmatics has the crucial function of allowing inference to the source's intentions. Crucial, too, is a link to stable attributes of the speaker: for example, to personality traits. Knowing about the speaker's attributes, the hearer might modify their interpretation of the speaker's intentions. To adapt Hilton's (1995) example, imagine an underwhelming reference for a job. How

---

[3] This is not to say that such biases of judgment and decision-making will necessarily disappear when pragmatics is factored in. The conjunction fallacy, for instance, is a robust effect: for discussion, see, for example, Jarvstad and Hahn (2011), and for experiments that control for pragmatic involvement, see Tentori and Crupi (2012). But, as the discussion of framing effects below shows, comprehensive pragmatic accounts are possible, and must be eliminated before judgment and decision biases can be proclaimed.

should we interpret this reference? If we know that the writer is generally generous with praise, producing glowing references, we will likely conclude that the reference is meant as a non-recommendation. By contrast, if we know that the writer is very cautious, or loath to give praise, then we will likely hesitate before concluding that the reference is bad.

The Attributional Model calls for an integration of pragmatic inference with information about specific sources. This integration remains, to the best of our knowledge, underexplored. However, researchers have followed Hilton's lead in assessing the impact of the experimenter's presumed intentions, especially in the psychology of judgment and decision-making. For instance, Sher & McKenzie (2006) explored how participants responded to seemingly equivalent descriptions of cups as half-full or half-empty. Participants inferred from the description 'half-full' that the experimenter intended to refer to a cup that had previously been empty; from the description 'half-empty', to one that had previously been full. More generally, Mandel (2014) has argued that key framing effects – attribute and risky-choice frames – rely on the pragmatics of number terms. For instance, participants famously prefer options that will save 200 out of 600 lives to those that will result in the loss of 400 out of 600 lives, even though these situations are, on some readings, mathematically equivalent (Tversky & Kahneman, 1981). Mandel (2014) has shown that such framing effects depend on participants assuming that the experimenter meant 'at least 200/400 lives', and that when the word 'exactly' is inserted, the framing effect disappears. It may well be that these much-vaunted equivalence framing effects – where the information content is supposedly equivalent but the persuasiveness differs – depend on pragmatic inference (for a discussion of equivalence frames, see the review of Levin, Schneider, & Gaeth, 1998).

Likewise, pragmatics seems to play a role in another key phenomenon, reasoning from conditionals ('if-then' statements). In the famous Wason selection task (van der Henst & Sperber, 2004; for the original exposition of the selection task, see Wason, 1966), participants are tasked with testing a rule of the form 'If P then Q'; the correct responses is checking 'P' and 'not Q'. Consider, for instance, the following set of items, adapted from van der Henst & Sperber (2004) :

Rule: If there is a 6 on one of the card, there is an E on the other.

Card 1: 6   Card 2: 7   Card 3: E   Card 4: G

The normative response is supposedly to turn over the cards '6' and 'G' (at least on the rather dubious interpretation of natural language if-then as the material conditional of propositional logic). However, participants tend to turn over the cards '6' and 'E'; only around 10% choose the normative response (van der Henst & Sperber, 2004). These findings can be qualified in two ways. Firstly, more participants show the normative response in a deontic scenario (Griggs & Cox, 1982; van der Henst & Sperber, 2004). For instance, when testing the rule 'If you are drinking beer (P), you must be 18 years old (Q)', participants would turn over cards for 'drinking beer' (P) and '16 years old' (not Q) (van der Henst & Sperber, 2004). Secondly, alternative accounts suggest that testing P and Q may, in fact, be more informative (Oaksford & Chater, 1994). More importantly, for present purposes, is that varying the pragmatics of the materials also varies participants' responses. Girotto et al. (2001) manipulated participants' performance on a deontic version of the task. Participants saw the following instructions:

Imagine that you work in a travel agency and that the boss asks you to check that the clients of the agency had obeyed the rule "If a person travels to any East African country, then that person must be immunized against cholera," by examining cards representing these clients, their destinations and their immunizations.

Participants then had to choose which of four cards to turn over, representing P (the person travels to an East African country), not P (the person does not travel to an East African country) and Q (the person is immunized against cholera) and not Q (no immunizations done). In this condition, the relevant goal is to prevent people travelling to East Africa unimmunized; the relevant tests are P and not Q. Participants did, indeed, tend to test cards corresponding to P and not Q. In another condition, the same participants read that the boss had been mistaken, and that cholera immunization is no longer required to visit East Africa. As a result, the boss is concerned that she has misinformed clients, causing them to follow an obsolete rule. In this condition, the relevant goal is to test whether people have been immunized unnecessarily; the relevant tests are P and Q. Participants did, indeed, tend to test cards corresponding to P and Q. Thus, Girotto *et al.* (2001) found evidence that pragmatics does indeed affect people's reasoning.

The preceding examples of the role of pragmatics in assessing experimentally human rationality, finally, are themselves simply expressions of a more fundamental pragmatic connection between content and source. As described above, pragmatics involves inferences about speaker intentions in the service of interpreting utterances. In the context of those inferences, hearers make a default assumption not only that communicated information should be relevant to them, but that it is so because speakers possess some (degree of) rationality. Among possible interpretations we choose (in the first instance) those that allow the speaker to be perceived as 'making sense' (on the so-called 'principle of charity' or 'rational accommodation' see e.g., Davidson, 1973; Quine, 1960, p. 59).

In summary, utterance comprehension intrinsically connects content and source at a basic level. It is by making assumptions about an argument's source that we understand the form and purpose of an argument. As we have seen, pragmatics has changed markedly since Grice, with a popular movement towards more streamlined theory but with wider scope. This raises the question of whether the streamlining of pragmatic theory can beget the streamlining of argumentation theory. We will return to this question below. Important though pragmatics may be in interpreting arguments, its role in evaluating the strength of arguments is less clear. The discussion in the following section moves towards a deeper, probabilistic notion of argument strength, but one that nevertheless reflects the importance of pragmatics through its link to more stable source characteristics such as source reliability (for extended discussion of this link, see McCready, 2014).

4. Source Reliability

As arguments occur in natural language, they are inherently subject to natural-language pragmatics. Pragmaticists after Grice have argued that pragmatics contributes even to the recovery of the literal proposition: that is, to understand what an argument even is, we need to use pragmatics. It follows from this account that, when they hear arguments, audiences must make considerable assumptions about a source. This account contrasts strongly with the dominant view on arguments and their sources in social psychology. In social psychology, fundamental aspects of communication are relegated to heuristic processes that can be turned on and off.

Social psychology treats arguments in the context of persuasion. In other words, social psychology is primarily interested not in when arguments are 'good' but when they 'work'. Though source characteristics feature heavily in the persuasion process, they have been given a decidedly 'second class' role in argument evaluation, at least by the dominant models. In particular, the Elaboration Likelihood Model (ELM) separates persuasion into two different routes: the central and peripheral routes (e.g. Petty & Cacioppo, 1984; Petty & Cacioppo, 1986). In the central route, people engage in sustained and critical processing of an argument's content. In the peripheral route, people evaluate an argument briefly or heuristically using source information or presentational features. Which route obtains is a function of the audience's motivation and ability to analyze the issue at hand, with central processing being engaged only under conditions of high personal involvement.

All of this sounds both surprising and problematic in light of the points raised in the preceding sections. In fact, it seems questionable whether it is even possible to understand, let alone be persuaded by, natural language arguments without at least some consideration of the source from which the argument or persuasive message is coming.

This is not to doubt that social psychology has obtained findings that seem roughly compatible with source processing assigned to a lesser, 'peripheral' route which is used only on occasion, as contrasted with an analytic route focused on content. However, in light of the implausibility (and impossibility) of strict separation between content and source, these data would seem in need of re-interpretation and process accounts of persuasion which need to be revised in light of more careful consideration of what, if anything, about sources can and cannot be ignored.

Not only does it seem difficult to ignore source considerations when evaluating arguments, it also seems normatively undesirable. Intuitively the reliability, veracity and competence of a reporting source seem informationally relevant (see also Schum, 1972). In keeping with this, sources feature more prominently in research into argumentation schemes (e.g. Walton, Reed, & Macagno, 2008) and Bayesian Argumentation (see, e.g., papers in Zenker, 2012) – both of which have overtly normative concerns.

When is information about sources relevant? There may be cases when the audience knows enough about a subject to evaluate an argument fully and ignore source information (Hahn et al., 2012). Perhaps, for example, when audiences encounter a deductive argument, they need not refer to source information as long as they accept the truth of the premises. Likewise, perhaps when audiences encounter an inductive argument, they need not refer to source information as long as they can independently corroborate the facts presented. One can, nevertheless, imagine many cases when source information will be relevant. Even in deductive arguments, we may need assurances about the truth of the premises before granting that the argument is sound. It remains possible that the degree of source involvement does and should differ across types of inference. When source information does contribute, however, to ignore it is to risk forming inaccurate beliefs.

One way to capture sources is to treat source-related content through argumentation schemes – an approach that has been popular also in the context of 'critical thinking' (e.g. Inch & Warnick, 2009). Argumentation schemes represent patterns of inference in common types of argument as a way to get a handle on argument strength. Argumentation schemes have yielded considerable insight into common arguments. Many sets of schemes are available (e.g., Garssen, 1997; Hastings, 1962; Kienpointner, 1992; Perelman & Olbrechts-Tyteca, 1969; Schellens, 1985), but probably the most extensive is Walton *et al.* (2008). According to Walton *et al.* (2008), schemes model arguments with defeasible reasoning: conclusions are held to be true or false but provisionally so; conclusions should be revised in light of new evidence. Defeasible reasoning contrasts with deductive reasoning, in which new information should not change a valid argument. Source content – indeed, argument content more generally – is represented in two ways. Firstly, content appears in the schemes themselves; secondly, it is probed in a corresponding set of critical questions. On this approach, hearers should judge an argument's strength by identifying the relevant argumentation scheme, and trying to find acceptable answers to the critical questions.

Walton et al. (2008) provide a set of a source arguments and attendant source fallacies (e.g. *argumentum ad hominem, ad populum, ad verecundiam*).

We can best assess the strengths and weaknesses of argumentation schemes by considering some examples. We will consider the argument from expertise (*argumentum ad verecundiam*) and the argument from popular opinion (*argumentum ad populum*), our analysis following that of Hahn and Hornikx (2015). Walton et al. (2008) provide the following scheme and set of critical questions:

> Source E is an expert in subject domain S containing proposition A
> E asserts that proposition A (in domain S) is true (false)
> A may plausibly be taken to be true (false).
> (Walton et al., 2008, p. 14)
>
> How credible is E as an expert source?
> Is E an expert in the field that A is in?
> What did E assert that implies A?
> Is E personally reliable as a source?
> Is A consistent with what other experts assert?
> Is E's assertion based on evidence?
> (Walton et al., 2008)

On further consideration, though, the list of critical questions balloons: Walton et al. (2008, pp. 92-93) distinguish some 21 questions that might be relevant in different contexts. While this list demonstrates the depth of the analysis, it also demonstrates what Walton et al. (2008) call the completeness problem. Simply put, we can always think of new critical questions to ask. The completeness problem argues for a way to integrate source information with broader argument content in a way that allows for uncertainty and for appropriately hedged degrees of belief. The argument from popular opinion demonstrates a further limitation to argumentation schemes. The scheme for this argument is as follows:

> S1: Everybody (in a particular reference group) accepts that A.
> Therefore, A is true (or you should accept A).
> S2: Everybody (in a particular reference group) rejects A.
> Therefore, A is false (or you should reject A).
>
> CQ1: Does a large majority of the cited reference group accept A as true?
> CQ2: Is there other relevant evidence available that would support the assumption that A is not true?
> CQ3: What reason is there for thinking that the view of this large majority is likely to be right?
> (Walton et al., 2008, p. 123)

Here there also seems to be some degree of question begging: CQ3 asks the same question that critical questions are supposed to address (Hahn & Hornikx, 2015). This problem of circularity indicates the need for supporting accounts of when such evidence can be diagnostic, can truly support a conclusion (for a discussion of Condorcet's Jury theorem in this context, see Hahn & Hornikx, 2015). In other words, it is not clear when and why argumentation schemes are normatively binding. That is, while the schemes focus the attention on crucial aspects of common arguments, they do not necessarily compel an audience to accept the argument. Although this point is compounded by the completeness problem, it would hold, nonetheless, even if we had a definitive set of critical questions for each scheme. We would also need a definitive set of standards for judging whether we have satisfactorily answered each question. Argumentation schemes do not offer such a set and, indeed, currently do not offer much insight into what satisfactory answers would look like.

Such a normative basis might, however, be provided through the use of probability theory to capture uncertainty, degrees of belief, and varying diagnosticity of evidence in order to provide an explicit treatment of argument strength. Such an approach has come to be known as Bayesian Argumentation (see, for example, Hahn, Harris, & Oaksford, 2012; Hahn & Oaksford, 2006, 2007, 2012; Zenker, 2012). Through the application of Bayesian probability theory, Bayesian Argumentation also provides strong links to social epistemology, in which there are rich seams of work on sources and testimony that draw on the Bayesian framework to explicate intuitions on these topics within epistemology and philosophy of science (e.g. Bovens & Hartmann, 2002; Olsson, 2011; Olsson & Vallinder, 2013; Zenker, 2012). Bayesian Argumentation extends the use of Bayesian probability theory from its now familiar use in the psychology of reasoning (Oaksford & Chater, 2007).

Bayesian Argumentation allows degrees of belief, which are modelled as subjective probabilities, and Bayes' rule allows calculation of the probability of a hypothesis or claim given the evidence (argument) for that claim. Bayesian Argumentation has much of the appeal of formal logic (in providing a well-founded mathematical formalism) while also handling the kinds of uncertain arguments that we typically encounter. It also provides a natural account of argument strength[4] and of how we should adjust our beliefs in the light of new evidence (for specific examples see e.g., Hahn & Oaksford, 2007). What is more, Bayesian inference can be justified more broadly as a normative system. For instance, Bayesian inference has been shown to be an optimal form of inference[5] (Leitgeb & Pettigrew, 2010a, 2010b; Rosenkrantz, 1992) and an effective strategy to avoid Dutch Book Arguments (for discussion, see Corner & Hahn, 2012; Hahn, 2014).

One project, here, then, is to recast argumentation schemes – such as the *argumentum ad hominem*, the appeal to expert opinion or the appeal to popular opinion – within the Bayesian framework and thus provide a normative basis for those schemes (see. e.g., Hahn et al., 2012; Oaksford & Hahn, 2012; Hahn & Hornikx, 2015, for Bayesian treatments of these schemes).

Consideration of the effects of variation in the strength of argument and content and variation in the reliability of a reporting source within a Bayesian framework (see e.g., Hahn et al., 2009) leads to a very different perspective than that adopted by dual-route models of persuasion within the social psychology literature. Specifically, the multiplicative nature of Bayes' rule implies statistical interactions between source reliability and evidence, not just additive effects. Hahn, Harris, and Corner (2009) found evidence of just such an interaction. They used experiments to explore evidence strength and source reliability, and found that when source reliability is high and evidence strong, there is an extra boost to argument strength.

---

[4] The likelihood ratio (or its logarithm) is a traditional and influential Bayesian measure of the evidential support of given data for a hypothesis of interest (Brössel, 2013, offers a recent survey). Also, there is experimental evidence that this measure captures people's judgments quite well at least in certain settings (see Tentori, Crupi, Bonini, & Osherson, 2007). From an argumentation perspective, however, it will not typically matter whether the likelihood ratio or other, alternative probabilistic measures of confirmation are chosen. What matters is that they afford quantitative measures of argument strength.

[5] Leitgeb & Pettigrew showed (see also D'Agostino & Sinigaglia, 2010, and Predd, Seiringer, Lieb, Osherson, Poor, & Kulkarni, 2009; Joyce, 1998) that, unless an agent A complies with a probabilistic representation of uncertain judgment and inference, there will always be an alternative (probabilistic) belief state which dominates A's in epistemic accuracy.

Closer consideration of how to model source reliability within the Bayesian framework has also drawn attention to other aspects of the relationship between arguments and their sources. In particular, sources are not only informative of argument content, but content may shape our views of the source. In particular, it seems intuitive that we may, in everyday life, use the content of what people say not just to revise our beliefs on the topic in question, but also to revise our beliefs about their reliability. Philosophers have, in recent years, been exploring the epistemological implications of using content to revise beliefs about source (see e.g., Bovens & Hartmann, 2003; Olsson, 2013; Schubert & Olsson, 2012). These results shed light not only on fundamental philosophical questions, but also on practical problems such as argument within the climate debate (Hahn, Harris & Corner, 2015) and suggest rich avenues for psychological research.

For example, Collins, Hahn, von Gerber & Olsson (2015) recently examined how argument content is used to draw inferences about reliability. A set of experiments manipulated source reliability and evidence strength, the latter being operationalized as a simple distinction between expected and unexpected information (i.e. high/low prior probability). Consider, for example, the following set of items. In the task on argument convincingness, participants first provided a rating of the initial claim, then of the claim repeated with source information:

Initial Claim: One of the best remedies against a severe cough is lots to drink, hot or cold.
Repeated Claim: Now imagine that Michael, who is a clinical nurse specialist, told you the following: "One of the best remedies against a severe cough is valium."

And, in the source-reliability task, a separate group of participants first provided a rating of the source, then of the source after reading their claim:

Initial source information: Michael is a drug addict.
Claim: Now imagine that Michael told you the following. "One of the best remedies against a severe cough is valium."

Reliable sources significantly increased the perceived convincingness of an argument; unreliable sources decreased it. Expected information significantly increased the perceived reliability of a source; unexpected information significantly decreased it. These data also provide evidence for source anti-reliability: highly unreliable sources can cause people to revise their beliefs in the opposite direction of what the source is claiming. In short, these data provide evidence for an intimate relationship between arguments (claims) and their sources. Not only does source information affect perceived argument strength; argument (claim) strength also affects perceived source reliability.

As we have seen, Bayesian models offer a subtle account of source reliability, and source reliability is a (relatively) stable attribute of discussants in argumentation. But source reliability is just one such attribute; there are doubtless many others. This raises the question of whether all such attributes can be handled in the same way. Recall the example of the reference for a job application, and generous versus the cautious referee. Both referees seem reliable as sources as long as we know of their attributes; nevertheless, we would interpret the evidence they offer in rather different ways. One of the strengths of the Bayesian framework is that all probabilistically relevant aspects can be handled in essentially the same way and thus inferentially integrated in probabilistic models.

This is illustrated with a final example from the recent empirical literature. Harris, Corner, and Hahn (2013) explored how pragmatic inferences can be cashed out in terms of

Bayesian inference. Specifically, they considered the example of an underwhelming reference for a maths course. If, other things being equal, one received a reference stating only that "James is polite and punctual", one would seem to be entitled to infer that the reference is unfavorable. The implication is a lack of recommendation. Formally, this is a type of argument from ignorance: the inference about James' math skills follows from what is *not* being said, not from the fact that punctuality is negatively correlated with mathematical ability (see Harris et al., 2013 for details). This was borne out in an experimental condition that probed the effects of being told that James was punctual and polite in addition to being informed of his mathematical ability in the context of a fictional university application. Where no further evidence about James is provided by the referee (other than that he is punctual and polite), that is, no evidence of his mathematical ability is given, the impact of that failure should nevertheless be moderated by beliefs about the specific characteristics of the source. Whereas one would expect information on his mathematical ability from an 'expert' source (James' maths teacher) the lack of such evidence from an inexpert source (James' personal tutor, a history teacher) is less surprising. In keeping with this, Harris et al. (2013) found that the weak reference led to decreases in the belief that James should be admitted to a university mathematics course only when it came from the maths teacher.

5. Integrating the Accounts

We have discussed, so far, rather different accounts of argumentation and the considerable evidence that they can call on. These accounts differ considerably in their approach. Procedural rules focus on arguments as a dialectical process. Argumentation schemes focus on content, treating arguments as inferential objects. Bayesian Argumentation focuses on arguments both as claims and as inferential objects, factoring in assumptions about sources. Whereas procedural rules and argumentation schemes are largely qualitative accounts, the Bayesian framework is inherently quantitative, even though it may be used to draw purely qualitative evaluations. And whereas procedural rules are well-suited to extended arguments, argumentation schemes and Bayesian Argumentation tend to focus on smaller fragments (though see Kadane & Schum, 1996, for a Bayesian analysis of all the evidence in the Sacco and Vanzetti trial). Since argumentation is a complex business, it is hardly surprising that there should be different accounts focusing on different aspects. It would be desirable both to integrate these accounts, and to simplify the theoretical frameworks involved. We have already given some indication of ways in which this might be achieved, but further considerations on this issue for future research seem an apt way to conclude.

We have suggested, above, that pragmatics could contribute more to argumentation theory when the developments since Grice are included. One way to simplify argumentation theory is to reduce the amount of things it has to explain. Perhaps some of the burden can be shifted to pragmatics, at least where we are concerned with mistaken reasoning rather than deliberate deception. Take, for example, the pragma-dialectic procedural rules. The rules draw on Gricean Maxims of Conversation: the Relevance Rule corresponds to the Maxim of Relation; the Usage Rule corresponds to the Maxim of Manner; and various rules (e.g. the Standpoint and Unexpressed Premise Rules) seem to echo the Maxim of Quality. Why not allow pragmatic principles to bear the load? We have also seen that pragmatics after Grice has tended to invoke fewer principles: most dramatically, Relevance Theory replaces Grice's maxims with an expanded notion of relevance, which is supposed to apply to pragmatics and to cognition more generally

(Sperber & Wilson, 1995). It is a tempting prospect that a more minimal set of pragmatic principles could account for good and mistaken reasoning.

It is also possible to reconcile pragmatic and Bayesian inference. Indeed, there is an emerging field of probabilistic semantics and pragmatics (for an introduction, see Goodman & Lassiter, in press) and the Harris *et al.* (2013) example of James and his mathematical ability provides a simple example.

In all of this, the notion of relevance is key. Relevance is a further point of contact between pragmatics and probabilistic reasoning. In Relevance Theory, relevance is still qualitatively defined: relevance arises out of a trade-off between positive cognitive effects and processing effort. But within probabilistic theories of reasoning there have long been suggestions that probability provides a natural vehicle for relevance, specifically in the form of the axioms of conditional independence (Hahn & Hornikx, 2015; Korb & Nicholson, 2011; Pearl, 1988). To clarify, Korb and Nicholson (2011) give the example of a simple causal model in which A causes B, which in turn causes C (i.e. A>B>C). Let us follow them in assuming that A is smoking, B is cancer, and C is dyspnea (shortness of breath). To quote,

> If we don't know whether [a] woman has cancer, but we do find out she is a smoker, that would increase our belief both that she has cancer and that she suffers from shortness of breath. However, if we already knew she had cancer, then her smoking wouldn't make any difference to the probability of dyspnea. That is, dyspnea is conditionally independent of being a smoker given the patient has cancer. (Korb & Nicholson, 2011, p. 39)

The notion of conditional independence thus captures dynamic aspects of relevance: how (probabilistic) relevance changes as a result of what one come to know. Certainly, much remains to be done in testing how well probabilistic relevance captures the relevance of Relevance Theory and argumentation. In other words, does probabilistic relevance map onto the kinds of inference that are, or should be, made in argumentation? But a probabilistic notion of relevance has the potential to tie together research in Bayesian inference, pragmatics and argumentation theory.

This is not to say that pragmatics and probabilistic reasoning can, together, say everything we might want to say about argumentation. Even if relevance proves useful in explaining, more minimally, various aspects of argumentation, it seems unlikely that it will account for the following points. Firstly, in pragmatics it can be relatively unimportant whether what is said is strictly true; hence in Relevance Theory, for instance, the Maxim of Quality is considered largely redundant (Clark, 2013). For instance, we do not expect utterances such as (8) and (9) to be strictly true:

(8) A hundred people showed up to my party.
(9) Aberdeen is five hundred miles from London.
(Clark, 2013, p. 70)

What is important is not the truth of such apparent claims but what the speaker meant (implicated) by them. In argumentation, truth seems to matter rather more. Indeed, argumentation theory has been partly motivated by the desire to avoid such counter-intuitive inferences as the paradoxes of material implication: for instance, from 'If P then Q' and 'Not P', in propositional logic we can infer 'Therefore Q' (Arthur, 2011). Secondly, relevance does not capture what we might call the ethical aspects of argumentation. However well relevance treats accidentally bad reasoning, it will not extend to deliberate bad, deceptive reasoning. There may be nothing inherently wrong, from the perspective of relevance, with committing an abusive *ad hominem* argument or threatening force (*argumentum ad baculum*) or deliberately misconstruing the opponent's standpoints or implicit premises (straw man fallacy). If we want to rule out such arguments, we will have to have recourse to procedural rules of some kind. To explain why such arguments might be effective, however, we can invoke relevance and Bayesian inference.

6. Conclusion

We tentatively suggest the following division of labour for a unified approach to argumentation. It seems plausible that a unified framework will have a common core of relevance, and probabilistic reasoning may help tie together relevance in communication and argumentation. Bayesian inference provides a natural vehicle for assessing the normative strength of arguments, and for prescribing appropriate belief revision. Bayesian inference can also be used to formalize insights from other fields, for instance, from argumentation schemes (for discussion, see Hahn & Hornikx, 2015). Pragmatics provides a natural vehicle for studying how people (mis-)understand arguments. Insights from pragmatics have already seeped into argumentation theory through procedural rules, but a fuller pragmatic treatment might reduce the burden on argumentation theory, and allow a shorter list of procedural rules. Procedural rules, in turn, provide a natural vehicle for explaining what we have called the ethical aspects of argumentation: why arguments can seem fair or unfair. Together these approaches can offer a rich account of argumentation.

References

Arthur, R. T. W. (2011). *Natural Deduction: An Introduction to Logic with Real Arguments, a Little History, and Some Humour*. Peterborough, Ontario: Broadview.
Bovens, L., & Hartmann, S. (2002). Bayesian networks and the problem of unreliable instruments. *Philosophy of Science, 69*(1), 29–72.
Breheny, R. (2006). Communication and folk psychology. *Mind & Language, 21*(1), 74–107.
Brössel, P. (2013). The problem of measure sensitivity redux. *Philosophy of Science, 80*, 378-397.
Christmann, U., Mischo, C., & Flender, J. (2000). Argumentational integrity: A training program for dealing with unfair argumentative contributions. *Argumentation, 14*(4), 339–360.
Christmann, U., Mischo, C., & Groeben, N. (2000). Components of the evaluation of integrity violations in argumentative discussions relevant factors and their relationships. *Journal of Language and Social Psychology, 19*(3), 315–341.
Clark, B. (2013). *Relevance Theory*. Cambridge: Cambridge University Press.
Clark, H. (1996). *Using Language*. Cambridge: Cambridge University Press.
Cohen, J. (1988). *Statistical Power Analysis for the Behavioral Sciences*. L. Erlbaum Associates.
Cohen, J. (1992). A power primer. *Psychological Bulletin, 112*(1), 155–159.
Corner, A., & Hahn, U. (2012). Normative theories of argumentation: are some norms better than others? *Synthese, 190*(16), 3579–3610.

D'Agostino, M., & Sinigaglia, C. (2010). Epistemic accuracy and subjective probability. In M. Suárez, M. Dorato, and M. Rèdei (Eds.), *Epistemology and Methodology of Science* (pp. 95-105). Berlin: Springer.

Eemeren, F. H. V., & Houtlosser, P. (1999). Strategic manoeuvring in argumentative discourse. *Discourse Studies, 1*(4), 479–497.

Eemeren, F. H. van, Garssen, B., & Meuffels, B. (2012). The disguised abusive ad hominem empirically investigated: Strategic manoeuvring with direct personal attacks. *Thinking & Reasoning, 18*(3), 344–364.

Garssen, B. J. (1997). *Argumentatieschema's in pragma-dialectisch perspectief: Een theoretisch en empirisch onderzoek.* Amsterdam: IFOTT.

Girotto, V., Kemmelmeier, M., Sperber, D., & van der Henst, J.-B. (2001). Inept reasoners or pragmatic virtuosos? Relevance and the deontic selection task. *Cognition, 81*(2), B69–B76.

Goodman, N., & Lassiter, D. (in press). Probabilistic semantics and pragmatics: Uncertainty in language and thought. In S. Lappin & C. Fox (Eds.), *Handbook of Contemporary Semantic Theory* (2nd ed.). Wiley-Blackwell.

Grice, H. P. (1957). Meaning. *The Philosophical Review, 66,* 377-388.

Grice, H. P. (1969). Utterer's meaning and intention. *The philosophical review,* 147-177.

Grice, H. P. (1989). *Studies in the Way of Words.* Cambridge, MA: Harvard University Press.

Griggs, R. A., & Cox, J. R. (1982). The elusive thematic-materials effect in Wason's selection task. *British Journal of Psychology, 73*(3), 407–420.

Hahn, U. (2014). The Bayesian boom: good thing or bad? *Cognitive Science, 5,* 765.

Hahn, U., Harris, A. J. L., & Corner, A. (2009). Argument content and argument source: An exploration. *Informal Logic, 29*(4), 337–367.

Hahn, U., Harris, A.J.L. & Corner, A.J. (2015). Public Reception of Climate Science: Coherence, Reliability, and Independence. Topics in Cognitive Science.

Hahn, U., Harris, A. J. L., & Oaksford, M. (2013). Rational argument, rational inference. *Argument & Computation, 4*(1), 21–35.

Hahn, U., & Hornikx, J. (in press). A normative framework for argument quality: Argumentation schemes with a Bayesian foundation. A normative framework for argument quality: Argumentation schemes with a Bayesian foundation. *Synthese.*

Hahn, U., & Oaksford, M. (2006). Why a normative theory of argument strength and why might one want it to be Bayesian? *Informal Logic, 26,* 1–24.

Hahn, U., & Oaksford, M. (2007). The rationality of informal argumentation: A Bayesian approach to reasoning fallacies. *Psychological Review, 114*(3), 704–732.

Hahn, U., & Oaksford, M. (2012). Rational argument. In K. J. Holyoak & R. G. Morrison (Eds.), *The Oxford Handbook of Thinking and Reasoning* (pp. 277–298). Oxford: Oxford University Press.

Hahn, U., Oaksford, M., & Harris, A. J. L. (2012). Testimony and argument: A Bayespian Perspective. In F. Zenker (Ed.), *Bayesian Argumentation* (pp. 15–38). Dordrecht: Springer.

Harris, A. J. L., Corner, A., & Hahn, U. (2013). James is polite and punctual (and useless): A Bayesian formalisation of faint praise. *Thinking & Reasoning, 19*(3-4), 414–429.

Hastings, A. C. (1962). *A reformulation of the modes of reasoning in argumentation.* Unpublished dissertation., Evanston, IL: Northwestern University.

Huang, Y. (2007). *Pragmatics.* Oxford: Oxford University Press.

Inch, E. S., & Warnick, B. H. (2009). *Critical thinking and communication: The use of reason in argument.* (6th ed.). London: Pearson.

Jarvsad, A., & Hahn, U. (2011). Source reliability and the conjunction fallacy. *Cognitive Science, 35*(4), 682-711.

Joyce, J.M. (1998). A non-pragmatic vindication of probabilism. *Philosophy of Science, 65*, 575–603.

Kadane, J. B., & Schum, D. A. (1996). *A Probabilistic Analysis of the Sacco and Vanzetti Evidence*. John Wiley & Sons.

Kienpointner, M. (1992). *Alltagslogik: Struktur und Funktion von Argumentationsmustern*. Stuttgart-Bad Cannstatt: Friedrich Fromman.

Korb, K. B., & Nicholson, A. E. (2011). *Bayesian Artificial Intelligence* (Second Edition). Boca Raton, Florida: CRC Press.

Leitgeb, H., & Pettigrew, R. (2010b). An objective justification of Bayesianism ii: the consequences of minimizing inaccuracy. *Philosophy of Science, 77*, 236–272.

Leitgeb, H., & Pettigrew, R. (2010). An objective justification of Bayesianism I: Measuring Inaccuracy. *Philosophy of Science, 77*(2), 201–235.

Levin, I.P., Schneider, S.L., & Gaeth, G.J. (1998). All frames are not created equal: A typology and critical analysis of framing effects. *Organizational Behavior and Human Decision Processes, 76*(2), 149–188.

Levinson, S. C. (1983). *Pragmatics*. Cambridge: Cambridge University Press.

Mandel, D. R. (2014). Do framing effects reveal irrational choice? *Journal of Experimental Psychology. General, 143*(3), 1185–1198.

McCready, E., (2014). *Reliability in Pragmatics*. Oxford Studies in Semantics and Pragmatics 4. Oxford: Oxford University Press.

Mischo, C. (2003). Cognitive, emotional, and verbal responses in unfair everyday discourse. *Journal of Language and Social Psychology, 22*(1), 119–131.

Oaksford, M., & Chater, N. (1994). A rational analysis of the selection task as optimal data selection. *Psychological Review, 101*(4), 608–631.

Oaksford, M., & Chater, N. (2007). *Bayesian Rationality: the Probabilistic Approach to Human Reasoning*. Oxford: Oxford University Press.

O'Keefe, D. J. (1977). Two concepts of argument. *Journal of the American Forensic Association, 13*, 121–128.

O'Keefe, D. J. (2009). Persuasive effects of strategic manoeuvring: Some findings from meta-analyses of experimental persuasion effects research. In F. H. van Eemeren (Ed.), *Examining argumentation in context: Fifteen studies ofn strategic manoeuvring* (pp. 285–296). Amsterdam: John Benjamins.

Olsson, E. J. (2011). A simulation approach to veritistic social epistemology. *Episteme, 8*(02), 127–143.

Olsson, E. J., & Vallinder, A. (2013). Norms of assertion and communication in social networks. *Synthese, 190*(13), 2557–2571.

Pearl, J. (1988). *Probabilistic Reasoning in Intelligent Systems*. San Mateo, CA: Morgan Kaufman.

Perelman, C., & Olbrechts-Tyteca, L. (1969). *The New Rhetoric: A Treatise on Argumentation*. (J. Wilkinson & P. Weaver, Trans.). Notre Dame: University of Notre Dame Press.

Petty, R. E., & Cacioppo, J. T. (1984). Source factors and the elaboration likelihood model of Persuasion. *Advances in Consumer Research, 11*, 668–672.

Petty, R. E., & Cacioppo, J. T. (1986). The elaboration likelihood model of persuasion. In L. Berkowitz (Ed.), *Advances in Experimental Social Psychology* (Vol. 19, pp. 123–205). Academic Press.

Predd, J.B., Seiringer, R., Lieb, E.J., Osherson, D., Poor, H.V., & Kulkarni, S.R. (2009). Probabilistic coherence and proper scoring rules. *IEEE Transactions on Information Theory, 55*, 4786-4792.

Quine, W. v. O. (1960). *Word and Object*. Cambridge, MA: MIT Press.

Rosenkrantz, R. . (1992). The justification of induction. *Philosophy of Science, 15*, 527–539.
Schellens, P. J. (1985). *Redelijke argumenten: Een onderzoek naar normen voor kritische lezers.* Dordrecht: Foris Press.
Schiffer, S. R. (1972). *Meaning.* Oxford: Oxford University Press.
Schreier, M., Groeben, N., & Christmann, U. (1995). "That's not fair!" argumentational integrity as an ethics of argumentative communication. *Argumentation, 9*(2), 267–289.
Schubert, S., & Olsson, E. J. (2012). On the coherence of higher-order beliefs. *The Southern Journal of Philosophy, 50*(1), 112–135.
Sher, S., & McKenzie, C. R. M. (2006). Information leakage from logically equivalent frames. *Cognition, 101*(3), 467–494.
Strawson, P. F. (1964). Intention and convention in speech acts. *The Philosophical Review, 73*(4), 439–460.
Tentori, K., & Crupi, V. (2012). On the conjunction fallacy and the meaning of and, yet again: a reply to Hertwig, Benz, and Krauss (2008). *Cognition, 122*(2), 123-34.
Tentori K., Crupi V., Bonini N., and Osherson D. (2007). Comparison of confirmation measures. *Cognition, 103*, 107-119.
Tversky, A., & Kahneman, D. (1981). The framing of decisions and the psychology of choice. *Science (New York, N.Y.), 211*(4481), 453–458.
Van der Henst, J.-B., & Sperber, D. (2004). Testing the cognitive and communicative principles of relevance. In I. Noveck & D. Sperber (Eds.), *Experimental Pragmatics* (pp. 229–279). London: Palgrave.
Van Eemeren, F., Garssen, B., & Meuffels, B. (2009). *Fallacies and Judgments of Reasonableness: Empirical Research Concerning the Pragma-Dialectical Discussion Rules* (Vol. 16). Dordrecht: Springer.
Van Eemeren, F. H., & Grootendorst, R. (1984). *Speech acts in argumentative discussions: A theoretical model for the analysis of discussions directed toward solving conflicts of opinion.* Dordrecht: Floris Press.
Van Eemeren, F. H., & Grootendorst, R. (1992). *Argumentation, communication, and fallacies: A pragma-dialectical perspective.* Hillsdale, N.J.: Lawrence Erlbaum Associates.
Van Eemeren, F. H., & Grootendorst, R. (1995). The Pragma-Dialectical Approach to Fallacies. In H. V. Hansen & R. C. Pinto (Eds.), *Fallacies: Classical and Contemporary Readings.* Philadelphia: Pennsylvania State University Press.
Van Eemeren, F. H., & Grootendorst, R. (2004). *A systematic theory of argumentation: The pragma-dialectical approach.* Cambridge: Cambridge University Press.
Walton, D. N. (1998). *Ad Hominem Arguments.* University of Alabama Press.
Walton, D., Reed, C., & Macagno, F. (2008). *Argumentation Schemes.* Cambridge: Cambridge University Press.
Wason, P. C. (1966). Reasoning. In B. M. Foss (Ed.), (Vol. 1). Harmondsworth: Penguin.
Zenker, F. (2012). *Bayesian Argumentation: The practical side of probability.* Dordrecht: Springer Science & Business Media.

Part III
Biases and Fallacies

Chapter 9

Don't Worry, Be Gappy! On the Unproblematic Gappiness of Alleged Fallacies

Fabio Paglieri

Istituto di Scienze e Tecnologie della Cognizione, CNR, Roma, Italy, fabio.paglieri@istc.cnr.it

Abstract. The history of fallacy theory is long, distinguished and, admittedly, checkered. I offer a bird eye view on it, with the aim of contrasting the standard conception of fallacies as attractive and universal errors that are hard to eradicate (section 1) with the contemporary preoccupation with "non-fallacious fallacies", that is, arguments that fit the bill of one of the traditional fallacies but are actually respectable enough to be used in appropriate contexts (section 2). Godden and Zenker have recently argued that reinterpreting alleged fallacies as non-fallacious arguments requires supplementing the textual material with something else, e.g. probability distributions, pragmatic considerations, dialogical context. Thus fallacies remain *gappy* on all accounts, and this is the hallmark of their failure. However, I argue that such gappiness is typically unproblematic, and thus no more flawed than enthymematic argumentation in general (section 3). This, in turn, calls into question the usefulness of the very notion of fallacy.

1. Introduction: The Standard Conception of Fallacies and Its Problems

This is how popular wisdom, aka Wikipedia, defines the notion of fallacy:

> A fallacy is the use of *poor, or invalid, reasoning* for the construction of an argument. A fallacious argument may be *deceptive by appearing to be better than it really is*. Some fallacies are committed intentionally to manipulate or persuade by deception, while others are committed unintentionally due to carelessness or ignorance. (my emphasis)[1]

Such definition is actually close to the scholarly conception of fallacies, at least as represented by the Stanford Encyclopedia of Philosophy:

> Two competing conceptions of fallacies are that they are false but popular beliefs and that they are *deceptively bad arguments*. These we may distinguish as the belief and argument conceptions of fallacies. […] There are yet other conceptions of what fallacies are, but the present inquiry focuses on the argument conception of fallacies. Being able to detect and avoid fallacies has been viewed as a supplement to criteria of good reasoning. The knowledge of them is needed to arm us against *the most enticing missteps we might take with arguments*—so thought not only

---

[1] Source: https://en.wikipedia.org/wiki/Fallacy (last consulted on September 23, 2015).

Aristotle but also the early nineteenth century logicians Richard Whately and John Stuart Mill. (my emphasis)[2]

This view of fallacies as enticing missteps is considered to be the standard conception in the Western tradition (Hitchcock, 2006), and indeed instances of it abounds in textbooks and scholarly articles. The following are just two notable examples:

> Fallacies are the *attractive nuisances* of argumentation, the *ideal types of improper inference*. They require labels because they are thought to be *common enough* or *important enough* to make the costs of labels worthwhile. (Scriven, 1987, p. 333, my emphasis)

> By definition, a fallacy is a *mistake* in reasoning, a mistake which *occurs with some frequency* in real arguments and which is characteristically *deceptive*. (Govier, 1987, p. 177)

In his recent monograph on fallacies, John Woods (2013) captures the defining features of this view using the acronym EAUI: on the standard conception, fallacies are thought to be *errors* (E) that are *attractive* (A), *universal* (U), and *incorrigible* (I) – in the sense that people are expected to persevere in fallacious reasoning, even after being shown the error of their ways. After Charles Hamblin's seminal work on fallacies (1970) revamped the interest in this subject, the EAUI conception have attracted substantial criticism (Massey, 1981; Finocchiaro, 1981; Hintikka, 1987; Hitchcock, 1995; van Eemeren & Grootendorst, 1995; Woods, 2013; Boudry, Paglieri, & Pigliucci, 2015), yet it remains prominent in the literature, as shown by the previous excerpts, and it still inspires the treatment of fallacies in many logic textbooks, where the most common, quick-and-dirty definition of fallacies is "an argument that seems to be valid but it is not".[3]

Instead of attempting to cover all the reasons why the EAUI conception is unsatisfactory, here I will only focus on its most glaring difficulties. A *prima facie* worry is that this view conflates a normative claim (that fallacies are errors) with various empirical assertions (that they are universal, attractive, and incorrigible). While this is not automatically a *faux pas*, the fact that such empirical claims appear hard, if not outright impossible to verify should raise a big red flag. How are we supposed to check whether an alleged fallacy is universal, attractive, and incorrigible? Even if we interpret the universality of these claims as just pertaining a majority of the human population, as charity demands, too many details remain unclear to allow for satisfactory verification. How many exceptions, i.e. people who do not commit a certain error, would disqualify that as a fallacy?[4] The EAUI conception remains utterly silent on these points. Unfortunately, as Paglieri argued with respect to Aristotle's definition of dialectical reasoning (2014), it is never good for a theory to incur a huge empirical debt without having the means to pay it off. This is even more true when defining something which is supposed to be of crucial importance for everyday practice, as it is certainly the case with fallacies.

---

[2] Source: http://plato.stanford.edu/entries/fallacies/ (version of May 29, 2015). Importantly, this entry is authored by Hans Hansen, a well-known fallacy scholar.

[3] As Maurice Finocchiaro once pointedly remarked (1981), many textbook cases of fallacies have the habit of existing *only* in textbooks. The most likely reason for this oddity is that when an argument is patently bad (e.g., "If an animal is a gorilla, then it is a primate. X is a primate, therefore X is a gorilla"), people do not find it attractive in the least, hence they do not use it, nor would fall for it if others were to try it.

[4] Similar worries apply even if one accepts the refined version of universality articulated by Woods (2013, p. 141), according to which fallacies on the standard conception are meant to identify the *typical ways* in which reasoners err, regardless of how frequently they do in fact err. Even *that* claim, however, is far from being proven, as Woods convincingly shows.

Similar considerations recently led Boudry, Paglieri and Pigliucci (2015) to formulate a dilemma in fallacy theory, which they propose to call "the Fallacy Fork". According to them, if fallacies are construed as demonstrably poor forms of reasoning,[5] then they have very limited applicability in real life, i.e. few actual instances: thus we get the normative part right (fallacies are indeed error), but we lose completely the empirical features attributed to them by the standard conception – those errors are neither especially attractive nor incorrigible for people, therefore certainly not universal in their occurrence. If, instead, our definitions of fallacies become sophisticated enough to capture real-life complexities, then they can no longer be held up as an effective tool for discriminating good and bad forms of reasoning, since we start talking of fallacious and non-fallacious instances of alleged fallacies – which is exactly what is commonplace in the literature nowadays (for a detailed exposition, see Walton, 1995). In section 2 I will discuss some of these attempts to articulate a theory of "non-fallacious fallacies", whereas in section 3 I will present evidence supporting the claim that people are not especially prone to obviously mistaken forms of reasoning – at least, not to the extent required by the standard conception of fallacies. For now, let us take stock of the Fallacy Fork main implication for the present argument: you can have the E without the AUI or the AUI without the E, but not both at once – hence the standard conception fails to provide a suitable account of fallacies.

Admittedly, the Fallacy Fork is a problem mostly for *structural accounts* of fallacies: that is, theories that describe fallacies as instantiation of flawed reasoning schemes. What makes the Fallacy Fork so pernicious for structural accounts is, quite simply, the abundance of available counterexamples to their proposed definition of a fallacy: arguments that instantiate the allegedly fallacious scheme, yet do not appear flawed, either intuitively or upon reflection. Take the *ad hominem* fallacy as a case in point, i.e. the "error" of accepting or dismissing a claim based on personal features of its source, and consider the following argument: "John's research is being funded by Marlboro, therefore his conclusions on the lack of negative side-effects of tobacco consumption are unreliable". The fact that the argument is deductively invalid is as obvious as it is irrelevant: the question is whether it is a bad argument also on weaker, more sensible standards. Here it is obvious that John's blatant conflict of interest gives excellent reasons to consider his claims as unreliable, insofar as they happen to accord with his personal convenience in the matter. However, the argument fits the structural definition of the *ad hominem* fallacy, even if it is perfectly reasonable to consider it good. Cleary, structural accounts of fallacies strive to remedy this problem by introducing further constraints on their definition of the fallacy in question: so, for instance, Brinton (1995) proposes that *ad hominem* arguments are admissible as long as they are directed at someone's advocacy of P, while they are fallacious when targeting P itself. Thus our example is admissible because the conclusion refers to John's advocacy (his claims are described as unreliable), whereas the argument would be fallacious if the conclusion pertained the facts of the matter (for instance, his claims were considered as false). This makes perfect sense in principle, but now the account is exposed to the second horn of the Fallacy Fork: many arguments have a partly testimonial character, and stand or fall with the credentials of the person advocating them. In our example, in practice considering John's claims to be unreliable is tantamount to consider them as probably false, since they end up being dismissed either way. Thus the notion of fallacy, once properly

---

[5] Crucially, "demonstrably poor" is much broader than "deductively invalid", and deliberately so. The fact that many perfectly sensible arguments are not deductively valid has now been remarked endlessly in the literature, so we should stop flogging the deductive horse – it is long dead. The problem of fallacy theory is a different one: namely, the fact that, when an argument is bad even according to non-deductive standards, laypeople are still smart enough to avoid buying it, thus making the alleged deceptiveness of fallacies something of a myth.

refined to avoid mislabeling good reasoning, ends up losing much of its practical significance.

However, fallacies have also been defined in non-structural terms, but rather *dialectically*, as infractions of some norms of proper dialectical engagement (Hamblin, 1970; van Eemeren & Grootendorst, 1995; Walton & Krabbe, 1995).[6] These approaches manage to provide a solid theoretical grasp on fallacies without incurring in the Fallacy Fork, yet they suffer instead from a labelling problem. In fact, even if the same names are used, it is hard to see how the original fallacy may constitute an instantiation of its dialectical definition: as Woods noted with respect to pragma-dialectics (PD), "the striking thing of the PD approach to fallacies is that whereas it show no reluctance to forgo the traditional concept of fallacy, it hangs on tightly to the traditional *list*. [...] This is puzzling, [because] with the possible exception of begging the question [...], which PD theorists may see as a violation of the unexpressed premise rule, none of the [traditional fallacies] is in any obvious way the violation of a PD rule. What PD rule does many questions violate? What does hasty generalization violate?" (2013, pp. 507-508). Thus, while in general I find dialectical theories of fallacies more viable than structural ones, I doubt these two approaches really focus on the same phenomena, and in what follows I will limit my analysis to a structural understanding of fallacies – which of course can be informed by dialectical considerations, yet sees fallacies as reasoning errors, rather than missteps in communication.

2. The Quest for Non-Fallacious Fallacies

Taken literally, the expression "non-fallacious fallacies" is clearly oxymoronic, in that it refers to an error that is not an error. This is not the sense in which I will use it here, nor in which it is used elsewhere: instead, non-fallacious fallacies stands as shorthand for "non-fallacious arguments that structurally match the definition of a certain fallacy". The adoption of this weird label is intended to emphasize the problems inherent in clinging to the term "fallacy", while dealing with patterns of reasoning. Even if argumentation scholars nowadays are fully aware that virtually every type of informal fallacy has its legitimate counterpart, the label "fallacy" is still very much in use, as discussed. While this habit is not logically contradictory, it is indeed conceptually confusing and rhetorically pointless: so, by the end of this chapter, I hope I will have started giving reasons to discontinue the practice, and send the term "fallacy" into retirement.

In the meantime, let us review some other examples of non-fallacious fallacies, the weird beast that has preoccupied fallacy theorists over the last 40 years or so. My first specimen is based on the *ad ignorantiam* fallacy, that is, drawing a conclusion based on absence of evidence to the contrary. As the old epistemological maxim goes, "absence of evidence is not evidence of absence". But now consider the two following arguments:

- My wife will not give me a present for my birthday, since I have received no indication to the contrary from her.
- My flight tomorrow from Schiphol will not be delayed, since I have received no indication to the contrary from the airline.

---

[6] This distinction between structural and dialectical accounts of fallacies is reminiscent of Walton's articulation of fallacies in two broad categories: *paralogisms* (fallacious premise-inference structures that fail to meet some necessary requirement of an argumentation scheme) and *sophisms* (fallacious tactics used to unfairly try to get the best of a speech partner in the course of a dialogical exchange; see Walton, 1995).

Although structurally both arguments instantiate the *ad ignorantiam*, they elicit very different intuitive appraisals: the first one is immediately perceived as extremely weak, if not outright erroneous, whereas the second will be good enough for anyone except the most skeptical person – in fact, we routinely reason that way the day before taking a trip. Nor is the difference particularly hard to account for: the presumptive weight of the absence of evidence for X depends on whether or not there is a justified expectation to obtain that evidence, whenever X is indeed the case. If that expectation is low or non-existent, then absence of evidence carries very little presumptive weight; but if the expectation is robust enough, then we are presumptively justified in concluding that X is unlikely, given that lack of evidence. This is precisely what supports our intuitions regarding these two arguments: whereas I have no reason to expect my wife to broadcast her plans for my birthday present (indeed, I have reasons to assume the opposite), I know as a fact that most airlines will make every possible effort to keep their customers well informed of any schedule change. This is why absence of evidence does suggest evidence of absence in the second case, but not in the first. Crucially, however, you do not need to appreciate the reason behind the difference to be able to detect it: laypeople would be perfectly capable of correctly rating the strength of the two arguments, even if they were unable to articulate a justification for it (empirical evidence to that effect will be reviewed in section 3, in case someone doubts it).

Nor is this phenomenon specific of the *ad ignorantiam*. If we now move to consider the *ad verecundiam*, i.e. accepting a certain conclusion on the say-so of some source, the same pattern arises. Consider the following two examples:

- Hollywood celebrities claim that homeopathy is effective, so it is likely to be effective.
- Medical practitioners claim that homeopathy is effective, so it is likely to be effective.

Again, the structure of both arguments is the same and fits the description of the *ad verecundiam* fallacy: however, the latter is patently stronger than the former, as any competent speaker immediately intuits. There is of course no big mystery behind such an intuition: here the different appraisal of the arguments mostly depends on considerations of expertise. While there is absolutely no reason to defer to Hollywood celebrities as experts on homeopathy, the competence of medical practitioners on this matter is manifest. Moreover, medical practitioners are typically assumed to be skeptical of homeopathic remedies, thereby making their endorsement of such remedies all the more convincing. Conversely, Hollywood celebrities are often thought as having a penchant for outlandish medical treatments, thus making their claims on homeopathy less likely to carry presumptive weights to people that are not already persuaded of its merits.

At this point, a defender of fallacy theory may consider retreating to a more entrenched position: that is, s/he may be willing to concede that non-fallacious fallacies abounds when it comes to informal fallacies, yet insists that no such thing exists, when it comes to formal fallacies, the real "bad boys" of logical textbooks. The prime examples of formal fallacies are, of course, *denying the antecedent* (DA) and *affirming the consequent* (AC). Given a conditional statement such as "If a bird is a flamingo, then it is pink", there are two correct reasoning patterns one can apply to it: *modus ponens* (from the conditional and the fact that "That bird is a flamingo", it is derived that "That bird is pink") and *modus tollens* (from the conditional and the fact that "That bird is not pink", it is derived that "That bird is not a flamingo"). DA and AC are their fallacious cousins: DA uses the negation of the antecedent to erroneously infer the consequent (from the conditional and the fact that "That bird is not a flamingo", it is derived that "That bird is not pink"), whereas AC mistakes the truth of the consequent as evidence for the truth of the antecedent (from the conditional and the fact that "That bird is pink", it is derived that "That bird is a flamingo").

The fact that DA and AC are deductively invalid is as obvious as it is uninformative. The real question, instead, is whether these reasoning patterns are unequivocal mistakes under *any* legitimate standard of inference, besides deductive validity. The short, non-technical answer to that question is that it depends on how many non-flamingo pink birds are out there – more generally, it depends on the likelihood of alternative reasons for the consequent, other than the antecedent. As Luciano Floridi pointed out, DA and AC can be coherently interpreted as "Bayesian 'quick and dirty' informational shortcuts [which] assume [...] that there are no false positives (double implication), or that, if there are, they are so improbable as to be disregardable (degraded Bayes' theorem)" (2009, p. 322; see also Stone, 2012). Take the following reasoning pattern: "When I have the flu, I have a fever. Today I have a fever, therefore I have the flu". This is certainly not deductively valid, but is it also blatantly mistaken on any other legitimate standard of inference? Ultimately, whether the argument is a fallacious instance of AC or an acceptable inference to the best explanation depends on the likelihood of alternative reasons for the observed symptoms, i.e. my fever: if this likelihood is low enough, in relation to the conditional probability of having a fever given the flu, the inference goes through without particular problems. Granted, it is still defeasible, and necessarily so, since no amount of probabilistic considerations can produce deductive validity out of thin air. But defeasibility, in and by itself, is not enough to brand it as rubbish on every normative standards. The same applies to our flamingo example: in a world where non-flamingo pink birds are rare enough, then for a bird being pink is a reliable (albeit defeasible) indicator of it being a flamingo (AC), whereas not being a flamingo drastically reduces the likelihood for that bird to be pink (DA).

More to the point, not only non-fallacious formal fallacies are possible, they might even be as widespread as non-fallacious informal fallacies. Compare the following:

- If an animal is a chimpanzee, then it is a primate. King Kong is a primate, therefore King Kong is a chimpanzee.
- When I have the flu, I have a fever. Today I have a fever, therefore I have the flu.

As usual, both arguments match the structural blueprint of a fallacy (AC, in this case): yet the first one is patent nonsense, whereas the second one sounds good enough under the appropriate circumstances, e.g. for a first, rough-and-ready diagnose of one's own medical condition. What matters, however, is the fact that their difference in strength is once again intuitively accessible to any competent speaker. Who in his/her right mind would fall for (or even use in the first place) the King Kong argument? In contrast, making an educated guess on a certain illness based on one of the symptoms that illness causes is a perfectly legitimate, albeit fallible, epistemic practice, assuming the probabilities provide enough support to justify such inference.

Now, of course the definition of the various fallacies, formal and informal, can be made precise enough to exclude non-fallacious instances: indeed, much of the recent efforts in fallacy theory have been directed to detail the conditions under which a certain argumentative pattern, previously labelled as fallacious in general, can be used legitimately, and when instead the accusation of fallaciousness still stands.[7] Yet comparing generic vs.

---

[7] Douglas Walton, for instance, authored a series of monographs on how to distinguish truly fallacious from spuriously fallacious uses of argument patterns traditionally considered to be fallacies in general: among others, *ad verecundiam* (1997), *ad hominem* (1998), *ad ignorantiam* (1999), *ad baculum* (2000). For similar efforts by other

specific definitions of fallacy brings us again into the grip of the Fallacy Fork. If fallacies are defined precisely, then they instantiate reasoning mistakes that are, however, not particularly common in everyday practice, nor especially attractive. If, on the other hand, fallacies are defined generically, then most of their instances turn out to be perfectly legitimate. Since the intuitive notion of a fallacy is meant to capture a mistake that occurs with alarming frequency, either way the expression fails to deliver on that intuition: the precise approach captures mistakes that are relatively infrequent, whereas the generic view encompasses argumentative patterns that are widespread but typically fine. Thus on both accounts the term fallacy is being misused, with no good reason to do so, other than a nod to historical tradition. Precise "fallacy" theories deal with real reasoning mistakes that are neither particularly frequent nor very appealing to arguers, hence should not be named "fallacies". And generic "fallacy" theory does not even identify mistakes, so the term "fallacies" is out of place there as well.

3. The Gappiness of Fallacies

There is, however, a way of trying to defend the legitimacy of the notion of fallacy in something akin to its standard conception. A brilliant effort in that direction has recently been made by David Godden and Frank Zenker (2015): these authors first provide a comprehensive and well researched review of recent attempts to show that formal fallacies, i.e. DA and AC, are ordinarily cogent – that is, "well-reasoned [on] a generic, theoretically-neutral, objective, normative standard of argumentative or inferential goodness" (p. 89). Then they argue *against* this position, by showing how, on all these attempts, "the acceptability of the conclusion of DA and AC arguments depends on factors not asserted by the stated conditional" (p. 89). Godden and Zenker think that such gappiness is what makes AC and DA problematic, since on their view AC and DA "fail to be cogent whenever they conspicuously fail to cite as reasons the conditions on which the acceptability of their conclusions properly depends" (p. 89).

Before criticizing their line of argument, I want to stress something on which we are in full agreement: all recent attempts to show AC and DA to be cogent (non-fallacious, in the sense I have been using it here) hinge on supplementing the interpretation of the stated conditional with something not included in its explicit formulation. In other words, Godden and Zenker are correct in emphasizing the gappy nature of formal fallacies – indeed, I think such gappiness is characteristic of *any* so called fallacy, formal or informal, even though their paper only focuses on AC and DA. What exactly is supposed to fill the gap depends on the theoretical approach being considered: on a Bayesian reading of formal fallacies (Korb, 2004; Hahn & Oaksford, 2007; Floridi, 2009; Stone, 2012; Hahn, Oaksford, & Harris, 2013; Zenker, 2013), the gap would be filled by probabilistic considerations; on a reconstruction of the inference in terms of Stalnaker's conditionals (Stalnaker, 1968), cogency would depend on similarity relations among possible worlds; on interpretations that invoke conditional perfection[8] as a way of justifying the rationality of AC and DA (Geis & Zwicky, 1971; Moldovan, 2009), what matters would be pragmatic indicators; and

---

scholars, see Mackenzie (1980), Massey (1981), Brinton (1995), Goodwin (1998), Woods (1998), Levi (1999), Hitchcock (2007), Mizrahi (2010) – among many others.
[8] Conditional perfection refers to the interpretative tendency to "perfect" a stated conditional to an unstated biconditional, due to pragmatic considerations. The standard examples involve promises or threats, e.g. a statement like "If you mow the lawn, I'll give you five dollars" is typically (and correctly) interpreted as shorthand for "If *and only if* you mow the lawn, I'll give you five dollars". This interpretation, clearly, makes alleged instances of AC and DA utterly unproblematic, since they are just applications of MP and MT to the other direction of the biconditional.

on dialectical theories of fallacies (Hamblin, 1970; van Eemeren & Grootendorst, 1995; Walton & Krabbe, 1995; Godden & Walton, 2004), filling the gap would require considering the contextual features of the dialogue. In spite of the key differences among these alternative approaches, it is always the case that the cogency of the alleged fallacy depends on something that is not explicitly mentioned in its utterance.

The crux of the matter, however, is whether or not this gappiness constitutes a significant flaw. Godden and Zenker think so, whereas I beg to differ. My argument against their conclusion can be summarized as follows:

1. Gappiness per se cannot be considered a capital sin, lest we throw away the baby of enthymematic reasoning with the bathwater of alleged fallacies.
2. What would make the gappiness of fallacies "bad" are the same criteria that would disqualify as illegitimate any other enthymeme: namely, that what is left implicit is controversial or unclear.
3. Thus whether the gappiness of fallacies is problematic or not ultimately rests on an empirical question, to wit, whether the "missing ingredient" in non-fallacious fallacies is typically controversial or unclear.
4. Extant evidence supports the claim that most alleged fallacies are gappy in unproblematic ways, vindicating the intuition that there is nothing wrong with non-fallacious fallacies, and thus making the continued use of the label "fallacy" problematic, for the reasons exposed above.

Point 1 should be rather self-evident: gappiness is a property of another well-known family of arguments, i.e. enthymemes, and there is an abundance of different theories on how their missing premise is to be reconstructed (Ennis, 1982; van Eemeren & Grootendorst, 1982; 1983; Gerritsen, 2001; Walton, 2001; 2008; Paglieri & Woods, 2011); even approaches that deny the standard characterization of enthymemes as argument with a missing premise (Hitchcock, 1985; 1998) make their validity, or lack thereof, dependent on something which is not explicitly stated in the argument itself, e.g. a Toulminian warrant (Toulmin, 1958), which would still count as a form of gappiness in the sense implied by Godden and Zenker. However, nobody thinks that all enthymemes are automatically flawed due to their gappy nature: indeed, the whole point of their analysis is to find ways of (a) interpreting their structure to (b) assess how good or bad they are *qua* arguments. If gappiness was enough of a fault to consider all gappy arguments fallacious, then all enthymemes would have to be considered fallacious; inasmuch as we deny the latter, we are also denying that gappiness per se is problematic. In fact, *prima facie* the gappiness of fallacies seems analogous to the gappiness of enthymemes, thus whether it is problematic or not will depend on the same features that matter for enthymemes: namely, whether what is left unstated is controversial or unclear (point 2). This is of course an empirical question, so I take point 3 to be also self-evident. As for how to respond to this empirical question, the experimental evidence I will shortly review suggests that the gappiness of fallacies is typically unproblematic (point 4).

Before discussing the experimental data, however, it is worth looking at how Godden and Zenker replies to this line of criticism in their paper (2015, pp. 121-126).[9] Basically,

---

[9] Full disclosure: I was one of the anonymous reviewers of their paper, whose comments prompted Godden and Zenker to include a rejoinder in the published article. Thus the present chapter should be seen as the public continuation of a debate started in private. As already mentioned, the fact that I still disagree with their position on DA and AC arguments does not subtract in any way from the value of their paper, especially as an excellent overview of various attempts to articulate the cogency of these alleged fallacies.

they concede point 1 and 3, while disputing 2 and 4. As I see it, their defense is two-pronged (the labels are mine):

- *The what argument* (contra 2): the flaw of fallacies is not gappiness per se, but rather *what* they are gappy about – in other words, their gappiness is different, and *worse*, than the gappiness of ordinary enthymemes.
- *The bias argument* (contra 4): evidence on how good people are at discriminating good vs. bad "fallacies" is undermined by results on *other reasoning biases* that plague our cognitive processes.

Let us first consider the what argument. This is how Godden and Zenker put it:

> We do not claim that the incogency of DA and AC arguments results from *that* something has been left unstated; rather the problem we identify concerns *what* has been left unstated. […] DA and AC arguments fail to assert the conditions on which the truth of their conclusions, and indeed the positive relevance of their stated premises, depend. These are properly construed as reasons, not background assumptions. Premises that are unproblematically supplemented to putatively enthymematic arguments are, minimally, ones that are *reasonably acceptable to* both arguer and audience, and also *accepted by* both arguer and audience. Only then can one take it for granted that these premises are not at issue—that they *go without saying*. Yet, with DA and AC arguments this does not seem to hold. The unstated conditions […], on which the cogency of DA and AC arguments depend, do not ordinarily go without saying. […] Indeed, because they give the very conditions on which the cogency of the given argument depends, they are precisely the kinds of claims that are, or should be, at issue, and should therefore be expected to be found among the stated premises of the argument. (2015, pp. 122-123, emphasis in the original)

The claim here is that what is missing in DA and AC arguments is qualitatively different from what is missing in ordinary enthymemes, and the crux of the difference is that the former fail to assert the conditions on which the truth of the conclusion and the relevance of the stated premises depend, whereas the latter are not supposed to do the same. Additionally, Godden and Zenker also claim that what is missing in DA and AC arguments does not typically go without saying, because it is either not reasonably acceptable by all parties, not actually accepted by them, or both. Unfortunately, I fail to see how textual evidence would support either of these claims. Compare the two following arguments:

- When I have the flu, I have a fever. Today I have a fever, therefore I have the flu.
- Socrates is a man, therefore he is mortal.

The former is a potentially acceptable instance of an AC argument, depending on the usual relevant assumptions on probabilities, whereas the latter is "the father of all enthymemes", i.e. the most frequently cited example of a valid enthymematic argument. What is missing in the first case is, roughly, (A) "The likelihood of alternative causes to my fever is low enough", whereas what is missing in the second case is something like (B), "All men are mortals". Now, as anyone can easily verify, *both* (A) and (B) support their respective argument in that they "assert the conditions on which the truth of their conclusions, and indeed the positive relevance of their stated premises, depend", as Godden and Zenker put it. In particular, if (B) was not the case, i.e. some men happened to be immortal, then the truth of the conclusion "Socrates is mortal" would no longer be established, and the relevance of the stated premise would also be in question – a lot or a little, depending on the percentage of immortal men over the whole population. On the face of it, the truth of the conclusion and the relevance of the stated premise depend on the missing element as much in enthymemes as they do in fallacies, contra Godden and Zenker. Besides, filling the gap appears to be equally unproblematic in both cases (against, contra Godden and Zenker): the fact that someone claiming to have an illness based on one of its symptoms is assuming other explanations to be unlikely, or at least significantly less likely than the proposed diagnosis, "goes without saying" as much as the reconstruction of the major premise in the Socrates enthymeme. Of course, the *truth* of the implicit assumption is much more questionable in the first case (i.e., alternative causes for a fever are unlikely) than in the second one (i.e., all men are mortal), based on background knowledge. But this has nothing to do with correctly *interpreting* the arguments, which is all that matters for the present discussion: given the enthymeme "Dumbo is an elephant, therefore he flies", we all interpret it (correctly) as relying on something like "All elephants fly", even if we know this assumption to be false. But as far as correct interpretation is concerned, what is missing is equally important, and equally unproblematic for arguers to individuate, for both enthymemes and alleged fallacies.

Now it is time to turn to the experimental record. I will first summarize some key findings on fallacy interpretation, and then discuss the results that, according to the bias argument proposed by Godden and Zenker, would call those findings into question. Ulrike Hahn, Mike Oaksford and others have demonstrated in recent years that people are intuitively good at discriminating between strong and weak instances of the same, potentially fallacious argument schemes, and they do so in ways consistent with a bona fide inference standard (Bayesian update) and showing sensitivity to rationally relevant factors, such as prior belief, argument strength, and nature of the evidence (Korb, 2004; Oaksford & Hahn, 2004; Hahn & Oaksford, 2006, 2007; Hahn, Harris, & Corner, 2009; Corner & Hahn, 2012; Harris, Hsu, & Madsen, 2012; Hahn, Oaksford, & Harris, 2012, 2013; Collins, Hahn, von Gerber, & Olsson, 2015). Since reviewing this still growing body of literature is not my aim here, I will just outline the gist of their methodology and findings with respect to a specific fallacy, i.e. *ad ignorantiam*, for the sake of illustration. Nonetheless, it is worth emphasizing that the same approach was applied with similar results to a garden variety of other alleged fallacies (e.g., *ad hominem*, *ad populum*, *petitio principii*, and *slippery slope*) and that preliminary theoretical considerations suggest that this kind of Bayesian explanation may be extended to virtually all the fallacies in the classic catalogue (Hahn & Oaksford, 2006).

Let us focus now on the *ad ignorantiam*. To study people's intuitive appraisal of this type of argument, participants are first presented with some ground conditional that they are asked to accept as valid, e.g. "If Drug A produces toxic effects in legitimate tests, then Drug A is toxic". Then the first manipulation is introduced, in terms of *argument type* (positive vs. negative), as follows:

- Drug A is toxic because a toxic effect was observed (positive argument).
- Drug A is not toxic because no toxic effects were observed (negative argument).

On a deductivist reading, the first argument is a valid instance of *modus ponens*, whereas the second argument is a blatant case of DA. Even from a non-deductivist perspective, the latter could still be regarded as instantiating a fallacious *ad ignorantiam*. Ordinary speakers, however, have different intuitions: whereas they typically consider the first argument to be the stronger one, they do not immediately regard the second one as flawed, but rather as a weaker yet potentially valuable inference, depending on other factors. One of these factors is the *amount of evidence* being considered, which can be easily manipulated as follows:

- Drug A is not toxic because no toxic effects were observed in one test (weak evidence).
- Drug A is not toxic because no toxic effects were observed in fifty tests (strong evidence).

The latter argument is clearly stronger than the former, and participants have no difficulty is intuiting the difference. The same is true for another factor that affects the credibility of the conclusion, namely, *prior belief* in it. This is manipulated by using short vignettes, like the following:

> Barbara: Are you taking digesterole for it?
> Adam: Yes, why?
> Barbara: Well, because I *strongly* believe that it does have side effects.
> Adam: It does have side effects.
> Barbara: How do you know?
> Adam: Because I know of one experiment in which they found side effects.

Now these vignettes can be used to manipulate all three factors at once (argument type, amount of evidence, and prior belief) and see how they affect people's intuitions. The previous vignette represents the combination of a positive argument with weak evidence and strong prior belief. The following instead conveys the opposite combination, i.e. a negative argument with strong evidence and weak prior belief (all other combinations are easy to obtain, of course):

> Barbara: Are you taking digesterole for it?
> Adam: Yes, why?
> Barbara: Well, because I *weakly* believe that it does not have side effects.
> Adam: It does not have side effects.
> Barbara: How do you know?
> Adam: Because I know of fifty experiment in which no side effects were found.

When participants are presented with similar vignettes and asked to what extent Barbara will accept Adam's argument, their responses (i) show sensibility to all three factors (ii) in ways that match almost perfectly the Bayesian prediction (Oaksford & Hahn, 2004; Hahn & Oaksford, 2007). Thus the relevant finding is not just that people have consistent intuitions on how to discriminate between actual fallacies and their non-fallacious counterparts, but also that such intuitions are grounded on a legitimate normative standard, namely, Bayesian update. Similar results, which have been established for a variety of alleged fallacies, lead me to claim that "filling the gap" in these instances is no more problematic that in any other case of enthymematic reasoning, contra Godden and Zenker's what argument.

However, this is where their bias argument is supposed to kick in, casting doubt on the relevance of the empirical findings I just briefly reviewed. This is how Godden and Zenker present their case:

> One cannot simply assume by fiat that arguers sensitively track the unstated cogency conditions on which DA and AC inferences depend. The empirical evidence purporting to show that arguers do so is, in our view, equivocal and inconclusive. Moreover, other empirical evidence suggests that arguers instead (or perhaps also) attend to various irrelevant aspects of reasoning problems, while neglecting relevant aspects. For example, studies of two-premise conditional reasoning show that reasoners respond to logically irrelevant aspects such as premise order (Girotto, Mazzocco and Tasso, 1997) and to whether antecedent or consequent conditions are negated (Evans and Lynch, 1973), resulting in the matching bias (Evans, 1998). Yet other studies show that reasoners fail to attend to logically relevant information. For example, one of the first cognitive biases to be named, the confirmation bias (Wason, 1966), regularly registers as being alive and well (Nickerson, 1998; Mendela et al, 2011). Similarly, base-rate neglect (Eddy, 1982) (which can be mitigated, though not eliminated, by presenting relevant information in a frequentist format (Gigerenzer and Hoffrage, 1995)) remains prevalent and recalcitrant (Barbey and Sloman, 2007). (2015, pp. 124-125)[10]

It is unfortunate that Godden and Zenker do not elaborate on their claim that the evidence in question is "equivocal and inconclusive": without some indications on why they consider this to be the case, it is hard to rebut the accusation in a meaningful way – especially since, as far as I can see, the evidence is not equivocal at all, but rather quite convergent on the point in question. As for it being inconclusive, well, it is certainly open to falsification, as any form of (good) empirical evidence is; yet it is not *especially* inconclusive, so I fail to see the relevance of this remark. Perhaps Godden and Zenker want to suggest that what makes this evidence problematic is the existence of those other reasoning biases they mention in this passage, i.e. matching bias, confirmation bias, and base rate neglect. If that is the case, then the point of course is whether or not such biases in fact undermine the quality of our folk intuitions on fallacies. Again, no argument is provided to support such conclusion, so perhaps Godden and Zenker think the facts speak for themselves. But when I started browsing some of the papers they mention, I found nothing speaking against the validity and robustness of the findings gathered by Hahn, Oaksford, and others.

By way of example, consider the paper on premise order by Girotto, Mazzocco and Tasso (1997): this presents a series of studies aimed at corroborating Johnson Laird's mental model theory of conditionals, and 6 out of the 8 studies deal only with valid forms of inference (MP and MT), whereas the last two studies conclude that "conditional fallacies [i.e., AC and DA] are *not* significantly affected by the premise order" (p. 1, my emphasis). Far from proving that ordering effects undermine our ability to evaluate potentially fallacious reasoning, this study shows that people are relatively immune from such effects when it comes to deal with possible fallacies – thus actually scoring a point *against* Godden and Zenker's claim. Or consider the matching bias, i.e. "a tendency to see cases as relevant in logical reasoning tasks when the lexical content of a case matches that of a propositional rule, normally a conditional, which applies to that case" (Evans, 1998, p. 45). Godden and Zenker cite a paper by Evans and Lynch (1973) as presenting evidence that would lead us to doubt people's folk intuitions on fallacy discrimination. However, the paper in question deals with a rather different topic: Evans and Lynch introduce a modification of the classic Wason's selection task, thus showing that the typical mistakes observed in this task do not depend on a tendency to look for confirmation in evaluating conditional statements (Wason's original hypothesis), but rather on the matching bias itself. While this is certainly interesting, it also suggests that these mistakes are superficial rather than logical: as Evans

---

[10] To avoid clogging the bibliography of this paper with unnecessary references, I have included in it only the articles cited by Godden and Zenker that are subsequently discussed in the present article. For all other references mentioned in this passage, I refer interested readers to the original essay by Godden and Zenker (2015).

and Lynch note, "the overall pattern, with matching bias cancelled out, gave no evidence for a verification bias, indicating instead that the logically correct values were most frequently chosen" (1973, p. 391). How is this relevant for establishing the value of our folk intuitions on fallacy discrimination? No obvious answer is forthcoming, since some alleged fallacy could be explained by invoking a matching bias (e.g., AC), whereas others could not (e.g., DA). More to the point, however, is the fact that this bias does not seem to have any import for our ability to discriminate between, say, a truly fallacious instance of AC and a *prima face* reasonable inference to the best explanation, since in both cases the minor premise (the affirmation of the consequent) would "match", in Evans' sense, the original conditional (major premise).

In light of these considerations, I regard the bias argument as being both underdeveloped, thus hard to assess properly and reply adequately, and based on empirical evidence that has no clear bearing on the matter under discussion, that is, whether or not our folk intuitions on fallacy discrimination are reliable enough. As such, it does not suffice to persuade me that the gappiness of fallacies is especially problematic, over and above the gappiness of enthymemes in general. Moreover, the claim that well-known biases of individual reasoning are worrisome in the dialogic context where alleged fallacies typically occur and are evaluated is at odds with the *argumentative theory of reasoning* recently proposed by Hugo Mercier and Dan Sperber, which suggests the opposite – biases are expected to be particularly widespread in solo reasoning, whereas they largely disappear when the appropriate social context is present. According to Mercier and Sperber (2011), reasoning did not evolve as a correction mechanism for mistaken intuitions or as a support system for individual decisions, contrary to the view championed by Kahneman and others (e.g., Kahneman, 2011), but rather as a tool to effectively engage in a special type of social activity – namely, argumentation. On this view, the function of reasoning is to find and evaluate reasons in dialogic contexts, not through solitary introspection. The evolutionary rationale of the theory has been extensively discussed elsewhere (Sperber et al., 2010; Mercier & Sperber, 2009, 2011; Mercier, 2013) and it is of secondary importance here. What matters is that the argumentative theory of reasoning offers a strikingly different interpretation of the findings mentioned by Godden and Zenker. Mercier and collaborators argue that, if the function of reasoning was indeed to correct mistaken intuitions, then the high incidence of errors observed in laboratory tasks would be hard to account for, especially considering tasks in which reaching the correct solution only requires the application of basic knowledge in logic (Wason, 1966), statistics (Tversky & Kahneman, 1982), probability theory (Tversky & Kahneman, 1983), or mathematics (Frederick, 2005). Moreover, according to Mercier and colleagues the reason why people on their own fail to achieve better performance in such tasks is not because they try to reason correctly and fail, but rather because reasoning itself leads them further into error, by strengthening, rather than questioning, their initial, mistaken intuitions, in line with the literature on motivated reasoning (Kunda, 1990).

In contrast, the very same features that make solitary reasoning so desperately flawed, e.g. myside bias (Nickerson, 1998; Stanovich & West, 2007; Mercier, 2010) and reasoning laziness (Kuhn, 1991; Perkins, Farady, & Bushey, 1991; Kahneman, 2011), make perfect sense in a social context, where unilaterally making your case as strong as possible is (i) personally advantageous and (ii) collectively counterbalanced by the critical scrutiny others will direct against your claims, so that (iii) directing your cognitive efforts towards the evaluation of the arguments of others, rather than your own, is just a sensible allocation of your limited resources, provided this laziness is selective, i.e. it affects self-evaluations and not the cross-examination of the arguments made by others (as shown in Trouche, Johansson, Hall, & Mercier, in press). In support of this interpretation, Mercier and collaborators cite the significant benefits of group discussion on performance in a variety of tasks that individuals fail to master alone, such as the Wason selection task (Moshman & Geil, 1998), in which group performance was on average four times higher than individual results (data analyzed in Mercier, Trouche, Yama, Heintz, & Girotto, 2015). Recently, similar results were reported also with respect to the Cognitive Reflection Task,[11] where simply exposing participants to each other solutions, with no opportunity for actual dialogue among them, was sufficient to rapidly converge on the correct answer (Rahwan, Krasnoshtan, Shariff, & Bonnefon, 2014). More generally, the benefits of group discussion have been observed for "problems or decisions for which there exists a demonstrably correct answer within a verbal or mathematical conceptual system" (Laughlin & Ellis, 1986, p. 177). The relevant fact seems to be the ability of the correct answer to assert itself: as long as one participant gets it right, other members converge on that solution, even if it is originally held only by a minority, or even by a single individual, and independently from how confident the original "truth-bearer" is (Trouche, Sander, & Mercier, 2014). The fact that people, including experts, systematically underestimate the benefits of group discussion on reasoning performance (Mercier et al, 2015) is unfortunate and contributes to explain the widespread picture of laypeople as "bad arguers", but it does not alter what the evidence suggests: provided with the appropriate social context in which argumentation is meant to take place, we do not argue (nor reason) nearly as poorly as the standard interpretation of so called "fallacies" would lead us to believe.

Finally, it is worth noting that argumentative fallacies and psychological biases are different classes of phenomena: while it may be productive to study their interplay (and I believe it is), the two labels should not be used as synonymous.[12] To be fair, I am quite sure Godden and Zenker would wholeheartedly subscribe to this sentiment, and I do not intend to accuse them of confusing these two concepts in their paper. Yet the fact that data on a garden variety of psychological biases are mentioned as being automatically relevant for the ongoing debate on argumentative fallacies, with no further reason offered, does raise a red flag, so I think a minor cautionary note is in order. To put it simply for the sake of brevity, a bias in psychology is defined as a cognitive heuristic, typically automatic and inflexible, that may or may not lead to error. In fact, the most heated debate on this subject hinges on whether biases should be regarded mostly as carriers of irrationality (in a nutshell, the position held by Kahneman and Tversky; e.g., Tversky & Kahneman, 1982, 1983; Kahneman, 2011), or rather as "simple heuristics that make us smart", i.e. useful strategies in our adaptive toolbox (the so called ecological rationality approach, championed by Gigerenzer and others; e.g., Gigerenzer & Selten, 2001; Gigerenzer, Hertwig, & Pachur, 2011). In contrast, a fallacy in argumentation theory is a mistake by definition, except when the term is used as a shorthand for "something that looks like a

---

[11] In the Cognitive Reflection Task (CRT from now on; see Frederick, 2005), participants are presented with three simple problems that elicit a strong, intuitive, yet mistaken response, which prevents most people from reasoning out the correct solution – even if such solution is well within their cognitive powers. Performance on the CRT is notoriously low even in highly educated and intellectually brilliant samples (see data in Frederick, 2005), and it is taken to measure the ability to inhibit an intuitive mistake, rather than general intelligence. By way of example, one of the three problems in the CRT is the following: "A bat and a ball cost $1.10 in total. The bat costs $1.00 more than the ball. How much does the ball cost?". Here the intuitive mistake is 10 c, whereas the correct response is 5 c.

[12] Matters are complicated by the fact that psychologists have appropriated the concept of fallacy and used it in their own work as, indeed, roughly equivalent to bias (e.g., Pohl, 2004). As a result, some well-known psychological biases have actually been popularized as fallacies: notable examples include the so called conjunction fallacy (the tendency to consider the probability of each conjunct as lower than the probability of their conjunction; Tversky & Kahneman, 1982) and the gambler's fallacy (the tendency to think that past random events influence the probability of future random events; Tversky & Kahneman, 1974). To avoid unnecessary confusion, I propose we reserve the label "fallacies" to the kind of argumentative phenomena analyzed in this paper, while reserving the name "biases" for their more recent psychological cousins. The two phenomena are related but far from identical.

fallacy but may be not", as in "non-fallacious fallacies" – for my reservations on this strange usage, see the end of section 2. At the same time, fallacies are not necessarily considered heuristic, and certainly there is no proof of their automaticity – in fact, upon reflection the only clear parallel in the two notions is between the inflexibility of biases and the alleged incorrigibility of fallacies. Thus treating fallacies as argumentative biases would require a drastic (and perhaps interesting) revision of current fallacy theory. While the effort may well bear useful fruits, it would also end up making the notion of fallacy redundant, precisely because it would be supplanted by the more flexible and value-neutral concept of bias. Thus, once again, our analysis suggests that fallacy, as a theoretical construct, may be due for retirement, at long last.

4. Conclusions

The fact that many arguments structurally identifiable as fallacies on the standard approach are not fallacious at all is nowadays firmly established in argumentation theory, and a variety of approaches have endeavored to spell out what demarcate a truly fallacious argument from its structurally similar but ultimately legitimate cousins (for a brief discussion, see section 2). This puts the notion of fallacy in a bit of a quandary, since its denotation is unambiguously pejorative, yet it ends up being used, more often than not, to talk about acceptable arguments – hence the bizarre expression "non-fallacious fallacies". Should the problem be merely terminological, a regimentation would still be in order, but perhaps not terribly urgent. What makes the matter important, however, are the practical implications of the continued use of the concept of fallacy, especially in education.

Indeed, fallacies still loom large in textbooks, both of logic (as a case in point, take the classic Copi, Cohen, & McMahon, 2010, now in its 14th edition) and of critical thinking (a recent authoritative exemplar is Tindale, 2007). While the latter typically offer a much more nuanced analysis of fallacies, along the lines discussed in this paper, the former popularize among students an outdated and misleading treatment of the subject matter. By way of example, consider the following argument, that Copi and colleagues (2010) propose as an example of fallacious DA:

> If Carl embezzled the college funds, then Carl is guilty of a felony. Carl did not embezzle the college funds. Therefore Carl is not guilty of a felony.

Only someone with logical blinders on (an expression I gladly borrow from Cohen, 2013, where he was commenting on another bad example from the same textbook) could fail to see that this argument, although obviously deductively invalid, also embodies a perfectly reasonable pattern of inference, that may very well be appropriate in a variety of contexts. In fact, in the most obvious context of application, i.e. legal proceedings, this is the *only* acceptable conclusion that could be drawn from the evidence, given that in most legal systems the presumption of innocence places the burden of proof on the prosecutor. This is especially pertinent since "being guilty of a felony" is, on the face of it, a legal concept, not just a mere statement of facts. In order to be guilty of a felony, Carl has not only to have committed the relevant deed, but also to have been proven responsible for it beyond reasonable doubt – which is precisely what is not achieved according to the argument's premises, therefore fully justifying the conclusion that Carl is not guilty of a felony.

This is not just nitpicking and taking cheap shots at a well-respected textbook. General concerns about the usefulness of fallacy theory to promote critical thinking education have been voiced by scholars who are fully committed to improve, rather than stifle, such educational aim (e.g., Hitchcock 1995; Feldman 2009).[13] The main reason behind these worries was empirical, since fallacy theory was noted to have little measurable impact on critical thinking proficiency. Now the argumentative theory of reasoning proposed by Mercier and collaborators justifies further concerns: insofar as the main negative side-effect of our cognitive inclinations is to make us exclude too much social information (since the communicative arms race makes people overly skeptical of each other's claims and agendas), priming us to be even more skeptical of arguments may actually be deleterious, rather than just useless – a provocative empirical hypothesis, which should be tested in future studies.

Thus teaching students that something like the embezzling argument is a fallacy does a severe disservice to critical thinking education. Not only because it trivializes and misrepresents the sophisticated understanding of fallacies characteristic of contemporary argumentation theory, but also because it perpetuates the view of fallacies as "silly mistakes that dumb people do". This is actually intrinsic to the very concept of fallacy, and gives yet another reason to discontinue its usage: a fallacy is something bad we do to ourselves, a pit we fall in of our own free will, a mistake of which we are responsible. As discussed, the alleged frequency with which this happens has been vastly exaggerated: as Woods put it, "if we attended to the empirical record of the reasoning behavior of individual agents, it would become quickly apparent that, with the [standard list of fallacies] as our guide, beings like us hardly ever commit *them*" (2013, p. 139, emphasis in the original).

So why keep insisting that we are so desperately prone to error? Most likely, to contain the unavoidable hubris that would overcome us if we were to realize that, all things considered, we are not that bad at reasoning, especially when we are free to reason in the appropriate social context. I happen to agree that such hubris would be dangerous and ultimately misguided: being on average decent reasoners does not make us infallible (far from it, in fact), so we must ensure people have a good grasp of their own limitations, as well as their own potential. But the fallibility caveat should be presented in a very different light, based on the results of our discussion. Fallacies, if we insist to keep teaching about them, should no longer be framed as attractive mistakes that dumb reasoners are prone to make, but rather as *points of vulnerability that sophisticated arguers may try to exploit* to their advantage, and possibly to our chagrin. The source of such vulnerability is, interesting enough, our very ability to discriminate between real fallacies and alleged ones. As Godden and Zenker correctly pointed out, that discrimination hinges on assumptions that are external to the textual material, and thus more malleable to strategic manipulation.

This perhaps offers a way of vindicating their reluctance to abandon the concept of fallacy, but it also requires a radical reinterpretation of their claim: the point is not that the gappiness of fallacies is especially problematic, but rather that, as any other kind of gappiness, it opens the way to the smart exploitation of our interpretative practices. The same holds, once again, for enthymemes in general, which indeed are known to be amenable to strategic tampering – advertising campaigns and political speeches offer a wealth of excellent examples. This is no reason to stop using enthymemes, nor to look at fallacy discrimination with special concern. But it does justify rethinking how we teach fallacies to students: instead of hammering them about mistakes that most likely they were not committing in the first place, we should train them to exert due diligence in assessing the arguments presented by their fellows, with an emphasis on how to spot and neutralize strategic derangements of enthymematic reasoning.

This will require a shift in how we conceive fallacies: from the proneness-to-error approach to the vulnerability-to-manipulation view. It may seem a small shift, but it is not,

---

[13] Yet another critique of the traditional pedagogy of fallacies was proposed by Hundleby (2010), but from a very different angle – namely, a concern for the exceedingly adversarial and authoritarian nature of that pedagogy.

especially in terms of the pedagogical methods used to teach about fallacies – or whatever we end up calling them. Luckily, some useful suggestions in this direction are to be found in the literature: e.g., Johnson and Blair well-established conception of critical thinking as "logical self-defense" (1977), and the more recent notion of epistemic vigilance (Sperber et al., 2010). This, I suggest, is the direction that fallacy theory ought to explore – in fact, possibly the *only* direction that might justify its continued success, hopefully in a new, improved incarnation.

References

Boudry, M., Pigliucci, M., & Paglieri, F. (2015). The fake, the flimsy, and the fallacious: Demarcating arguments in real life. *Argumentation, 29(4)*, 431-456.
Brinton, A. (1995). The *ad hominem*. In H.V. Hansen & R.C. Pinto (Eds.), *Fallacies: Classical and contemporary readings* (pp. 213-222). University Park: Penn State University Press.
Cohen, D. (2013). Virtue, in context. *Informal Logic, 33(4)*, 471-485.
Copi, I., Cohen, C., & McMahon, K. (2010). *Introduction to logic*. 14th edition, New York/London: Routledge.
Ennis, R. (1982). Identifying implicit assumptions. *Synthese, 51*, 61-86.
Evans, J. St. B.T. (1998). Matching bias in conditional reasoning: Do we understand it after 25 years? *Thinking and Reasoning, 4*, 45-82.
Feldman, R. (2009). Thinking, reasoning, and education. In H. Siegel (Ed.), *The Oxford handbook of philosophy of education* (pp. 67-82). Oxford: Oxford University Press.
Finocchiaro, M. (1981). Fallacies and the evaluation of reasoning. *American Philosophical Quarterly, 18(1)*, 13-22.
Floridi, L. (2009). Logical fallacies as informational shortcuts. *Synthese, 167*, 317-325.
Frederick, S. (2005). Cognitive reflection and decision making. *Journal of Economic Perspectives, 19(4)*, 25-42.
Geis, M., & Zwicky, A. (1971). On invited inferences. *Linguistic Inquiry, 2*, 561-566.
Gerritsen, S. (2001). Unexpressed premises. In F. H. van Eemeren (Ed.), *Crucial concepts in argumentation theory* (pp. 51-79). Amsterdam: Sic Sat.
Gigerenzer, G., & Selten, R. (Eds.) (2001). *Bounded rationality: The adaptive toolbox*. Cambridge: The MIT Press.
Gigerenzer, G., Hertwig, R., & Pachur, T. (Eds.) (2011). *Heuristics: The foundations of adaptive behavior*. New York: Oxford University Press.
Godden, D., & Walton, D. (2004). Denying the antecedent as a legitimate argumentative strategy: A dialectical model. *Informal Logic, 24*, 219-243.
Godden, D., & Zenker, F. (2015). Denying antecedents and affirming consequents: The state of the art. *Informal Logic, 35(1)*, 88-134.
Goodwin, J. (1998). Forms of authority and the real *ad verecundiam*. *Argumentation, 12(2)*, 267-280.
Govier, T. (1987). *Problems in argument analysis and evaluation*. Dordrecht: Foris.
Hahn, U., & Oaksford, M. (2006). A Bayesian approach to informal argument fallacies. *Synthese, 152*, 207-236.
Hahn, U., & Oaksford, M. (2007). The rationality of informal argumentation: A Bayesian approach to reasoning fallacies. *Psychological Review, 114*, 704-732.
Hahn, U., Harris, A. J. L., & Corner, A. (2009). Argument content and argument source: An exploration. *Informal Logic, 29(4)*, 337-367.

Hahn, U., Oaksford, M., & Harris, A. J. L. (2012). Testimony and argument: A Bayesian Perspective. In F. Zenker (Ed.), *Bayesian argumentation* (pp. 15-38). Dordrecht: Springer.

Hahn, U., Oaksford, M., & Harris, A. J. L. (2013). Rational inference, rational argument. *Argument & Computation, 4*, 21-35.

Hamblin, C. (1970). *Fallacies*. London: Methuen

Harris, A.J.L., Hsu, A.S., & Madsen, J.K. (2012). Because Hitler did it! Quantitative tests of Bayesian argumentation using ad hominem. *Thinking & Reasoning, 18(3)*, 311-343.

Hintikka, J. (1987). The fallacy of fallacies. *Argumentation, 1(3)*, 211-238.

Hitchcock, D. (1985). Enthymematic arguments. *Informal Logic, 7(2-3)*, 83-97.

Hitchcock, D. (1995). Do the fallacies have a place in the teaching of reasoning skills or critical thinking? In H.V. Hansen & R.C. Pinto (Eds.), *Fallacies: Classical and Contemporary Readings* (pp. 319-327). University Park: Penn State University Press.

Hitchcock, D. (1998). Does the traditional treatment of enthymemes rest on a mistake?. *Argumentation, 12*, 15-37.

Hitchcock, D. (2006). Informal logic and the concept of argument. In D. Jacquette (Ed.), *Philosophy of logic. Handbook of the philosophy of science, vol. 5* (pp. 101-129). Amsterdam: Elsevier.

Hitchcock, D. (2007). Why there is no *argumentum ad hominem* fallacy. In F. H. van Eemeren & B.Garssen (Eds.), *Proceedings of the Sixth Conference of the International Society for the Study of Argumentation* (Volume 1, pp. 615-620). Amsterdam: Sic Sat.

Hundleby, C. (2010). The authority of the fallacies approach to argument evaluation. *Informal Logic, 30(3)*, 279-308.

Johnson, R., & Blair, A. (1977). *Logical self-defense*. Toronto: McGraw-Hill Ryerson.

Kahneman, D. (2011). *Thinking, fast and slow*. New York: Farrar, Straus and Giroux.

Korb, K. (2004). Bayesian informal logic and fallacy. *Informal Logic, 24*, 41-70.

Kuhn, D. (1991). *The skills of arguments*. Cambridge: Cambridge University Press.

Kunda, Z. (1990). The case for motivated reasoning. *Psychological Bulletin, 108(3)*, 480-498.

Laughlin, P., & Ellis, A. (1986). Demonstrability and social combination processes on mathematical intellective tasks. *Journal of Experimental Social Psychology, 22*, 177-189.

Levi, D. S. (1999). The fallacy of treating the *ad baculum* as a fallacy. *Informal Logic, 19(2-3)*, 145-159.

Mackenzie, P. T. (1980). *Ad hominem* and *ad verecundiam*. *Informal Logic, 3(3)*, 9-11.

Massey, G. (1981). The fallacy behind fallacies. *Midwest Studies in Philosophy, 6(1)*, 489-500.

Mercier, H. (2010). The social origins of folk epistemology. *Review of Philosophy and Psychology, 1(4)*, 499-514.

Mercier, H. (2013). Our pigheaded core: How we became smarter to be influenced by other people. In K. Sterelny, R. Joyce, B. Calcott & B. Fraser (Eds.), *Cooperation and its evolution* (pp. 373-398). Cambridge: MIT Press.

Mercier, H., & Sperber, D. (2009). Intuitive and reflective inferences. In J. St. B. T. Evans & K. Frankish (Eds.), *In two minds: Dual processes and beyond* (pp. 149-170). New York: Oxford University Press.

Mercier, H., & Sperber, D. (2011). Why do humans reason? Arguments for an argumentative theory. *Behavioral and Brain Sciences, 34(2)*, 57-74.

Mercier, H., Trouche, E., Yama, H., Heintz, C., & Girotto, V. (2015). Experts and laymen grossly underestimate the benefits of argumentation for reasoning. *Thinking & Reasoning, 21(3)*, 341-355.

Mizrahi, M. (2010). Take my advice—I am not following it: *Ad hominem* arguments as legitimate rebuttals to appeals to authority. *Informal Logic, 30(4)*, 435-456.

Moldovan, A. (2009). Pragmatic considerations in the interpretation of denying the antecedent. *Informal Logic, 29*, 309-326.

Moshman, D., & Geil, M. (1998). Collaborative reasoning: Evidence for collective rationality. *Thinking & Reasoning, 4(3)*, 231-248.

Nickerson, R. (1998). Confirmation bias: A ubiquitous phenomena in many guises. *Review of General Psychology, 2*, 175-220.

Oaksford, M., & Hahn, U. (2004). A Bayesian approach to the argument from ignorance. *Canadian Journal of Experimental Psychology, 58*, 75-85.

Paglieri F., & Woods J. (2011). Enthymematic parsimony. *Synthese, 178*, 461-501.

Paglieri, F. (2014). Accepted by whom? On the empirical roots of Aristotle's dialectic. *Revue Internationale de Philosophie, 270*, 393-402.

Perkins, D., Farady, M., & Bushey, B. (1991). Everyday reasoning and the roots of intelligence. In J. Voss, D. Perkins & J. Segal (Eds.), *Informal reasoning and education* (pp. 83-105). Hillsdale, NJ: Lawrence Erlbaum Associates.

Pohl, R. (Ed.) (2004). *Cognitive illusions: A handbook on fallacies and biases in thinking, judgement, and memory*. Hove/New York: Psychology Press.

Rahwan, I., Krasnoshtan, D., Shariff, A., & Bonnefon, J.-F. (2014). Analytical reasoning task reveals limits of social learning in networks. *Journal of the Royal Society Interface, 11(93)*, 20131211.

Scriven, M. (1987). Fallacies of statistical substitution. *Argumentation, 1*, 333-349.

Sperber, D., Clément, F., Heintz, C., Mascaro, O., Mercier, H., Origgi, G., & Wilson, D. (2010). Epistemic vigilance. *Mind and Language, 25(4)*, 359-393.

Stalnaker, R. (1968). A theory of conditionals. *Studies in logical theory, American Philosophical Quarterly monograph, no. 2* (pp. 98-112). Oxford: Blackwell.

Stanovich, K., & West, R. (2007). Natural myside bias is independent of cognitive ability. *Thinking & Reasoning, 13(3)*, 225-247.

Stone, M. (2012). Denying the antecedent: Its effective use in argumentation. *Informal Logic, 32*, 327-356.

Tindale, C. (2007). *Fallacies and argument appraisal*. Cambridge: Cambridge University Press.

Toulmin, S. (1958). *The uses of argument*. Cambridge: Cambridge University Press.

Trouche, E., Johansson, P., Hall, L., & Mercier, H. (in press). The selective laziness of reasoning. *Cognitive Science*, doi:10.1111/cogs.12303

Trouche, E., Sander, E., & Mercier, H. (2014). Arguments, more than confidence, explain the good performance of reasoning groups. *Journal of Experimental Psychology: General, 143(5)*, 1958-1971.

Tversky, A., & Kahneman, D. (1974). Judgment under uncertainty: Heuristics and biases. *Science, 185(4157)*, 1124-1131.

Tversky, A., & Kahneman, D. (1982). Evidential impact of base rates. In D. Kahneman, P. Slovic & A. Tversky (Eds.), *Judgment under uncertainty: Heuristics and biases* (pp. 153-160). Cambridge: Cambridge University Press.

Tversky, A., & Kahneman, D. (1983). Extensional versus intuitive reasoning: The conjunction fallacy in probability judgment. *Psychological Review, 90(4)*, 293-315.

van Eemeren, F., & Grootendorst, R. (1982). Unexpressed premises: Part I. *Journal of the American Forensic Association, 19(2)*, 97-106.

van Eemeren, F., & Grootendorst, R. (1983). Unexpressed premises: Part II. *Journal of the American Forensic Association, 19(4)*, 215-225.

van Eemeren, F., & Grootendorst, R. (1995). The pragma-dialectical approach to fallacies. In H.V. Hansen & R.C. Pinto (Eds.), *Fallacies: Classical and contemporary readings* (pp. 130-144). University Park: Penn State University Press.
Walton, D. (1995). *A pragmatic theory of fallacy*. Tuscaloosa: University of Alabama Press.
Walton, D. (1997). *Appeal to expert opinion: Arguments from authority*. University Park: Penn State University Press.
Walton, D. (1998). *Ad hominem arguments*. Tuscaloosa: University of Alabama Press.
Walton, D. (1999). The appeal to ignorance, or *argumentum ad ignorantiam*. *Argumentation, 13(4)*, 367-377.
Walton, D. (2000). *Scare tactics: Arguments that appeal to fear and threats*. Dordrecht: Kluwer.
Walton, D. (2001). Enthymemes, common knowledge, and plausible inference. *Philosophy and Rhetoric, 34(2)*, 93-112.
Walton, D. (2008). The three bases for the enthymeme: A dialogical theory. *Journal of Applied Logic, 6(3)*, 361-379.
Walton, D., & Krabbe, E. (1995). *Commitment in dialogue: Basic concepts of interpersonal reasoning*. Albany: SUNY Press.
Wason, P. C. (1966). Reasoning. In B. Foss (Ed.), *New horizons in psychology: I* (pp. 106–137). Harmandsworth: Penguin.
Woods, J. (1998). *Argumentum ad baculum. Argumentation, 12(4)*, 493-504.
Woods, J. (2013). *Errors of reasoning. Naturalizing the logic of inference*. London: College Publications.
Zenker, F. (Ed.) (2013). *Bayesian argumentation: The practical side of probability*. Berlin: Springer.

Chapter 10

Reliable Debiasing Techniques in Legal Contexts? Weak Signals From a Dark Corner of the Social Science Universe

Frank Zenker[1] & Christian Dahlman[2]

Lund University, Sweden; [1] Department of Philosophy & Cognitive Science, frank.zenker@fil.lu.se;
[2] Law Faculty, christian.dahlman@jur.lu.se

Abstract. Debiasing techniques seek to make a positive impact on personal or procedural features of reasoning and decision-making. This paper discusses theoretical and empirical considerations regarding the effectiveness of such techniques in legal contexts, addressing the case for prescriptive ameliorative interventions from a historical perspective, and commenting on the rise and popularization of research on human biases and heuristics. After reviewing a set of analytical distinctions that help understand why the results of empirical research on the effectiveness of debiasing techniques in legal contexts have remained mixed, we provide examples of such techniques, state reasons why research has delivered less than some could have hoped for, and briefly comment on methodological improvements to the *status quo*.

1. Introduction

How to improve decision-making is a pertinent question whenever judgments cannot be avoided. The decisions that judges and juries must reach virtually every day provide a case in point, especially when these bear heavily on the fates of individual and collective agents. Since biased reasoning and decision-making is (rightly) thought to occur also in legal contexts (see, e.g., Langevoort, 1998 for a review: English *et al.*, 2006; Guthrie *et al.*, 2001; Irwin & Real, 2010; cf. Mitchell, 2002), no further argument is required that it *ought* to be reduced. Rather, empirical knowledge is wanted how to reliably achieve such a reduction.

Prior to presenting the rather narrow range of debiasing techniques that have been empirically studied with respect to their potential of positively impacting personal or procedural features particularly of legal reasoning and decision-making, we turn to theoretical and empirical considerations regarding their effectiveness. We review the case for prescriptive ameliorative interventions from a historical perspective (Sect. 2), explain the rise and popularization of research on heuristics and biases (Sect. 3), provide useful analytical considerations regarding debiasing techniques (Sect. 4), and present the rather mixed results of empirical inquiries into the effectiveness of debiasing techniques in legal contexts (Sect. 5). Identifying reasons why such research has rather not delivered what one could hope for, we briefly comment on methodological improvements to the *status quo*

(Sect. 6), and close with general conclusions (Sect. 7). Although the debiasing techniques discussed here pertain to the legal context, the focal research and its implications also pertain to non-legal contexts.

2. The Case for Ameliorative Prescriptive Intervention

According to an ancient *topos,* human beings are imperfect—*errare humanum est* ("to err is human"). Owed to their first nature or the untrained state, agents regularly accept and act upon *seeming* truths. Leaving the umbrella at home because the weather seemed fine in the morning makes for an everyday example; of course, the issue extends well beyond such trivialities. Being subject to passions and appetites, moreover, humans have long been known to endorse "vain opinions, flattering hopes, false valuations, imaginations as one would, and the like," or so Francis Bacon's essay *Of Truth* (1625) echoes this classical idea in the Renaissance, even speaking of "a natural though corrupt love of the lie itself" (ibid.).

Bacon denounced most of the Aristotelian scholarship that was then "peddled" in the predecessors to the modern university. He sought no less than new foundations for reasoning and knowledge in order to make human belief-forming processes more sensitive to the empirical, explicitly intending these as a counter-weight to the broadly negative human dispositions that his *idolatry* had identified (see Lord *et al.*, 1984). Comprising four idols—of the cave, the tribe, the theater, and the marketplace—Bacon had thus provided an early modern taxonomy of what contemporary authors refer to as *biases*. As we will see, his taxonomy wasn't much worse than what current scholarship offers. Far from being a contemporary idea, then, it had been, and continues to be, recognized with respect to both reasoning and behavior that coming closer to perfection requires prescriptive ameliorative intervention onto man's original state, that is, efforts at enculturating the human savage.

While "Aristotelian logic had still seen itself primarily as a tool for training [only] 'natural' abilities at reasoning, later logics [modern ones, that is, being developed from the 17th century onwards] proposed *vastly* improving meagre and wavering human tendencies and abilities" (Hintikka & Spade, 2012: 12). Indeed, by partially succeeding in controlling such wavering tendencies, the development of modern logics "pushed" human reasoning abilities past their then natural state. It would nonetheless be mistaken to view the abstract formal accounts that these logics constitute as having arisen for their own sake. This holds particularly for probability-theory as it developed in the 18th century, and derivatively also for decision-theory and risk-analysis in the 19th and 20th century, respectively. Rather, such accounts were and remain driven by the desire to improve the practical effects of human reasoning and decision-making that are collectively referred to as outcomes.[1]

Besides offering a descriptive apparatus that facilitates an improved understanding of various empirical aspects of the world, these formal accounts invariably rest upon normative standards which, however, are neither necessarily unequivocal nor readily applicable without further assumptions. These standards generally inform how humans *should* act in given contexts so as to avoid broadly suboptimal forms of reasoning and decision-making. Deriving from current best expert-knowledge in order to guide real-world

---

[1] It can hardly be viewed as a coincidence, for instance, that the formal understanding of probability-as-chance thrived among those who played monetary reward-games such as roulette (Hacking, 2001). Similarly, decision-theory developed in parallel to the rise of, and to this day remains intimately connected to, stock and bond markets. Particularly risk-analysis may be viewed as an ameliorative intervention in response to various new and improved technological possibilities that 20th and 21st century science made available—notably nuclear energy generation and genetic modification—and expectably continues to provide in the long run (Weingart, 1999).

decision-making of practical relevance, any such formal account thus contains principally revisable norms such as the maximization of choice utility through the exclusion of dominated strategies in decision theory, or particular constraints such as the complement-rule $P(p)=1-P(non\ p)$ in Pascalian probability.

Underlying the development of normative standards, of course, is the insight that suboptimal reasoning and decision-making processes may in particular cases lead to suboptimal outcomes. Such accounts therefore provide reasons *why* a particular type of reasoning and decision-making is suboptimal, irrespective of whether a given token in fact produces a less good outcome than that which would have been obtained had the relevant normative standard been correctly applied. One should here bear in mind that the decision-theoretic role of *heuristics* is to abbreviate a given normative standard. But since suboptimal reasoning processes do not inevitably lead to suboptimal outcomes, heuristics cannot plausibly be viewed as their sole cause.

On this background, the informational value of a *prescriptive ameliorative intervention* always constitutes an efficient cause whose effect shows as an improved (or perfect) alignment between the relevant normative standard, on one hand, and the actual reasoning and decision-making processes *cum* its decision-making outcome, on the other. This immediately entails that interventions are required *only if* natural dispositions alone do not reliably produce this alignment, while interventions fail when, upon being deployed, the alignment remains too imperfect with respect to the reasoning-process, its result, or both.

Notice, too, that further assumptions are required before the application of a given normative standard licenses *but one* correct reasoning process or decision-making outcome. Whether fulfilled, or not, these assumptions are at any rate *not* delivered by the normative standard itself, and may include necessary or contingent features of the decision-making context such as standardized institutional procedures, but also motivational and cognitive capacities of agents who enact such contexts. Consequently, a *successful* ameliorative intervention makes such assumptions true.[2]

In the legal domain, interventions include elements of the jurisprudential procedure, of which some have become so sedimented that few might readily be interpreted as intending to impact natural dispositions (Arlen & Tontrup, 2014; Farnsworth, 2003; Jolls & Sunstein, 2003; 2006; Pi *et al.*, 2013). A vivid example is the dictum *audi alteram partem* ("you must listen to the other side"). Widely endorsed in both Greek and Roman law to counter the arbitrary one-sidedness of legal decision-making (Kelly, 1964), this dictum reappears as the total evidence principle in inductive reasoning, upon which the sciences at large depend.

We return to empirical aspects pertaining to the effectiveness of such interventions in Sect. 5, and now provide further background on biases by drawing on related research from psychology as well as medical and cognitive science.

3. How Heuristics and Biases Came to Be Viewed as a Bad

Although not under this exact term, systematic empirical research into human biases is roughly as old as empirical psychology itself. Some early work did admittedly remain anecdotal, but *empirical* demonstrations of distortions or selective filtering in perception,

---

[2] We thus take the position of the ameliorist, while those being disinclined to accept that particular cases of reasoning and decision-making *should* have been any different than they were take a panglossian position. By contrast, apologists recognize that a particular normative standard *should* have been deployed, but seek to explain why that standard *could* not have been deployed in the particular case (see Stanovich, 1999, for these distinctions). Of course, rather than 'debias', the term 'rebias' will sound more appropriate to those who hold that the last word on a given normative standard has not been spoken yet.

memory, judgment, belief, and choice had been available as early as MacDougall (1906). Similar effects had earlier provided "bread and butter" to the phenomenological tradition, from which 20th-century psychology and cognitive science, the latter arising in the 1980s, sought distance by endorsing the experimental paradigm, on the one hand, and by making ever heavier use of statistical inference methods on the other.

Selective filtering had already then been well-distinguished from "*distorted evaluations of our necessarily selective perceptions of the world*" (Hahn & Harris, 2015, 44), echoing the distinction between obtaining evidence and assessing its informational value. Until well into the 1970s, moreover, most researchers refrained from drawing overly pessimistic conclusions with respect to human rationality, despite Wason's (1960; 1968) infamous card-turning experiment having established 'confirmation bias' as a technical term. Indeed, the rather upbeat message had then been that 'man as intuitive statistician' (Peterson & Beach, 1976), although regularly falling short of judgments that *perfectly* match an applicable normative standard, nevertheless tends in the right direction, yet remains overly conservative—that is, *under*-confident(!)—for instance when evaluating the diagnosticity of statistical evidence. Marr & Poggio (1967) developed their rational analysis research program in a similar vein, which was further advanced by Anderson (1990). Its main premise was that, when viewed as (just) another cognitive organism, man adapts optimally to a given reasoning-task *as perceived*. Since such tasks are regularly misunderstood, this itself provides evidence of man's bounded rationality (Simon, 1982).

This optimistic view regarding man's rationality would soon change. By the 1980s, science journalism duly reported the opposite message, using language and drawing conclusions that were

> only modestly more apocalyptic than those appearing in scholarly journals both inside and outside psychology. [...] From [U.S.] coast to coast, researchers seem to agree that people of all stripes are seriously deficient in their decision-making abilities. (Lopes, 1991, p. 65).

Based on citation analysis, Lopes traces the probable cause of this change in tone to a series of publications by Daniel Kahneman and Amos Tversky that culminated in a widely cited summary article in *Science* (Tversky & Kahneman, 1974) and provided the "cornerstone for what today is called the biases and heuristics literature" (Lopes, 1991, 67).[3]

Obtained in laboratory studies that implemented the *strong inference* method (Platt, 1964), the heuristics and biases program invariably communicates the results of posing decision-theoretic and probabilistic reasoning-tasks that leave experimental participants with but one of two possible answers: a right and a wrong one. Obtaining the wrong response from a vast majority of subjects then allows for the immediate inference that subjects reason in violation of the applicable normative standard. Since countless such studies have meanwhile been conducted and published by authors who mostly seem to be content with repeating the "apocalyptic" message, it presently fails to make for news

---

[3] Several authors (e.g., Sandri, 2009; Hahn & Harris, 2014) observe that this apocalyptic message had not been a *central* part of the original research endeavors, a contention that Tversky and Kahneman express as follows: "it soon became apparent that although errors of judgments are but a method by which some cognitive processes are studied, the method has become a significant part of the message" (Kahneman & Tversky, 1982, 124). This suggests that the message changed, as it typically does when research results "travel" (Rehg, 2009) from the scientific to the public context but also to other scholarly contexts, as an effect of boundary work that feeds back into the context of origin (Goodwin & Honeycutt, 2009). Those familiar with popularizations such as Kahneman (2011) presumably agree that, exceptions permitting, science-journalism has in this case anticipated, but also provided a partial cause to, the scientific message becoming almost as dumb as the popular one.

anywhere. Notably, use of the strong inference method did reorient the focus of this research, away from the reasoning-*outcome*—which, as we saw, had been interpreted to tend in the right direction—and towards the reasoning-*process* that generates this outcome.

Despite the criticism that this method has received (see below), experimental results were, and until today remain, standardly interpreted as strong evidence in support of the hypothesis that participants rely on deploying heuristics in place of more effortful reasoning-processes that relevant normative standards demand. Biases are perhaps best thought of as the subjective or internal antecedent to deploying these heuristics. Unlike biases, which are considered latent, moreover, heuristics can in principle be externalized; they thus become intersubjectively appreciable (even objectifiable) in addition to the behavioral manifestations that go along with the reasoning-outcome. Being latent, in contrast, in the absence of specific tests, biases are assumedly closed to reflective access. The literature, however, is less than clear on what sets a bias apart from a heuristics; both terms are often used interchangeably.

To give an example of a particularly (in)famous reasoning-task that has led to a plethora of empirical replications, but also to conceptual work seeking to explain why it constitutes a stable effect, consider what has become known as the 'Linda problem', the 'conjunction fallacy', or the 'conjunction error'. In this case, the vast majority of experimental participants are said to deploy an *availability* heuristic, rather than rely on probability theory. The heuristic seemingly misleads participants when experimenters task them with comparing the probability of at least two evidence-statement, S and S*, which are presented along with a textual description of a person, called Linda, that experimenters also provide. Importantly, the semantic content of S* has been manipulated to be *more representative* of Linda-as-described than the content of S, on one hand, while S* also repeats the content of S, on the other. Since the content of S* is thus more specific than that of S, the former sentence is a logical conjunction of S and additional information—hence the term 'conjunction fallacy'. Typically the vast majority, though hardly ever all participants, find the more specific and more representative content expressed by the sentence S* to be *more probable* than the state of affairs expressed by S. It is therefore immediate that participants engage in reasoning-processes that fail to align with the Kolmogorov axiomatization of probability theory, the second axiom of which dictates that more specific content be equally or less, but *never more* probable than less specific content, in virtue of S* describing a subclass of the events that S describes. (We return to the presuppositions of similar experiments below.)

Fig. 1 provides a list of biases typically discussed in the literature, which originates in medical research.[4] The primary intention is to motivate checklists that address various errors, mishaps, and failures of medical diagnosis which biases are through to effect.

---

[4] Over the years, researchers have coined various more or less creative terms for allegedly *new* biases. Genuinely statistical and decision-theoretic ones excepted (which, normatively speaking, could be identified as biases no sooner than that the relevant normative standard had historically been developed; see Sect. 2), most such terms can be mapped onto predecessors terms from the classical or the early modern period, where they were subsumed under the category-label 'fallacies' (Hamblin, 1970; Hansen & Pinto, 1995; Woods, 2014). For better or worse, this label stuck despite the standard view of fallacies-as-reasoning-errors having gradually given way to a more nuanced dialectical treatment since the middle years of the 20th-century. But this latter view has largely failed to enjoy reception among those *not* having kept up with developments in rhetoric and communication studies. The sheer number of allegedly novel terms such as 'sunk-cost bias', '*status quo*' bias', or 'gender bias', moreover, has led to the impression that enumerating *all* biases would make for a very long list. This list has more recently also come to include forms of social stereotyping, now often referred to as *implicit biases*, in reference to an empirical test that is said to reveal them (Project Implicit, 2015). But, again, also these terms simply reference various ways in which individuals may be discriminated against—for instance on the basis of age, gender, sexual orientation, origin, religious or political persuasion—and so are as old as the discriminatory practices themselves.

Cognitive Biases and Failed Heuristics Addressed by Diagnostic Checklists

Bias or heuristic	Definition*	Role of checklist
Anchoring	The tendency to perceptually lock on to salient features of the patient's presentation too early in the diagnostic process and failing to adjust this impression in light of later information.	Prompt physician to consider diagnoses other than the initially favored one.
Availability	The disposition to judge things as being more likely or frequently occurring, if they readily come to mind.	Prompt physician to consider diagnoses other than those that readily come to mind.
Base-rate neglect	The tendency to ignore the true prevalence of a disease, either inflating or reducing its base rate and distorting Bayesian reasoning.	Remind physician of the relative prevalence of diseases in primary care for the patient's complaint.
Premature closure	The decision-making process ends too soon; the diagnosis is accepted before it has been fully verified. "When the diagnosis is made, the thinking stops."	Prompt physician to reopen the diagnostic process and consider alternative diagnoses before discharging the patient.
Representativeness restraint	The physician looks for prototypical manifestations of disease (pattern recognition) and fails to consider atypical variants.	Prompt physician to consider causes for the symptoms other than the ones that readily fit the pattern.
Search satisficing	The tendency to call off a search once something is found.	Prompt physician to consider additional causes of the complaint after something is found.
Unpacking principle	The failure to elicit all relevant information in establishing a differential diagnosis.	Prompt physician to ask questions that might confirm or rule out alternative diagnoses.
Context errors	The critical signal is distorted by the background against which it is perceived.	Encourage physician to rethink assumptions and maintain objectivity.

Figure 9. Typical biases listed in conjunction with a characterization, rather than a definition, and the role of checklist to mitigate their influence in the context of medical diagnosis (source: Ely et al., 2011, adapted from Croskerry, 2009b; creative common license).

Such checklists primarily serve medical practitioners when presented with emergency cases, and may be viewed as debiasing techniques in the sense of decision-making props (discussed below). But their purpose crucially differs from debiasing techniques in legal contexts, where biases seem to persist although the time constraints on decision-making are far less pressing. Nonetheless, also judges work under time constraints. As every judge is assigned cases by the court's head judge, judges regard themselves as being overburdened, and often enough feel pressed to decide issues faster than they wish.

Larrick's (2008, 319 f.) typology of *errors* may also be used to categorize biases and their outcomes. It analytically reduces biases to one or more of the following overlapping categories: (i) psychophysically-based errors such as reference point effects, e.g., overweighing recent information vis-à-vis all information received; (ii) association-based errors, e.g., deciding on the basis of available evidence rather than searching for more complete or representative evidence; and (iii) strategy-based errors, e.g., the use of inferior strategies or decision rules such as 'a bird in the hand is worth two in the bush.'

With a view to interventions that could correct biased reasoning, a theoretically grounded explanation *why* a heuristic regularly replaces a normatively correct reasoning process is harder to find. Even as participants are assumedly able to *potentially* reason in normatively adequate ways,

> [p]roponents of the heuristics and biases tradition [...] (Kahneman & Tversky 1996) [did and continue to] maintain that [...] heuristics 'can be assessed experimentally' and that testing the hypothesis that probability judgments are mediated by these heuristics [for instance, in the case of the Linda problem] 'does not require a theoretical model' (ibid.). (Samuels, Stich & Bishop, 2002, p. 32)

Until today, besides rather loose theoretical ideas on dual systems, dual processes, or dual types of human reasoning, little in the way of a theoretically well motivated why-explanation has become available.

In the early 20th century, similar ideas have been voiced in the work of William James and Sigmund Freud, for instance (see Osman, 2004); the distinction between intuition and deliberation itself had been available at least since the classical age. Contemporary 'dual-systems accounts of reasoning' originally sought to separate intuitive from reflective reasoning, but have meanwhile come to operate with diverse clusters of properties related to human consciousness, evolutionary considerations, functional characteristics, and individual differences in reasoning (Evans, 2008). This widening in scope, however, has provided little more than handy terms for interdisciplinary (mis-)use among a scholarly community that by and large appears to be more comfortable with hypotheses-generating and hypothesis-testing research, at the broad expense of constructing theories from which differentiable empirical hypotheses may be deduced for predictive purposes. (We return to this aspect in Sect. 6).

Mutatis mutandis, this critical remark equally pertains to scholarly views that make the regular deployment of heuristics an *ecologically rational* affair in view of the conceptual truth that natural environments are uncertain, rather than risky. Heuristics are said to have evolved as adaptations (or exaptations) to natural environments, that is, as effects of *successfully* resisting the "survival uncertainty" that natural contexts present to the human organism (Gigerenzer *et al.*, 1999; Gigerenzer & Todd, 2009). This view emphasizes that, by definition, whenever the environmental initial and boundary conditions are right, then deploying heuristics delivers reasoning-outcomes[5] that are as good as, and in a special sense better than, reasoning-outcomes delivered by deploying decision theory or probability theory. After all, non-vague probabilities, and hence well-defined risks, can be properly calculated *only if* the space of possible outcome-events is finite. But genuinely uncertain environments fail this condition in virtue of featuring so-called 'unknown-unknowns'.

On this view, *optimal* reasoning and decision-making is a matter of the best fit between a task, a context, and various resources. But this can at best provide a partial vindication for deploying heuristics, because the relevant notion of optimality is bound to the specifics of one or another context. Much of this research is therefore devoted to describing the exact environmental features that let the heuristics constitute a *contextually optimal* problem-solution strategy.

Particularly in view of agents' limited cognitive resources and time-constraints, deploying a heuristics can thus be viewed to result from trading-off against more effortful, reflective reasoning (Gigerenzer & Sturm, 2012). Such considerations ultimately invoke the economics of occasion-cost, and put an explanation sketch in place why a heuristic has evolved—namely through an evolutionary adaptation to natural environments (rather than risky ones). But any explanation why a heuristic is deployed *now* remains empty if it fails to differ from the trivial, though perhaps correct, hypothesis that the heuristics is all the agent had readily in stock. Given a task T and a context C, even if a heuristics H may provide an optimal result R, this sheds light onto the question *why* H is deployed in view of a T*-task and a C*-context only on the assumption that agents regularly mistake T* for T or C* for C, that is, *mis*identify the relevant elements.

4. Debiasing: How to Tame the Critters of the Mind?

Among most researchers working on dual process accounts of reasoning and decision-making, for fuller accounts of which see Osman (2004) and Evans (2008), the

---

[5] Some scholars distinguish a heuristic from a reflective mode of inferencing by strictly reserving the term 'reasoning' for the reflective mode. Rather than take a stance on the viability of this distinction, we note that these scholars are consequently opposed to calling the outcome of deploying a heuristic a '*reasoning* outcome'.

consensus is that there are two major sources: innate, hard-wired biases that developed in our evolutionary past, and acquired biases established in the course of development and within our working environments [and that b]oth are associated with abbreviated decision-making in the form of heuristics. (Croskerry, Singhai & Mamede, 2013, ii58)

Croskerry *et al.* (2013) provide an analytical schema of the processes and the tasks arising in medical reasoning when transitioning from patient presentation to diagnosis (Fig. 2).

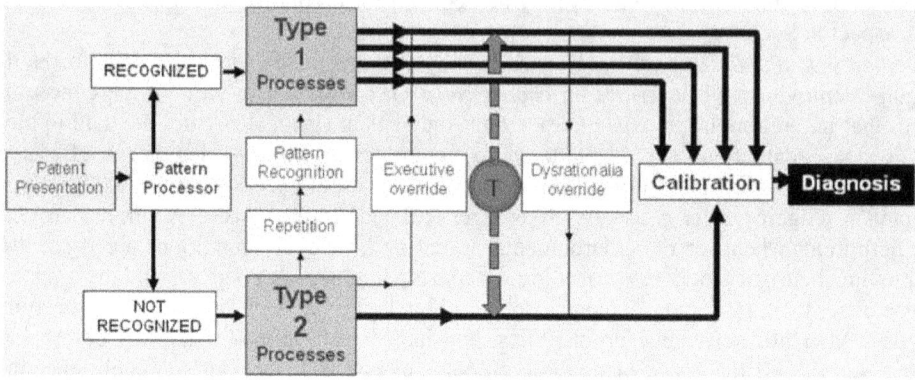

Figure 2. Dual process model for decision making (source: Croskerry, 2009a); 'executive override' signifies just this, the term 'dysrationalia' denotes the inability or unwillingness to reason in effortful ways, and 'calibration' the degree to which the intuitive and the deliberative diagnoses align.

The schema readily transfers to other domains concerned with the assessment of evidence, broadly conceived, including the legal context, of course. Being a dual-process schema, it prominently features a toggle function (above abbreviated by T)—seemingly introduced on the basis of practitioners' observational self-reports—which represents that decision-makers may "move forth and back between [intuitive] Type 1 and [deliberative] Type 2 processes" (ibid., ii60). Information processing here occurs in terms of recognized or unrecognized patterns that trigger an intuitive or an effortful reasoning-response, respectively.

To successfully overcome a bias, that is to *debias*, in order to significantly mitigate or completely avoid the effects on a decision-making outcome that a bias, once triggered, is thought to cause in contexts where a normatively (more) adequate decision-making outcome is possible (see Sect. 2), requires changing the intuitive reasoning-process *whenever doing so is apt*. Aligning the outcome to the applicable normative standard in order to achieve what in broad terms are accuracy-gains in information processing thus presupposes the following conditions to be jointly fulfilled (Fig. 3 below).

These five conditions—(i) awareness of the bias, (ii) motivation to correct it(s effects), (iii) awareness of the bias's direction and magnitude, (iv) ability to apply an appropriate debiasing technique, and (v) successful debiasing (Wilson & Brekke, 1991; see Croskerry et al., 2013)—are listed in implicational order. Conditions occurring downwards in the diagram, although being fulfilled, may *not* suffice to ameliorate the bias's effects on the decision-making outcome unless *all* conditions occurring upwards are also fulfilled. For instance, the *ability* to deploy an appropriate debiasing technique (condition v) is ineffective, analytically speaking, without the *motivation* to do so (condition i). So if the model is any good, then a debiasing technique should be more likely to fail if its successful deployment presupposes conditions that are in fact false. Effective debiasing techniques will therefore need to leave, or make, *all* relevant conditions fulfilled; efficient techniques avoid wasting resources. Notice that these considerations pertain to empirical realities although it may be less clear how to ascertain that the five conditions are (not) fulfilled.

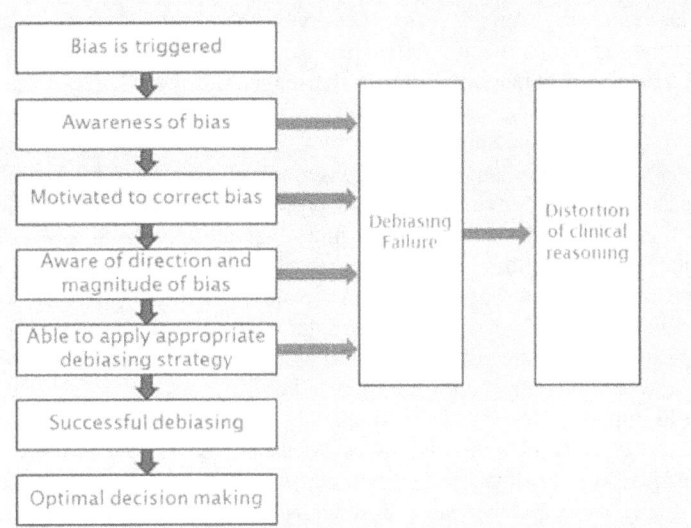

Figure 3. Successive steps in cognitive debiasing (Croskerry *et al.*, 2013; adapted from Wilson & Brekke, 1991).

One may debate the methodological implications of the final step in Fig. 3, leading from successful debiasing to optimal decision-making. The main insight from research on the ecologically rational use of heuristics teaches that the reasoning-outcome alone does not uniquely presuppose this final step (Betch & Held, 2012). The intuitive and the reflective mode may well differ with respect to their underlying reasoning processes, and yet deliver the same outcome. Therefore, even fully debiased agents, as it were, may still reach suboptimal decision-making outcomes *although* they do in fact engage in apt reasoning processes that are licensed by a normative standard. So they may be "innocent" of committing the strategic reasoning-error ('selection-error') that deploying a heuristic instantiates whenever deploying the heuristic yields *vastly* different outcomes than those deemed normatively correct. Rather, it is easy to see that debiased agents may still commit 'run-errors', that is, errors *within* an otherwise optimal reasoning process which would yield the decision-making outcome under normal conditions. That "faulty" process, however, need not arise in virtue of bias; it may rather owe its suboptimal outcome to more mundane matters such as fatigue, distraction, or species-typical cognitive limits, particularly on memory and perception, which can thus be cited as non-normal conditions.

Pace this important caveat, the leading idea is that long term successful debiasing at a personal level requires *decoupling* intuitive from quasi-automatic responses. Whether or not through having deployed a debiasing technique, once agents *can* engage in a deliberative mode of reasoning, this very mode may become quasi-automatic, provided the *former* intuitive mode is lastingly overridden. Only then is the seemingly paradoxical claim meaningful that, in the long run, expert decision-makers can command expert-knowledge *in intuitive ways*. This, of course, differs from the claim that experts are *never* subject to bias. Experts too, particularly judges, jury members, but also attorneys *are* subject to biases, and are thus also engaged in biased decision-making (see Sect. 5).

Leaving non-biased forms of outcome-deviance such as run-errors to the side, each debiasing technique will seek to improve agents' reasoning in ways that let outcome-optimal modes of reasoning become self-sustaining—in the long run and in the best cases. These techniques thus present ameliorative interventions onto what, in broad strokes, are *personal* characteristics of reasoning and decision-making, seeking to change one's "reasoning-character" in lasting ways. They may therefore be understood as cultural techniques, in the sense that a more developed culture provides better average-individual-chances at developing a personality that, in the longer run, ever leaves traits of a bygone age behind.[6]

It is widely accepted that human agents may in the shorter run be *nudged* (Thaler & Sunstein, 2008) into better decisions, through readily deployable formal procedures, conducive decision-making environments and props (catchphrase: on a buffet, place fruit first and muffins last), as well as selective hints that literally "ring in" on one's mobile phone. But nudges do generally not require agents to be *personally* aware of a bias, nor to command substantial amounts of debiasing-motivation. Therefore, they present little more than a work-around to the challenges that the internalist conditions (i) to (iii), above, pose. If such shepherding results in better decision-making, then this is also because nudges are by definition attuned to agents' *ready* abilities, rather than seeking to further these. This lets nudges differ in important respects from intendedly enculturating debiasing techniques. The evidence for the effectiveness of nudges as debiasing techniques is therefore harder to assess, for nudges do not primarily intend to educate; they merely assist agents in choice-behaving as if they were. But whenever a nudge is absent or—as in having hit the snooze-button on one's alarm clock—has been put off-duty, then agents should rather do as they please. It is in this sense only that nudging can be viewed as a form of *liberal* paternalism.

A final step in preparing for an overview of extant debiasing techniques in legal contexts consists in a methodological remark. Starting perhaps with Cohen's (1981) classic paper, it has regularly been noticed that the claims supported by data obtained through the strong inference method (see Sect. 2) incur substantial presuppositions. As we saw, this inference runs *from* the observation that participants solve a task incorrectly *to* the claim that the normative standard is not (correctly) deployed which experimenters had antecedently coordinated to a given task. But the task being presented to participants does in each case come with particular task-instructions. Therefore, the observation can

---

[6] Disagreement persists, of course, as to the desirable properties of such cultures and how they develop—questions we cannot adequately address here. The biases currently sought to be ameliorated nonetheless reach beyond primarily cognitive shortcomings. Furthermore implicated are various forms of suboptimal *social* decision-making. These may, for instance, be observed in the differentially inclusive praxis of hiring and promotion referred to as 'direct reciprocal altruism', 'nepotism', or 'clientelism'. The detrimental socio-economic effects of the false conservatism that even political conservatives may view in this praxis are rather well-documented; they bear out in a sluggish cultural, organizational, and personal development. Its lasting ubiquity across cultures and ages has led some political scholars to accept, rather than explain, this praxis on the basis of evolutionary biology (Fukuyama, 2011; 2014). Notice that, despite lip-service to the contrary being regularly paid and laws against such practices being in place among most modern nations, it stands to reason that agents who command a power-status which enables them to "hand out" favors may well be personally motivated to avoid nepotism. But they may for various reasons remain unable to do so at the *social level* where nepotism plays out, partly because those in lower power positions, who expect to be favored in return for subservient behavior, provide a continuous over-supply of temptation (ibid.), and partly because they tend to overestimate the personal risk of taking measures against it in view of falsely estimating the preferences of others to do the same (Bicchieri, 2006). Generally, such social biases appear to be more effectively treatable by interventions that target not the personal level, but instead provide decision-making props such as blinding-screens for hiring orchestra musicians, anonymized picture-free CVs (see Kenyon & Guillaume, 2014), or hiring committees whose members are *not exclusively* selected by those who expect to interact with a new colleague. We return to such props as elements of a conducive decision-making infrastructure in Sect. 6.

meaningfully support the intended conclusion—that, owed to the presence of a bias, heuristics mediate the manner in which *the* task is solved, as opposed to how it should be solved—*only if* participants felicitously interpret the task, that is, correctly understand the instructions. So participants and experimenters must successfully coordinate on a shared understanding of the task, which requires a shared meaning (see Stenning & van Lambalgen, 2008). If semantic coordination remains imperfect, however, then to observe suboptimal reasoning and decision-making behavior is ambiguous between supporting, on one hand, the intended claim and, on the other, the hypothesis that one deals with systematic *experimental* error rather than systematic *human* error that is partially or fully owed to biases.

If so, then *but one* among the assumptions listed in Fig. 4 in fact pertains to cases where individuals may be personally biased, namely when what the debiasing technique addresses originates primarily with a faulty judge—an *imperfect but perfectible* individual—rather than coming about through a misunderstood task, or an unfair one, or a mismatch between judge and task.

Assumption	Strategies
Faculty tasks	
Unfair tasks	Raise stakes
	Clarify instructions
	Dispel doubts
	Use better response modes
	Discourage second guessing
	Ask fewer questions
Misunderstood tasks	Demonstrate alternative goal
	Demonstrate semantic disagreement
	Demonstrate impossibility of task
	Demonstrate overlooked distinction
Faulty judges	
Perfectible individuals	Warn of problems
	Describe problem
	Provide personalized feedback
	Train extensively
Incorrigible individuals	Replace them
	Recalibrate their responses
	Plan on error
Mismatch between judges and task	
Restructuring	Make knowledge explicit
	Search for discrepant information
	Decompose problem
	Consider alternative situations
	Offer alternative formulations
Reeducation	Rely on experts
	Educate from childhood

Figure 4. Assumptions and debiasing methods, here called 'strategies', according to Fischhoff (1982; reprinted in Fischhoff, 2010); for 'Faculty tasks' read: 'Faulty tasks'.

Clearly, unfair tasks should be made fair and misunderstood tasks should be clarified, while—by definition—incorrigible individuals cannot improve their performance. Since all ways of correcting a mismatch between judge and task must improve the availability of relevant information and provide broadly educational support, moreover, it is less clear whether simply making such information available in readily digestible form alone can plausibly count as an intervention that ameliorates a bias. This goes to show that there are a number of reasons *other than personal or social bias* why tasks may be solved incorrectly.

Originally laid out by Fischhoff (1982), the assumptions in Fig. 4 also provide evidence for a lack of analytical rigor. In a rather misleading use of the phrase 'artificial bias of individual judgement', for instance, Sandri (2009) correctly observes that it can be "difficult [in practice] to draw a clear dividing line between a misleading [task-]description which artificially biases the individual judgement and a description which frames the individual interpretation of the situation faced" (Sandri, 2009, 110). This irresponsibly conflates 'bias as a partial cause of a suboptimal decision-making outcome' with the 'sub-optimality of the outcome', whether artificially induced or not. But as we saw, one can unproblematically speak of personal bias and its effects *only if* the task is clear, and if the resources that normatively correct decision-making requires are in fact available.

Of course, this does not excuse suboptimal decision-making whenever it occurs. But it makes plain what the claim entails that 'suboptimal reasoning and decision-making owed to a bias is sought to be ameliorated through a specific technique', as opposed to understanding the various ways in which reasoning and decision-making processes can be improved although they do *not* involve personal biases. Analytically, this distinction is clean. It turns "dirty" only because it is in practice hard to separate the presence of personal bias from conditions of reasoning that produce suboptimal decisions-making outcomes, or from conditions that may interact with a bias and so can *jointly* produce a suboptimal outcome. Recall that a suboptimal outcome underdetermines whether it arose as the effect of a bias alone, of a bias plus other quasi-facilitating conditions, or in the absence of bias. Just as value-choices do not—without producing scholarly non-sense—*require* the interpretation that these are owed to biases (Weinstein, 2002), some suboptimal decision-making outcomes may thus have nothing to do with biases, neither personal nor social ones.

This pedestrian insight appears to be underappreciated in the literature, also because the term 'bias' is regularly used in the ambiguous manner to which its appropriation by scholars outside of the term's domain of origin typically leads. Phrased more positively, *only if* the 'scientific hygiene' of an empirical study is secured can one unproblematically interpret participants' suboptimal decision-making behavior as evidence of bias.[7]

5. Debiasing Techniques: Tested, Tried, and Failing About Half the Time

We now turn to extant research on debiasing techniques in legal contexts, keeping in mind that 'bias' should be understood in the loose sense that currently dominates the literature. As Hahn & Harris (2014, 42) put it, "[a] reader venturing into the psychological literature about human biases soon realizes that the word 'bias' means many things to many people." Correspondingly, 'debiasing technique' broadly denotes all "strategies designed to reduce the magnitude of judgment errors" (Arkes, 1998, 449), including errors that may not be owed to personal biases but to non-conducive environmental, institutional, or cultural conditions, as well as to constraints such as fatigue, and possibly the condition formerly known as stupidity (Cipolla, 1987), now more politely called 'severe cognitive limitations'.

As we saw in Sect. 4, among the four necessary conditions for successful debiasing to occur, conditions (i) and (iii) are cognitive ones, citing awareness of the bias and of its direction; condition (ii) cites the motivation to become unbiased; and condition (iv)

---

[7] Such hygienic considerations have led to studying them "in terms of the four essential elements of any behavior: the organism, the stimulus being evaluated, the response mode for expressing preferences, and potentially distracting contexts" (Fischhoff, 2010, 728).

demands an appropriate measure or strategy to be deployable. Despite the suffix 'able', we view (iv) as a partly technological condition rather than a solely cognitive one; even a "cognitively perfect" agent would require being provided with the right technique(s). So cognition and motivation must meet with the right tool, unless a fully motivated cognitively perfect agent could never be subject to biases—an assumption one cannot endorse, for even the best actual cognizer has *some* capacity limit.

We will meet this triad—cognition, motivation, and technology—further below as the top level of the perhaps most basic contemporary taxonomy of debiasing techniques that is structured according to the primary goal that a technique pursues (Larrick, 2008). By then, the following should have become painfully clear: in the absence of having specified one or more of the necessary debiasing-conditions as that which a particular debiasing technique seeks to ameliorate, and in the absence moreover of having validated these conditions as plausible partial causes of poor decision-making that are responsive to acts of ameliorative prescriptive intervention, one cannot generate plausible empirical hypotheses regarding the effectiveness of a debiasing technique given a task, a context, and a sample of agents. Nor can one reasonably expect that deploying some such technique will meet with success other than by sheer luck, that is, thanks to conditions one did *not* control. After all, if several conditions C_1 to C_n jointly account for, and so can be thought to individually or jointly give rise to, a suboptimal decision-making outcome O_B, then an effective debiasing technique T must treat all conditions that would still give rise to O_B if any of C_1 to C_n remain untreated.

A slightly more complex way of putting the matter is to additionally recognize that O_B may be determined not only by these conditions, but also with respect to the specific biases B_1 to B_n as well as further broadly contextual conditions that may collectively give rise to O_B. After all, there is no reason to doubt that the *same* outcome can be brought about by more than one bias, that is, be "multiply determined" (Larrick, 2008), just as the same optimal decision-making outcome is potentially an effect of deploying either a heuristic or a normatively more virtuous standard, or the effect of deploying one of several ways of applying the normative standard.[8]

The literature mostly reports cases where a particular debiasing technique T_i barely worked half-way, so it effectively improves the decision-making outcome only in some cases, compared to a control group where a different or no such technique is deployed, but proves ineffective in other cases where it is deployed. However, already for decision-making outcomes that are determined by a single bias, if one understands O_B as a function of any or all of C_1 to C_n being *unfulfilled*, and if T_i successfully treats only C_i but remains ineffective with respect to C_j, then obtaining such mixed-results is but a matter of due course. Metaphorically, these are cases where one "cannot cure a disease by dancing around the patient." Worse yet, in other cases one cannot "extinguish fire by pouring oil onto it," so rather than ameliorate the reasoning-process, some interventions may make it worse, normatively speaking.

---

[8] For instance, if the task is 'add 79 and 16', one may add 16 to 80 and then subtract 1. Alternatively, one may add 10 to 70, then 9 to 6, and finally 80 to 15. This yields 95 either way; none of these algorithms is less *correct*, if perhaps aesthetically different. In the suboptimal case, by contrast, one's overconfidence in the ability to lose five pounds of body mass, say, or a lack of refined aesthetic judgment might both see one purchase a pair of jeans that fits someone else better. As for heuristics, finally, one might purchase the same computer that one's neighbor has recently acquired and reports to deliver good performance, thus relying on an availability heuristic, and merely search for the best prize. Alternatively, one can spend the time required to compare the available options, subjectively prize the features one in fact (should) want, read consumer reports, etc., and yet end with the neighbor's model as the best candidate. The comparative triviality of these examples does not affect the underlying principle, which is non-trivial.

The literature generally offers such mixed results, and the tendencies to not properly report statistically non-significant experimental results expectably lets the true number of failed interventions be much higher (file-drawer problem). There is no reason to suspect that such cases are typical only of empirical studies on debiasing techniques in legal contexts; the same would hold elsewhere, too. Extant studies moreover display a certain absence of awareness that complexities such as the above could have been anticipated. Provided the necessary resources are in place, they can be addressed in methodologically sound ways often enough by increasing the sample size, and by undertaking efforts at studying cohorts other than the confused and inexperienced which the mixed quality of higher education lets adolescents be too often (see Henrich *et al.*, 2010).

A central difficulty, which we can only briefly address, is that hardly any two such studies may be unproblematically compared with respect to the biases treated, the debiasing technique employed, the experimental set-up used, or qualities of the participants recorded. This indicates a grave shortcoming that has besieged empirical social science research for decades, but had first been widely recognized in medical research (see Ioannidis 2005a,b; Witte & Zenker, 2015). Partly because of a lack of coordination among (competing) researchers and research groups, and despite the fact that meta-reviews are prized highly as indispensable instruments to accumulate research results, hardly any study is in fact open to an informative meta-analysis. What follows should therefore be read with tons of salt.

Before turning to empirical methodologies that could improve such data, so that a meaningful meta-review becomes possible after all (Sect. 6), we now give a brief overview of findings from extant empirical studies on the effectiveness of debiasing techniques in legal contexts. Readers should bear "the largely uncharted frontier of debiasing" (Lilienfeld *et al.*, 2009, 391) in mind; the legal context merely makes the absence of well-founded empirical knowledge especially pressing.

Following Larrick (2008), virtually all extant debiasing techniques may be classified according to whether the strategy that a given technique pursues is in the widest sense a *motivational*, a *cognitive*, or a *technological* one. As these terms suggest, motivational and cognitive strategies operate at the individual level (even when they are deployed to groups of agents), while technological strategies can also operate at the group or institutional level. Respectively, they seek to provide personal incentives to deploy a more desirable mode of reasoning and decision-making in contexts where doing so would be apt; to personally equip subjects with the requisite knowledge, skill, or information that suffices to correctly deploy such reasoning-processes; and to provide environmental props that facilitate the deployment of these processes as a matter of standardized procedures that agents enact in the course of institutionalized due procedures. In brief, debiasing techniques thus target individual dispositions and abilities as well as the structure of collective environments.

The following might come as a sobering surprise to those teaching the heuristics and biases literature as part of formal education, notably in critical thinking instruction: judging on the basis of experimental results it may not be completely impossible in particular cases, but it generally appears to be a waste of time and resources to 'tell people that they are potentially subject to biases'. In terms of debiasing effects, the same by and large holds for instructing agents to reason harder, better, longer, deeper, to concentrate, or to just put oneself to it. While brief non-technical tutorials on biases have reportedly been effective in reducing people's tendency to fall prey to them (Lilienfeld *et al.*, 2009, 393), teaching meta-cognitive rules—i.e., those that engage cognition about cognition—such as 'consider the opposite' appear to be more promising (Willingham, 2007). But these can go significantly beyond a non-technical level of instruction. Similarly, the speculative hopes that Duthil Novaes (2012, 221-248) voices for formal logic instructions to have debiasing effects do, in the same book, meet with a technical level of instruction. Logic instruction at less technical levels, however, has proven to be ineffective in transferring to *all* relevant domains where such skills would improve the reasoning-outcome (Lilienfeld *et al.*, 2009).

In the legal context especially, a judge's instructions that a jury should for various reasons disregard or discount on evidence is similarly ineffective (Arkes, 1998; Wissler & Sacks, 1985). This is perhaps expectable; after all, task-instruction for the Linda problem (see Sect. 3) or Wason's (1968) card-turning task also (implicitly) require of participants to disregard background information, but regularly meet with no success. Attempting to *generally* raise people's awareness of biases thus probably constitutes an ineffective way of improving their reasoning and decision-making, that is, such techniques do rather not ameliorate. Instead, they may in particular cases worsen the effects of a bias, particularly with respect to hindsight bias, confirmation bias, overconfidence, and belief persistence (Arkes, 1998; Fischhoff *et al.*, 1977; Kurtz & Garfield, 1978; Kenyon, 2014; Wood, 1978).

Standard explanations why similar instructions remain ineffective or even become counterproductive have remained vague, and tend to fail more demanding standards of why-explanation. (Of course, theoretical shortcomings do not make the empirical data less real.) Explanation candidates for the impotency of instructions range from the postulation of self-defense and immunization mechanisms that are automatically deployed when such instructions are perceived as the *ad personam* criticism they are, over invoking null-explanations such a 'bias blindspot'—which offers the name of the *explanandum* in place of an *explanans* (see Pronin *et al.*, 2002; 2007)—to a presumably non-empirical because non-testable explanation known as the Dunning-Kruger effect. It postulates the *principled* incompetence of the incompetent to comprehend their own incompetence, and to normally go along with a similar incompetence to comprehend the competence of others. "If you're incompetent, you can't know you're incompetent. [...] the skills you need to produce a right answer are exactly the skills you need to recognize what a right answer is" (Dunning *et al.*, 2003). Owing to the severe consequences of their decisions (e.g., sending someone to jail), *that* judges at times commit errors may thus require that they distance themselves from the idea of committing errors, for constant awareness thereof may be experienced as a burden. These considerations alone would put a stop to all personal and perhaps also to most environmental debiasing efforts whenever they require *some* amount of self-awareness, and so could only leave replacing individuals from their organizational positions if their non-malleable personal condition causes greater social or personal harm to others.[9]

In addition to being pathologically incorrigible and hence unreceptive to a debiasing technique, or receptive in a direction other than aimed at, Lilienfeld *et al.* (2009) identify the following barriers to debiasing. These loosely map onto the cognitive and motivational ones among the necessary conditions for debiasing identified above (Sect. 4), but also include affective states as a mediator: debiasing may be perceived as irrelevant to personal welfare (lack of motivation); present cognitive abilities remain domain-specific and do not transfer to new contexts (lack of cognition); "individual and cultural differences in personality, cognitive styles, and developmental level may predict the efficacy of debiasing efforts (Stanovich & West, 2000)" (ibid., 394) (differential individual cognition); mediated by strong affects, ideological extremism may strongly dispose to perceiving confirmatory

---

[9] Compare Fischhoff's category 'incorrigible individuals', above, along with the broader implications of footnote 6. The need for amelioration is perhaps more pressing in bureaucratic organizations such as state-run agencies than in private businesses, the more so when "firing" is a seeming cultural impossibility. After all, since positive external incentives to ameliorate may be either lacking or constitute a sad value-proposition, bureaucratic careers are regularly made not by promotion due to excellence, but because one has over many years merely avoided brute error. The process allegedly continues until individuals reach the organizational position that reflects the highest level of their own personal incompetence (Peter & Hull, 1996).

evidence only (lack of fuller cognition owed to affect). Effective debiasing techniques will have to address such barriers.[10]

It may by now have become clear that effective debiasing techniques look ever more unlikely to be 'one pill fits all'-treatments. Rather, as is the case for cutting-edge medical research, the success of ameliorative prescriptive interventions expectably varies from one individual or group to the next. In terms of cost and effort, such research hence incurs all the challenges of personalized medicine, since an ameliorative intervention must be customized to the constraints of those individuals and groups that require it, to the moment at which decisions are made, and to the larger context in which such actions occur.

Among the more promising techniques reported in the literature is 'consider the opposite' (Lord *et al.*, 1984; Mumma & Wilson, 1995; Mussweiler *et al.*, 2000), which we did meet above as a meta-cognitive strategy and had already met in Sect. 1 as the principle *audi alteram partem*. This does in some cases reportedly ameliorate instances of hindsight bias as well as the effects of overconfidence and selective availability owed to anchoring effects. For mixed experimental evidence in legal scenarios, see Kamin & Rachlinski (1995); a more upbeat result is reported in Arkes *et al.* (1988). Similarly, Babcock *et al.* (1997) report the positive effect of mitigating self-serving bias in litigation negotiations subsequent to explicit instruction to consider the weaknesses of one's own case.

However, if whatever *the* opposite to consider may be must yet be generated from imagination, rather than being made available in full detail, for instance by a defense attorney, then the same strategy may be less likely to work. Presumably, sheer lack of imagination, or lack of motivation to probe it, could lead agents to conclude (incorrectly) on the basis of the few alternatives thus generated—of which some may be 'out of the way'-possibilities—that the original belief was not problematic after all. It is easy to see that the role of the defense attorney may be viewed as a counter-weight to a lack of imagination, for instance by providing plausible scenarios that put the defendant, at the relevant time, at places other than the crime scene, say, or explanations for the presence of the evidence presented by the prosecution even if the defendant was innocent (aka the probability of the evidence conditional on the falsity of the hypothesis).

The strategy of playing 'devil's advocate' amounts to a similar idea under a different name (see Büyükkurt & Büyükkurt, 1991). A particular version thereof is known as 'reframing', where one presents the *same* state of affairs from a different point of view, for instance as a gain-framed rather than a loss-framed message in order to counteract loss-aversion (see Hodginson *et al.*, 1999). Particularly with respect to statistical evidence, for instance, it is well-documented that reframing information such as 'the probability of x is 5%' to a frequency-format such as '5 in 100 cases exhibit property x' improves the

---

[10] This is not made easier whenever individuals see no personal gain from being debiased because there is no gain. In environments where suboptimal decision-making has become the norm, for instance, substantial incentives might be required at an organizational level, as well as plain courage at a personal level, to have being debiased pay personally. This provides a seemingly paradoxical twist of the same economic logic for decision-making that boundedly rational actors allegedly do *not* deploy widely (see Jolls & Sunstein, 2005, 13; Bicchieri, 2006): absent proper incentives, biased individuals who fail to personally command significant levels of motivation to improve may have *no* valid reasons to debias, and they might know, too. As Lilienfeld *et al.* (2009) observe, the issue acquires a cultural aspect and may implicate religious backgrounds, or their absence. For the kinds of incentives that could support more optimal modes of decision-making—ranging from the mere display of explicit and unmistakable peer-respect to significant monetary rewards, but also including equally explicit and unmistakable peer-criticism—are differentially distributed in group-oriented homogeneous contexts as opposed to individualistic ones. This implicates the family over social groups to societies, hence the micro, meso, and the macro level of social structures. Forms of quasi-punishment which are deployed in response to acts of good decision-making that nevertheless starkly deviate from the organizational normal condition, of course, can only worsen the case for mounting a sound means-ends argument in favor of personal debiasing efforts.

accuracy of judgment, although providing the less initiated with a more detailed explanation of the task will normally also be required (Lopes, 1987; Gigerenzer & Hoffrage, 1995; Cosmides & Tooby, 1996).

The same holds for the strategy of 'providing supporting reasons' for a decision as a measure to increase accountability, which in the legal case falls upon the judge or the jury. But this may fail in contexts where the differentially pronounced ability to state reasons can lead to concluding that the original decision was better than it is—for, as it were, so few reasons speak against it (see Koriat *et al.*, 1980; Zenker *et al.*, 2015). Generally, individuals seem naturally inclined to produce supporting reasons, though the ability to do so again varies significantly. Readily generating counter-reasons to one's own belief normally requires training. (To readily test this, readers may try to construct a hypothetical case for the legitimacy of terrorist attacks; see Lilienfeld *et al.*, 2009.) A specific variant is the 'Socratic method' where guided questioning seeks to elicit more accurate responses from participants, for instance about a range of prior cases similar to that under consideration (Büyükkurt & Büyükkurt, 1991). In the legal context, the witness's cross examination is an obvious example, but notice that so-called "leading questions" may not be allowed in all relevant contexts (Federal Rules of Evidence, 611).

The above strategies generally seek to make countervailing evidence *more salient* than it is (Arkes *et al.*, 1988; Arkes, 1998), and may thus lead to debiasing effects on future judgments, for instance when participants are informed about their objective performance-score on a task they were highly overconfident about. Such debiasing effects are thus brought about by humbling the agent.

Also Seamone's (2006) proposal for developing judicial mindfulness is an awareness-raising technique, but remains empirically untested. As part of both initial and ongoing legal training, it recommends implementing various strategies taken from areas as diverse as clinical psychology, creative thinking, and improvisational theatre. This includes role reversal, Brecht's alienation effect (*Verfremdungseffekt*), various brain-teasing and breathing exercises, but also classical techniques from phenomenological research such as focusing, or therapeutic ones such as role enacting and journal writing. All strategies seek to improve awareness of self and others also at an affective level.

Besides such broadly personal interventions, various sources (see Kamin & Rachlinski, 1995) report mixed results from deploying environmental props that are aimed at reducing decision-makers reliance on cognition, particularly on memory which is easily depleted. This includes the use of formal models such as Bayesian ones, but also causal maps. These tools, however, can reportedly be severely detrimental to the decision-making outcome whenever such models are applied, as they often are, without detailed knowledge of the underlying formalities, and so risk using the right tools in the wrong way (see, e.g., Pundik, 2012). Generally, there is a vast gap between those in the full know about the complexity of such models and those who rely on them in their decision-making (Petersen & Zenker, 2015). As should be clear, possessing the right tool does not alleviate the need to develop the skills necessary to use it well.

Next to such props, more procedural forms of interventions include separating the decision-making process into distinct phases, as is again already the case in legal proceedings. Wiggins & Breckler (1990) report ameliorating effects from bifurcation, that is, from creating an additional process-separation between juries being instructed and coming to a decision on the amount of damages paid in negligence cases, for instance. Kamin & Rachlinksi (1995) experimentally demonstrate such cases to be highly subject to hindsight bias, but to be ineffectively treated by merely instructing participants to imagine outcomes alternative to the actual course of events.[11]

6. Implications for Future Research

As was indicated in Sect. 5, the available empirical evidence on the effectiveness of debiasing techniques in legal contexts suffers from methodological drawbacks. To start, the number of studies alone remains embarrassingly low. Compared to experimental "confirmations" of the effects of biases on reasoning and decision-making, moreover, research on debiasing compares at a rather saddening ratio of 1:7.5, based on a PsycInfo query from June 2008 that compares citations for 'debias' with those for 'cognitive bias', as reported in Lilienfeld *et al.* (2009); for the same terms, Google trends reports a ratio of 1:4.6 for the period from 2004 until January 31st, 2015.

But merely intensifying the research endeavors or increasing the typically low sample size, while not unhelpful, can hardly suffice. Rather, the experimental study of debiasing techniques should adopt crucial lessons learned during the ongoing replication crisis in the medical sciences (Ioannidis, 2005a, b), which had soon after arisen also in psychology (Spellmann, 2012; Pashler & Wagenmakers, 2012; Sturm & Mühlberger, 2012). We cannot hope to do full justice to this important issue here. But the crisis strongly indicates that significantly improved research coordination is required at levels unprecedented in the social sciences. Better coordination between (competing) research teams could ensure that empirical studies will at least be comparable with respect to the sampled population and the test-situation created, that is, the tasks, the task-instructions, and the debiasing techniques. The current normal condition in the empirical social sciences lets such samples differ too widely as to be readily comparable in methodologically sound ways.

Strangely enough, conducting the near-exact replication attempts that are required to compare and extrapolate from data were, until recently, seemingly not considered worth the effort. For most practical purposes, this more or less remains the case to date also because career-incentives do not let researchers gain comparable amounts of credit for replication-research as the community grants for allegedly "discovering" an new phenomenon—the stranger, the better (Witte & Zenker, 2015). This novelty-driven research is sustained by a general lack of appreciation for theoretical efforts at providing something similar to covering laws, let alone theories that allow reliable predictions. Particularly the popularity of "two-systems talk" (see Sect. 4), where a handy term re-describes the data rather than offering a theoretically well-motivated explanation of it, suggests that the social sciences generally, and empirical social psychology in particular, thrive on collecting seemingly interesting effects, rather than attempting to subsume them under theoretical accounts.

As this is unlikely to change in the short-term, immediate measures for future research include verbatim protocols to collect self-reports that reflect participants' reasoning. These may allow insight into the reasoning processes that participants do in fact deploy in experimental contexts, but are currently not a standard instrument. On the assumption that *some* aspects of allegedly intuitive processes can be introspectively accessed, verbalized, and deliberated upon, such protocols could also go some way towards understanding which among the documented errors, fallacies, and systematic mishaps might not be owed to a bias, but could rather be accounted for by unfamiliarity with executing the relevant

---

[11] An anonymous reviewer has suggested that we provide readers with a summarizing chart of the debiasing techniques discussed, the biases (not) mitigated, and their relevance to the methodological considerations raised. We believe, however, that such a chart would risk signaling a *false* state of knowledge, all warnings notwithstanding.

normatively correct reasoning process. At the latest since the work of Sigmund Freud, of course, the ability to accurately self-report the mind's inner working has (rightly) been under doubt, which may partly explain why verbatim protocols are not in fashion.

Finally, there are reasons to pay increased attention to a particular data-pattern that is systematically obtained in these reasoning experiments. Typically around five percent among a given sample appear to deploy the "right" reasoning process; at any rate, they provide the experimenter-intended reasoning outcome. As Stanovich & West (2000) argue, this pattern may reflect individual differences in reasoning insofar as a strong correlation obtains between giving "correct" responses and scoring high on the scholarly aptitude test (SAT). The commentaries to Stanovich & West's article demonstrate that this interpretation is debatable. But this small subgroup of high-scorers nevertheless aligns *much* better to the experimenter-assumed standard. Structured interviews might provide an adequate instrument to ascertain why a minority of the sample regularly differs markedly.

7. Conclusion

This paper has addressed the case for prescriptive ameliorative intervention from a historical perspective, and commented on the rise and popularization of research on human biases and heuristics. As debiasing techniques seek to positively impact personal or procedural features of reasoning and decision-making, we discussed both theoretical and empirical considerations regarding the effectiveness of such techniques in legal contexts.

Our review of pertinent analytical distinctions should have served to better understand why empirical results on the effectiveness of such techniques in legal contexts have remained mixed. Overall, few examples of tested techniques are available; the few tested ones suffer from low sample sizes and further await proper replication attempts. This goes some way towards explaining why research has delivered less than one could have hoped for. More positively, we have suggested that effective debiasing techniques must simultaneously address aspects of cognition, motivation, and technology in ways that bear stronger resemblances to personalized medicine than to the currently typical forms of university level instruction.

Stressing the role of individual differences in reasoning, it also became clear that being personally biased and being potentially debiasable—whatever exactly these terms mean—also requires conducive (institutional) environments. After all, agents who profit from the *status quo* in terms of social status and related economic power may see no valid reasons to debias because, for them, there seemingly is no reason. Future work may therefore consider also the larger socio-political implications of debiasing. In the meantime, the legal context continues to provide a particularly important area for the much wider-reaching question how suboptimal decisions and choice actions can be reliably improved.

Acknowledgements

Previous versions of this paper were presented at Lund University, Sweden, the University of Windsor, Canada, Rochester Institute of Technology, USA, and Copenhagen University, Denmark. The authors would like to thank audience members for useful comments and discussion that helped improve this manuscript, particularly Christian Kock, Clarence Sheffield Junior, Lori Buchanan, and Sune Holm Petersen. We thank the editor, Fabio Paglieri, and an anonymous reviewer for useful comments. The authors gratefully acknowledge funding for the research project "Judges without bias" from the Ragnar Söderberg Foundation.

References

Arlen, J., & Tontrup, S. (2015). Does the endowment effect justify legal intervention? The debiasing effect of institutions. *The Journal of Legal Studies*, 44(1), 143-182.

Anderson, J. R. (1990). *The Adaptive Character of Thought*. Hillsdale, NJ: Erlbaum.

Arkes, H.A. (1998). Principles in Judgment/Decision Making Research Pertinent to Legal Proceedings. *Behavioral Sciences & the Law*, 7(4), 429–456.

Arkes, H.R., Faust, D., Guilmette, T.J., & Hart, K. (1988). Eliminating the hindsight bias. *Journal of Applied Psychology*, 73(2), 305–307.

Bacon, F. (1625). Of Truth. In: Eliot, C.W. (Ed.), *Essays, Civil and Moral*. (Vol. 3, Part 1). New York: P.F. Collier & Son.

Babcock, L., Loewenstein, G., & Issacharoff, S. (1997). Creating convergence: Debiasing biased litigants. *Law & Social Inquiry*, 22(4), 913-925.

Betsch, T., & Held, C. (2012). Rational decision making: balancing RUN and JUMP modes of analysis. *Mind and Society*, 11, 69–80.

Bicchieri, C. (2006). *The Grammar of Society: The nature and dynamics of social norms*. Cambridge, UK: Cambridge University Press.

Büyükkurt, B.K., & Büyükkurt, M.D. (1991). An experimental study on the effectiveness of three debiasing techniques. *Decision Sciences*, 22(1), 60–73.

Cipolla, C.M. (1987). The Basic Laws of Human Stupidity. *Whole Earth Review*, 2–7.

Cohen, J.L. (1981). Can human irrationality be experimentally demonstrated? *Behavioral and Brain Sciences*, 4, 317–370.

Cosmides, L., & Tooby, J. (1996). Are humans good intuitive statisticians after all? Rethinking some conclusions from the literature on judgment under uncertainty. *Cognition*, 58(1), 1–73.

Croskerry, P. (2009a) A universal model of diagnostic reasoning. *Academic Medicine*, 84, 1022–1028.

Croskerry, P. (2009b). Cognitive and affective dispositions to respond. In: Croskerry, P., Cosby, K., Schenkel, S., & Wears, R. (Eds.), *Patient Safety in Emergency Medicine* (pp. 219–227). Philadelphia, PA: Lippincott Williams & Wilkins.

Croskerry, P., Singhal, G., & Mamede, S. (2013). Cognitive debiasing 1: Origins of bias and theory of debiasing. *BMJ Quality & Safety*, 22 (Suppl 2), ii58–ii64.

Croskerry, P., Singhal, G., & Mamede, S. (2013). Cognitive debiasing 2: impediments to and strategies for change. *BMJ Quality & Safety*, 22 (Suppl 2), ii65–ii72.

Dunning, D., Johnson, K., Ehrlinger, J., & Kruger, J. (2003). Why people fail to recognize their own incompetence. *Current Directions in Psychological Science*, 12(3), 83–87.

Dutilh Novaes, C. (2012). *Formal Languages in Logic: A Philosophical and Cognitive Analysis*. Cambridge, U.K.: Cambridge University Press.

English, B., Mussweiler, T., & Strack, F. (2006). Playing dice with criminal sentences. The influence of irrelevant anchors on experts' judicial decision making, *Personality and Social Psychology Bulletin*, 32(2), 188–200.

Ely, J., Graber, M., & Croskerry, P. (2011). Checklists to reduce diagnostic errors. *Academic Medicine*, 86, 307–13.

Evans, J.S.T.B. (2008). Dual-Processing Accounts of Reasoning, Judgment, and Social Cognition. *Annual Review of Psychology*, 59, 255–278.

Farnsworth, W. (2003). The Legal Regulation of Self-Serving Bias. *UC Davis Law Review, 37*, 567–603.
Fischhoff, B. (1982). Debiasing. In: Kahneman, D., Slovic P., & Tversky, A. (Eds.), *Judgment under Uncertainty: Heuristics and Biases* (pp. 422–444). New York: Cambridge University Press.
Fischhoff, B. (2010). Judgment and decision making. *WIREs Cognitive Science, 1*, 724–735.
Fischhoff, B., Slovic, P., & Lichtenstein, S. (1977). Knowing with certainty: The appropriateness of extreme confidence, *Journal of Experimental Psychology: Human Perception and Performance, 3*(4), 552–564.
Fukuyama, F. (2011). *The Origins of Political Order*. New York, NY: Farrar, Straus and Giroux.
Fukuyama, F. (2014). *Political Order and Political Decay*. New York, NY: Farrar, Straus and Giroux.
Gigerenzer G., Todd, P.M., & the ABC Research Group (1999). *Simple Heuristics That Make Us Smart*. New York, NY/Oxford, UK: Oxford University Press.
Gigerenzer, G., & Brighton, H. (2009). Why biased minds make better inferences. *Topics in Cognitive Science, 1*, 107–143.
Gigerenzer, G., & Hoffrage, U. (1995). How to improve Bayesian reasoning without instruction: Frequency formats. *Psychological Review, 102*, 684–704.
Gigerenzer, G., & Sturm, T. (2012). How (far) can rationality be naturalized? *Synthese, 187*, 243–268.
Goodwin, J., & Honeycutt, L. (2009). When science goes public: From technical arguments to appeals to authority. *Studies in Communication Sciences, 9*(2), 125–36.
Guthrie, C., Rachlinski, J.J., & Wistrich, A.J. (2001). Inside the Judicial Mind. *Cornell Law Review, 86*, 777–830.
Hacking, I. (2001). *An Introduction to Probability and Inductive Logic*. Cambridge University Press.
Hahn, U., & Harris A.J.L. (2014). What does it mean to be biased: motivated reasoning and rationality. *Psychology of Learning and Motivation, 61*, 41–102.
Hansen, H.V., & Pinto, R. (Eds.). (1995). *Fallacies—Classical and Contemporary Readings*. University Park, PA: Penn State University Press.
Hamblin, C. (1970). *Fallacies*. London: Methuen.
Henrich, J., Heine, S., & Norenzayan, A. (2010). The Weirdest People in the World? *Behavioral and Brain Sciences, 33*(2–3), 61–135.
Hintikka, J.J., & Spade, P.V. (2012). History of Logic. *Encyclopedia Britannica*. http://www.britannica.com/EBchecked/topic/346217/history–of–logic.
Hodgkinson, G.P., Bown, N.J., Maule, A.J., Glaister, K.W., & Pearman, A.D. (1999). Breaking the frame: an analysis of strategic cognition and decision making under uncertainty. *Strategic Management Journal, 20*(10), 977–985.
Irwin, J. F. & Real, D.L. (2010). Unconscious Influences on Judicial Decision-Making: The Illusion of Objectivity. *McGeorge Law Review, 43*, 1–20.
Ioannidis, J.P.A. (2005a). Contradicted and initially stronger effects in highly cited clinical research. *Journal of the American Medical Association, 294*(2), 218–228.
Ioannidis J.P.A. (2005b). Why Most Published Research Findings Are False. *PLoS Medicine, 2*(8), e124.
Jolls, C., & Sunstein, C.A (2005). Debiasing through law. *NBER Working Paper* 11738, 1–49. http://www.nber.org/papers/w11738.
Jolls, C., & Sunstein, C.A. (2006). The law of implicit bias. *California Law Review, 94*(4), 969–996.
Kahneman, D., & Tversky, A. (1982). On the study of cognitive illusions. *Cognition, 11*, 1123–141.

Kahneman, D. & Tversky, A. (1996). On the reality of cognitive illusions: A reply to Gigerenzer's critique. *Psychological Review, 103*, 582–591.

Kahneman, D. (2011). *Thinking, Fast and Slow*. New York, NY: Farrar, Strauss and Giroux.

Kamin, K. & Rachlinski, J. (1995). Ex Post ≠ Ex Ante: Determining Liability in Hindsight. *Law and Human Behavior, 19*(1), 89–104.

Kenyon, T. (2014). False polarization: debiasing as applied social epistemology. *Synthese, 191*(11), 2529–2547.

Kelly, J.M. (1964). Audi Alteram Partem. *Natural Law Forum, 9*, 103 -110.

Kenyon, T., & Guillaume, B. (2014). Critical Thinking Education and Debiasing. *Informal Logic, 34*(4), 341–363.

Koriat, A., Lichtenstein, S., & Fischhoff, B. (1980). Reasons for confidence. *Journal of Experimental Psychology: Human learning and memory, 6*(2), 107–118.

Kurtz, R.M., Garfield, S.L. (1978). Illusory correlation: A further exploration of Chapman's paradigm. *Journal of Consulting and Clinical Psychology, 46*(5), 1009–1015.

Larrick, R.P. (2008) Debiasing. In: Koehler, D.J., & Harvey, N. (Eds.), *Blackwell Handbook of Judgment and Decision Making* (pp. 316–337). Malden, MA: Blackwell Publishing Ltd.

Langevoort, D. C. (1998). Behavioral Theories of Judgment and Decision Making in Legal Scholarship: A Literature Review. *Vanderbilt Law Review, 51*, 1499–1540.

Lilienfeld, S.O., Ammirati, R., & Landfield, K. (2009). Giving debiasing away: Can psychological research on correcting cognitive errors promote human welfare? *Perspectives on Psychological Sciences, 4*(4), 390–398.

Lopes, L.L. (1987). Procedural debiasing. *Acta Psychologica, 64*, 167–185

Lopes, L.L. (1991). The rhetoric of irrationality. *Theory & Psychology, 1*(1), 65–82.

Lord, C.G., Lepper, M.R., & Preston, E. (1984). Considering the opposite: A corrective strategy for social judgment. *Journal of Personality and Social Judgment, 47*(6), 1231–1243.

Macdougall, R. (1906). On secondary bias in objective judgments. *Psychological Review, 13*, 97–120.

Marr, D., & Poggio, T. (1976). From understanding computation to understanding neural circuitry. *Neurosciences Research Program Bulletin, 15*, 470-488.

Mitchell, G. (2002). Why law and economics' perfect rationality should not be traded for behavioral law and economics' equal incompetence. *Georgetown Law Journal, 91*, 67–167.

Mumma, G.H., & Wilson, S.B. (1995). Procedural debiasing of primacy/anchoring effects in clinical-like judgments. *Journal of Clinical Psychology, 51*(6), 841–853.

Mussweiler, T., Strack, F., & Pfeiffer, T. (2000). Overcoming the inevitable anchoring effect: considering the opposite compensates for selective accessibility. *Personality and Social Psychology Bulletin, 26*(9), 1142–1150

Osman, M. (2004). An evaluation of dual-process theories of reasoning. *Psychonomic Bulletin & Review, 11*(6), 988–1010.

Pashler, H., & Wagenmakers, E.-J. (2012). Editors' introduction to the special section on replicability in psychological science: A crisis of confidence? *Perspectives on Psychological Science, 7*, 528–530.

Peter, L.J, & Hull, R. (1969). *The Peter Principle: Why Things Always Go Wrong*. New York: William Morrow and Company.

Peterson, C.R., & Beach, L.R. (1967). Man as an intuitive statistician. *Psychological Bulletin, 68*, 29–46.

Petersen, G., & Zenker, F. (2015). *Lies, damn lies, and models*. Submitted manuscript available from the author.

Pi, D., Parisi, F., & Luppi, B. (2013). Biasing, Debiasing, and the Law. *Minnesota Legal Studies Research Paper*, 13-08.

Platt, J.R. (1964). Strong inference. *Science, 164*, 347–353.

Project Implicit (2015). https://implicit.harvard.edu/implicit/index.jsp

Pronin, E., Lin, D., & Ross, L. (2002). The bias blind spot: Perceptions of bias in self versus others. *Personality and Social Psychology Bulletin, 28*, 369–381.

Pronin, E., & Kugler, M. (2007). Valuing thoughts, ignoring behavior: The introspection illusion as a source of the bias blind spot. *Journal of Experimental Social Psychology, 434*, 565–578.

Pundik, A. (2012). Was it wrong to use statistics in R v Clark? A case study of the use of statistical evidence in criminal courts. In: Zenker, F. (Ed.), *Bayesian Argumentation* (pp. 87–109). Dordrecht: Springer.

Rehg, W. (2009). *Cogent Science in Context: The Science Wars, Argumentation Theory, and Habermas*. Cambridge, MA: MIT Press.

Samuels, R., Stich, S., & Bishop, M. (2002). Ending the rationality wars: How to make disputes about human rationality disappear. In: Elio, R. (Ed.), *Common Sense, Reasoning and Rationality* (pp. 236–268). New York: Oxford University Press.

Sandri, D. (2009). *Reflexivity in Economics: An Experimental Examination on the Self-Referentiality of Economic Theories*. Dordrecht: Springer.

Seamone, E.R. (2006). Understanding the person beneath the robe: Practical Methods for neutralizing harmful judicial biases. *Willamette Law Review, 42*(1), 1–76.

Simon, H.A. (1982). *Models of Bounded Rationality*. Cambridge, MA: MIT Press.

Spellman, B.A. (2012). Introduction to the Special Section on Research Practices. *Perspectives on Psychological Science, 7*, 655–656.

Stanovich, K.E. (1999). *Who is rational? Studies of individual differences in reasoning*. Hillsdale, NJ: Erlbaum.

Stanovich, K.E., & West, R.F. (2000). Individual differences in reasoning: Implications for the rationality debate? *Behavioral and Brain Sciences, 23*, 645–726.

Stenning, K., & Lambalgen, M. van (2008). *Human Reasoning and Cognitive Science*. Cambridge, MA: The MIT Press.

Sturm, T., & Mülberger, A. (2012). Crisis discussions in psychology – New historical and philosophical perspectives. *Studies in History and Philosophy of Biological Sciences, 43*, 425–433.

Thaler, R.H., & Sunstein, C.R. (2008). *Nudge: Improving Decisions about Health, Wealth, and Happiness*. Yale University Press.

Tversky, A., & Kahneman, D. (1974). Judgment under uncertainty: Heuristics and biases. *Science, 185*, 1124–1131.

Wason, P.C. (1960). On the failure to eliminate hypotheses in a conceptual task. *Quarterly Journal of Experimental Psychology, 12*, 129–140.

Wason, P. (1968). Reasoning about a rule. *Quarterly Journal of Experimental Psychology, 20*, 273–281.

Weingart, P. (1999). Scientific expertise and political accountability: paradoxes of science in politics. *Science and Public Policy, 26*(3), 151–161.

Weinstein, I. (2002). Don't believe everything you think: Cognitive bias in legal decision making. *Clinical Law Review, 9*, 783–834.

Wiggins, E.C., & Breckler, S.J. (1990). Special verdicts as guides to jury decision making. *Law & Psychology Review, 14*, 1–36.

Willingham, D. (2007). Critical thinking: Why is it so hard to teach? *American Educator, 31*(2), 8–19.

Wilson T.D., & Brekke, N. (1994). Mental contamination and mental correction: unwanted influences on judgments and evaluations. *Psychological Bulletin, 116*, 117–142.

Wissler, R.L., & Saks, M.J. (1985). On the inefficacy of limiting instructions: When jurors use prior conviction evidence to decide on guilt. *Law and Human Behavior, 9*(1), 37–48.

Witte, E.H., & Zenker, F. (2015). *From discovery to justification: outline of an ideal research program in empirical psychology*. Unpublished manuscript available from the author.

Wood, G. (1978). The knew-it-all-along-effect. *Journal of Experimental Psychology: Human Perception and Performance, 4*, 345–353.

Woods, J. (2014). *Reasoning Errors: Naturalizing the Logic of Inference* (2nd ed.). London: College Publications.

Zenker, F., Dahlman, C., Bååth, R., & Sarvar, F. (2015). Giving Reasons *Pro et Contra* as a Debiasing Technique in Legal Decision Making. In Mohammed, D., & Lewinski, M. (Eds.), *Proceedings of the First European Conference on Argumentation, Lisbon, June 2015*. London: College Publications.

Chapter 11

The Biased Use of Argument Evaluation Criteria in Motivated Reasoning: Does Argument Quality Depend on the Evaluators' Standpoint?*

Hans Hoeken[1] & Mariecke van Vugt

Utrecht Institute for Linguistics OTS, Utrecht University, Utrecht, The Netherlands, [1] j.a.l.hoeken@uu.nl

Abstract. People without a background in argumentation theory possess several criteria to distinguish strong from weak arguments. The fact that people have these criteria does not imply that they will use them to objectively assess the quality of an argument. Research on motivated reasoning suggests that people take a more critical stance toward arguments that go against their opinions compared to arguments that are in accordance with these opinions. In this study, the question was addressed whether people employ criteria to evaluate arguments in a biased way. Forty participants were told that they would take part in a debate and either had to defend the claim that mixed schools (that is, schools attended by children with different ethnic backgrounds) were desirable or the claim that these were undesirable. All participants received sixteen (strong and weak) arguments and were asked to prepare themselves for the debate while thinking aloud. Analysis of the think aloud protocols showed that people almost exclusively used criteria to boost the quality of arguments supporting their claim while disqualifying arguments that went against it. These results provide important insights into the nature of motivated reasoning because they show how people deploy argument criteria in this process.

1. Introduction

Mercier and Sperber (2011) claim that there is a big difference between situations in which people have to evaluate the acceptability of a claim and its supporting arguments and situations in which they themselves have to defend a certain claim or already have an opinion on the issue at hand. In the former situation, people are hypothesized to assess the quality of the arguments in a relatively objective way whereas in the latter situation, they are predicted to evaluate the arguments in a more biased manner. Biased evaluation implies that arguments in accordance with their opinion or preference are perceived as stronger whereas counterarguments are considered weaker, regardless of the arguments' objective

---

*A version of this paper has been published in Dutch: Hoeken, H., & Vugt, M. van (2014). Het bevooroordeelde gebruik van argumentatieschema-specifieke criteria. *Tijdschrift voor Taalbeheersing, 36*, 87-105. ISSN: 1573-9775, DOI: 10.5117/TVT2014.1.HOEK. The study reported here was conducted by the second author as part of her master thesis for the master Communication & Persuasion at the Radboud University, the Netherlands. The first author served as her supervisor.

characteristics. Mercier and Sperber explain this asymmetry by stating that this leads to an optimal division of labor: Supporters of a certain claim have to develop arguments in favor of their claim, opponents have to come up with counterarguments, and the ultimate judge of whose arguments are the most convincing ones is the audience. If the audience members do not already have an opinion on the issue, they are well-equipped to assess and weigh the reasonableness of the arguments provided by the debaters.

Schellens and De Jong (2004) state that for people to evaluate argument quality, they need to possess argument scheme specific criteria. That is, to assess the quality of an argument from authority, other criteria are relevant than to assess the quality of an argument from analogy or an argument from example. Recent studies have shown that lay people without special training in the field of argumentation theory possess and apply a number of such criteria when distinguishing strong from weak arguments. These findings appear to support Mercier and Sperber's (2011) claim that people are relatively well versed in argument evaluation. In this paper, we focus on the second part of their prediction, namely that people will interact with arguments in a biased way when they have a preference for a claim. In the next paragraphs, we will discuss studies on the extent to which people are capable of evaluating arguments as well as the research on motivated reasoning. The latter research shows how people go about when evaluating arguments that go against their own opinions.

1.1 Evaluating Arguments Employing Argument Scheme Specific Criteria

According to dual-process models of the persuasion process, such as the Elaboration Likelihood Model (ELM: Petty & Cacioppo, 1986) or the Heuristic-Systematic Model (HSM: Chaiken, 1987), people can form (or change) their attitudes as a result of a careful evaluation of arguments. If people arrive at an attitude in such a way, this attitude is believed to be more stable, less susceptible to counter persuasion attempts, and a better predictor of attitude related behavior (Petty, Haugtvedt, & Smith, 1995). These models state that people need to be both motivated to and capable of carefully evaluating arguments in order for this to occur. Scrutinizing arguments takes time and effort and people will only be willing to invest these if they find it important to have a correct attitude. People will be more inclined to carefully evaluate the arguments if accepting the claim has far reaching consequences (e.g., when buying a house), compared to decisions with only minor implications (e.g., when buying toothpaste). In general: the more people consider an issue as relevant to their own and/or their loved ones' well-being, the more they will be inclined to set out for a careful evaluation of the arguments.

Being motivated is not sufficient in itself; people also need to be capable of evaluating arguments. Schellens and De Jong (2004) have shown what capacities this requires. They analyzed twenty Dutch public information brochures for the general public in which certain behaviors were either discouraged (e.g., smoking, gambling) or encouraged (e.g., using sun tan lotion, exercising). They have shown that for a careful evaluation of the arguments, people need to take three steps. First, in the majority of the analyzed brochures, there are no explicit arguments. Arguments are presented implicitly in the guise of factual information. For instance, in a brochure on exercising, positive consequences of exercising are presented without explicitly stating that these consequences are arguments in support of the claim that exercising is good. As a result, people should be able to identify which information serves as an (implicit) argument. Second, Schellens and De Jong (2004) show that different types of argument are used in the brochures, such as arguments from authority, arguments from example, and arguments from analogy. For a critical evaluation of such arguments, argument scheme specific criteria are needed. For instance, when scrutinizing an argument from analogy, the similarity between the two cases that are being compared is important, whereas the number and representativeness of examples is at stake when evaluating an argument from example. A careful evaluation of arguments therefore requires two additional skills: identification of the argument scheme and the possession of and ability to apply scheme specific criteria.

People appear to be able to distinguish strong from weak arguments. In many experiments, argument quality has been manipulated and the attitude after reading the message has been measured. Meta-analyses have revealed that, in general, people are more convinced by messages containing strong arguments than by messages containing weak arguments (Carpenter, 2015; Johnson, Smith-McLallen, Killeya, & Levin, 2004; Park, Levine, Kingsley Westerman, Orfgen & Foregger, 2007). Both the meta-analyses and the individual studies reported on in these meta-analyses, do not shed light on the criteria people use to distinguish strong from weak arguments. The manipulation of argument quality is usually based on the intuitions of individual researchers rather than on theory guided normative criteria. As a result, strong arguments differ on many dimensions from their weak counterparts. For instance, Van Dijk-Van Enschot, Hustinx, and Hoeken (2003) have shown that Petty and Cacioppo's (1986) manipulation of argument quality has led to their strong arguments differing on various dimensions compared to their weak arguments. Different consequences are referred to in the strong arguments compared to the weak ones, and the evidence in support of the likelihood that the consequences will occur, does also differ. As a result, it is not possible to link differences in claim acceptance or attitudes to specific characteristics of these arguments. Therefore, it is not possible to identify the criteria participants in these studies have employed to distinguish strong from weak arguments.

There are two lines of research that provide insights into the extent to which people are capable of applying argument specific criteria. Van Eemeren, Garssen, and Meuffels (2009) report on a series of experiments in which they showed that arguments that do not meet the standards of reasonableness were perceived as less reasonable by ordinary language users. For instance, in one of their studies, they had participants rate the reasonableness of an argumentative move within a conversational context. One of the conversational partners attacked the credibility of the other person. In one condition, this could be considered a reasonable move whereas in other conditions, it could be considered as an instance of an 'ad hominem' fallacy. In the former condition, participants rated the move as more reasonable compared to the reasonableness of the moves in the latter conditions. This research shows that people can distinguish a fallacious move from a non-fallacious one.

In a different line of research, it is studied to what extent people are more convinced by arguments that meet specific argument scheme criteria than by those that do so to a lesser extent. For instance, in case of an argument from example, the number of examples is relevant. Several studies have shown that a claim is accepted more strongly when supported by a statistical summary of a number of cases compared to a single case (Hoeken & Hustinx, 2009; Hornikx & Hoeken, 2007). Even adding one additional example led to a stronger claim acceptance (Hoeken, Šorm, & Schellens, 2014). In the latter study, it was also shown that the representativeness of the example in the argument, had a positive impact on claim acceptance. These studies thus show that people are sensitive to two normative criteria for the quality of an argument from example: number and representativeness.

A similar sensitivity to normative criteria is found for the argument from analogy. The most important criterion for this type of argument is the extent to which the two cases that are being compared can be considered similar. Both Hoeken and Hustinx (2009) and Hoeken, Timmers, & Schellens (2012) have shown that the more similar the two cases are, the more people are willing to accept the claim supported by the argument from analogy.

Sensitivity to argument scheme relevant criteria also has been documented for the argument from authority (Hoeken et al., 2012, 2014; Hornikx & Hoeken, 2007), the argument from cause to effect (Hoeken et al., 2014), and the argument from consequences (Hoeken et al., 2012). For all of these argument schemes it was found that people accepted the claim more strongly, if the argument met relevant, normative criteria.

1.2 Biased Evaluation of Arguments

The fact that people possess relevant criteria for argument evaluation does not guarantee that they will use those to objectively assess the merits of the argument at hand. The concept of motivated reasoning (Kunda, 1990; Molden & Higgins, 2005; Westen, Blagov, Harenski, & Hamann, 2006) refers to the biased evaluation of arguments: arguments that are in line with one's opinion are evaluated less critically than those that go against it. It is an important part of the so-called confirmation bias phenomenon (see, for a review, Nickerson, 1998). This broader phenomenon not only includes the biased evaluation of arguments but also the selective seeking of information that corroborates one's opinion.

Mercier and Sperber (2011, p.76) state that "people are good at assessing arguments and are quite able to do so in an unbiased way" but also attach a precondition for such an unbiased assessment: "provided they have no particular axe to grind". Mercier and Landemore (2012, p. 251) claim that "the confirmation bias mostly affects the production, and not the evaluation of arguments". The production of arguments is guided by a persuasive goal, that is, to have another person accept one's standpoint (Mercier, 2012, p. 317). As a result, the production is geared to finding arguments that will support the individual's standpoint (or rebut the opponent's arguments). Argument evaluation, on the other hand, "aims at distinguishing good arguments from bad ones, and hence genuine information from misinformation" (Mercier & Sperber, 2011, p. 72). However, if people do have an axe to grind, that is, they already have an opinion on a topic, the evaluation of arguments may display motivated reasoning as well.

Lord, Ross, and Lepper (1979) were one of the first to provide empirical evidence for the existence of motivated reasoning. They had participants read two studies on the deterrence effect of capital punishment. In one study, the results appeared to provide support for this effect whereas the other study's results failed to do so. Some of their participants were in favor of capital punishment whereas others were against it. Those in favor of capital punishment were much more critical about the quality of the study that did not find positive effects of capital punishment on crime rates compared to the study reporting such an effect. Exactly the opposite pattern of results was obtained for the opponents of capital punishment. In addition, despite the fact that both groups were presented exactly the same information, the supporters felt more certain about the positive effect of capital punishment and the opponents about the absence of such an effect. Several other studies replicated these findings (e.g., Ditto & Lopez, 1992; Edwards & Smith, 1996).

In the studies described above, participants were already in favor of, or against, capital punishment before they entered the experiment. As a result of this difference in attitude, supporters and opponents probably also differed with respect to their knowledge on the issue. This prediction follows from the existence of a confirmation bias, which would lead those in favor of capital punishment to seek out and remember information that is in line with their opinion on the capital punishment's deterrence effect whereas the opposite will hold for the opponents of capital punishment. One of the strategies that people apply when evaluating arguments is coherence checking. That is, people activate their previously held beliefs and check to what extent the new information is accordance with these beliefs (Mercier & Sperber, 2011, p. 60). If incoherencies come to the fore, people have to choose between revising their previous beliefs or rejecting the new information. If an opponent of capital punishment reads a study reporting deterrence effects of the death penalty, incoherencies are evoked and he or she is well-equipped to bring up counterarguments against the assumptions and interpretation of this study; the same holds for supporters of capital punishment who can use their reservoir of facts and beliefs to tackle the study reporting the absence of a deterrence effect. In sum, the difference in knowledge accompanying the difference in attitude may have made it more easy for both supporters and opponents of the death penalty to criticize the studies reporting results that go against their views.

Jain and Maheswaran (2000) conducted two experiments in which they manipulated instead of observed the participants' standpoint. That is, in the first stage of the experiment, participants received either positive or negative information about a telephone answering machine. In this way, prior attitude was manipulated instead of observed. Because the experiments were conducted immediately afterwards, the participants could not seek out and store information relevant to their preferences. As a result, they differed in prior attitude, but the differences with respect to prior knowledge were limited. In both experiments, it was found that participants generated more counterarguments when confronted with information that went against their initial attitude compared to when new information was in line with that attitude. So, even if people's preference is manipulated instead of observed, it still leads to a biased evaluation of subsequent arguments that go against that preference. Still, the participants differed in their knowledge about the product which may have enabled them find more flaws in the arguments that went against their attitude.

The results reported by Jain and Maheswaran (2000) do not enable an assessment of the extent to which people use argument scheme specific criteria to critically evaluate arguments. Klaczynski, Gordon, and Fauth (1997) provided some insight into this issue. They studied to what extent participants employed normative criteria when evaluating arguments that went against their interests. Students received arguments in support of the claim that the teaching program they were in either provided excellent or very meager prospects at the job market. The arguments were not very strong. For instance, claims about the job prospects of the program as a whole were backed up by the case of a single alumnus who was either very successful or very unsuccessful in getting a job. Participants were asked to rate the arguments and to provide an explanation of their ratings. The results showed that arguments were rated as stronger when they were in support of positive prospects on the job market whereas they were rated as weaker when they were in support of negative prospects. Analyses of the explanations revealed that participants employed argument scheme specific criteria when evaluating these arguments. For instance, they referred to the danger of hasty generalization when the meager job prospects claim was supported by the experiences of a single alumnus. In four experiments, Klaczynski *et al.* showed that people thought longer, deeper and more critical when confronted with arguments that went against their vested interest compared to when the arguments were in favor of their interests. Apparently, people use normative criteria to refute displeasing arguments.

Nienhuis, Manstead, and Spears (2001) showed that under certain conditions people who have to defend a claim, do evaluate arguments in an objective manner. In three experiments, they had participants read a text in which a certain claim was defended, for instance, "hard drugs should be legalized". There were two versions of the text: one containing strong and one containing weak arguments. Some participants were told that in a subsequent part of the study, they would have to convince another participant of the claim's acceptability and that they themselves would be evaluated with respect to their task performance. After they had read the text, participants were asked to write down all thoughts they had while reading the text. Participants who had read the text containing strong arguments generated more thoughts that were favorable with respect to the issue than those who had read the text containing weak arguments. The favorableness of the thoughts listed by participants who had not received this instruction did not differ as a result of the quality of the arguments in the text. In addition, the attitude toward the claim of the former group was more positive after reading the version with the strong arguments. An important difference between the Nienhuis et al. study and the one by Klaczinsky et al. (1997) was that in the former study participants did not have a strong opinion on the issue whereas in the latter study they immediately had one (i.e., it was about their chances on the job market). This raises the question to what extent motivated reasoning is mainly evoked when people have a vested interest in an issue.

1.3 Research Questions

Several studies have shown that people possess and use normative criteria to distinguish strong from weak arguments (see, e.g., van Eemeren et al., 2009; Hoeken & Hustinx, 2009; Hoeken et al., 2012, 2014; Hornikx & Hoeken, 2007). In addition, research has shown that people respond more critically toward arguments that go against their existing opinions or vested interests (Ditto & Lopez, 1992; Edwards & Smith, 1996; Lord et al., 1979). Klaczinsky et al. (1997) provided evidence that people use normative criteria when evaluating arguments in a biased fashion. In these studies, the participants already held these opinions or had a vested interest when they entered the experiments. That is, they were already in favor (or against) the death penalty and they had an interest of being in a program that would give them excellent opportunities on the job market. As a result of these existing preferences, they may have also differed in their knowledge about these issues which may have made them better equipped to spot the weaknesses in (strong) arguments that ran counter their opinion and repair the flaws in weak arguments in favor of their position. Jain and Maheswaran (2001) manipulated instead of observed the participants' opinion. However, their manipulation consisted of providing participants with different information which may also have enabled participants to spot weaknesses or incompatibilities between the first batch of information and the second. In addition, it is unclear whether the participants in the Jain and Maheswaran study employed normative criteria when evaluating the arguments in a biased way. Finally, Nienhuis et al. (2001) have shown that people who have to defend a claim are not necessarily blind to differences in argument quality. In their study, however, participants were only exposed to arguments in support of the claim they had to defend.

In this paper, a study is reported in which we had participants evaluate arguments ostensibly as a preparation for a discussion in which they had to defend a certain claim. This instruction resembles the one used by Nienhuis et al. (2001) but in this study we provided participants with normatively weak and strong arguments in favor of the claim as well as weak and strong arguments against the claim the participant had to defend. In addition, we asked participants to verbalize their thoughts while preparing for the debate in order to assess whether or not they employed argument specific criteria when evaluating the arguments. This set up enabled us to address the question to what extent people respond more critically towards arguments that go against the claim they have to defend compared to arguments that are in line with that claim. This general question can be subdivided in a number of more specific questions. First, it is interesting to assess how people evaluate arguments in such a context. It is unclear whether they employ specific criteria in this process. Therefore, the first question is:

1. Do people use argument scheme specific criteria while evaluating arguments in preparation of a debate?

Research on motivated reasoning has shown that people pay more attention to arguments that go against their opinion compared to arguments that are in line with it (see, e.g., Klaczynski *et al.*, 1997). It is unclear whether more attention implies a more frequent use of criteria. This raises the second question:

2. To what extent do people deploy argument scheme specific criteria more often when evaluating arguments that go against the opinion they are instructed to defend compared to arguments that are in line with it?

Nienhuis *et al.* (2001) showed that people distinguish strong from weak arguments when they expect to have to convince someone else of an opinion; research on motivated reasoning claims that people are more likely to seek weaknesses in arguments against their opinion compared to arguments that are in line with it. This discrepancy leads to the third question:

3. To what extent do people deploy argument scheme specific criteria to distinguish strong from weak arguments in an objective way or to what extent they use them in a biased way, that is, disqualify arguments that go against the opinion they have to defend and boost the quality of arguments that are in line with that opinion?

With this study, we hope to shed more light into the exact nature of motivated reasoning and the conditions under which it is likely to occur.

2. Method

2.1 Materials

Sixteen arguments were developed relevant to the issue of "mixed schools". This issue pertains to the question as to whether it is desirable or undesirable that pupils in primary schools come from (many) different ethnic backgrounds. Eight arguments supported the claim that mixed schools are desirable while the other eight were in support of the claim that they are undesirable. For each of these sets of eight arguments, half did meet argument scheme specific criteria, which made them strong from a normative point of view, whereas the other half did not meet these criteria, which made them weak from a normative point of view. In summary, there were four strong and four weak arguments in favor of mixed schools, as well as four strong and four weak arguments against mixed schools.

Four different argument schemes were employed: argument from analogy, argument from authority, argument from example, and argument from cause to effect. Argument quality was manipulated by having arguments meet to a stronger or lesser extent criteria for which it had been established in previous research that people are sensitive to. For the argument from analogy, cases were selected that either were more similar to the Dutch situation at hand (e.g., "In Sweden, the introduction of mixed schools has led to better results") or less similar (e.g., "In France, mixing people with different ethnic backgrounds on the work floor has led to better results."). For the argument from authority, sources were selected that had no vested interest in the issue (e.g., Peter den Boer, renowned child pedagogic) or ones who had such an interest (e.g., Annieck ten Haven, director of a mixed school). For the argument from examples, the representativeness of the example was manipulated (e.g., "Lisa de Bruin performs better since she moved to a mixed school" vs. "Metab Rakkech performs better since she moved from an all black school to a mixed school"). Finally, for the argument from cause to effect, the plausibility of the causal relation was manipulated. A more plausible relation would be "Mixing will lead to a reduction of social and cultural differences" compared to "Mixing will lead the management of the school to pay attention to the educational policy". The quality of arguments in favor of mixed schools as well as those against mixed schools were manipulated in this way.

2.2 Research Design and Procedure

Participants were approached by the second author to ask whether they were willing to take part in a study. If the participant agreed, the experimenter told the participant that the study was about the way in which people prepare for a debate and how they debate. Next, they received the sixteen arguments about the issue of mixed schools. Participants were randomly assigned to either the condition in which they were told that they had to defend the claim that mixed schools are desirable or the condition in which they had to defend the opposite claim. The experimenter explained that in the first stage of the study, the participant would have to prepare for the debate by evaluating the provided arguments while thinking aloud. This part of the session would be recorded. After transcribing the recording, the data would be anonymized. Upon completing this part, the participants were told the aim of the study and it was explained to them that there would be no debate. Any remaining questions were answered. A session lasted 30 minutes on average.

2.3 Participants

Forty participants took part in the study. Age varied from 21 to 60 with a mean of 35 years. Slightly more women (22) than men (18) participated. Level of education varied from vocational studies (13), applied university (19), to a completed master's degree (8). There were no significant differences between the two conditions with respect to the participants' age, gender, and level of education (p's > .52). At the end of a session, participants were asked whether they already had an opinion on the issue of mixed schools. The majority (28) held a neutral stance towards the issue, five held a strong favorable attitude, six held a slightly favorable attitude on the issue; only one participant held a (slightly) negative attitude towards mixed schools. Participants holding a prior attitude were almost equally distributed over the conditions.

2.4 Data Processing

The think aloud protocols were analyzed from the perspective of whether participants used argument scheme specific criteria and if they did, whether that led to a positive or a negative qualification of the argument's quality. For each participant, it was established how often he or she applied a criterion to qualify an argument as strong and how often application led to qualify an argument as weak. For each of the four sets of (four) arguments (i.e., in favor & strong, in favor & weak, against & strong, against & weak), the number of criterion based evaluations by a participant were computed. The resulting scores were analyzed using a 2 (Instruction: defend desirability, defend undesirability) x 2 (arguments in favor of claim to be defended, arguments against) x (strong arguments, weak arguments) x 2 (criterion application leads to positive qualification or leads to negative qualification) Analysis of Variance was conducted with the final three factors being within-participants factors.

3. Results

The first question was whether participants would use argument scheme specific criteria when evaluating the arguments. This proved to be the case for each and every participant. With a minimum of 4 and a maximum of 13, the average number of criteria used by a participant was 6.58 (SD = 1.92). First, the results of the quantitative analyses of the data will be presented, followed by presenting quotes from the think aloud protocols which reveal how participants employed these criteria. Table 1 contains the mean scores (and standard deviations) for the application of criteria for the different conditions.

	Supportive of claim		Opposing of claim	
	Strong	Weak	Strong	Weak
Positive Evaluation				
In favor	1.30 (0.73)	1.35 (0.93)	0.05 (0.22)	0.00 (0.00)
Against	1.50 (0.76)	0.80 (0.70)	0.00 (0.00)	0.00 (0.00)
Negative Evaluation				
In favor	0.00 (0.00)	0.00 (0.00)	1.90 (0.72)	2.15 (1.14)
Against	0.00 (0.00)	0.00 (0.00)	2.20 (0.77)	1.90 (0.91)

Table 1. The means and standard deviations for the number of times criteria for the evaluation of arguments were applied as a function of the claim the participant had to defend (in favor, against), argument relation with claim (supporting claim to be defended, opposing claim to be defended), argument quality (strong, weak), and evaluation of argument (positive, negative) (Minimum = 0, Maximum = 4).

Research question 2 was about whether people would employ argument scheme specific criteria to a greater extent when evaluating arguments that went against the claim they had to defend compared to arguments that were in line with that claim. This proved to be the case (F (1, 38) = 40.89, p < .001, η^2 = .52). When the arguments went against the claim to be defended, participants applied normative criteria more often (M = 1.03, SE = .050) compared to evaluating arguments that supported this claim (M = 0.62, SE = .049). Participants were also more likely to deploy criteria to evaluate an argument as weak (M = 1.01, SE = .051) than to evaluate an argument as strong (M = 0.63, SE = .050; F (1, 38) = 36.34, p < .001, η^2 = .49). The nature of the claim (mixed schools are desirable vs. mixed schools are undesirable) had no effect on the number of criteria they applied (F (1, 38) < 1), nor was there a main effect for argument quality (F (1, 38) = 1.79, p = .19).

Research question 3 was about whether participants would use normative criteria to assess argument quality in an objective or in a biased way. If they had used them in an objective way, participants would use normative criteria mainly to classify strong arguments as strong and weak arguments as weak. This would result in an interaction between argument quality and evaluative judgment. However, this interaction was not significant (F (1, 38) = 1.79, p = .19). If, on the other hand, participants used normative criteria in a biased way, they would use them to qualify arguments that went against the claim they had to defend as weak, and the arguments that were in line with that claim as strong. In that case, the interaction between relation to claim (in favor, against) and evaluative judgment should become significant. This interaction was indeed highly significant ($F(1, 38) = 445.24$, $p < .001$, $\eta^2 = .92$) and explained almost all of the variance in the data. Without exception, criteria were deployed to qualify arguments in support of the claim to be defended as strong ($M = 1.24$, $SE = .098$), and – with one exception - to qualify arguments as weak that went against the claim to be defended ($M = 2.04$, $SE = .10$).

Two other effects were significant as well. First, participants who had to defend the claim that mixed schools are desirable deployed normative criteria in equal numbers for strong and weak arguments whereas those who defended the claim that mixed schools are undesirable were more likely to use these criteria to evaluate strong arguments. This resulted in a significant interaction between claim (mixed schools are desirable, undesirable) and argument quality ($F(1, 38) = 4.96$, $p = .03$, $\eta^2 = .12$). Given that application of criteria could lead to both a positive or a negative evaluation of the argument, this effect is not relevant for the research questions. Finally, the four way interaction also proved significant ($F(1, 38) = 5.47$, $p = .03$, $\eta^2 = .13$). Three-way analyses of variance were conducted separately for people who had to defend the claim that mixed schools are desirable and people who had to defend the opposite claim. For the former group, no three-way interaction arose ($F(1, 19) < 1$) whereas it did for the latter group $F(1, 19) = 5.76$, $p = .03$, $\eta^2 = .23$). Given the small sample size, this interaction should be interpreted with caution, but it appears that the opponents deployed criteria more often to evaluate strong supporting arguments as strong than to evaluate weak supporting arguments as strong whereas they used criteria equally often to evaluate contradicting arguments as weak, regardless of argument quality.

The think aloud protocols provided more detailed information on how participants used criteria to evaluate arguments. A good example of how participants went about in tailoring the arguments to their purpose, is provided by the different evaluations of the same argument from analogy. One of the arguments in support of the desirability of mixed schools was that Sweden had favorable experiences with mixed schools whereas one of the counterarguments was that Russia had unfavorable experiences with mixed schools. Given that most people regard Sweden as more similar to the Netherlands than Russia, the first argument should be considered stronger than the second. Participants who had to defend the claim that mixed schools are desirable made remarks such as "If you look at how things work in other countries, then that will work for the Netherlands as well. Sweden makes a good comparison for the Netherlands" and "Sweden is in many respects comparable to the Netherlands, so it is a good comparison". When evaluating the Russia argument, participants made remarks such as "The organization of the educational system in Russia is so saddening, that whatever their experience is, it says nothing about the Netherlands". So in both cases, the criterion of "sufficient similarity" is used to evaluate these arguments. Participants who had to defend the opposite claim, also applied the same criterion but reached opposite evaluations. For the Sweden argument, they pointed to dissimilarities: "Sweden is not a good example because there are far fewer people with a different ethnic background, therefore the mix will also be less extreme" and "Sweden is very different from the Netherlands. Just because mixed schools are effective in Sweden does not imply that it will work in the Netherlands as well". They also used this criterion to evaluate the Russia argument as strong: "Russia is a country with many ethnic differences and therefore a perfect gauge for the Netherlands" and "It has been proven that it doesn't work in Russia so why would it work in the Netherlands? Russia has, similar to the Netherlands, many different cultures".

A similar strategic employment of criteria was also observed for the other argument schemes. For instance, for the argument from authority, proponents reacted very differently compared to opponents when evaluating the argument that a renowned child pedagogic held a positive opinion on mixed schools. Whereas proponents were positive ("The fact that a child pedagogic says something about it, that it's good, that proves something. It's not a lay man who says so."), opponents questioned the source's authority ("On what grounds does he claim that? He can be a child pedagogic, but that doesn't make him automatically an expert on the issue of mixed schools"). When a director of a mixed school expressed her opinion on the desirability of mixed schools, opponents pointed out that she had a vested interest in expressing that claim: "Director of a mixed school, so she's definitely biased. She will never say that a mixed school is bad. You never say about your own school that it performs worse." Opponents, on the other hand, qualified this argument as strong, not through focusing on the reliability criterion but by focusing on the criterion of relevant expertise: "A director has knowledge about the results and thus knows a lot about the effects of a mixed school. She can provide a well-founded statement about mixed schools."

4. Conclusion and Discussion

Without exception, all participants in this study used argument scheme specific criteria to evaluate arguments. Previous research has shown that people spend more time evaluating arguments that went against their initial opinion than arguments that were in line with it (Klaczynksi *et al.*, 1997). The results of our study replicated that finding and in addition showed that this additional attention also implied the more frequent application of normative criteria. Finally, the results clearly provided a pattern consistent with the concept of motivated reasoning: rather than using normative criteria to distinguish strong from weak arguments, participants used them to disqualify arguments that went against the claim they had to defend and, to a lesser extent, glorify the arguments that were in line with this claim. The results of the think aloud protocols showed how people who had to defend different claims used the exact same criterion on the exact same argument to reach opposite conclusions about the argument's quality.

This study provides an importation addition to our understanding of motivated reasoning for four reasons. First, we manipulated the claim the participants had to defend instead of using already established opinions. As a result, differences in prior knowledge could not serve as an alternative explanation for the results. Second, we provided participants with both supporting and counterarguments. Because participants differed systematically with respect to the claim they had to defend, the exact same argument was a supporting argument in one condition while being a counterargument in the other condition. As a result of this design, we could establish for each argument how it was treated when considered as a supporting argument and when considered as a counterargument. Third, the quality of the arguments was manipulated systematically employing criteria for which it had been established in previous research that lay people could use them. This enabled a much more controlled sample of arguments for participants to react upon. Finally, by having participants express their thoughts verbally while evaluating the arguments, we were able to assess online the extent to which participants used normative criteria and identify which specific criteria they used.

As cited in the introduction, Mercier and Landsmore (2012) predict that the confirmation bias is more likely to occur in the production of arguments than in their evaluation. In our study, participants were evaluating arguments instead of producing them, and still their evaluation pattern revealed a strong confirmation bias. However, this finding does not go against the predictions of Mercier and Sperber's (2011) argumentative theory of reasoning. Mercier (2012) describes how people may go about when finding arguments to defend their position in a discussion. From his description it becomes clear that there is a lot of evaluation going on when arguments are produced: when looking for arguments to use in a discussion, potential arguments are evaluated in a serial manner. This evaluation process is geared by a strong confirmation bias in the sense that "positive consequences of one's proposition, as well as negative consequences of one's interlocutor's proposition, often make good arguments, while the converses don't" (Mercier, 2012, p. 319). In our study, participants did not have to generate arguments themselves; these were presented to them. However, Mercier (2012) suggests that the confirmation bias manifests itself in the evaluation of the arguments, not in their generation. The evaluation patterns of the participants in our study confirm Mercier's prediction.

Participants were instructed to defend a claim on an issue the vast majority did not have strong opinions on. This context may have created a mindset in which they considered the task a game without real-life consequences. As a result of this playful mindset, their evaluation patterns may have become more extreme and less representative of how they would act under normal circumstances. It would therefore be interesting to see how participants would evaluate these same arguments when instructed to form their own opinion on this issue, an opinion that they would have to defend in a debate with another person. Under those conditions, it is more likely that participants would use the criteria to distinguish strong from weak arguments. Mercier and Sperber (2011) claim that we have immediately intuitions about the validity of a belief. These intuitions may subsequently guide our evaluation of the arguments. To assess the validity of this prediction, one could unobtrusively manipulate the participants' attitude towards an issue, for instance, through conditioning or priming and subsequently assess whether participants evaluate the arguments in the same biased way as reported in this study.

In conclusion, the results of this study provide more insight into the exact nature of motivated reasoning by revealing that people employ normative criteria in a biased way. The study's design, having participants evaluate systematically manipulated arguments while thinking aloud, provides interesting opportunities to further explore the conditions under which motivated reasoning occurs as well as assess predictions that can be derived from Mercier and Sperber's (2011) argumentative theory of reasoning. Apart from the scientific implications, this study has also practical consequences. It is believed that attitudes and opinions based upon a careful evaluation of relevant arguments yield a more stable attitude that is more predictive of subsequent behavior. Implicitly, the idea is that argument quality will make the difference. This study reveals that people can be very creative in assessing the quality of arguments.

References

Carpenter, C. J. (2015). A meta-analysis of the ELM's argument quality x processing type predictions. *Human Communication Research, 41(4)*, 501-534.

Chaiken, S. (1987). The heuristic model of persuasion. In M. P. Zanna, J. M. Olson, & C. P. Herman (Eds.), *Social influence: The Ontario symposium* (Vol. 5, pp. 3-39), Hillsdale, NJ: Erlbaum.

Ditto, P. H., & Lopez, D. F. (1992). Motivated skepticism: Use of differential decision criteria for preferred and nonpreferred conclusions. *Journal of Personality and Social Psychology, 64(4)*, 568-584.

Edwards, K., & Smith, E.E. (1996). A Disconfirmation Bias in the Evaluation of Arguments. *Journal of Personality and Social Psychology, 71(1)*, 5-24.

Hoeken, H., & Hustinx, L. (2009). When is statistical evidence superior to anecdotal evidence? The role of argument type. *Human Communication Research, 35,* 491-510.

Hoeken, H., Šorm, E., & Schellens, P.J. (2014). Arguing about beliefs: Lay people's criteria to distinguish strong arguments from weak ones. *Thinking & Reasoning, 20(1),* 77-98.

Hoeken, H., Timmers, R., & Schellens, P.J. (2012). Arguing about desirable consequences: What constitutes a convincing argument? *Thinking & Reasoning, 18(3),* 394-416.

Hornikx, J., & Hoeken, H. (2007). Cultural differences in the persuasiveness of evidence types and evidence quality. *Communication Monographs, 74(4),* 443-463.

Jain, S.P., & Maheswaran, D. (2000). Motivated reasoning: A depth-of processing perspective. *Journal of Consumer Research, 26(4),* 358-371.

Johnson, B. T., Smith-McLallen, A., Killeya, L. A., & Levin, K. D. (2004). Truth or consequences: Overcoming resistance with positive thinking. In E. S. Knowles & J. A. Linn (Eds.), *Resistance and persuasion* (pp. 215-233). Mahwah, NJ: Erlbaum.

Klaczynski, P., Gordon, D., & Fauth, J. (1997). Goal-oriented critical reasoning and individual differences in critical reasoning biases. *Journal of Educational Psychology, 89,* 470-485.

Kunda, Z. (1990). The Case for Motivated Reasoning. *Psychological Bulletin, 108(3),* 480-498.

Lord, C.G., Ross, L., & Lepper, M.R. (1979). Biased Assimilation and Attitude Polarization: The Effects of Prior Theories on Subsequently Considered Evidence. *Journal of Personality and Social Psychology, 37(11),* 2098-2109.

Mercier, H. (2012). Looking for arguments. *Argumentation, 20,* 305-324.

Mercier, H., & Landemore, H. (2012). Reasoning is for arguing: Understanding the successes and failures of deliberation. *Political Psychology, 33(2),* 243-258.

Mercier, H., & Sperber, D. (2011). Why do humans reason? Arguments for an argumentative theory. *Behavioral and Brain Sciences,* 34, 57–111.

Molden, D. C., & Higgins, E. T. (2005). Motivated thinking. In K. J. Holyoak & R. G. Morrison (Eds.), *The Cambridge Handbook of Thinking and Reasoning* (pp. 295-317). Cambridge: Cambridge University Press.

Nickerson, R. S. (1998). Confirmation bias: A ubiquitous phenomenon in many guises. *Review of General Psychology, 2*(2), 175-220.

Nienhuis, A. E., Manstead, A. R., & Spears, R. (2001). Multiple motives and persuasive communication: Creative elaboration as a result of impression motivation and accuracy motivation. *Personality and Social Psychology Bulletin, 27*(1), 118-132.

Park, H. S., Levine, T. R., Kingsley Westerman, C. Y., Orfgen, T., & Foregger, S. (2007). The effects of argument quality and involvement type on attitude formation and attitude change: A test of dual-process and social-judgment predictions. *Human Communication Research, 33,* 81-102.

Petty, R. E., & Cacioppo, J. T. (1986). *Communication and persuasion. Central and peripheral routes to attitude change.* Berlin: Springer.

Petty, R. E., Haugtvedt, C. P., & Smith, S. M. (1995). Elaboration as a determinant of attitude strength: Creating attitudes that are persistent, resistant, and predictive of behavior. In R. E. Petty, & J. A. Krosnick (Eds.), *Attitude strength: Antecedents and consequences* (pp. 93-130). Mahwah, NJ: Erlbaum.

Schellens, P. J., & Jong, M. D. T. (2004). Argumentation schemes in persuasive brochures. *Argumentation, 18,* 295-323.

Van Eemeren, F. H., Garssen, B., & Meuffels, B. (2009). *Fallacies and judgments of reasonableness: Empirical research concerning the pragma-dialectical discussion rules.* Dordrecht: Springer.

Van Enschot – Van Dijk, R., Hustinx, L., & Hoeken, H. (2003). The concept of argument quality in the Elaboration Likelihood Model. In F. H. van Eemeren, J. A. Blair, C.

A.Willard, & A. F. Snoeck Henkemans (Eds.), *Anyone who has a view. Theoretical contributions to the Study of Argumentation* (pp. 319-335). Dordrecht: Kluwer.

Westen, D., Blagov, P.S., Harenski, K., & Hamann, S. (2006). Neural Bases of Motivated Reasoning: An fMRI Study of Emotional Constraints on Partisan Political Judgment in the 2004 U.S. Presidential Election. *Journal of Cognitive Neuroscience, 18(11)*, 1947-1958.

Chapter 12

Evidence Quality Variations and Claim Acceptance: an Experimental Investigation of the Role of Distraction and Dilution[1]

Jos Hornikx

Centre for Language Studies, Radboud University, Nijmegen, the Netherlands, j.hornikx@let.ru.nl

Abstract. Studies on persuasive arguments have generally found that claims supported by high-quality evidence are better accepted than claims supported by low-quality evidence. However, an experiment by Hoeken and Hustinx (2007) demonstrated that this effect was only observed in short texts (a claim with evidence), but not in longer texts (where information unrelated to the evidence was added at the end of the text). The present experiment was conducted to examine whether this effect of text length could be explained by distraction (the additional text at the end distracts the reader) or by dilution (the additional text makes the fragment less diagnostic for claim evaluation). Participants ($N = 629$) read two texts with a claim supported by high-quality or low-quality (anecdotal, statistical, or expert) evidence. The text was presented in one of the three versions: (1) short, (2) long with additional information at the end, or (3) new in comparison to Hoeken and Hustinx (2007) – long with additional information at the start. The data found support for the distraction explanation. An effect of evidence quality on claim acceptance was observed in two conditions: in the short text, and in the longer text with additional information at the start. The effect of evidence quality was not found in the longer text with additional information at the end.

1. Introduction

People are more likely to accept claims when they are supported by strong arguments (e.g., Carpenter, 2015; O'Keefe, 2013; Park, Levine, Kingsley Westerman, Orfgen, & Foregger, 2007). In their Argumentative Theory of Reasoning, Mercier and Sperber (2011) expect people to be highly capable of distinguishing strong from weak arguments. One of the ways in which the quality of arguments can be defined is through the notion of the argument scheme, which is "a more or less conventionalized way of representing the relation between what is stated in the argument and what is stated in the standpoint" (Van Eemeren & Grootendorst, 1992, p. 96). For argumentation schemes, critical questions have been formulated that serve as criteria to assess an argument's quality (e.g., Kienpointner, 1992; Walton, Reed, & Macagno, 2008). For the argument from authority, for example, one question relates to the source's expertise, and another to the source's credibility (Walton,

---

[1] This chapter is a modified version of a paper published in Dutch (Hornikx, 2014). The author wishes to thank the publisher, Amsterdam University Press, for permission to use this paper for the current chapter.

1997). The idea is that an argument that respects such criteria is normatively strong, and that an argument that does not respect one or more criteria is normatively weak of weaker. For instance, a given argument from authority is of higher quality (normatively stronger) when the expert has a higher level of expertise and is more credible (on the normative status of this approach, see Hahn & Oaksford, 2012).

In line with the expectation of Mercier and Sperber (2011), empirical studies investigating the persuasiveness of evidence that is normatively strong or normatively weak have shown that claim acceptance is higher when the evidence provided is normatively strong (respecting critical questions from the related argumentation scheme) than when it is normatively weak (not respecting one or more critical questions from the related argumentation scheme). One caveat in this conclusion relates to the length of the text including the evidence: evidence quality has been shown to matter only for short texts (consisting only of a claim and supporting evidence), but not for longer texts (Hoeken & Hustinx, 2007). The present paper reports on an experiment examining two potential explanations for this interaction between evidence quality and text length: distraction and dilution. The current study may generate better insights into the conditions under which laypeople assess claim acceptance on the basis of the quality of arguments provided.

2. Evidence Quality and Text Length

2.1. Evidence Quality and Claim Acceptance

Researchers have been interested in the relationship between the quality of evidence and the acceptance of claims supported by the evidence. In a number of experiments, participants have been exposed to claims supported by high-quality and low-quality evidence. In Hornikx and Hoeken (2007, Study 2), for instance, 20 claims were presented with statistical evidence (which relies on a large number of observations) or with expert evidence (which relies on the expertise of a source) to Dutch and French participants. The quality of statistical evidence was manipulated on the basis of the sample size (small, large), and the quality of expert evidence was manipulated on the basis of whether or not the field of expertise of the source corresponded to the claim's topic. For Dutch participants, claim acceptance was higher after high-quality than after low-quality evidence; for French participants, however, this effect of evidence quality was absent. The effect of expert evidence quality on claim acceptance found for the Dutch participants has been replicated in studies conducted in the Netherlands (Hornikx & Ter Haar, 2013, Study 1), India (Hornikx & De Best, 2011), and Germany (Hornikx & Ter Haar, 2013, Study 1). The effect of statistical evidence was also observed in studies conducted in the Netherlands (Hornikx & Ter Haar, 2013, Study 1), but not in Germany (Hornikx & Ter Haar, 2013, Studies 1 and 2). Finally, effects of the quality of anecdotal evidence (which relies on a single observation) on claim acceptance have been reported in Hoeken and Hustinx (2009, Study 3). In that study, conducted with Dutch participants, high-quality anecdotal evidence resulted from the similarity between the case in the anecdotal evidence and the case in the claim; low-evidence quality was the result of a dissimilarity between the two cases. Analyses showed that high-quality anecdotal evidence was found to be more persuasive than low-quality anecdotal evidence.

In the studies discussed above, low-quality evidence differed from the high-quality evidence only in one critical question, such as the similarity between two cases. Two studies made comparisons between high-quality evidence on the one hand and different variations of low-quality evidence on the other. Hoeken, Timmers, and Schellens (2012) investigated anecdotal and expert evidence. The low-quality anecdotal evidence presented a dissimilar case (just as in Hoeken & Hustinx, 2009) or a case that was similar on a characteristic that was irrelevant to the claim in question. Both variations were found to be less persuasive than the high-quality counterpart. For expert evidence, the researchers developed five variations of low quality, such as when the expert had only moderate expertise in the field, or when the expert had a vested interest in the claim. For three of the five comparisons, high-quality evidence resulted in higher claim acceptance than low-quality evidence. While the claims in Hoeken *et al.* (2012) were related to the desirability of measures or behavior (e.g., 'The increased consumption of fruit drinks is a good thing'), the claims in Hoeken, Šorm and Schellens (2014) concerned the probability that measures or behavior resulted in specific effects (e.g., 'Obligatory driving lessons for people over 70 can reduce their fear in traffic'). Expert evidence had four different low-quality manipulations, causal evidence (which relies on an explanation of the relationship described in the claim) had three low-quality variations, and anecdotal evidence had two different variations of a low-quality manipulation. Across the three types of evidence, nine comparisons were made between high-quality and low-quality evidence, and in seven cases high-quality evidence resulted in higher claim acceptance than low-quality evidence. These results underline the impact of argument quality for claim acceptance.

2.2. Evidence Quality in Longer Texts

The studies presented under section 2.1 used claims with evidence without any other context to examine effects of evidence quality. This methodological choice resulted in high internal but low ecological validity: findings may not hold for evidence quality in realistic, longer texts. Only a limited number of studies have used longer texts to investigate the impact of evidence quality on claim acceptance. In Hoeken and Van Wijk (1997), participants read one longer text about tax increases in a Dutch city. The high-quality (low-quality) anecdotal evidence consisted of a similar (dissimilar) city where the increase had led to beneficial effects. While the manipulation of high versus low quality was found to be successful, the two quality variations were equally effective in terms of beliefs, attitudes, and voting behavior. The experiment conducted by Hornikx and Houët (2009) aimed at higher ecological validity by presenting a realistic municipal letter to actual inhabitants of the municipality that was said to send the letter. Again, the high-quality (low-quality) anecdotal evidence consisted of a city similar (dissimilar) to the municipality. Moreover, in half of the letters this city was said to be taken as example from a large sample of cities where the benefits had been observed. Only in the letters with this statistical evidence was the attitude towards the proposed measure higher for the high-quality than for the low-quality condition. In other words, the effect of evidence quality seems more pronounced in material that only consisted of claims with evidence than in material that embedded these in a longer text.

Hoeken and Hustinx (2007) provide empirical support for the relationship between text length and evidence quality. Their participants judged short texts (claims with normatively strong or weak anecdotal evidence) and longer texts (claims with normatively strong or weak anecdotal evidence, and with additional information irrelevant to the evidence). An interaction effect between evidence quality and text length was reported: an effect of evidence quality on claim acceptance was found in the short texts but not in the longer texts. Hoeken and Hustinx (2007) investigated 16 different claims – allowing some level of generalization to other claims. However, as they remarked themselves, they only included one type of evidence: "Whether laypeople are sensitive to other distinctions in argument quality is unclear" (2007, p. 630). The first goal of the present study therefore is to reexamine the interaction effect between evidence quality and text length for anecdotal, statistical, and expert evidence. The effect is expected to occur independently of the type of evidence:

H1 For anecdotal, statistical, and expert evidence, high-quality evidence leads to stronger claim acceptance than low-quality evidence in the absence of additional information but not in the presence of additional information.

2.3. Distraction or Dilution as Explanation?

People's sensitivity to the quality of arguments is generally said to depend on people's motivation and capacity to scrutinize the message that contains arguments (Elaboration Likelihood Model; Petty & Cacioppo, 1986; Petty, Rucker, Bizer, & Cacioppo, 2004). Hoeken and Hustinx (2007, p. 628) attribute their findings to the participants' low capability in the longer text condition: "It could be that the additional text distracts the participants' focus on the arguments." In the short texts, the "absence of context may have helped the participants to focus their attention completely on the argument itself thereby increasing the chance that they notice the differences in quality" (2007, p. 629). The two text conditions in Hoeken and Hustinx (2007) differed in two ways: text length and the *position* of evidence in the text. In the short text, the evidence was the last information participants read before indicating claim acceptance; in the longer text, the non-diagnostic additional information served as last information.

Distraction seems a plausible explanation for their findings, but an alternative may be the dilution effect (Nisbett, Zukier, & Lemley, 1981; Tetlock, Lerner, & Boettger, 1996), also known as the nondiagnosticity effect (Troutman & Shanteau, 1977). In one of the five studies presented in Nisbett *et al.* (1981), for instance, participants were asked to predict behavior of other people, such as how many movies these people had seen lately. Diagnostic information relevant to that prediction was given (i.e., whether they were premedical students or English majors), and – in half of the cases – additional non-diagnostic information was presented (e.g., about their religious background and spare time activities). Participants' judgements were found to be dependent on the diagnostic information, but much less so when non-diagnostic information was also presented in the material. The additional information diluted the total information that was available about the students in the text, and lowered the impact of the diagnostic information.

In Hoeken and Hustinx (2007), both dilution and distraction may explain the effects that were observed. A third condition would be needed to reveal which explanation holds, namely a condition that consists of a longer text that ends with evidence. If dilution is the explanation, an effect of evidence quality would occur in the short text condition, but not in the longer text conditions as the largest part of the text is non-diagnostic information (regardless or whether that information is positioned at the beginning of at the end of the text). If distraction is the explanation, an effect of evidence quality would also occur in this third condition (as there is no distraction between the evidence presented and the claim acceptance measure) and in the short text condition, but not in the longer text condition that ends with non-diagnostic information. Figure 1 shows the expected effects in the three conditions.

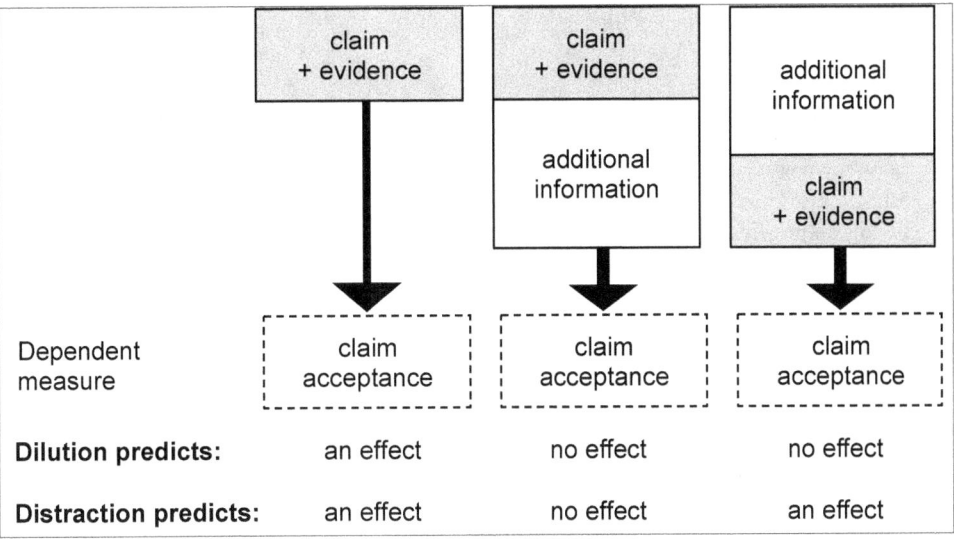

Figure 1. Dilution and distraction as predictors of an effect of evidence quality in the three conditions

The second, and most important goal of the present study is to investigate whether the interaction between evidence quality and text length can be explained by dilution or distraction:

RQ1 Does distraction or dilution explain the interaction between text length and evidence quality on claim acceptance?

3. Method

An experiment was designed in which participants were presented with two different texts including a claim with evidence. The texts in the different conditions varied in the quality of the evidence, the type of evidence, and the structure of the text.

3.1. Material

In order to select appropriate claims, 20 Dutch students (age: $M = 22.90$, $SD = 2.02$; 70% female) rated the 16 claims used in Hoeken and Hustinx (2007) on 7-point probability scales. Two claims were selected that scored around the midpoint of the scale (cf. Hornikx & Hoeken, 2007); in a Dutch translation, they read 'A longer wine list will increase drinking sales in restaurants' ($M = 3.45$), and 'Driving schools will see their registration rise when they paint their learner cars in pronounced colours' ($M = 4.05$).

For each of the two claims, three different structures were designed, based on the short text in Hoeken and Hustinx (2007).[2] The factor Structure was a combination of the absence/presence of non-diagnostic text, and – if present – of the position of this addition (beginning or end of the text). In the short text, an introductory sentence was followed by a claim and by supporting evidence. In the longer texts, non-diagnostic information consisting of 139 words was added before the short text, or after. Example (1) shows an

---

[2] The author wishes to thank Hans Hoeken and Lettica Hustinx for sharing their material.

English translation of the condition with the longer text starting with the non-diagnostic information:

> (non-diagnostic addition) Bistros are popular in the Netherlands. You can find them in villages and cities. Sometimes they were founded years ago, and still have the same owner. In other cases, they are relatively recent, such as in new housing estates, bringing to the neighborhood the necessary atmosphere that is often lacking. Local authorities are very interested in bistros. They believe these bistros are important for the vividness and livability of the areas. In most bistros, popular dishes are on the menu, such as soups, salads, satay, and spare ribs. For their turnover, the weekends are crucial for bistros. On special occasions, such as local events, they can be very busy. However, bistros are having a hard time in the Netherlands. (introductory sentence) Bistro 'Het Hommeltje' in Heerlen is a profitable bistro where customers can eat a lot of food at a reasonable price. Nevertheless, the consumption of drinks is fairly low. (claim) A good possibility to increase drinking sales is to present a longer wine list. (evidence) For bistro 'Den Dikke Dragonder' in Kerkrade, which targets the same type of customers, a longer wine list has increased drinking sales.

The evidence in the example was high-quality anecdotal evidence: the case in the evidence (bistro Den Dikke Dragonder) was a bistro similar to the case in the claim (bistro Het Hommeltje). The low-quality manipulation of anecdotal evidence, also borrowed from Hoeken and Hustinx (2007), consisted of presenting the case of the Da Vinci restaurant, proud owner of a Michelin star, which also saw its turnover increase after introducing a longer wine list. For each of the three conditions, high-quality and low-quality evidence were created for statistical and for expert evidence. Statistical evidence provided information about a large number of cases. Following the manipulation in Hornikx and Hoeken (2007), high-quality statistical evidence reported about a large sample size and a high percentage: 'A Dutch study among 104 restaurants has shown that a longer wine list increased drinking sales for 74% of those restaurants'. The low-quality evidence reported sales increases for 36% of the 28 bistros in the study sample. For expert evidence, the quality was dependent on the vested interest of the expert (cf. Hoeken et al., 2014). In both cases 'Dr Glastra argues that a longer wine list increases drinking sales in restaurants'. In the high-quality variant, he was described as a person 'who has a PhD in food and beverage management and who currently is a professor of retail marketing at Rotterdam University', and in the low-quality variant, he was described as a person 'who has a PhD in food and beverage management and who currently is sales director of wine merchant Colaris in Weert'. In total, there were 18 versions of each of the two texts, differing in Structure, Evidence Type, and Evidence Quality.

3.2. Participants

A total of 629 participants took part in the experiment, of whom 53.6% were female. The Dutch participants were on average 32.32 years of age ($SD = 14.21$; range: 15-84), and their highest education level ranged from primary school (1%) to a Master's degree (37%). The participants were randomly assigned to the 18 conditions of the material. Between these conditions, no differences were observed in the participants' mean age ($F(17, 611) = 1.60$, $p = .06$), gender distribution ($\chi^2(17) = 12.67$, $p = .76$), or educational level ($\chi^2(68) = 67.51, p = .49$).

3.3. Instrumentation

The questionnaire included a series of questions on 7-point scales that were identical for the two texts on bistros and driving schools: claim acceptance, distraction, motivation to read, and issue involvement.

The main dependent measure was the acceptance of the claim, which was repeated after the text ('question 1'), and which was followed by three items ('very improbable – very probable', 'very unbelievable – very believable', and 'very unreasonable – very reasonable' (text 1 about bistros: $\alpha = .94$; text 2 about driving schools: $\alpha = .95$).

Distraction was measured with three items. Likert scales followed three items (inspired by the fluency scale of Lee, Keller, & Sternthal, 2010): 'It was easy to answer question 1', 'For question 1, I was able to easily recall the topic of the text', and 'I had to think hard before I could answer question 1'. As the three items were not reliable (text 1: $\alpha = .58$; text 2: $\alpha = .65$), only item 1 and 3 were taken together (text 1: $r(627) = .43$, $p < .001$; text 2: $r(629) = .39$, $p < .001$). Perceived text comprehension was included as an additional measure of distraction: 'The text about the bistros / driving schools was: difficult – easy, complex – simple, unclear – clear' (text 1: $\alpha = .85$; text 2: $\alpha = .92$).

The questionnaire also checked participants' motivation to answer question 1 with Likert scales after the items: 'I found it interesting to answer question 1' and 'It was fun answering question 1' (inspired by the engagement scale of Lee et al., 2010; text 1: $\alpha = .83$; text 2: $\alpha = .87$). Involvement with the topics of the texts was measured with three of the four items developed in Wegman (1994), and adapted to these texts: 'To what extent do bistros / driving schools preoccupy you personally?', 'Do you ever think about bistros / driving schools?', and 'How important you feel bistros / driving schools are to you?' (text 1: $\alpha = .88$; text 2: $\alpha = .91$). The questionnaire ended with questions about participants' age, gender, and highest educational level.

3.4. Design

The experiment had a 3 (Structure: short, long starting with evidence, long ending with evidence) x 3 (Evidence Type: anecdotal, expert, statistical) x 2 (Evidence Quality: low, high) x 2 (Text: bistros, driving schools) design. Text was a within-subject factor: each participant responded to the two different texts. The other factors were between-subject factors. This means that participants responded to one of the 18 structure x quality x type conditions (for each of the two texts).

3.5. Procedure and Statistical Tests

Participants were approached individually to take part in one of the conditions of the study. When they agreed, they were randomly assigned to one of the conditions. The study was introduced as being about their judgements about bistros and driving schools. Participation took between 10 and 15 minutes.

H1 was addressed by examining the effect of evidence quality for the two conditions used in Hoeken and Hustinx (2007): short texts, and longer texts starting with evidence. RQ1 was addressed with two contrast analyses, one for distraction and one for dilution (see Van den Bercken & Voeten, 2002). For distraction, the longer text starting with evidence was contrasted to the other two texts; for dilution, the two longer texts were contrasted to the short text. No significant interactions were found between Text and the other factors; this means that effects that were (non) significant for the first text, were also (non) significant for the second text. Because of these non-significant interactions, data were collapsed over the factor Text; all means and standard deviations in the results section are based on data of the two texts together[3].

4. Results

4.1. Preliminary Analyses

Significant correlations were observed between claim acceptance and motivation ($r(629) = .14$, $p < .001$), and between claim acceptance and involvement ($r(629) = .26$, $p < .001$). Therefore, motivation and involvement were used in the GLM as covariates. The pattern of results was identical in analyses with and in analyses without these covariates.

4.2. Hypothesis and Research Question

The experiment was conducted to examine whether distraction or dilution could explain the interaction between evidence quality and text length. There was a main effect of Evidence Quality on claim acceptance ($F(1, 609) = 20.82$, $p < .001$, $\eta^2 = .03$): claim acceptance was higher after high-quality evidence ($M = 4.34$, $SD = 1.06$) than after low-quality evidence ($M = 3.89$, $SD = 1.13$). This main effect was not qualified by an interaction with Evidence Type ($F(2, 609) = 1.22$, $p = .30$). RQ1 was addressed with two contrast analyses examining the interaction between Evidence Quality and Structure. The first contrast analysis supported the distraction explanation ($F(1, 621) = 20.57$, $p < .001$, $\eta^2 = .03$); the second contrast analysis did not support the dilution explanation ($F(1, 621) < 1$)[4]. Figure 2 shows that evidence quality had an effect on claim acceptance in the short texts and in the longer texts ending with evidence, but not in the longer text starting with evidence.

For each Structure type, the effect of Evidence Quality on claim acceptance was measured (see also Figure 2). An effect of Evidence Quality was observed for the short texts ($F(1, 201) = 7.24$, $p < .01$, $\eta^2 = .04$) and for the longer texts ending with evidence ($F(1, 203) = 15.90$, $p < .001$, $\eta^2 = .07$), but not for the longer texts starting with evidence ($F(1, 201) = 1.13$, $p = .29$). This result seems to suggest that the effect obtained in Hoeken and Hustinx (2007) for anecdotal evidence was replicated here for three types of evidence. However, the interaction between Structure and Evidence Quality was not significant when only shorts texts and longer texts starting with evidence were considered ($F(1, 404) = 1.49$, $p = .22$). H1 was not supported.

The overall interaction between Structure and Evidence Quality may be further explained by participants' fluency of judging claim acceptance or their perceived text comprehension. There was no main effect of Structure on fluency ($F(2, 626) = 1.62$, $p = .20$), but there was an effect of Structure on perceived comprehension ($F(2, 626) = 6.26$, $p < .01$, $\eta^2 = .02$).

---

[3] There were no significant interactions between Text x Structure ($F(2, 611) < 1$), Text x Evidence Type ($F(2, 611) = 1.30$, $p = .27$), Text x Evidence Quality ($F(1, 611) < 1$), Structure x Evidence Type ($F(4, 611) < 1$), Evidence Quality x Evidence Type ($F(2, 611) = 1.39$, $p = .25$), Structure x Evidence Quality ($F(2, 611) < 1$), or Structure x Evidence Quality x Evidence Type ($F(4, 611) < 1$).
[4] The contrast analysis for distraction was significant for the first text ($F(1, 621) = 15.83$, $p < .001$, $\eta^2 = .02$) and for the second ($F(1, 621) = 9.48$, $p < .01$ $\eta^2 = .02$). The contrast analyses for dilution were not significant for the first and second text (each text: $F(1, 621) < 1$).

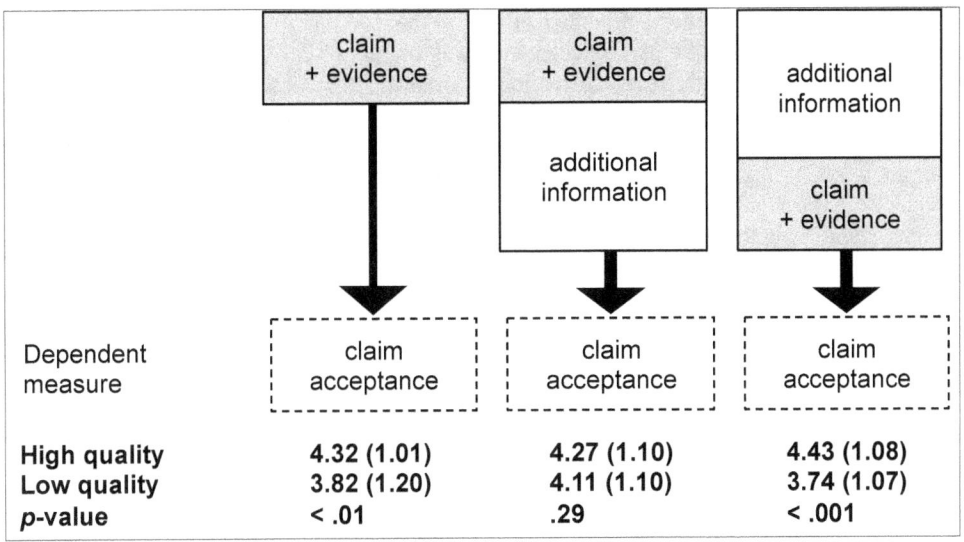

Figure 2. The effect of evidence quality in the three conditions (*SD* in brackets after *M*)

A post-hoc test with Sidak correction showed that perceived comprehension was higher for the short text ($M = 5.95$, $SD = 0.91$) than for the longer text starting with evidence ($M = 5.62$, $SD = 1.01$). However, including perceived comprehension as covariate in the contrast analyses did not alter the results (distraction: $F(1, 620) = 19.79$, $p < .001$, $\eta^2 = .03$; dilution: $F(1, 620) < 1$). Therefore, perceived comprehension did not have strong power in explaining the interaction between Structure and Evidence Quality.

Structure	Type	High Quality			Low Quality		
		M	*SD*	*n*	*M*	*SD*	*n*
short	total	4.32	1.01	104	3.82	1.20	105
	anecdotal	4.30	0.95	35	3.87	1.05	35
	statistical	4.56	0.92	34	3.86	1.21	35
	expert	4.01	1.12	35	3.74	1.34	35
long, evidence at the end	total	4.43	1.08	104	3.74	1.07	107
	anecdotal	4.46	1.01	35	3.84	0.94	35
	statistical	4.63	0.94	35	3.77	1.22	37
	expert	4.19	1.26	34	3.62	1.05	35
long, evidence at the start	total	4.27	1.10	104	4.11	1.10	105
	anecdotal	4.40	1.18	35	4.13	1.27	35
	statistical	4.48	0.98	34	4.27	0.99	35
	expert	3.92	1.08	35	3.93	1.03	35

Table 1. Persuasiveness of evidence in function of structure, type, and quality

For the sake of completeness, the other effects in the ANCOVA are also reported. A main effect of Evidence Type on claim acceptance was found ($F(2, 609) = 6.59$, $p < .001$, $\eta^2 = .02$). A post-hoc test with Sidak correction showed that anecdotal ($M = 4.17$, $SD = 1.09$) and statistical evidence ($M = 4.26$, $SD = 1.09$) generated a higher claim acceptance than expert evidence ($M = 3.92$, $SD = 1.15$). Finally, the following effects were not significant: the main effect of Structure ($F(2, 609) < 1$), the interaction between Structure and Evidence Type ($F(4, 609) < 1$), and the interaction between Evidence Type, Evidence Quality, and Structure ($F(4, 609) < 1$). Table 1 gives the claim acceptance scores in function of the three factors.

5. Conclusion and Discussion

According to Mercier and Sperber (2011), laypeople are good at differentiating between strong and weak arguments. Empirical studies on people's sensitivity to high-quality and low-quality evidence underline this idea. Hoeken and Hustinx (2007), however, argued that this sensitivity is only displayed when people judge short texts with claims and evidence. The present study was designed to examine whether distraction or dilution is able to explain people's insensitivity to evidence quality in the case of longer texts. Support was found for distraction as an explanatory factor: when additional information is present between the evidence and the acceptance measure, people do not consider the quality of the evidence when indicating their claim acceptance. The present study demonstrated that adding non-diagnostic information in itself does not hinder an effect of evidence quality. That is, in longer texts starting with non-diagnostic information and ending with evidence, an effect of quality on claim acceptance was found. The current experiment could not replicate the interaction observed in Hoeken and Hustinx (2007) between evidence quality and text length (short, long starting with evidence), although an effect was found for the short text but not for long text starting with evidence.

The present study contributes to existing research by examining a longer text in which evidence was positioned at the end or at the beginning, allowing the examination of the roles of distraction and dilution, and by extending the types of evidence that were studied. This study contrasted two processes, distraction and dilution, each of which generated different predictions of claim acceptance in one particular condition. Both processes, however, share the characteristic of adding non-diagnostic information to the claim and evidence. Results show that additional information can indeed hinder an effect of evidence quality on claim acceptance in longer texts ending with evidence – predicted by both distraction and dilution. Nevertheless, this additional information does not always dilute: with the non-diagnostic information at the beginning of the text, participants' claim acceptance was found to be sensitive to the quality of evidence. The experimental conditions, for two different texts, used only one length for the manipulation of non-diagnostic information, and one length for the different manipulations of evidence. Future research may vary in the relative proportion of diagnostic and non-diagnostic information to address the question as to how limited the proportion of non-diagnostic information can be to obscure effects of evidence quality on claim acceptance.

Another avenue for future studies is located in participants' involvement. The texts presented in the current study were, on purpose, neutral to the participants. The participants were therefore unlikely to scrutinize the evidence that is presented in the text in order to observe weaknesses in the presented arguments. The current experiment shows that people are insensitive to evidence quality when non-diagnostic information follows. The question is whether people are more sensitive to the quality of evidence presented when they are involved in the subject of the text? Research addressing this question may improve our understanding of the limits of when argument quality matters.

The results of the current study were based on three different types of evidence, and on two different texts, giving the results some level of robustness. It should be noted that the same effects were found for the second text, when participants had already been exposed to the first experimental condition and the different items. This first text did not help them to be more sensitive to evidence quality in the second text. This suggests that for laypeople, although they may be sensitive to the quality of the arguments that are presented (see Hoeken & Hustinx, 2009; Hoeken et al., 2012, 2014; Hornikx & Hoeken, 2007), a small additional paragraph is enough to distract them from using this quality to assess how likely they find a claim. For persuasion practice, this result may imply that persuaders who do not have strong arguments to underline their claims or who are not sure about the quality of their arguments may still generate successful texts. For researchers, the results of the current study may stimulate further research on people's acceptance of claims supported by arguments differing in quality.

References

Carpenter, C. J. (2015). A meta-analysis of the ELM's argument quality x processing type predictions. *Human Communication Research, 41*, 501-534.

Hahn, U., & Oaksford, M. (2012). Rational argument. In K. Holyoak, & R. Morrison (Eds.), *The Oxford handbook of thinking and reasoning* (pp. 277-300). Oxford, UK: Oxford University Press.

Hoeken, H. & Hustinx, L. (2007). The influence of additional information on the persuasiveness of flawed arguments by analogy. In F. H. Van Eemeren, J. A. Blair, C. A. Willard, & B. Garssen (Eds.), *Proceedings of the sixth conference of the International Society for the Study of Argumentation* (pp. 625-630). Amsterdam: Sic Sat.

Hoeken, H., & Hustinx, L. (2009). When is statistical evidence superior to anecdotal evidence in supporting probability claims? The role of argument type. *Human Communication Research, 35*, 491-510.

Hoeken, H., Šorm, E., & Schellens, P. J. (2014). Arguing about the likelihood of consequences: Laypeople's criteria to distinguish strong arguments from weak ones. *Thinking and Reasoning, 20*, 77-98.

Hoeken, H., Timmers, R., & Schellens, P. J. (2012). Arguing about desirable consequences: What constitutes a convincing argument? *Thinking and Reasoning, 18*, 394-416.

Hoeken, H., & van Wijk, C. (1997). De overtuigingskracht van anekdotische en statistische evidentie. *Taalbeheersing, 19*, 338-357.

Hornikx, J. (2014). Het effect van evidentiekwaliteit op de beoordeling van standpunten: De rol van toegevoegde tekst. *Tijdschrift voor Taalbeheersing, 36*, 107-125.

Hornikx, J., & de Best, J. (2011). Persuasive evidence in India: An investigation of the impact of evidence types and evidence quality. *Argumentation and Advocacy, 47*, 246-257.

Hornikx, J., & ter Haar, M. (2013). Evidence quality and persuasiveness: Germans are not sensitive to the quality of statistical evidence. *Journal of Cognition and Culture, 13*, 483-501.

Hornikx, J., & Hoeken, H. (2007). Cultural differences in the persuasiveness of evidence types and evidence quality. *Communication Monographs, 74*, 443-463.

Hornikx, J., & Houët, T. (2009). De overtuigingskracht van normatief sterke en normatief zwakke anekdotische evidentie in het bijzijn van statistische evidentie. In W. Spooren, M. Onrust, & J. Sanders (Eds.), *Studies in taalbeheersing, volume 3* (pp. 125-133). Assen: Van Gorcum.

Kienpointner, M. (1992). *Alltagslogik: Struktur und Funktion von Argumentationsmustern*. Stuttgart / Bad Cannstatt: Friedrich Frommann.

Lee, A. Y., Keller, P. A., & Sternthal, B. (2010). Value from regulatory construal fit: The persuasive impact of fit between consumer goals and message concreteness. *Journal of Consumer Research, 36*, 735-747.

Mercier, H., & Sperber, D. (2011). Why do humans reason? Arguments for an argumentative theory. *Behavioral and Brain Sciences, 34*, 57-111.

Nisbett, R. E., Zukier, H., & Lemley, R. E. (1981). The dilution effect: Non-diagnostic information weakens the implications of diagnostic information. *Cognitive Psychology, 13*, 248-277.

O'Keefe, D. J. (2013). The relative persuasiveness of different forms of arguments from consequences: A review and integration. In C. T. Salmon (Ed.), *Communication yearbook* (Vol. 36, pp. 109-135). New York: Routledge.

Park, H. S., Levine, T. R., Kingsley Westerman, C. Y., Orfgen, T., & Foregger, S. (2007). The effects of argument quality and involvement type on attitude formation and attitude change: A test of dual-process and social-judgment predictions. *Human Communication Research, 33*, 81-102.

Petty, R. E., Rucker, D. D., Bizer, G. Y. & Cacioppo, J. T. (2004). The Elaboration Likelihood Model of persuasion. In J. S. Seiter & G. H. Gass (Eds.), *Perspectives on persuasion, social influence, and compliance gaining* (pp. 65-89). Boston: Allyn & Bacon.

Petty, R. E., & Cacioppo, J. T. (1986). *Communication and persuasion: Central and peripheral routes to attitude change*. New York: Springer.

Tetlock, P. E., Lerner, J., & Boettger, R. (1996). The dilution effect: Judgmental bias, conversational convention, or a bit of both? *European Journal of Social Psychology, 26*, 915-935.

Troutman, C. M., & Shanteau, J. (1977). Inferences based on nondiagnostic information. *Organizational Behavior and Human Performance, 19*, 43-55.

Van den Bercken, J., & Voeten, M. (2002). *Variantie-analyse: De GLM-benadering*. Groningen: Wolters-Noordhoff.

Van Eemeren, F. H., & Grootendorst, R. (1992). *Argumentation, communication, and fallacies: A pragma-dialectical perspective*. Hillsdale, NJ: Lawrence Erlbaum.

Walton, D. N. (1997). *Appeal to expert opinion: Arguments from authority*. University Park, PA: Pennsylvania State University Press.

Walton, D. N., Reed, C., & Macagno, F. (2008). *Argumentation schemes*. Cambridge: Cambridge University Press.

Wegman, C. (1994). Factual argumentation in private opinions: Effects of rhetorical context and involvement. *Text, 14*, 287-312.

Chapter 13

Does Expertise Favor the Detection of the Metaphoric Fallacy?

Francesca Ervas[1], *Antonio Ledda*[2] *& Antonio Pierro*[3]

University of Cagliari, Cagliari, Italy, [1] ervas@unica.it; [2] antonio.ledda@unica.it; [3] antonio.pierro@gmail.com

Abstract. The paper aims at clarifying whether and to what extent expertise plays a role in the detection of ambiguity fallacies, such as *quaternio terminorum*, where a metaphor is the middle term in one of the premises (*metaphoric fallacy*). We tested a group of (N=40) non-experts adults and a group of (N=40) experts adults (scholars having a strong training in philosophical logic), by using a series of verbally presented arguments, having the structure of quaternio terminorum and containing either a lexical ambiguous or a metaphorical middle term. The experimental results of the study show that non-experts tend to judge sound *quaternio terminorum* with lexicalized metaphors as middle terms, when the conclusion of the argument is far from being patently false. Nonetheless, metaphorical middle terms seem to have also an effect on experts' intuitions on fallacious argument with plausible conclusion. However, this effect is mitigated by expertise.

1. Introduction

Following Van Eemeren and Grootendorst (2004, p. 158), "A standard definition of a fallacy that was accepted until recently is that of 'an argument that seems to be valid but that is not valid'. During the last few decades, however, argumentation theorists have raised several important objections to this definition: 'Seems' involves an undesirable amount of subjectivity; 'validity' is incorrectly presented as an absolute and conclusive criterion; […] These objections explain why nowadays it is preferred in some quarters to give a broader definition in which a fallacy is regarded as a deficient move in an argumentative discourse or text".

However, in everyday communication, it is unquestionable that there are psychological aspects that play a role at least in the perception of a fallacy, and *a fortiori* so for people bearing no normative notion of argument and of validity. This should be particularly evident in case of those fallacies *dependent on language* (*in dictione* in the Latin tradition), to be distinguished from those outside language, *extra dictionem* (for a detailed account see Hamblin, 1970). Fallacies in this class arise from ambiguity in the words or sentences in which they are expressed. Among them are of particular interest those whose ambiguity stems from the presence of a metaphor. A sentence where a metaphor occurs is literally false. However, the context of its usage might create the perception that it is true, or at least plausible. In other words, from a literal point of view a metaphor is false, but from a non-literal point of view it may appear plausible. Indeed, a metaphor is a very special kind of trickery whose most direct, and substantial effect should be regarded at a semantic/pragmatic level. For instance, with the words "Juliet is the sun" Romeo utters that Juliet is the warmth of his world, that his day begins with her, that only in her nourishment can he grow, etc. This statement is, of course, literally false, but metaphorically taken it may appear, at least, evocative, if not true. The relationships between metaphor and truth are controversial and widely discussed in the literature. Since a detailed account would lie outside of the scope of the present work, we refer the interested reader e.g. to Davidson (1984).

In (2015), Fischer pointed out that a range of varied disciplines have demonstrated the productive usage of metaphors in deductive reasoning: physics (Hesse, 1996), biology (Keller, 1995), psychology (Gentner & Grudin, 1985) and problem solving (Keefer *et al.*, 2014; Thibodeau & Boroditsky, 2011). At the same time, metaphors are governed by heuristic rules that never guarantee the preservation of truth, thus giving rise to systematic fallacies (Fischer 2011, 2014). This might be the reason why metaphors are highly persuasive by nature. Thus, the persuasiveness of an argument would eventually be influenced by the presence of (different kinds of) metaphors. In fact, there is a change in the mode of the inferential processing involved in metaphor comprehension, which changes with the degree of lexicalization. Non-lexicalized metaphors (or live metaphors) might require imagination to grasp the intended metaphoric meaning. Highly lexicalized metaphors might instead be acquired as a highly salient meaning (Giora, 2003) in the linguistic community, which have a status similar to literal meanings of a polysemic word and therefore might go unnoticed as figurative meaning. This fact suggests that the degree itself of lexicalization of a metaphor may affect the persuasiveness of an argument. In this study, however, we will discuss mainly the case of lexicalized metaphors, and we leave this other topic open for future research.

Expanding on an idea from (Ervas *et al.*, 2015), we aim to understand the effect of metaphors in the narrow context of arguments with the standard syllogistic structure (cfr. Section 2.1). In particular, we discuss the cases in which the middle term in the premises may assume different meanings because it possesses two different literal meanings (the case of *homonymy* and *polysemy*), or because it has both a literal and a non-literal meaning (the case of *lexicalized metaphor* and *live metaphor*). If the middle term assumes different meanings in the two premises, then the syllogism is a fallacy, usually referred to as *quaternio terminorum* or fallacy of the four terms.

The leading-questions of the research were:

1. Is there any difference in the detection of the fallaciousness of a syllogism, in case the middle is either an homonym, or polysemic, or a lexicalized metaphor?
2. Does expertise play a crucial role in the detection the fallaciousness of a syllogism whose middle term is used metaphorically in one premise and literally in the other (*metaphoric fallacy*)?

In order to provide a (tentative) answer to these questions, we set up an experimental design to clarify whether and to what extent expertise has an effective role in the detection of *quaternio terminorum*. And, moreover, whether such a role is related to the type of ambiguity of the middle term.

Our working hypotheses were that the persuasiveness of the syllogism should be proportional to the degree of semantic overlapping between the different meanings of the middle term (*Metaphoric Effect*). Furthermore, we claimed that the expertise should favor not only the disambiguation process of the premises but also the detection of the metaphoric fallacy (*Expertise Effect*). The results suggest the presence of a metaphoric effect in non-experts' detection of strong argument and standard *quaternio terminorum*, in particular in case of a plausible conclusion. Indeed, we will see that the expertise effect is sufficient in case of strong argument and standard *quaternio terminorum*. However, this is not so for fallacious arguments with plausible conclusion.

The paper is structured as follows. In the first section we provide the basic notions required to define *quaternio terminorum* and lexical ambiguity. In the second section, we present the design of an experimental study that aims to understand whether, and to what extent, expertise favor the detection of a lexical ambiguity fallacy, especially in case of metaphorical middle terms. In the third section, we present the methodology of the experimental study, and describe the groups of participants involved, the materials presented to the participants, and the procedure. In the fourth section we present the results of the experimental study. Finally, in the fifth section, we discuss the results in relation to the aims of the paper.

2. Basic Notions

In this section, we introduce all the basic notions from argumentation theory and lexical pragmatics used in the experimental study.

2.1. Argumentation Theory

A *sentence* is a declarative statement to which a truth-value (true or false), in classical logic, can be assigned (Govier, 1999; Copi & Cohen, 2014). In the context of this paper, by an *argument* or a *syllogism* we mean a set of three sentences: two premises (*major* and *minor*, respectively) and a conclusion. The conclusion in turn involves three terms: the subject, the predicate of the conclusion, and a third term (*middle term*) connecting the subject of the first premise to the predicate of the second.

Argumentation theory discusses the conditions under which a conclusion follows from true premises. To this effect, the concepts of *validity* and *soundness* are required (Walton, 2005, 2010):

- an argument is *valid* if its conclusion is true, whenever its premises are true;
- an argument is *sound* if it is valid and all its premises are true.

Validity and soundness are crucial in evaluating arguments. Nonetheless, these notions seem too tight for ordinary communication. Due to this reason, two weakened versions of these concepts may be expedient: *strength* and *goodness* (Bonissone, 1987; Borwein & Bailey, 2008; Epstein & Kernberger, 2006, chap. 3).

- an argument is *strong* if it is very *likely* that its conclusion is true, whenever its premises are true. In other words, premises provide reasons that support the probable truth of the conclusion. We call *weak* an argument which is not strong.

To spell out in full this classification of the arguments, some scholars distinguish between deductive and inductive standards. If, from the truth of the premises, the conclusion necessarily follows, the argument is said to be *deductive*. Conversely, if the truth of the premises does not necessarily establish the truth of the conclusion, but nonetheless their truth provides good reasons to believe the conclusion to be true, then the argument is *inductive* (Govier, 1980).

- an argument is *good* if it is *strong* and all its premises are *plausible* (We say that a statement is plausible if it seems worthy of approval or probable).

In this article we will deal with the strength and the goodness of those arguments whose middle term potentially call for (at least) two distinct meanings, as mentioned in the introduction.

Quaternio terminorum or *fallacy of the four terms* (Copi & Cohen, 2014, pp. 230-231; Smiley, 1973, pp. 136-154) is a well-known case of a fallacious argument based on the *ambiguity* of its middle term, which has different meanings in the two premises (Dunbar, 2001; Fearnside & Holter, 1959; Hamblin, 1970; Kroeger, 2005, para. 3.1; Quine, 1960, para. 27-31). If the middle term assumes a different meaning in each premise, then a syllogism contains indeed a fourth, hidden term, that causes the fallacy. Here is a simple example:

(P_1) Mick Jagger loves rock
(P_2) Rock is solid mineral material
(C) Mick Jagger loves solid mineral material.

2.2. Lexical Pragmatics

The most common lexical ambiguities are *homonymy* and *polysemy*. Homonymy refers to terms with different literal meanings bearing no semantic relation. Polysemy, instead, refers to terms with literal meanings conveying semantic overlaps (Lyons, 1977; Taylor, 2003). A typical example of homonymy is the term 'bank', which may refer to a financial institution in «Mike went to the bank to apply for a credit card», and to a riverside in «John and Yoko run on the bank». An example of polysemy is the term 'letter', which means both 'symbol of the alphabet' and 'written communication'. In the case of homonymy, the meanings of the different occurrences do not share any property, while in case of polysemy some properties are shared because of the semantic overlap between the two meanings. This overlap is induced by different phenomena, among which are metaphors. Metaphors have indeed been considered in connection with polysemy in cognitive semantics (Lakoff & Johnson, 1980, p. 248), because they are considered one of the most important way of creating new meanings (Bartsch, 2002).

At a linguistic level, metaphors are words with multiple meanings. Take the word 'grasp', which can mean: 'to hold on', but also 'to apprehend', 'to understand' and 'to grip'. All these meanings are lexicalized (i.e. stored in a lexicon); we do not perceive them as proper metaphorical usages (Black, 1993, p. 25). They are conventional uses (indeed, we find them in dictionaries!), that scholars call *lexicalized metaphors* or *dead metaphors*. A dead metaphor is a lexical item with a conventional figurative meaning different from the original (or some previous meaning in the chain of semantic change). It should be observed that, in order to understand a dead metaphor, we do not need to refer to its original meaning. For example, the term 'star' is a dead metaphor and both the lexical entries:

1. a fixed luminous point in the night sky that is a large, remote incandescent body like the sun;
2. a famous or exceptionally talented performer in the world of entertainment or sports;

occur in dictionaries.

Instead, *live metaphors* testify a creative usage of language, neither to be found in common usage nor already classified in dictionaries. They involve a creative comprehension process. However, in the conceptual theory of metaphor (Lakoff & Johnson, 1980), live metaphors are also supposed to be as much alive as the conventional and vital conceptual metaphors in which they are considered to be grounded in. For example, the conceptual metaphor LIFE IS A JOURNEY gives rise to several conventional meanings: as «I do not know which path to take», but also unconventional, creative utterances, as for instance in the poem "The Road Not Taken" (1920) by Robert Frost.

3. Design and Predictions

The study was designed to clarify whether and to what extent expertise plays a role in the detection of *quaternio terminorum*. And, moreover, whether such a role is related to the type of ambiguity of the middle term.

3.1. Working Hypotheses

In *quaternio terminorum*, middle terms could have two different literal meanings, i.e. homonymy and polysemy cases, or else, a literal meaning and a non-literal meaning, i.e. lexicalized metaphor and live metaphor cases. For convenience, we termed *metaphoric fallacy* a *quaternio terminorum* whose middle term is used *metaphorically* in one premise and *literally* in the other. How could the detection of *quaternio terminorum* be in such cases? Does expertise influence the detection of *quaternio terminorum*?

Our working hypotheses were that:

1. the persuasiveness of the syllogism should be proportional to the degree of semantic overlapping between the different meanings of the middle term (*Metaphoric Effect*);
2. the expertise should favor not only the disambiguation process of the premises but also the detection of the metaphoric fallacy (*Expertise Effect*).

As to the first claim, our prediction was that *quaternio terminorum* identification should rely mainly on the nature of the middle term, and therefore on the degree of partial semantic overlapping between the different readings of a middle term (degree of shared semantic properties). In particular, we presumed that arguments featuring homonymous words (e.g. 'bank') as middle terms would be more easily recognized as fallacious than arguments featuring polysemous terms (e.g. 'letter') or dead metaphors (e.g. 'star') as middle terms. This might be due to the fact that different meanings of a homonymous term would be clearly divergent, whilst the meanings of a polysemous or metaphorical term partly overlap (Ervas & Ledda, 2014; Gentner, Ratterman, Forbus, 1993; Gick & Holyoak, 1983). The disambiguation process involved in the disambiguation of a homonymous middle term, such as 'bank', would require the suppression of one of its two literal meanings; namely, the irrelevant one (Gernsbacher, 1990). In this case, the detection of a *quaternio terminorum* calls for the suppression of one of the two literal meanings in the first premise and vice versa in the second premise.

Interestingly enough, it has been pointed out in (Gernsbacher & Faust, 1991) that a suppression process is involved in both disambiguation and metaphor interpretation.

Nonetheless, in the process of disambiguation of a homonym the irrelevant meaning disappears significantly more quickly, when compared to the process of polysemy and metaphor interpretation, which requires more demanding attentional resources to suppress the corresponding literal meaning (Rubio Fernandez, 2007). Indeed, we supposed that arguments featuring polysemous or metaphorical terms were more persuasive. As an effect of their being lexicalized, premises with dead metaphors may appear true even if they are literally false. Consequently, we expected that *quaternio terminorum* with dead metaphors as middle terms would have been the most convincing. Instead, we surmised that premises with live metaphors would have been easily recognized as literally false, because the figurative meaning of a live metaphor clearly differs from its literal meaning, thus making the ambiguity easily detectable.

As to the second claim, it seemed reasonable to expect that a higher level of competence in the field, besides significantly improving the disambiguation procedure required in the general detection of a *quaternio terminorum*, could also be relevant in the identification of metaphoric fallacies. Indeed, we assumed that a specific knowledge of the abstract structure of a syllogism would attenuate, up to a certain degree at least, the persuasive effect that we would have expected from *quaternio terminorum* featuring polysemous or metaphorical terms. In order to test these hypotheses, we tried to single out and quantify the relevant differences between

1. a group of (N=40) *non-experts* adults,
2. a group of (N=40) *experts* adults (scholars having a strong training in philosophical or mathematical logic),

in the detection of the fallacy of the four terms. We tested both groups using a series of verbally presented arguments, with the structure of *quaternio terminorum*, and containing either a (literal) lexically ambiguous or a (non-literal) metaphorical middle term.

3.2. Sets of Arguments

The experimental design comprised four groups of ambiguous middle terms, classified as follows:

1. homonymy (H),
2. polysemy (P),
3. dead (lexicalized) metaphor (DM),
4. live metaphor (LM).

From now on, with H, P, DM, and LM we shall denote the classes of arguments featuring homonymous terms, polysemous terms, dead (lexicalized) metaphors, and live metaphors, respectively. We planned a set of arguments with H, P, DM, LM middle terms, having the structure of a *quaternio terminorum*: for different x, y, z

Premise 1: x verb y;
Premise 2: y verb z;
Conclusion : x verb z.

In order to spell out the perception of the strength of an argument, we asked participants to evaluate whether the conclusion *follows* from the premises.

Specifically, the test comprised 3 kinds of randomized arguments having as middle terms H, P, DM, LM:

1. Six *standard quaternio teminorum* x (H, P, DM, LM) with True premises/False conclusion.

A DM example follows:

(P1) Clooney is a star
(P2) A star is a celestial body
(C) Clooney is a celestial body

It should be noted that a patently false conclusion could lead non-experts to judge an argument weak without any real understanding of the overall argument. Hence, we had to make sure that non-expert participants wouldn't have considered an argument weak just because of its false conclusion. To this aim we devised a set of weak arguments whose middle term had different meanings in the premises, as in standard *quaternio terminorum* case, but whose conclusions were plausible:

2. Six *quaternio teminorum* × (H, P, DM, LM) with True premises/Plausible conclusion.

A LM example follows:

(P1) The old age is a dinner
(P2) A dinner is quite long
(C) The old age is quite long

Since the middle term possesses different meanings in the two premises, we had to make sure that non-expert participants would have been able to distinguish between middle terms needing disambiguation in the two premises, and middle terms unambiguously used. Therefore, we also prepared a set of strong arguments with true premises, true conclusions and a (potentially ambiguous) middle term used with the same meaning in both premises:

3. Six *strong arguments* × (H, P, DM, LM) with True premises/True conclusion.

Arguments in this class are strong, and should to be distinguished from strong arguments with potentially ambiguous (but unambiguously used) middle terms. In fact, arguments within this set do not involve any term that may call for distinct references in the two premises: they are completely devoid of any disambiguation effort.

We included this set of arguments in order to check participants' capacity to distinguish between weak arguments (standard *quaternio terminorum*) and strong arguments with unambiguously used middle terms (which do not require any suppression process).

An H example follows:

(P1) BNP Paribas is a bank
(P2) A bank is a financial institution
(C) BNP Paribas is a financial institution

Finally, to fully understand whether non-expert participants would have been able to distinguish between a *clearly strong* and a *clearly weak* argument, we planned a set of distractors having the structure of a standard *quaternio terminorum* featuring no ambiguous terms, divided in two exclusive subsets:

1. Twenty-five *clearly strong arguments* with True premises/True conclusion;
2. Twenty-five *clearly weak arguments* with True premises/False conclusion.

The role of distractors was twofold:

1. reflecting non-experts' understanding and commitment to the task they were assigned;
2. emphasizing non-experts' capacity to distinguish between a strong and a weak argument, without any explicit instruction.

Finally, we accepted only those participants who were able to distinguish between strong and weak arguments (acceptance threshold: more than 90% of correct answers).

4. Method

We tested premises and conclusions of arguments in three norming studies. The method included a primary selection of a set of middle terms (= 206 nouns), that could possibly be used to form middle terms belonging to the four categories H, P, DM, LM, according to their number of letters and frequency in the GRADIT (De Mauro, 2000). In the first norming study, we tested both emotional (positive and negative) meaning and familiarity of the selected terms. We eliminated terms with definite emotional meanings and insufficient familiarity. We used unambiguous terms to build live metaphors and then devised the arguments according with the middle terms we have chosen. In the second norming study, we tested the premises of the arguments separately. This was to make sure that participants attribute either the same meaning to the middle terms in case of strong arguments, or different meanings in case of *quaternio terminorum*. In the third norming study, we tested premises and conclusions separately in order to understand whether they were perceived as either true or false.

The results of the norming studies have shown that the majority of sentences featuring a dead metaphor (83%) are perceived as true (even if they are literally false) whilst the majority of sentences featuring a live metaphor (79%) are perceived as false. The results of the norming studies induced us to exclude, from the experimental design, LM middle terms. Indeed, they are not comparable with the other categories of middle terms since they are widely perceived as false. Therefore, the pilot study had a 3x3 experimental design: 3 argument structure conditions x3 (H, P, DM) middle term conditions.

4.1. Participants

All participants were native Italian speakers, and had normal/corrected vision. They signed informed consent agreements to volunteer for this study, which complies with APA ethical standards. The norming studies involved 209 participants (balanced across age and gender). All of these participants were undergraduate students at the University of Cagliari (Faculty of Education, Languages and Engineering) and did not participate in the experimental study.

The experimental study involved:

1. Non-experts (N = 40, balanced across age and gender), who were undergraduate students at the Faculty of Humanities, University of Cagliari, attending the first year BA Program in Communication Science. We made sure that participants were naïve, i.e. that they had not previously attended any logic course.
2. Experts (N = 40, balanced across age and gender), who were scholars in philosophical logic, history of logic, mathematical logic, philosophy of science, mathematics education, philosophy of language, linguistics.

4.2. Materials

The printed materials of the experimental study consisted of a set of 54 arguments and 50 distractors in Italian. The 54 arguments comprise 18 arguments with *quaternio terminorum*'s structure for homonymy (H), polysemy (P) and dead metaphor (DM) middle term condition. For each group of ambiguous middle terms, 6 arguments were *standard quaternio terminorum* (with true premises/false conclusion), 6 arguments were *strong arguments* (with true premises/true conclusion) and 6 arguments were *quaternio terminorum with plausible conclusion* (with true premises/plausible conclusion); see English Examples in Table 1 and Table 1a. in Appendix for examples of validated arguments in Italian.

In the norming studies, we tested the premises of the arguments separately in order to make sure that participants attribute either the same meaning to the middle terms in case of strong arguments or different meanings in case of *quaternio terminorum*. If the middle term is interpreted with the same reference in the two premises, a *quaternio terminorum* would be even valid, being an instance of *ex absurdo quodlibet sequitur*. Or else, as in the case of "A bank is at water's edge", interpreting the term "bank" as a "financial institution" which could (eventually) be placed next to a river, we would obtain an argument with a true premise and a *plausible premise*, together with a plausible conclusion. These cases, however, are rule out by the fact that we asked participants to assess on the *strength of an argument* (see Section 2.1). To this effect, *all the arguments in the experiment possess clearly true premises only*.

5. Results

We performed a t-test to determine the statistical significance. As regards distractors, in more than 90% of cases, the participants answered «Yes, the conclusion follows from the premises» in presence of a clearly strong argument (Non-experts' group: mean difference = 37.92; $p < 0.001$; Experts' group: mean difference = 38.32; $p < 0.001$), and «No, the conclusion does not follow from the premises» in case of a clearly weak argument (Non-experts' group: mean difference = 37.28; $p < 0.001$; Experts' group: mean difference = 38.48; $p < 0.001$). No significant difference was found among groups.

Results show a significant difference of performance between the 'strong arguments' (TP/TC) condition and the set of 'standard *quaternio terminorum*' (TP/FC) condition in both non-experts' (Figure 1) and experts' (Figure 2) groups. This significant difference is due to a greater number of correct responses ('YES') to the 'strong arguments' condition compared to the number of correct responses ('NO') to the 'standard *quaternio terminorum*' condition for homonymous middle term (Non-experts' group: mean difference = 11.67; $p < 0.001$; Experts' group: mean difference = 4.34; $p < 0.01$) and polysemous middle term (Non-experts' group: mean difference = 12.5; $p < 0.001$; Experts' group: mean difference = 4,33; $p = 0.035$) conditions. In case of dead metaphors as middle term, this difference is still significant in non-experts' group, but it is not significant in experts' group. (Non-experts' group: mean difference = 12.17; $p < 0.001$; Experts' group: mean difference = 1.00; $p = 0.60$). In both cases of strong arguments and standard *quaternio terminorum* no significant difference was found among homonymy, polysemy and dead metaphors conditions in both groups. Overall, experts perform significantly better than non-experts ($p = 0.0004$) and this is due to the greater number of errors in the non-experts' group.

	Standard quaternio terminorum	***Strong arguments***	***Quaternio terminorum with plausible conclusion***
Homonymous middle terms (H)	(P1) A dog's cry is a *bark*. (P2) A *bark* is a tree covering. (C) A dog's cry is a tree covering.	(P1) A folder is *a file*. (P2) *A file* is a document collection. (C) A folder is a document collection.	(P1) A financial building is a *bank*. (P2) A *bank* is at water's edge. (C) A financial building is at water's edge.
Polysemous middle terms (P)	(P1) An incision is a *cut*. (P2) A *cut* is a dress shape. (C) An incision is a dress shape.	(P1) F is a *letter*. (P2) A *letter* is an alphabet symbol. (C) F is an alphabet symbol.	(P1) A jog is a *run*. (P2) A *run* is a race. (C) A jog is a race.
Dead (lexicalized) metaphors as middle terms (DM)	(P1) Clooney is a *star*. (P2) A *star* is a celestial body. (C) Clooney is a celestial body.	(P1) A movement's chief is a *head*. (P2) A *head* is a leader. (C) A movement's chief is a leader.	(P1) A hot place is an *oven*. (P2) An *oven* is part of a kitchen. (C) A hot place is part of a kitchen.

Table 1. Argument's types, combined with H, P, DM middle term conditions.

The 50 distractors comprise 25 clearly strong and 25 clearly weak arguments, as in the English examples in Table 2 (see Table 2a. in Appendix for example of validated arguments in Italian).

Strong arguments	Weak arguments
(P1) Linda is a *girl*. (P2) A *girl* is a female. (C) Linda is a female.	(P1) Antonio hates *wine*. (P2) *Wine* is a drink. (C) Antonio is a drink.
(P1) Leo is a *child*. (P2) A *child* is a human being. (C) Leo is a human being.	(P1) Mary eats an *orange*. (P2) An *orange* is a fruit. (C) Mary is a fruit.

Table 2. Examples of distractors (clearly strong arguments and clearly weak arguments).

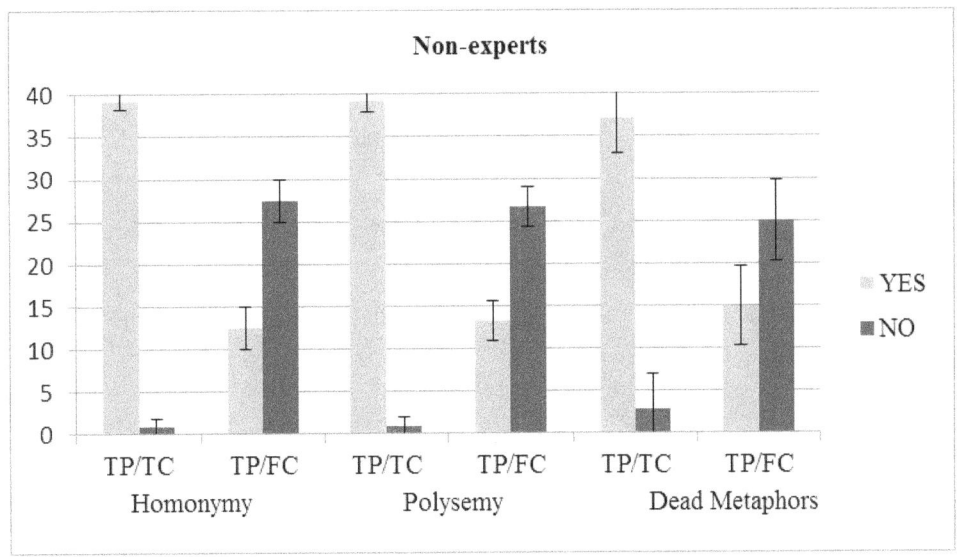

Figure 1. Non-experts' group: mean of answers for strong (TP/TC) and weak (TP/FC – standard *quaternio terminorum*) arguments with H, P, DM middle terms.

As to the non-experts' group, in case of *quaternio terminorum* with plausible conclusion (TP/PC) condition, no significant difference was registered between incorrect answers ('YES') and correct answers ('NO') in both homonymy (mean difference = -5.66; $p = 0.25$) and polysemy condition (mean difference = 9; $p = 0.13$). However, a significant difference was observed between incorrect answers ('YES') and correct answers ('NO') in dead metaphors condition (mean difference = 17.34; $p = 0.015$) (see Figure 3). As to the experts' group, in case of *quaternio terminorum* with plausible conclusion (TP/PC) condition, a significant difference was found between incorrect answers ('YES') and correct answers ('NO') in both homonymy (mean difference = -22.7; $p < 0.001$) and polysemy condition (mean difference = -12.33; $p = 0.018$). However, no significant difference was observed between incorrect answers ('YES') and correct answers ('NO') in dead metaphors condition (mean difference = 0.67; $p = 0.86$) (see Figure 4). Even in case of dead metaphors, overall experts' performance is significantly better than non-experts' ($p = 0.03$).

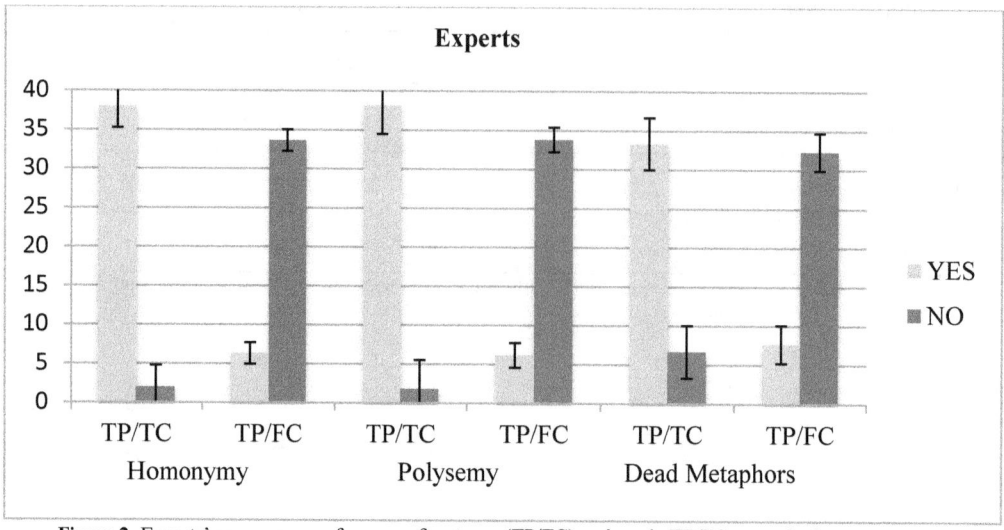

Figure 2. Experts' group: mean of answers for strong (TP/TC) and weak (TP/FC – standard *quaternio terminorum*) arguments with H, P, DM middle terms.

Legenda for figures 1 and 2:
TP = true premises;
TC = true conclusion;
FC = false conclusion;
PC = plausible conclusion.

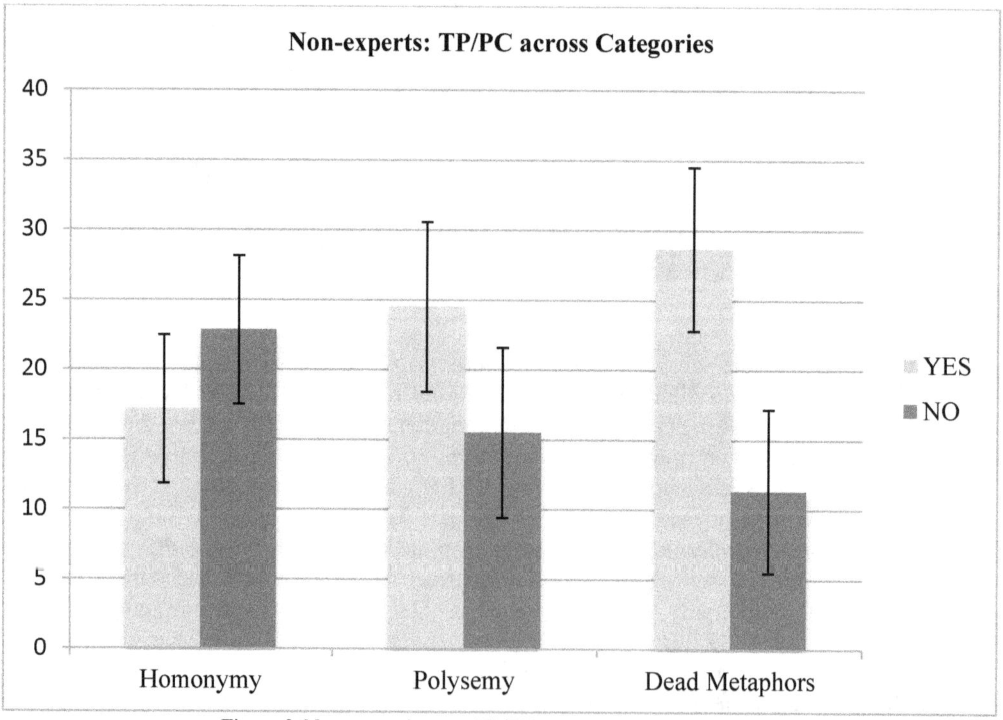

Figure 3. Non-experts' group: TP/PC cases across categories.

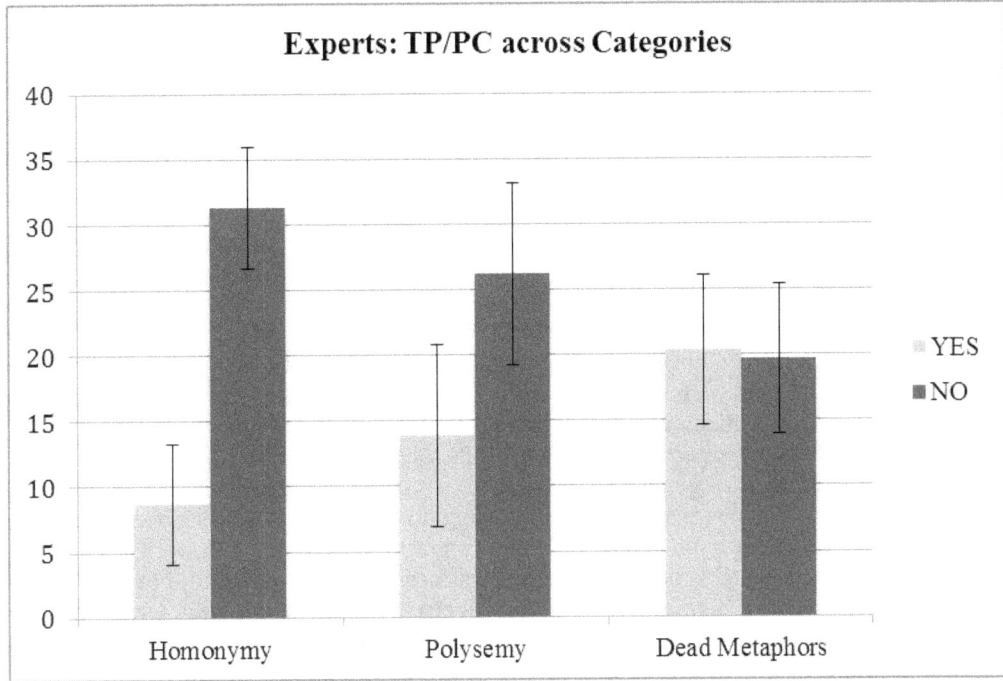

Figure 4. Experts' group: TP/PC cases across categories.

Furthermore, in the TP/PC condition, a certain trend towards significance was found while comparing the number of correct answers ('NO') in the homonymy condition and the number of correct answers ('NO') in the polysemy condition (mean difference = 7.33; $p = 0.08$), in the non-experts' group. Instead no significant difference was registered while comparing the number of correct answers ('NO') in the homonymy condition and the number of correct answers ('NO') in the polysemy condition (mean difference = 5.17; $p = 0.19$), in the experts' group.

A significant difference was observed in the number of correct answers ('NO') within the homonymy condition and the dead metaphors condition in both non-experts' (mean difference = 11.51; $p = 0.005$) and experts' (mean difference = 11.67; $p = 0.005$) groups. No significant difference was recorded between the number of correct answers ('NO') in the polysemy condition and the dead metaphor condition in both non-experts' (mean difference = 4.17; $p = 0.29$) and experts' (mean difference = 6.49; $p = 0.14$) groups.

6. Discussion

The study aimed to investigate whether the persuasiveness of a syllogism might depend on the nature of the lexical ambiguity of the middle term, and, in particular, on the figurative meaning of the middle term - as in the case of metaphors (Metaphoric effect). Moreover, the study was designed to test whether, in the detection of ambiguity fallacies, the expertise might favor not only the disambiguation process of the premises, but also the detection of the metaphoric fallacy (Expertise effect).

The results from the experimental study suggest that, in general, both expert and non-expert participants can evaluate correctly the strength of an argument in presence of non-ambiguous middle terms, as in case of distractors. On the one hand, the results reveal an effect of lexical ambiguity: especially in case of metaphors, fallacious arguments are harder to be detected by both groups (see § 5.1). On the other hand, the results reveal an effect of expertise: experts' group performs significantly better than non-experts' group, overcoming or mitigating the effect of lexical ambiguity (see § 5.2).

6.1. Metaphoric Effect

Both experts and non-experts detected good arguments, also in the presence of potentially ambiguous middle terms with the same meaning in both premises (as in the case of "strong arguments", see table 3 and 4). Indeed, the majority of participants are indeed able to identify strong arguments even in presence of potentially ambiguous middle terms. They are also able to single out standard *quaternio terminorum* with true premises and false conclusion. However, it should be observed that, compared to the identification of strong arguments with potentially ambiguous middle terms, their performance is slightly inferior in both groups. Why arguments with univocally used middle terms were easier to detect compared to fallacious arguments?

In our view, arguments with no lexical ambiguity are easier to detect since they do not require any disambiguation process. On the contrary, a *quaternio terminorum* requires a process of disambiguation to single out the existence of the four different meanings within the logical structure. Both experts and non-experts are asked to suppress one of the two meanings of the middle term, which means something in the first premise and something else in the second, to detect the ambiguity fallacy. Disambiguating a homonymous middle term would require the suppression of one of its two literal meanings, namely the irrelevant one in the premise (Gernsbacher, 1990; Gernsbacher & Faust, 1991). Processing a lexical form, such as 'bank', requires the activation of two different and unrelated lexical entries, and the suppression of the irrelevant one by default on the basis of the so-called *pre-semantic* or *narrow context* (Perry, 1997, 2001), i.e. the sentential context. Contrary to homonyms, in which completely independent meanings only are involved, polysemous words do not exhibit a list of possible meanings to select. In fact, in the case of polysemy, the selection of the relevant meaning requires a process of modulation and a broader context (Bach, 2012; Carston, 2002; Perry, 1997, 2001; Recanati, 2004). For instance, in the sentence «The President cut the taxes», we pick up a more specific concept 'cut' than the conventional concept encoded by the polysemous term 'cut'. This ad hoc concept is relative to the sentential context and differs from the ad hoc concept we pick up in the sentence «I cut the grass» (Falkum, 2011). The process of modulation might be a *lexical narrowing* when the concept expressed by the usage of a term conveys a more restricted interpretation than the linguistically-encoded concept, or a *lexical broadening* when the concept expressed by the usage of a term conveys a more general interpretation than the linguistically-encoded concept.

Middle terms used in a metaphorical sense have figurative meanings in the one premise that depart from their literal ones in the other. How might the detection of *quaternio terminorum* be in such cases? A metaphorical extension is a lexical broadening, where some properties of the linguistically-encoded concept are selected to grasp the intended, contextually relevant *ad hoc* concept. In this perspective, metaphors are explained as a local, on-line pragmatic modulation of the encoded lexical meaning that results in an *ad hoc* concept, which conveys the figurative meaning. For instance, to understand the sentence «Ned Ludd was the *head* of the movement», we need to select the properties of the linguistically-encoded concept 'head' required to grasp the intended ad hoc concept, i.e. to be the leader in a certain field. Other irrelevant properties of the linguistically-encoded concept 'head' (for instance, to be a traditional story) are suppressed (Glucksberg, Newsome & Goldvarg, 2001; Rubio Fernandez, 2007).

Metaphor interpretation process seems to have a major effect in the case of fallacious arguments with a plausible conclusion. Indeed, as expected, the most interesting results are from the set of *quaternio terminorum* with true premises and plausible conclusion (see Tables 5 and 6), which forces participants to evaluate the whole argument and avoid a separate reading of the single conclusion. In case the middle term is a dead metaphor, we observe a very significant difference between the number of correct answers given in the dead metaphors middle term condition and in the homonymous middle term condition in both experts' and non-experts' groups ($p = 0.005$). On the contrary, dead metaphors middle terms condition is not significantly different from polysemous middle terms condition in both groups. This may happen because they share a similar process of modulation, and therefore a similar mechanism of suppression of the irrelevant information, as previously argued.

In any case, a significant difference ($p = 0.015$) between non-expert participants missing and recognizing the fallacy is observed in dead metaphors middle term condition, but not in homonymous and polysemous middle term conditions. In the case of dead metaphors as middle terms, this seems to suggest that non-experts are more prone to judge strong a fallacious argument. Even experts, who could detect a fallacy in both homonymy and polysemy cases, appear more insecure in the case of dead metaphors as middle terms. Therefore, the results seem to suggest that a metaphoric effect alters participants' perception of the strength of an argument, influencing their intuitions about the logical connection of the entire argument.

6.1. Expertise Effect

Experts generally perform significantly better than non-experts in every condition in exam. In the case of standard *quaternio terminorum* with metaphors as middle terms, expertise helps participants to prevail over the metaphoric effect. The results of the experimental study suggest indeed some difficulties in recognizing the fallacy, especially in case of dead metaphors middle term condition in non-experts' group ($p < 0.001$). Despite that, this is not the case for the experts' group ($p = 0.60$). Indeed, in case of dead metaphors as middle terms, the results indicate that, for experts, detecting a standard *quaternio terminorum* is not harder than detecting a strong argument. Moreover, expertise favors the detection of the ambiguity fallacy in case of fallacious arguments with plausible conclusion, that forces the participants to evaluate the whole argument. Experts are indeed prone to judge fallacious these arguments in both homonymous and polysemous middle term conditions. Even in case of fallacious arguments with plausible conclusion and metaphorical middle term, experts perform better than non-experts ($p = 0.03$). In case of fallacious arguments with plausible conclusion, this suggests that the "metaphoric effect" is mitigated by expertise.

These results can be explained by distinguishing between the truth conditions of a literal sentence, and the intuitive truth conditions assigned by a speaker in specific contexts in which a sentence is used (Ervas & Ledda, 2014). While we can assign truth conditions to premises where a literal (homonymous or polysemous) middle term occurs, we usually assign intuitive truth conditions to premises containing a metaphorical middle term. In case of metaphorical middle terms, non-experts perceive premises as true, assigning them intuitive truth conditions. On the contrary, experts might perceive such premises as literally false. According to Contextualism (Carston, 2002; Recanati, 2004, 2010; Sperber & Wilson, 1986), the 'falsehood' of a metaphor is just an expert's "myth", and an attempt to judge the metaphor under some sort of truth conditions, the literal ones, that cannot explain the nature of metaphor in everyday contexts (Clark, 1994). In this perspective, there is no literal meaning in people's intuitions: when non-experts read a premise where a metaphor occurs, they assign intuitive truth conditions to the sentence, thus immediately considering the sentence in which it occurs plausible or even true. This could be the reason why non-experts judged premises featuring a dead metaphor as true, thus compromising the evaluation of the strength of the overall argument, when they are forced to judge whether the conclusion follows from the premises.

However, on the non-contextualist side, it could be claimed that dead metaphors are just perceived as true because they are lexicalized (Stern, 2006; Szabo, 2012). Proper, live metaphors would still be perceived as false, as the classical view predicts (Grice, 1989; Katz, 1972; Katz & Fodor, 1963; Searle, 1985). This is one of the main reasons why we could not compare live metaphors with other cases of lexical ambiguities in this experimental study. The experimental literature has shown that the interpretation process of novel metaphors diverges from that of conventional metaphors (Blasko & Connine, 1993; Thibodeau & Durgin, 2008). Indeed, live metaphor comprehension involves iconic representations of concepts or imagery (Carston, 2010; Indurkhya, 2007) and the contextual information in a sentence would be too narrow to produce their typical imagistic effect (Lai, Curran & Menn, 2009). In a narrow context, dead metaphors are instead perceived as true even though they are literally false. The case of lexicalized metaphors is indeed very interesting because, as the experimental literature shows, they are processed as fast as literal meanings (Giora, 2003) and participants show some difficulty in rejecting them as literally false (Glucksberg, 2003). This might be the main reason why 'common' dead metaphors make fallacious arguments persuasive, while expertise in the detection of formal fallacies might mitigate the metaphoric effect.

7. Conclusion

The results of our study emphasize that, for both experts and non-experts, it is more difficult to detect a *quaternio terminorum* in comparison with a strong argument. We argue that a reason for this phenomenon is that the disambiguation process required to single out the different meanings of the middle term in the premises ask for additional attentional resources. Especially in case of dead metaphors, the experimental evidences seem to indicate that non-experts have some difficulties in detecting the fallacy of the four terms, whilst expertise helps in prevailing over the metaphoric effect. Furthermore, the results highlight a tendency between non-experts in judging strong *quaternio terminorum* with lexicalized metaphors as middle terms. This is indeed so when the conclusion of the argument is far from being patently false. As a matter of fact, the metaphoric effect shows its influence when non-experts are asked to verify the connection between the premises and the conclusion of an argument. A motivation for this could be that non-experts assigned intuitive truth conditions to premises in which a dead metaphor occurs. This might have led them to consider true premises that are literally false. It is certain that expertise reduces the metaphoric effect in case of dead metaphors: experts could assign literal truth conditions to lexicalized metaphors. In spite of that, a minor metaphoric effect is still in place in case of fallacious arguments with plausible conclusions.

In this article we considered standard syllogisms only. This is due to the fact that their regular form makes them highly manageable and easy to set up within an experimental framework, even though experimental constraints might make them similar to textbook examples. However, this work is part of a wider project, which plans an extension of the experiment to arguments involving much wider contexts. It has indeed been shown that additional semantic information coming from the context may foster more stable representations, the so-called "context availability effect" (Glucksberg-Estes, 2000). As already experimentally tested, «the amount of attentional resources involved in interpreting a metaphorical expression would be determined by the combination of these two factors: the degree of familiarity of the metaphorical interpretation and the strength of the contextual bias» (Rubio Fernandez, 2007: 366). Moreover, future research will also consider participants' reaction times in the evaluation of arguments, with homonymous and polysemous middle terms. These data will be precious to understand better the influence different forms of lexical ambiguities may have in the detection of fallacies.

Finally, the experimental study excluded live metaphors as middle terms. In fact, the results from the norming studies indicate that premises in which a live metaphor occurs are perceived as patently false. This made them incomparable to the other group of arguments with true premises. We hypothesize that the detection of an argument, whose middle term is a live metaphor depends on the broadness of the context provided. This could explain why, within the narrow context involved in the premises, participants interpreted live metaphors as literally false. According with this perspective, live metaphors may appear 'less persuasive' when compared with dead metaphors: indeed, they are rarely perceived as true in a premise of an argument. In case of live metaphors, both groups justified the evaluations of the arguments in the norming studies, by reporting new analogies between the meanings of the middle term. In point of fact, previous literature suggested not only that the presence of (live) metaphors render unsound an argument, but also that metaphors induce a more creative style of reasoning, in a number of scientific disciplines ranging from physics to philosophy (Hesse, 1974, Goodman, 1976, Searle, 1979, Kuhn, 1979, Blackburn, 1984). Further research is required to understand when and why live metaphors could elicit a more creative and productive argumentation style, thus favoring analogical reasoning. Under this respect, metaphors should not be interpreted as a trap leading to fallacies, but as a helpful means for both experts' and non-experts' creative thinking.

Acknowledgements

This work is the outcome of a collaborative effort. However, for the specific concerns of Italian academy, Francesca Ervas is responsible for sections 3.1, 4, 6, 7; Antonio Ledda for sections 1, 2, 3.2 and Antonio Pierro for the section 5. We warmly thank Elisabetta Gola and Giuseppe Sergioli for the stimulating discussions on this experimental study which is part of the wider project "Argomentazione e Metafora. Effetti della comunicazione persuasiva nel territorio sardo" (RAS, L. //2007). Francesca Ervas gratefully acknowledges the support of the Sardinia Regional Government for the financial support (P.O.R. Sardegna F.S.E. Operational Programme of the Autonomous Region of Sardinia, European Social Fund 2007-2013 – Axis IV Human Resources, Objective l.3, Line of Activity l.3.1). Antonio Ledda gratefully acknowledges the support of the Italian Ministry of Scientific Research within the FIRB project "Structures and dynamics of knowledge and cognition", Cagliari-F21J12000140001.

References

Bach, K. (2012). Context Dependence. In M. García Carpintero, M. Kölbel (Eds.), *Continuum Companion to the Philosophy of Language* (pp. 153-184). London: Continuum.
Bartsch, R. (2002). Generating Polysemy: Metaphor and Metonymy. In R. Dirven, R. Porings (Eds.), *Metaphor and Metonymy in Comparison and Contrast* (pp. 49-74). Berlin-New York: Mouton de Gruyter.
Black, M. (1993). More about Metaphors. In A. Ortony (Ed.), *Metaphor and Thought* (pp. 19-41). Cambridge: Cambridge University Press.
Blackburn, S. (1984). *Spreading the Word*. Oxford, Oxford University Press.
Blasko, D., & Connine, C.M. (1993). Effects of Familiarity and Aptness on Metaphor Processing. *Journal of Experimental Psychology: 19*, 295-308.
Bonissone, P. (1987). Plausible Reasoning: Coping with Uncertainty in Expert Systems. In S.C. Shapiro (Ed.), *Encyclopedia of Artificial Intelligence* (pp. 854-863). New York: John Wiley.
Borwein, J., & Bailey, D. (2008). *Mathematics by Experiment: Plausible Reasoning in the 21st Century*. Wellesley, MA: AK Peters.
Carston, R. (2002). *Thoughts and Utterances: The Pragmatics of Explicit Communication*. Oxford: Blackwell.
Carston, R. (2010). Metaphor: Ad Hoc Concepts, Literal Meaning and Mental Images. *Proceedings of the Aristotelian Society, 110*(3), 295-321.
Clark, S.R.L. (1994). The Possible Truth of Metaphor. *International Journal of Philosophical Studies, 2*(1), 19-30.
Copi I. M., Cohen C., & McMahon, K. (2014). *Introduction to Logic*. Harlow: Pearson.
Davidson, D. (1984), What metaphors mean *Inquiries into Truth and Interpretation*. Oxford: Oxford University Press.
De Mauro, T. (2000). *Grande dizionario italiano dell'uso*. Torino: UTET.
Dunbar, G. (2001). Towards a Cognitive Analysis of Polysemy, Ambiguity and Vagueness. *Cognitive Linguistics, 12*, 1-14.
Epstein, R., & Kernberger, C. (2006). *Critical Thinking*. Belmont, CA: Thomson Wadsworth.
Ervas, F., & Ledda, A. (2014). Metaphors in *Quaternio Terminorum* Comprehension. *Isonomia, 4*, 179-202.
Ervas, F., Gola, E., Ledda, A., & Sergioli, S. (2015). Lexical Ambiguity in Elementary Inferences: An Experimental Approach. *Discipline Filosofiche, 22*(1), pp. 1-24.
Falkum, I. (2011). *The Semantics and Pragmatics of Polysemy: A Relevance-Theoretic Account*. PhD thesis: University College London.
Fearnside, W., & Holter, W.B. (1959). *Fallacy: The Counterfeit of Argument*. Upper Saddler River, NJ: Prentice Hall.
Fischer, E. (2011). *Philosophical Delusion and its Therapy*. New York: Routledge.
Fischer, E. (2014). Philosophical Intuitions, Heuristics, and Metaphors. *Synthese, 191*, 569-606.
Fischer, E. (2015). Mind the Metaphor! A Systematic Fallacy in Analogical Reasoning. *Analysis, 75*, 67-77.
Gentner, D., & Grudin, J. (1985). The Evolution of Mental Metaphors in Psychology. *American Psychologist, 40*, 181-192.
Gentner, D., Ratterman, M., & Forbus, K. (1993). The Roles of Similarity in Transfer: Separating Retrievability from Inferential Soundness. *Cognitive Psychology, 25*, 524-575.

Gernsbacher, M.A. (1990). *Language Comprehension as Structure Building*. Hillsdale, NJ: Lawrence Erlbaum.

Gernsbacher, M.A., & Faust, M. (1991). The Role of Suppression in Sentence Comprehension. In G.B. Simpson (Ed.), *Understanding Word and Sentence* (pp. 97-128). Amsterdam: Elsevier.

Gick, M.L., & Holyoak, K.J. (1983). Schema Induction and Analogical Transfer. *Cognitive Psychology*, 15, 1-38.

Giora, R. (2003). *On our Mind: Salience, Context, and Figurative Language*. New York: Oxford University Press.

Glucksberg, S., & Estes, Z. (2000). Feature Accessibility in Conceptual Combination: Effects of Context-induced Relevance. *Psychonomic Bulletin and Review*, 7, 510-515.

Glucksberg, S. (2003). The Psycholinguistics of Metaphor. *Trends in Cognitive Science*, 7(2), 92-96.

Glucksberg, S., Newsome, M.R., & Goldvarg, Y. (2001). Inhibition of the Literal: Filtering Metaphor Irrelevant Information During Metaphor Comprehension. *Memory and Symbol*, 16, 277-294.

Goodman, N. (1976). *Languages of Art*. Indianapolis: Hackett.

Govier, T. (1980). More on Inductive and Deductive Arguments. *Informal Logic Newsletter*, 3, 7-8.

Govier, T. (1999). *The Philosophy of Argument*. Newport: Vale Press.

Grice, H.P. (1989). *Studies in the Way of Words*. Cambridge, MA: Harvard University Press.

Hamblin, C.L. (1970). *Fallacies*. London: Methuen.

Hesse, M. (1966). *Models and Analogies in Science*. Notre Dame: University of Notre Dame Press.

Hesse, M.B. (1974). *The Structure of Scientific Inference*. London: Macmillan.

Indurkhya, B. (2007). Creativity in Interpreting Poetic Metaphors. In T. Kusumi. (Ed.), *New Directions in Metaphor Research* (pp. 483-501). Tokyo: Hitsuji Shobo.

Katz, J.J. (1972). *Semantic Theory*. New York: Harper & Row.

Katz, J.J., and Fodor, J.A. (1963). The Structure of a Semantic Theory. *Language*, 39(2), 170-210.

Keefer, L.A., Landau, M.J., Sullivan, D., & Rothschild, Z.K. (2014). Embodied Metaphor and Abstract Problem Solving: Testing a Metaphoric Fit Hypothesis in the Health Domain. *Journal of Experimental Social Psychology*, 53, pp. 12-20.

Keller, E.F. (1995). *Refiguring Life: Metaphors of Twentieth-century Biology*. New York: Columbia University Press.

Kroeger, P. (2005). *Analyzing Grammar. An Introduction*. Cambridge: Cambridge University Press.

Kuhn, T.S. (1979). Metaphor in Science. In A. Ortony (Ed.), *Metaphor and Thought* (pp. 409-419). Cambridge: Cambridge University Press.

Lai, V.T., Curran, T., & Menn, L. (2009). Comprehending Conventional and Novel Metaphors: An ERP Study. *Brain Research*, 1284, 145-155.

Lakoff, G., & Johnson, M. (1980) *Metaphors We Live by*. Chicago: Chicago University Press.

Lyons, J. (1977). *Semantics*. Cambridge: Cambridge University Press.

Perry, J. (1997). Indexicals and Demonstratives. In B. Hale, C. Wright (Eds.), *Companion to the Philosophy of Language* (pp. 586-612). Oxford: Blackwell.

Perry, J. (2001). *Reference and Reflexivity*. Stanford, CA: CSLI Publications.

Quine, W.V.O. (1960). *Word and Object*. Cambridge, MA: MIT Press.

Recanati, F. (2004). *Literal Meaning*. Cambridge: Cambridge University Press.

Recanati, F. (2010). *Truth-Conditional Pragmatics*. Oxford: Oxford University Press.

Rubio Fernandez, P. (2007). Suppression in Metaphor Interpretation: Differences between Meaning Selection and Meaning Construction. *Journal of Semantics, 24*, 345-371.

Searle, J. (1979). Metaphor. In A. Ortony (Ed.), *Metaphor and Thought* (pp. 83-111). Cambridge: Cambridge University Press.

Searle, J.R. (1985). *Expression and Meaning: Studies in the Theory of Speech Acts*. Cambridge: Cambridge University Press.

Smiley, T. (1973). What is a Syllogism? *Journal of Philosophical Logic, 2*, 136-154.

Sperber, D., & Wilson, D. (1986). *Relevance: Communication and Cognition*. Oxford: Blackwell.

Stern, J. (2006). Metaphor, Literal, Literalism. *Mind & Language, 21*, 243-279.

Szabo, Z. (2012). The Case for Compositionality. In M. Werning, W. Hinzen, E. Machery (Eds.), *The Oxford Handbook of Compositionality* (pp. 64-80). Oxford: Oxford University Press.

Taylor, J.R. (2003). *Linguistic Categorization*. Oxford, Oxford University Press.

Thibodeau, P.H., & Boroditsky, L. (2011). Metaphors We Think With: The Role of Metaphor in Reasoning. *PLoS ONE, 6*, e16782.

Thibodeau, P., & Durgin, F.H. (2008). Productive Figurative Communication: Conventional Metaphors Facilitate the Comprehension of Related Novel Metaphors. *Journal of Memory and Language, 58*(2), 521-540.

Van Eemeren, F.H., & Grootendorst, R. (2004). *A Systematic Theory of Argumentation. The pragma-dialectical approach*. Cambridge: Cambridge University Press.

Walton, D.N. (2005). *Fundamentals of Critical argument*. Cambridge: Cambridge University Press.

Walton, D.N. (2010). Why Fallacies Appear to be Better Arguments Than They Are. *Informal Logic, 30*(2), 159-184.

Appendix

	Standard quaternio terminorum	*Strong arguments*	*Quaternio terminorum with plausible conclusion*
Homonymous middle terms (H)	(P1) Fabrizio Corona è un *narciso*. (P2) Un *narciso* è un fiore. (C) Fabrizio Corona è un fiore.	(P1) Belen è una *valletta*. (P2) Una *valletta* è una showgirl. (C) Belen è una showgirl.	(P1) Talete è un *saggio*. (P2) Un *saggio* è una lettura piacevole. (C) Talete è una lettura piacevole.
Polysemous middle terms (P)	(P1) Un lingotto è una *barra*. (P2) Una *barra* è una lineetta. (C) Un lingotto è una lineetta.	(P1) Un viaggiatore ha una *mappa*. (P2) Una *mappa* è una carta geografica. (C) Un viaggiatore ha una carta geografica.	(P1) Il Papa ha un *seggio*. (P2) Un *seggio* ha un elettorato. (C) Il Papa ha un elettorato.
Dead (lexicalized) metaphors as middle terms (DM)	(P1) Il comico è una *sagoma*. (P2) Una *sagoma* è una linea di contorno. (C) Il comico è una linea di contorno.	(P1) L'amicizia è un *supporto*. (P2) Un *supporto* è un aiuto. (C) L'amicizia è un aiuto.	(P1) Una risata è un *farmaco*. (P2) Un *farmaco* serve alla salute. (C) Una risata serve alla salute.

Table 1a. Validated examples in Italian for each argument condition combined with H, P, DM middle term conditions.

Strong arguments	Weak arguments
(P1) Il comodino è un *mobile*. (P2) Un *mobile* è un oggetto. (C) Il comodino è un oggetto.	(P1) Carla ha una *sedia*. (P2) Una *sedia* ha quattro gambe. (C) Carla ha quattro gambe.
(P1) Brad Pitt è una *persona*. (P2) Una *persona* è un essere umano. (C) Brad Pitt è un essere umano.	(P1) L'opinionista ha un'*opinione*. (P2) Un'*opinione* è un'idea. (C) L'opinionista è un'idea.

Table 2a. Validated examples of distractors in Italian (clearly strong arguments and clearly weak arguments).

Chapter 14

Face the Consequences! Strategic Maneuvering With the *Argumentum ad Consequentiam*

Bart Garssen

Speech Communication, Argumentation Theory and Rhetoric, University of Amsterdam, Amsterdam, the Netherlands, b.j.garssen@uva.nl

Abstract. The argumentum *ad consequentiam* is a fallacy in which the arguer points to the positive or negative consequences of holding a particular belief in order to show that this belief is true or false. The question is how this obviously unreasonable type of argumentation can be given a more reasonable appearance. The *ad consequentiam* fallacy mimics other reasonable types of argumentation: pragmatic argumentation and *ad absurdum* argumentation. It is argued that in different variants of the *ad consequentiam* fallacy its resemblance to reasonable argument forms is exploited. This mimicry explains how the *argumentum ad consequentiam* may seem more reasonable.

1. Introduction

In an interview with evolutionary biologist Richard Dawkins, Wendy Wright, president of the Concerned Women of America, argues the following in her defense of creationism:

1. What a person believes about how human beings are created shapes what they believe about human beings. And if we believe that human beings were created out of love, that is by a loving creator, and has given each one of us not only a material body but a spirit and a soul we are more likely to treat others with respect and dignity.

Dawkins replies that evolution theory is a fact, and that you might as well be concerned with gravity or the Milky Way. Apparently he is worried about the quality of Wright's argument. She seems to argue than creationism is true because as a consequence of our belief in creationism it is more likely that we treat others with respect and dignity. Wright defends a factual standpoint by pointing out the desirable consequences of this standpoint, and by doing so she commits a fallacy that is called the *argumentum ad consequentiam*.

In the *argumentum ad consequentiam* an illicit step from a normative premise to a descriptive standpoint is made. The mistake made in this fallacy is quite obvious even to untrained arguers. This is the finding of an empirical research project concerning the judgments of ordinary arguers of the reasonableness of fallacious and non-fallacious discussion contributions (van Eemeren, Garssen & Meuffels, 2009). In this empirical investigation it was found that the *argumentum ad consequentiam* was judged as a very unreasonable discussion move (2009, pp. 176-179). It should be noted that in the experiments conducted in this project only clear and straightforward examples of the fallacy were used. When the fallacy appears in daily life it is harder to detect.

The question arises why this fallacy is harder to detect in argumentative discourse. The extended pragma-dialectical argumentation theory provides an excellent way of dealing with this problem. The aim of this paper is to show that the *ad consequentiam* resembles other reasonable types of argumentation and that arguers can hide the unreasonableness of this fallacy by way of certain presentational techniques. In this paper I will first give a pragma-dialectical account of *the argumentum ad consequentiam*, then I will give an analysis of this fallacy as a specific type of derailment of strategic maneuvering and describe some presentational strategies used to conceal the fallacious character of *ad consequentiam*.

2. The *Argumentum Ad Consequentiam* According to the Pragma-Dialectical Argumentation Theory

In the pragma-dialectical argumentation theory fallacies are regarded as violation of rules for critical discussion. The *argumentum ad consequentiam* is a violation of the argument scheme rule: standpoints may not be regarded as conclusively defended by argumentation that is not presented as based on formally conclusive reasoning if the defense does not take place by means of appropriate argument schemes that are applied correctly (Van Eemeren & Grootendorst, 2004, p. 194).

An argument scheme – the central notion of this rule – characterizes the way in which the acceptability of the premise is transferred to the standpoint. The pragma-dialectical typology of argument schemes consists of three schemes each with a number of subtypes and variants: (1) symptomatic or 'token' argumentation, where there is a relation of concomitance between the premise and the standpoint; (2) comparison or 'similarity' argumentation, where the relation is one of resemblance; and (3) instrumental or 'consequence' argumentation, where there is a causal relation between the premise and the conclusion.

In each of these schemes a unique argumentative principle is used to relate the argument and the standpoint in such a way that the transference of acceptability is in principle possible – given of course that the assumptions made in the premises are correct. The argument schemes in this typology differ from each other because each scheme comes with different evaluative criteria.[1]

The argument scheme rule consists of two parts: 1) an argument scheme should be appropriate and 2) the argument scheme should be correctly applied. According to the latter part of the rule, the argument scheme should be applied. After the protagonist put forward an argument using a certain type of argument scheme, the antagonist is entitled to ask critical questions pertaining to this particular scheme. The protagonist is in turn allowed to reply to the critical questions. This leads to additional argumentation that is put forward to cope with this criticism. However, if the protagonist does not succeed in coping with the critical questions adequately, he uses the argument scheme in an incorrect way and violates the argument scheme rule.

In principle, the appropriateness of an argument scheme is a matter of intersubjective agreement. In the opening stage of a critical discussion the parties come to an agreement

---

[1] Other aspects of the argumentation, such as the type of standpoint or the type of premise, are important – as we will see – but they are not necessary for distinguishing between them.

about the material starting points, the way the discussion is to be conducted and about the argument schemes that are permissible in the discussion. A protagonist violates the argument scheme rule if he uses a scheme that has been banned from the discussion earlier in this intersubjective procedure.

This happens when, for example, two people discuss the esthetic value of a piece of art and one of them uses an argument by authority (which is a variant of symptomatic argumentation). This arguer may get criticized because the other party finds this type of argument not appropriate in a discussion about esthetics: referring to professor such and such who thinks that this particular art piece is very important does not contribute to this type of discussion and can therefore be considered inappropriate (Garssen, 2006, p. 108).

In other cases, the argument scheme involved is inappropriate irrespective of the intersubjective decisions of the arguers, of instance because of an illicit combination of proposition types. These types of argument schemes are always without exception inappropriate means of defense. *Ad consequentiam* argumentation constitutes an inappropriate argument scheme, because the combination of a descriptive standpoint and a normative premise always leads to an inapplicable scheme (Van Eemeren and Grootendorst, 1992, p. 162). This means that this fallacy should be regarded as violation of the first part of the rule. In *ad consequentiam* argumentation the transference of acceptability is not possible because the acceptability of a descriptive standpoint is independent of the values that are attached to the causal effects of the state of affairs that is described in the standpoint. Asking critical questions does not make sense because it has already become clear that the type of reasoning that is used is inadmissible. In fact, it is not possible to tell what critical questions need to be asked in case of an *argumentum ad consequentiam*.

In other approaches to argumentation other conceptions of *ad consequentiam* are developed, that involve a slightly different kind of conception of this type of reasoning. Walton (1999) for instance, takes the following definition by Rescher as a starting point: this type of argumentation may be broadly characterized as the argument for accepting the truth (or falsity) of a proposition by citing the consequences of accepting that proposition (or of not accepting it) (Walton, 1999, p. 252).

This definition is troublesome because of an ambiguity. In the first part, "accepting the truth (or falsity) of a proposition," can mean: "to hold something to be true" or "to explicitly approve of something." If we choose the last possibility, there is no *ad consequentiam* fallacy at all: the standpoint counts as an incentive to accept something or to refrain from question a certain standpoint. This is for instance the case in following example, used by Walton:

2. The United States had justice on its side in waging the Mexican war of 1848. To question this is unpatriotic, and would give comfort to our enemies by promoting the cause of defeatism (1999, p. 252).

In this example the arguer does not point to any consequences of the proposition that is expressed in the standpoint, but to the consequences of taking a stance.

This view of the *argumentum ad consequentiam* is not entirely in line with the pragma-dialectical conception of *ad consequentiam*. Pointing to the consequences of uttering a certain standpoint takes place in the confrontation stage and is a violation of the freedom rule: "discussants may not prevent each other from advancing standpoints or from calling standpoints into question" (Van Eemeren and Grootendorst, 2004, p. 190). *Ad consequentiam*, however should be seen as a violation of a rule that belongs to the argumentation stage of the critical discussion. The fact that these two fallacies constitute two different rule violations is a strong indication that we are dealing with two different types of fallacies.

In conclusion, if someone tries to prevent the opponent in the discussion from putting forward a standpoint, then the move counts as a violations of the freedom rule. If someone points to the negative (or positive) consequences of the content of a certain utterance in order to defend a certain claim, then the move counts as a violation of the argument scheme rule. In order words, a distinction should be made between the consequences of putting forward a standpoint (illocutionary consequences) and the consequences of including the elements in the contents of an utterance (propositional consequences). From a pragma-dialectical perspective only the latter type of move should be considered an *argumentum ad consequentiam*. However, as we shall see in section 4, pointing at the consequences of advancing a standpoint may be an interesting mode of strategic maneuvering when it comes to explaining the deceptive character of the *argumentum ad consequentiam*.

3. *Ad Consequentiam* and Strategic Maneuvering; Structural Resemblance with Other Types of Arguing

The question remains how it is possible that a seemingly absurd mode of reasoning is actually used in argumentation. Why would arguers portray themselves as being unreasonable by openly deviating from the discussion rules? The answer to this question can be found in extended pragma-dialectical theory. Here, a rhetorical component of effectiveness has been added to and integrated within the dialectical framework of classical, standard pragma-dialectics (van Eemeren, 2010).

In their aim to be effective, discussants will maneuver strategically in such a way that they will try to achieve their dialectical goal – keeping to the rules of critical discussion – while simultaneously trying to realize their rhetorical goal: winning the discussion by having their standpoint accepted by the other party. Balancing these two objectives of dialectical resolution-oriented reasonableness and rhetorical effectiveness and trying to reconcile the simultaneous pursuit of these two aims, which may be at times at odds, the arguers make use of what can be called *strategic maneuvering* (van Eemeren, 2010).

In itself there is nothing wrong with wanting to win a discussion, but trying too hard can lead to a derailment: if arguers allow their commitment to having a reasonable exchange be overruled by their eagerness for achieving effectiveness, their strategic maneuvering has been derailed. Viewed from this perspective, fallacies are derailments of strategic maneuvering that involve violations of critical discussion rules. By violating the rules for critical discussion the argumentative move they have made hinders the process of resolving a difference of opinion on the merits and so their strategic maneuvering must be condemned as fallacious.

Derailments of strategic maneuvering may easily escape attention of the interlocutors because deviations of the rules of critical discussion are often hard to detect since none of the parties in the discussion will be keen on portraying themselves as being unreasonable – if only because this will make their contribution ineffective in the end. So arguers will most likely try to stick to the established dialectical means for achieving rhetorical objectives which are possibly at odds with the dialectical rationale for a certain discussion rule, and "stretch" the use of these means so much that the fallacious maneuvering is also covered (van Eemeren, 2010, p. 140). As a consequence, derailments of strategic maneuvering can be very similar to sound instances of strategic maneuvering.

Each fallacy has, in principle, sound counterparts that are manifestations of the same mode of strategic maneuvering (van Eemeren, 2010, p. 199). This also goes for the *argumentum ad consequentiam*. Hence, the first step in the analysis of the fallacy is to identify the potentially reasonable counterparts of this mode of strategic maneuvering. In the *ad consequentiam* the arguer points at consequences of what is stated in the standpoint: the standpoint is true (or false) because the consequences are good (or bad). This happens for instance in the following straightforward example:

3. The research on the bell-curve cannot be right, because adhering to the claim that it is right would lead to discrimination of African Americans.

The arguer points to the negative *causal* effects of accepting the proposition that is denied in the standpoint.

There are several instances of reasonable argumentation where the standpoint is defended by pointing at consequences. For instance, in pragmatic argumentation, which is a subtype of causal argumentation, the standpoint recommends a certain course of action (or discourages a certain course of action) and the argumentation consists of summing up the favorable or unfavorable consequences of adopting that course of action.

The pragma-dialectical characterization of the argument scheme of pragmatic argumentation is as follows:

1	Standpoint:	Action X should be carried out
1.1	Because:	Action X will lead to positive result Y
(1.1')	And:	(Actions of type X that lead to positive results of type Y must be carried out)

Ad consequentiam argumentation is similar to pragmatic argumentation because of the causal claim that is made in both types of argument. The acceptability of pragmatic argumentation depends among other things on its conditional prediction (carrying out this plan will lead to the effects mentioned in the argument). In pragmatic argumentation the standpoint concerns a normative or inciting proposition, while in *ad consequentiam* argumentation the standpoint concerns a descriptive proposition. In this respect the *argumentum ad consequentiam* can be seen as a derailment of pragmatic argumentation.[2]

The consequences that are pointed at in the *ad consequentiam* fallacy do not always need to be causal. In the following example the consequences are logical in nature:

4. You can't agree that evolution is true, because if it were, we would be descendants of the apes and that thought is terrible.

This example differs from the bell-curve argument. For one no causal relation is assumed between the argument and the standpoint: the fact that we are descendants of the apes is not presented as an effect of the evolution theory.

In example (4) the arguer does not point to the causal consequences of adhering to the denial of the standpoint, but rather to its logical consequences. We have to accept the standpoint because of the undesirable logical consequences of the denial of the standpoint. In *ad consequentiam* argumentations of this type the premise does not point to a falsehood

---

[2] In another way Bonnefon (2012, p. 380), also makes a connection with consequential arguments and *ad consequentiam* fallacies. He is particularly interested in cases in which preferences are used to infer beliefs, as in "She doesn't like French cuisine, therefore she prefers not to go to a French restaurant." In this example a preference is mentioned in the premise but this does not make the premise itself normative; in fact, the premise is factual.

or contradiction of what follows from the denial of the standpoint but to a value judgment (the thought that humans are decedents from the apes is terrible).

In this case the arguer exploits the reasoning scheme of the *reductio ad absurdum*. In a *reductio ad absurdum* the standpoint is defended by making it clear that a falsehood or contradiction follows from the denial of the standpoint. This becomes clear after changing the argument in the example, so that we end up with a regular *reduction ad absurdum*:

5. The evolution theory cannot be right, because if it were right, that would mean that we are descendants of the apes, and that is not true.

In a *reductio ad absurdum* the standpoint is defended by showing that what logically follows from the opposite of the standpoint is obviously untrue, while in the *ad consequentiam* fallacy the standpoint is defended by showing that what logically follows from the opposite of the standpoint is undesirable.

In sum, we need to make a distinction between two variants of *ad consequentiam* fallacies: a variant in which pragmatic argumentation is exploited (table 1) and a variant in which the logical form of *reductio ad absurdum* is exploited (table 2).

Reasonable pragmatic argumentation	*ad consequentiam* I
positive variant	
Plan X should be carried out, *because*	Statement X is true, *because*
Plan X leads to positive result Y	Statement X leads to positive result Y
and Actions of type *X* that lead to positive results of type *Y* must not be carried out	*and* Statements of type x that lead to positive results of type Y are true
negative variant	
Plan X should not be carried out, *because*	Statement X is false, *because*
Plan X leads to negative result Y	Statement X leads to negative result Y
and Actions of type *X* that lead to negative results of type *Y* must not be carried out	*and* Statements of type x that lead to negative results of type Y are false

Table 1. *Ad consequentiam* as an exploitation of pragmatic argumentation

Reasonable *reductio ad absurdum* (modus tollens)	*ad consequentiam* II
positive variant	
X has to be true, *because*	X has to be true, *because*
If X were not true, Y would be true	If X were not true, Y would be true
and Y is not true	*and* Y is undesirable
negative variant	
X cannot be true, *because*	X cannot be true *because*
If X were true, Y would be true	If X would be true, Y would be true
and Y is not true	Y is undesirable

Table 2. *Ad consequentiam* as an exploitation of *reductio ad absurdum*

The fact that we are dealing with two different forms also becomes clear if we compare the two fallacious types and their reasonable counterparts more closely. *Ad consequentiam* as an exploitation of pragmatic argumentation features a change in the standpoint: the standpoint in the pragmatic counterpart is normative (plan X should be carried out) while in the *ad consequentiam* form it is descriptive. This is different from the *ad absurdum* variant where the argument is descriptive in the reasonable form and normative in the unreasonable form.

Apart from these formal differences between the two types, there are also noticeable differences in formulation. Obviously, the standpoint of the fallacious *ad absurdum* variant will be formulated using phrases such as, "it cannot be the case that," or "this just has to be true," because these are used with *ad absurdum* reasoning. The premise in the causal variant will point to causal effects of what is stated in the standpoint. Therefore, phrases like 'that will make x look bad' are likely to be used.

4. Concealing the *ad Consequentiam* Fallacy

The fact that the *ad consequentiam* fallacy bears structural resemblances to reasonable argumentative moves may already provide some explanation how it is possible that the *ad consequentiam* fallacy appears to be reasonable. By way of presentation devices, the unreasonableness of the fallacy may be further concealed.

The fallaciousness of the logical (*ad absurdum*) variant of the *ad consequentiam* can camouflaged by presenting the premise that is used in defense of the standpoint in a special way. This premise is normative and, as we have seen, in the *reductio ad absurdum* variant it is factual. A way to conceal the unreasonable character of the logical variant of the *ad consequentiam* is to choose the wording of the premise in such a way that the charitable reader chooses an interpretation of the argumentation that comes closer to the *ad absurdum* variant.

The arguer could for instance use the word 'absurd' instead of 'terrible' to qualify the consequences. The original example was formulated as follows: "You can't agree that evolution is true, because if it were, we would be descendants of the apes and that thought is terrible". Example (6) looks less unreasonable because of the phrasing of the last part of the premise:

6. You can't agree that evolution is true, because if it were, we would be descendants of the apes and that is absurd.

This choice is strategically advantageous because 'absurd' can mean 'terrible', but also 'untrue'. In other words, the speaker makes use of an unclarity or ambiguity do get away with the *ad consequentiam* fallacy. The general strategy is to present the *argumentum ad consequentiam* as if it is an argument based on the *ad absurdum* reasoning scheme.

When presenting the causal variant of *ad consequentiam* it is important to conceal the nature of the standpoint, since if it is too obvious that the standpoint is descriptive, the fallacious character of this type of reasoning will come to light too easily. This is illustrated in the following example:

7. It is very careless to say that cancer is chiefly a matter of irrepressible cell growth that can only be stopped by means of quasi-heroic interventions by doctors. This is even deceitful because this unproven theory deprives the patient of his self-curing capacities that can clean up the cancer from within.

In this argument the arguer defends the (descriptive) standpoint that cancer is not just a matter of irrepressible cell growth. The argument however seems to be put forward in defense of another (normative) standpoint: it is better not to tell patients that cancer is a matter of irrepressible cell growth. The use of the word 'careless' in the standpoint makes both interpretations possible.

The *argumentum ad consequentiam* looks like another mode of strategic maneuvering: pointing out that advancing a certain standpoint has harmful consequences. This mode of strategic maneuvering is different from the pointing to consequences of putting forward a standpoint (see section 2). Pointing at the illocutionary consequences of advancing a standpoint is a mode of strategic maneuvering that belongs to the confrontation stage. This type of move is fallacious when it is violates the freedom rule (the rule for the confrontation stage): "Discussants may not prevent each other from advancing standpoints or from calling standpoints into question". (van Eemeren and Grootendorst, 2004, p. 190). When a critic charges an arguer with advancing a standpoint that has harmful consequences he violates the freedom rule.

Fallacies that are typically committed in the confrontation stage are associated with pointing at consequences of advancing a standpoint. An arguer may for instance point at the negative consequences for the opponent if the opponent sticks to his standpoint. In this case the argumentum *ad baculum* is committed. If an arguer points at the tiresome consequences for him or herself he or she may commit an *ad misericordiam* fallacy. The opponent it silenced by a technique of restricting his or her freedom by his or her feelings of compassion (Van Eemeren and Grootendorst, 1992).

Generally speaking, in these cases the protagonist is forced to withdraw his standpoint because of the negative consequences of maintaining the standpoint. In both cases the freedom rule is violated.[3]

Pointing at the consequences of advancing a standpoint can only be seen as violation of the freedom rule if the opponent is prevented from advancing standpoints or from calling standpoints into question. However, in a discussion a speaker may in all reasonableness warn against openly venting a certain opinion. He may defend this warning by pointing at the negative consequences of openly venting this opinion for a certain audience or group. This advice or warning is not directly related to the current discussion and hence it cannot be seen as a violation of the freedom rule. This type of move is also not an *argumentum ad consequentiam* because the arguer does not claim that the assertion is true or false. The arguer only advices the opponent not to advance a certain standpoint in a particular situation. By reformulating the original example (7), we can make it seems as if the arguer just advices not to advance a standpoint.

8. It is unwise to say that cell growth can only be stopped by means of quasi-heroic inventions by doctors because this unproven theory deprives the patient of his self-curing capacities that can clean up the cancer from within.

Whether this is true or not, when formulated in this way it may be seen as just an advice. But it can still also be read as argumentation defending a factual standpoint, although literary speaking, no factual standpoint is put forward. Again, an ambiguity is introduced. A similar technique is used in the following example:

---

[3] Van Laar, (2006) argues that there may be exceptions to this rule when the consequences are dialectically relevant. "For instance, the critic may allege that advancing the standpoint, in these problematic circumstances, or presented in this controversial manner, will probably lead to the termination of the discussion, without the difference of opinion having been resolved."

9. Rationality and analytic faculty cannot be called male attributes. If we do regard them as such, we give men an unwarranted advantage in job applications and promotion.

The standpoint in this example is not that rationality and analytic faculty *are* male attributes. It seems to be a matter of choice: *we should not call them male attributes*. The latter is not a factual standpoint, but a practical one. Presenting the standpoint as a factual statement would immediately reveal the fact that the arguer is trying to defend a factual statement by pointing at the negative consequences of this statement. In this case, the strategy for concealing the fallaciousness of the causal *ad consequentiam* is presenting the standpoint as a policy statement.

I hope to have shown in this paper that *ad consequentiam* argumentation is structurally similar to reasonable types of argumentation: pragmatic argumentation and *ad absurdum* argumentation. The arguer can strategically conceal the treacherous character of the *ad consequentiam* by making use of certain presentational devices so that this fallacy looks like these reasonable counterparts. In case of the logical variant of *ad consequentiam* the arguer can make it seems as if he points out the fact that what logically follows from a statement is not true, when in fact he actually points at the negative consequences of these consequences. In case of the causal variant of the *ad consequentiam*, the arguer may present his standpoint not as a factual statement but as a policy statement. It is a matter of empirical research to show to what extent these techniques are effective in practice.

References

Bonnefon, J. F. (2012). Utility conditionals as consequential arguments: A random sampling experiment. *Thinking and Reasoning, 18*, 379–393.

Eemeren, F. H. van, Garssen, B., & Meuffels, B. (2009). *Fallacies and judgments of reasonableness. Empirical research concerning the pragma-dialectical discussion rules*. Dordrecht: Springer.

Eemeren, F. H. van & Grootendorst, R. (1992). *Argumentation, communication and fallacies*. A pragma-dialectical perspective. Hillsdale, New Jersey: Lawrence Erlbaum.

Eemeren, F. H. van & Grootendorst, R. (2004). *A Systematic theory of Argumentation*. The Pragma-Dialectical Approach. Cambridge: Cambridge University Press.

Garssen, B. (1997). *Argumentatieschema's in pragma-dialectisch perspectief. Een theoretisch en empirisch onderzoek*. Amsterdam: IFOTT.

Garssen, B. (2006). *Beweringen met nare consequenties: twee varianten van het argumentum ad consequentiam*. In B. J. Garssen & A. F. Snoeck Henkemans (Eds.), *De redelijkheid zelve. Tien pragma-dialectische opstellen voor Frans van Eemeren* (pp. 107-115). Amsterdam: Rozenberg.

Laar, J. A. van (2006) Don't say that! *Argumentation, 20*, 495-510.

Walton, D. (1999) Historical Origins of Argumentum ad Consequentiam. *Argumentation, 13*, 251-264.

Part IV
Communication and Persuasion

Chapter 15

The Psychological Approach to Interpersonal Argumentation in the U.S. Argumentation Community[1]

Dale Hample

Department of Communication, University of Maryland, College Park MD, 20742, USA, dhample@umd.edu

Abstract. The development of the U.S. approach to studying interpersonal arguing is examined from an historical perspective. The intellectual climate leading into the 1970s is summarized, and then Joseph Wenzel's influence at the University of Illinois is explained. Historical and current trends in this scholarly tradition are summarized.

1. Introduction

This chapter surveys the history and intellectual development of psychological approaches to understanding interpersonal argumentation, as these developed in the discipline of communication in the United States. Although some very early points of interest are now detectable in retrospect (e.g., Yost, 1917), this orientation only became self-aware in the 1970s and 1980s. Argumentation studies first had to break off from the rhetorical and debate traditions that had previously monopolized the study of argument in the discipline of communication, and did so partly by concentrating on the idea of arguing as an interpersonal exchange. Once the legitimacy of seeing arguing as a facet of interpersonal communication was accepted, scholars could easily examine not only face-to-face encounters but also various processes and elements that underlie public behaviors.

These "processes and elements" came to include cognitive processes involved in message production and reception as well as various traits that were found to predispose ordinary actors toward various understandings of argument. This paper surveys these research emphases historically in an effort to sketch the development of a minority position in both argument studies and in the discipline of communication. Both in the beginning and in the present, all this work has struggled for attention, even in the open-minded environment of argumentation studies.

---

[1] I am grateful to Susan L. Kline for checking and supplementing my memories of our experience at the University of Illinois. Remaining mistakes are mine.

2. Background to the 1970s

The 1970s was a key decade for the development of interpersonal and psychological argumentation studies. Young scholars inherited traditions of argumentation theory and practice, but were also immersed in the radical changes taking place in the discipline at large. In effect, and perhaps without much self-conscious awareness that they were doing so, they fused traditional rhetorical and pedagogical argumentation work with the new, energizing interests in interpersonal communication that were controversially gaining a secure place in departments across the country.

2.1. The Inheritance

The formal birth of the modern American discipline of communication took place on 28 November 1914, when 17 members of the National Council of Teachers of English finally decided that they had little future in English departments, and voted to leave the NCTE to establish their own association, which several name changes later is now the National Communication Association (A. Weaver, 1959). They were teachers of public speaking, housed in English departments where they were regarded with less respect than literary critics or composition teachers. They taught public speaking, and as the decades went on, they added debate, oral interpretation, small group discussion, and radio broadcasting to their curricula. All these were performance courses.

Entering the 1970s, doctoral education in communication was far broader in scope than it is today. Graduate coursework was similar in most Ph.D.-granting institutions, the majority of which were in the Midwestern region of the U.S. Doctoral students would take a seminar in British public address, one or two seminars in American public address (probably broken at the Civil War if there were two), at least one course in classical rhetorical theory (if two, one would be for the Greek writers and the other for the Romans; the medieval and Renaissance periods were not widely known), a methodology course in rhetorical criticism, a course in radio/TV (probably emphasizing production rather than theory), and a course in speech and hearing sciences (in which they would learn the phonetic alphabet and basic information about stuttering, hearing loss, and similar matters). Very commonly, a doctoral student would also take a course in either theater or oral interpretation. Contemporary rhetorical theory courses were added after the Second World War. These would cover the British period (Campbell, Blair, and Whately), and eventually would include American theorists such as Richards, Weaver, and Burke. Argumentation courses were rare. If they were offered, they would generally take the form of instruction in forensics – how to coach debate and individual events, perhaps even with a little performance by the graduate students.

Most of the point of rhetorical theory prior to the 1970s was to inform critical analysis. Marie Hochmuth Nichols brought Kenneth Burke (Hochmuth, 1952) and I.A. Richards (Hochmuth, 1958) to the field's attention in the 1950s, but prior to that, criticism was neo-Aristotelian or Ciceronian. The critic would choose a Great Man, identify a Great Speech, and then analyze the ethos, logos, and pathos present in the text, along with style and the other canons of rhetoric. In the postwar period, critics and rhetoricians began to have more choices about their intellectual commitments. Burke and Richards both produced rich theories of symbols and rhetoric, but neither expressly generated an argumentation theory. Richard Weaver (1953) did, but his clearly politicized applications of it gave many scholars pause. Finally, Toulmin's *Uses of Argument* was championed by Brockriede and Ehninger (1960), and Perelman and Olbrechts-Tyteca's *New Rhetoric* by Dearin (1969). From that point on, what we would today recognize as sophisticated modern argumentation theories found a place in doctoral seminars' reading lists, and this encouraged scholars to hunt out more writers, including Henry Johnstone, Maurice Natanson, Jürgen Habermas, and others from outside the field. Within the discipline, some writers began to try to create argumentation theory themselves, and the *Journal of the American Forensic Association* (now called *Argumentation and Advocacy*) began its slow evolution from a journal interested mainly in pedagogy and forensic competition to one trying to define the cutting edge of argumentation theory in the U.S.

Scientific research, mainly aimed at understanding speech and hearing problems and the necessary psychological characteristics of a good orator or speech teacher, had a small presence in the field from its beginning (Cohen, 1994, ch. 5; Thompson, 1967, ch. 2). But after World War II, graduate students began to take a greater interest in social science. The persuasion research of the Hovland group was unmistakably relevant to the discipline and some people wanted to push forward with those experimental and quantitative methods (Cohen, 1985; see Thompson, 1967). Hovland's work, along with the postwar development of attitude theories in the field of psychology (consistency theory, dissonance theory, attribution theory, and the theory of reasoned action) made 'social science' nearly synonymous with 'psychological research' for young scholars interested in persuasion or face-to-face influence. Early work on interpersonal communication was also dominated by psychological studies (rather than those in counseling or psychotherapy). Quality work in political science and sociology, often dealing with propaganda or the effects of mass media, was available during this period, but it was read mainly by communication scholars interested in the mass media and not by interpersonally-oriented researchers.

This interest was sometimes met with fierce resistance by rhetorical scholars, and during the 1960s and 1970s more than one important department was unpleasant to work in because of controversies about types of research. A state of the discipline conference was convened in 1970, and its product was *The Prospect of Rhetoric* (Bitzer & Black, 1971), a book that betrays little notice of the social science work that was already struggling for presence. But during the 1970s, most doctoral departments made their peace, and most had groups of people doing both humanistic and social scientific work. The social science research was focused on persuasion, with attention to interpersonal communication coming next, but later.

Argumentation in the discipline was still mainly the province of logic, debate practice, and rhetorical theory in the 1970s (Cox & Willard, 1982), but several important publications diverging from those influences appeared in this period. McCroskey (1969) reported a series of studies concerning evidence and its persuasive effects. Thompson (1967) published a summary of social scientific research concerning public address, including a section on invention (pp. 50-65). Miller and Nilsen (1966) edited a book of original essays on argumentation, and some of the field's leading social scientists contributed to the volume (e.g., G. R. Miller, John Waite Bowers, Robert N. Bostrom). Graduate students during this period would encounter argumentation theory in the context of logic and rhetorical theory and would find the greatest development there. But social science voices were audible, if one wanted to listen.

2.2. Wenzel at Illinois

With all this as background, our story now focuses considerably, on one teacher in one place in one decade. The University of Illinois was one of the most prominent communication departments, and had a rich history of remarkable faculty: Karl Wallace, Marie Hochmuth Nichols, and Richard Murphy, among others. This faculty graduated many doctoral scholars, including Wayne Brockriede in 1954 and Joseph Wenzel in 1963.

Wenzel returned to Illinois as a debate coach and assistant professor a year or two after graduating, and spent the remainder of his career there. For all practical purposes, the interpersonal approaches to argumentation began in his classrooms.

Wenzel himself was a fairly traditional rhetorician early in his career (e.g., Wenzel, 1974). He was intrigued by Wallace's teaching about Aristotle, and Brockriede and Ehninger's interest in Toulmin. He himself was one of the first in the discipline to feature Habermas in his thinking (Wenzel, 1979). But as a former debater and coach, he always understood that although arguments appeared in public speeches, they were also part of ordinary human interactions. In his classes, he would sometimes illustrate points by reference to public orations, but just as often by simple examples of spouses or friends talking with one another.

Wenzel was one of those rare teachers who could energize students without trying to press them into his own mold. In the early to mid-1970s, these students were in his classes, often together: Charles Willard, Daniel O'Keefe, Barbara O'Keefe, Sally Jackson, Scott Jacobs, Susan Kline, Brant Burleson, and myself. Just a few years later, Pamela and William Benoit would study with Barbara O'Keefe at Wayne State University. None of us ended up doing the sort of work that Wenzel had done when we met him, although Willard was perhaps closest to him intellectually and Kline and Burleson directly picked up Wenzel's interest in Habermas. Much of what Wenzel assigned us to read was straight out of the rhetorical and philosophical traditions that I have already mentioned, but some was not. I clearly remember one seminar in which we read Toulmin (not merely *Uses of Argument*), Perelman (not only the *New Rhetoric*), and Habermas. Besides this material, closest to Wenzel's heart, he also made sure we were exposed to the work of some cognitive psychologists who studied people's performance on syllogism problems. I remember that several of us were quite impressed with Henle (1962). Her paper made a point quite congenial to us, namely that people solve syllogisms as they understand them, not as the experimenter thought they were. With that qualification in mind, we concluded that people are quite capable of deductive thought, in spite of the counter-evidence we found when we gave exams in undergraduate classes on argumentation.

But Wenzel was not our only teacher. Among the other faculty at the time was the newly-hired Jesse Delia, who propounded a new approach to interpersonal communication. Most of what had been done in the area of interpersonal communication to that time was derivative from personality psychology and tended to develop ideas variable-to-variable in surveys or experiments. Delia found his inspiration in Kelly's construct theory, and developed it into constructivism, one of the few original communication theories of interpersonal communication at that time. This was a cognitive approach to social life, because it emphasized the importance of the constructs people apply in perceiving others (Delia, 1977). All of us took Delia's courses, too. Besides studying constructivism, we also learned the more common approaches to interpersonal communication, which often involved traits (e.g., dogmatism, locus of control, Machiavellianism) that our field had imported from psychology. Many of us took Burke seminars from either Marie Hochmuth Nichols or Wayne Brockriede (who was on a sabbatical leave; Burleson had also studied with him as an undergraduate), and this led the budding social scientists to appreciate symbols and motives at a deeper level. Another line of influence on us was in the sociology department, where Norman Denzin was championing qualitative methods. Not all of us took those classes, but for those who did (e.g., Jackson and Jacobs), this opened the door to conversation analysis and the work of Harvey Sacks and Harold Garfinkel. Kline had already learned speech act theory from Fred Kauffeld before transferring to Illinois. Larry Grossberg gave lectures on Habermas, which strengthened the work Wenzel had done with us. Finally, the Illinois psychology department was one of the nation's best, and we took classes from Harry Triandis, Robert S. Wyer, and particularly from Martin Fishbein.

Fishbein actually used a book by the communication scholar Gary Cronkhite as a text in one of his graduate courses on attitude change, and the very name of his theory – the theory of reasoned action – announced clear connections to argumentation.

Looking back, we were all full of several active ingredients: rhetorical approaches to argumentation, some acquaintance with formal logic, the modern theorists Wenzel was teaching us, cognitive and other psychological approaches to persuasion and interpersonal communication, philosophical analyses of speech acts, micro-sociological work on conversation and interpersonal communication, and training in critical, quantitative, and qualitative research methods. The work that was soon to appear was circulating among us. We were aware of the instruction in conversation analysis that Jacobs and Jackson brought back from their visit to UCLA, and we watched Willard (e.g., 1983, 1989) begin his work bridging rhetorical and interpersonal approaches to argumentation. In addition – and I think this is quite important – we were all intercollegiate debaters. Wenzel, Willard, and I coached debate at Illinois. Jackson, Jacobs, and both O'Keefes debated for us. Kline and Burleson had debated elsewhere. We were all interested in argumentation in the first place because we saw it as a direct outgrowth of the skills that had preoccupied us as undergraduates. And debate is face-to-face. No one ever had to explain to us that arguing could be an interpersonal process.

3. The Genesis of the Psychology of Argument

Our work in the early part of the 1970s was mostly confined to Wenzel's classrooms, but by the end of the decade some important papers were in print and others were on the verge of publication. First I will continue my people-centered story, and then move on to a more conceptual summary of the psychological argumentation scholarship that began in the last quarter of the 20th century.

3.1. More Community Biography

Probably the most important paper any of us ever wrote in service to interpersonal theories of argument was Daniel O'Keefe's (1977) explanation of the two senses of 'argument.' One of these, argument$_1$, was pretty much the rhetorical tradition's target for criticism. This was something a person *makes* and its product was generally a monologic speech or written text. But the important news was the other sense of 'argument,' argument$_2$, a thing that people *have*. So two friends deciding where to have lunch *have* an argument. This, straightforwardly, was interpersonal argument. Significantly, it was its own thing from the beginning. We were not instructed to imagine two orators dramatically elocuting at one another over topics such as dishwashing. We were aimed instead at observing ordinary interpersonal exchanges, and noticing the public reasoning in them. O'Keefe's paper was in response to another early monument in this story, Brockriede's (1975, p. 179) dramatic claim that "arguments are not in statements but in people." Brockriede's paper connected with Wenzel's classroom insistence on distinguishing between the 'product' and the 'process' of arguing (later he published this, adding 'procedure' to the mix: Wenzel, 1980) and with O'Keefe's own philosophical leanings (e.g., D. O'Keefe, 1975) to produce an original ordinary language analysis of argumentation. This remains a foundational paper, and has been found by international communities as well as the American one. Later, I argued that we should also study argument$_0$, the cognitions involved in producing and understanding arguments (Hample, 1985). So between us, we had suggested that arguments were personal (Brockriede), interpersonal (O'Keefe), and cognitive (Hample). All of us understood that we were placing interpersonal argumentation next to, but not within, rhetorical and forensic orientations.

O'Keefe's paper was published in *Journal of the American Forensic Association*. This was the argumentation and debate community's main journal, but it was not regarded as very important by the discipline at large. O'Keefe's essay appeared in a special issue on argumentation theory, and this issue might be an historical mark in the journal's evolution from a debate journal to an argumentation one. American argumentation scholars' reception of O'Keefe's paper – it ranged from bemusement to acceptance to enthusiasm – gave us entrée to the (one) argumentation journal and, very importantly, to the biennial conference on argumentation held in Alta, Utah for the first time in 1979. The Alta conference actually published more papers than *JAFA* did, and the argumentation community was more personal than it was printed. Presenting an interpersonal paper at Alta inevitably meant that your bibliography was in the audience. Acceptance at Alta, in the argumentation divisions of our associations, and by *JAFA* met with no special resistance within the argumentation and debate communities, even though they were still (and even now) dominated by rhetoricians and coaches.

Prominence in the discipline at large was another story. In the 1970s many communication departments still offered tenure track positions to debate coaches, but this was also the decade that began to teach most leading departments that this was a questionable investment of faculty lines, if the return were only to be measured in publications. At our national association's meetings, there were periodic rumblings about the number of forensics associations that programmed panels and about how trimming those affiliates would leave more program slots for more important research, but these impulses never quite got to the floor for a vote. So finding respect in the mainstream (where key journals, jobs, and teaching assignments are controlled) was an uphill fight.

Still, there was some success. Jackson and Jacobs (1980) published their theory of conversational argument in the discipline's most prestigious journal, and Burleson and Kline (1979) broke into that same venue with their Habermas paper. I published my first experiment on a cognitive theory of argument in the field's leading social science journal (Hample, 1978) and was later able to place an early study of argument editing in another such journal (Hample & Dallinger, 1987). But most of our explicitly labeled argumentation research was at Alta or in *JAFA*. It was important, I think, that many of us were placing other sorts of research in the best journals: on constructivism, persuasion, comforting, conversation, message production, and interpersonal communication. This gave credibility to our argumentation work as well.

3.2. Lines of Inquiry

Finally, it is time to move to a conceptual inventory of the American cognitive and psychological research on interpersonal argumentation. I must first be clear about one thing: there is not, and never has been, anything like a unified American school of interpersonal argumentation. Even in Wenzel's classes, we wrote very different papers, and we continued on our separate paths. People well outside Wenzel's circle made many important contributions. A few key scholars have never even been to Alta. This lack of cohesion contrasts with other major movements in argumentation studies. Although I am only on the fringes of these traditions, it seems to me that there is an essential unity within pragma-dialectics and within informal logic. Perhaps theorists aligned with those orientations think that there are raging internal divisions, but it does not look like that to outsiders. I suppose that those scholars might consider that we are somehow unified by adherence to some basic propositions, but I cannot imagine what those might be, beyond the simple recognition that people argue face to face.

Nonetheless, as we move through the developments I summarize next we can see a central theme. People, alone or in contact, were the central preoccupation (in contrast to syllogisms, argument schemes, inferences implemented in software, or some other possibilities). Even those who preferred to study argument texts without interrogating the arguers were looking at individualized (or mutually emergent) lines of action and expression, uniquely personal productions. None of us ever thought that expressing argumentative content in formal propositions was a useful way to describe what happens when people argue. Categorizing impulses, such as the theorizing or counting of fallacies, were left to others without regret. We wanted to know what arguers' motives were (did they orient to the topic or to the possibility of personal attacks?), the development of arguing skill from childhood to adulthood, how we could intervene with children to improve their arguing ability, the ways serial arguments influence the course and prospects of our most important personal relationships, the manner in which arguments could most efficiently be used to persuade people, the nature of people's argumentative reactions to changes in their immediate circumstances, and, as a kind of summarizing impulse, how arguers understand arguing. These research topics often resulted in quite distinct lines of work, but in retrospect they were all centered on the individual or the dyad. All these seemed to be obvious and engaging research topics and they are all essentially person- or relationship-centered matters. Textual, psychological, cognitive, or interactive, all our approaches were aimed at enlarging our understanding of the human experience of arguing.

3.2.1. Psychologism, Text, and Externalization

The first matter that needs to be acknowledged is the fact that not everyone involved in the study of interpersonal argumentation was immediately open to psychological approaches. Conversation analysis, for instance, consists in close scrutiny of discourse, and finds intention in the illocutionary act and identifies effects by the interlocutor's conversational response. In defining this mode of inquiry, Jacobs and Jackson emphasized that they were providing an alternative to psychologism (Jacobs & Jackson, 1989). By interrogating records of conversation, they could discover the structure and function of interpersonal arguing without needing to ask respondents what they thought or how they felt (Jackson & Jacobs, 1980). The text of a conversation is publically observable, and whatever can be learned from that has a certain solidity that people's marks on Likert scales do not.

A parallel impulse was expressed by the pragma-dialecticians, with whom Jackson and Jacobs later worked (van Eemeren, Grootendorst, Jackson, & Jacobs, 1993). One of the basic principles of pragma-dialectics is "externalization," by which these scholars mean that argumentations are public texts, and the arguments must be found there and only there (van Eemeren & Grootendorst, 1983). Pragma-dialectics (introduced to the U.S. in an Alta panel organized by Wenzel and by his publication of some early work during his editorship of JAFA: van Eemeren & Grootendorst, 1982) concerns itself with critical discussions, face-to-face efforts to resolve disputes. The sojourning of this tradition in the U.S. reinforced the idea that interpersonal arguing ("critical discussions") could be studied without using psychological constructs or data.

Although firmly held, these reservations about psychological work were methodological, not philosophical. The commitments to text simply carved out what these researchers would study, and what they would not. As the research developed and moved into neighboring domains, all these scholars proved themselves willing to take a quantitative psychological approach. Where needed, Jackson and Jacobs have always been ready to do such work (e.g., Jackson, Jacobs, Burrell, & Allen, 1986). Jacobs' (1999) normative pragmatics certainly has lost no interest in analysis of the message text, but in reconstructing the arguments that people actually experience, he appreciated that

> When people interpret a message, they construct a context of assumptions and inferences that make sense of what was said and of what was not said but could have been said, and that make sense of how and when all of it is said. The words are not the message. The words and sentences are simply part of an assembly of *cues* that people use to construct the message. It is the context of interpretive assumptions and inferences that is the message. And it is the message that has argumentative functions. (p. 398; italics orig.)

The pragma-dialecticians, too, have been willing to move beyond externalization when their research questions required it. To find out how fallacies are registered, for instance, they collected self-reports of people's judgments about the fallacies' reasonableness (van Eemeren, Garssen, & Meufels, 2009). The best known coding system for public argumentative exchanges, originated in good part by a number of Illinois people including David Seibold, Renee Meyers, and Dale Brashers (e.g., Canary & Seibold, 2010), was developed over the years by scholars praised for their psychological investigations as well.

In short, the occasional opposition to psychological approaches that one can find in some of the early work on interpersonal argumentation was never an objection to the reality of psychological experience, or even a disparagement of its importance. It was simply a way of marking what the researchers wanted to analyze in their scholarship. The other material I cover in this section never hesitated over this matter and enthusiastically dove into self-reports and recollections to answer its research questions.

3.2.2. *Argumentativeness and Verbal Aggressiveness*

Amid the widespread interest in conversation analysis, the 1980s also saw the development of a clearly distinct line of research dealing with two psychological traits, argumentativeness and verbal aggressiveness. These originated at Kent State University under the leadership of Dominic Infante, a former debater and coach who was never really part of the personal community of argumentation scholars. His research group, prominently including Andrew Rancer, Charles Wigley, and Theodore Avtgis, has never had much presence at Alta or in the argumentation journals. In fact, the first key paper was not even published in a communication journal (Infante & Rancer, 1982), although the second one was (Infante & Wigley, 1986). Others have carried this research into the common places for argumentation research, such as the conferences at Alta and Amsterdam, journals such as *Argumentation and Advocacy* and *Argumentation*, and mainstream sites in the discipline at large.

Argumentativeness and verbal aggressiveness are both aggressive impulses, but they differ in their targets. Argumentativeness refers to the motivation to present a controversial view or to attack another person's reasons, evidence, position, or case. Verbal aggressiveness is the impulse to attack the other person's character, identity, habits, or nature. Argumentativeness has been shown to have constructive consequences in a variety of settings, and verbal aggressiveness to be corrosive. These traits have become two of the most commonly researched topics in communication. Infante and Rancer (1982) has been cited more than 400 times in journal articles, and Infante and Wigley (1986) about 550 times. Offspring papers are numerous, and the current summary of this work (Rancer & Avtgis, 2014) is necessarily book length.

Argumentativeness and verbal aggressiveness are theorized to assert themselves in reaction to situational circumstances. These two traits have several complementary and simultaneous characters. They influence how people understand a given situation (substantive or personal; constructive or destructive). They define thresholds for particular behaviors (how much stimulus is required to call out hostile insults). They help describe the profile of someone's personality (aggressive or quiet). Several of these points will be taken up momentarily. These two simple self-report measures have been applied in many nations, and in a variety of settings in the U.S., including classrooms, organizations, families, and the mass media (see Rancer & Avtgis, 2014).

3.2.3. Children

Delia's constructivism always had a pointed interest in the development of interpersonal abilities through the lifespan, and it is no accident that study of children's argumentative ability was begun by Barbara O'Keefe and one of her students, Pamela Benoit (B. O'Keefe & Benoit, 1982). Susan Kline (Kline & Carlson, 1989; Kline & Oseroff-Varnell, 1993) also explored children's abilities. Both O'Keefe and Kline, of course, studied and published with Delia. Constructivism made many well-regarded contributions regarding children's communication, concentrating on the person-centeredness of their messages, the children's developing perceptiveness about others, and how their family experiences affected their later abilities.

However, perhaps the most important work on the explicit topic of children's arguments has been done by someone outside the discipline, Deanna Kuhn and her research teams (e.g., Kuhn, 2005; Kuhn, Katz, & Dean, 2004; Kuhn & Udell, 2003). Kuhn has gone into urban elementary and middle schools in the U.S., many of them in challenging neighborhoods, and has demonstrated that through careful scaffolding of one argument skill onto the last, that it is possible to instill substantially better arguing skills in children. These are probably the most successful of all the interventions aimed at improving people's arguing skills.

A very limited amount of work has been done at the other lifespan extreme (Congalton & Olson, 1993), and so it cannot be said that the American argumentation community has been thoroughly attentive to the development of arguing ability through people's lives. However, research on both constructivism and Kuhn's scaffolding has been very informative in regard to children.

3.2.4. Serial Arguments

In 1985, Robert Trapp (who studied with Brockriede at the University of Denver) made a serendipitous discovery. He and a colleague were engaged in a qualitative study of how arguments affect interpersonal relationships. Interviewing a dozen dyads, they stumbled across the fact that nearly all the arguments being reported to them had occurred before within that dyad (Trapp & Hoff, 1985). They termed this phenomenon serial arguing.

The Trapp and Hoff paper was widely assigned in classes, but it fell to someone outside the argumentation community, Michael Roloff (a former debater), to initiate systematic research on serial arguing (e.g., K. Johnson & Roloff, 1998; Malis & Roloff, 2006). He and his students have explored demand/withdraw tactics in these recurring episodes, and have also done essential work to connect serial arguing with relational satisfaction as well as both physical and mental health outcomes. Jennifer Bevan and her colleagues (e.g., Bevan, Finan, & Kaminsky, 2008; Bevan, Hale, & Williams, 2004) have accomplished the important task of determining what goals lead people to participate in serial arguments, and how those goals relate to arguing tactics and relational outcomes. I have contributed to this research myself, by carrying the topic into specialized settings, such as classrooms, organizations, and inter-ethnic relationships (Hample & Allen, 2012; Hample & Cionea, 2012; Hample & Krueger, 2011).

Together we have established clear links between motivating goals (e.g., to hurt the other), tactics (e.g., demand/withdraw), and outcomes (e.g., relational satisfaction). Research seems to show that the frequency with which episodes occur has little effect on relational satisfaction, but arguers' estimates of whether the argument is resolvable are quite consequential. We have also discovered that serial arguments are common, and go some distance in describing the ethos and content of important interpersonal relationships.

3.2.5. Persuasive Effects

Due to the rhetorical and persuasion elements of our heritage, it was natural to wonder what persuasive effects arguments have. This interest ran against the grain of some theory. Effectiveness, while a legitimate avenue for argument research (Wenzel, 1980), was often contrasted with soundness. The idea was that a sound argument *deserved* to be effective whether it was or not, and this conveniently left persuasion in the hands of mainstream researchers who were not much oriented to argumentation. However, several strands of argumentation research nonetheless took up the issue of effectiveness.

Daniel O'Keefe did several meta-analyses on the effects of various argument structures on attitude change (e.g., D. O'Keefe, 1997, 1998; summarized in D. O'Keefe, 2015). He found, for example, that omitting the conclusion or any other key structural element of a public argument lessened its effectiveness. Rather than advising people to leave their arguments enthymematic to be spontaneously completed by receivers, O'Keefe's research showed that making things explicit and unmistakable was a better tactic.

Another approach to this general question was my own cognitive model of argument (Hample, 1978; recently refreshed in Hample & Richards, in press). This is a theory of belief processing based on the law of total probability. The idea is that people's adherence to a claim can be predicted by their rational registration of two sets of subjective probabilities. One set pertains to the message syllogism, a summary of the argument actually expressed. The model says that adherence to a claim is partly predicted by the subjective probability that the evidence is true, times the probability that the claim is true given that the evidence is true. The other predictor is a parallel syllogism summarizing everything pertinent that is not in the message: the subjective probability that the evidence is false, times the probability that the claim is true even though the evidence is false. This model has a logical/rational basis in the two syllogisms as well as in its loose appropriation of the law of total probability. The equations produce very accurate estimates of how probable people regard a conclusion, and they control substantial amounts of variance (often half of it) in adherence.

This research, along with some outside studies (e.g., Park, Levine, Westerman, Orfgen, & Foregger, 2007), makes it clear that argumentative content predicts (if not controls) persuasive effects. Combined with the continuing research on the importance of evidence (e.g., Reynolds & Reynolds, 2002), all this work makes it clear that argumentation is foundational to persuasion.

3.2.6. Situation

All interpersonal arguments are concrete and situated, and argumentation researchers have identified several features of the situation that affect arguing. Among the key elements of the situation are the other arguer and the argument topic.

One of the earliest of these investigations was Infante (1987). He collected people's argumentativeness scores and then asked them whether or not they would like to debate an opponent they viewed on videotape. When their willingness to argue was predicted solely by their argumentativeness scores, a modest positive correlation appeared. But when Infante also included people's private estimates of the likelihood they would prevail over the videotaped opponent, predictions were dramatically more accurate, increasing from about 7% of the variance in intention to argue to about 40%.

Likelihood of winning an argument has also emerged as an important predictor of willingness to engage in an argument, in a more recent research program stimulated by Paglieri's (2009; Paglieri & Castelfranchi, 2010) analysis of the risks of arguing. Projecting Paglieri's theory into a cost benefit model of argument engagement has produced quite accurate estimates of when people do and do not intend to argue (e.g., Hample, Paglieri, & Na, 2012). The most important predictors of engagement are people's estimates of whether they will win the argument, along with their sense that such an argument would be appropriate or not in that situation.

Another situational feature that must be mentioned is the sort of issue people are arguing about. A. Johnson (2002) introduced a key distinction, between public and personal topics. Personal topics are internal to the arguers' relationship and concern matters such as affection, shared tasks, and relational identity. Public topics are external and deal with matters like capital punishment, city governance, and sports. Extensive research (reviewed in A. Johnson, Hample, & Cionea, 2014) has shown that personal topics are weightier and more involving, while public arguments are more entertaining and have less need to be resolved.

Finally, Jackson, Jacobs, and several of their former students at the University of Arizona have developed another approach to the connections between situations and arguing. They have studied how particular circumstances can be designed in order to promote better arguing, and how design failures impede proper public reasoning (e.g., Aakhus, 2007; Jackson, 1998; Weger, 2001). This approach regards the situation as malleable rather than given, and shows how to intervene into circumstances to generate the sorts of arguing that would be desired.

Collectively, all this scholarship modifies the personality research dealing with argumentativeness and other measures by showing that situational matters need to be included in our theories. Arguing behaviors are not solely trait-determined (i.e., consistent in any situation), nor are they person-independent (i.e., dictated by circumstances). Allowing for an interaction of person and setting is necessary to predict how and when people will argue.

3.2.7. Arguers' Understanding of Arguing

An essential part of the movement away from rhetorical theory toward interpersonal argumentation was focused attention on the people doing the arguing. For many, this meant study of arguers' traits and attitudes toward the argument's topic. But for some, this led naturally to the question of how ordinary people understood the activity of arguing. At first, these investigations were somewhat global (e.g., Benoit & Benoit, 1990; Martin & Scheerhorn, 1985) and produced the finding that people associate 'argument' with hostility and closed-mindedness, whereas collaboratively conducted arguments are called 'discussions.' A similar friction in vocabulary for argumentation is presently being noticed in various cross-cultural studies that require the translation of various argumentation instruments into other languages (see section 2.3.8).

However, later research became more detailed about people's beliefs. Rancer, Baukus, and Infante (1985) identified a number of possible beliefs people might have about arguing. These had to do with various characteristics or affordances of argumentation: hostility, process, dominance, conflict, self-image, learning, and skill. They found that highly argumentative people had more positive beliefs. For instance, they felt that arguing was an opportunity for learning, for reducing conflict, and enhancing self-image. A. Johnson (2002) extended this work by applying it to types of argument issues. She found that public issue arguments were consistently thought to be more enjoyable and to have more positive effects on self-image, when contrasted to personal topics.

A different research program deals with argument frames, which are offered as a summary of how people understand the general project of arguing with another person (Hample, 2005, ch. 2). Some frames are self-oriented (people say they argue in order to obtain some benefit, display some identity, assert dominance over the other person, or to play), some are other-oriented (not blurting, being cooperative, and being civil), and they all culminate in a measure that contrasts scholars' understandings with other common ones (e.g., is arguing an alternative to violence, or an incitement to it?). Several of the frames have been singled out for special study. Playful arguing was found to be more aggressive than enjoyable (Hample, Han, & Payne, 2010), blurting was typical of people who are reactive and aggressive (Hample, Richards, & Skubisz, 2013), and arguing to display identity was pleasant and did not require controversy for its initiation (Hample & Irions, in press).

While this research is interesting in its own right, little work has been done to show how different beliefs about arguing actually affect interactional behaviors. No doubt that will soon be a topic for careful study.

3.2.8. Cross-Cultural Argumentation Research

Whether I had decided to focus on the American argumentation community or not, it would remain the case that nearly all the research on interpersonal arguing was conducted in the U.S., primarily using undergraduate respondents. This certainly does not represent most of the people in the world (Arnett, 2000; Henrich, Heine, & Norensayan, 2010), and so the question naturally arises, How much of this research can be exported to other cultures?

Several researchers, notably Stephen Croucher, an American working at the University of Jyväskylä in Finland, have taken up this challenge, and have studied key matters in other nations. Argumentativeness and verbal aggressiveness have been the most common topics of investigation (e.g., Croucher, 2013; Croucher, et al., 2009; for a summary, see Rancer & Avtgis, 2014, ch. 7). Argument frames and the tendency to personalize conflicts have also been explored in nations other than the U.S. (e.g., Hample & Anagondahalli, 2015; Xie, Hample, & Wang, in press). Translation of U.S. instrumentation into other languages has proved to be challenging, particularly when 'argument,' 'arguing,' 'arguer,' and similar terms need to be re-expressed. A proper summary of this work would need to proceed variable by variable and nation by nation, and this is not the place for that. Most generally, researchers have found that many of the associations that appear in the U.S. are also in evidence elsewhere, but that some are not. While not very specific, this conclusion shows the importance of testing (rather than postulating) American theories elsewhere in the world. If the present paper is a chiefly a summary of American interpersonal arguing practices, this justifies developing Chinese, or Indian, or Japanese theories of face-to-face arguing as well.

However, this research has not really been very 'cultural.' In fact, nations have generally been used as proxies for cultures. This is quite unsatisfactory because few nations are culturally homogenous, and this fault in the research record will eventually need to be repaired. Others have more defensibly investigated cultural arguing practices, and we can find possible models in Katriel's (1986) ethnography of Israel, Fitch's (1998, ch. 3) study of Columbia, or Ellis' (e.g., 2012) investigations of Israeli-Palestinian exchanges. Hazen (e.g., 2007) has long been a voice for proper cultural investigations of argumentation. Happily, interpersonal research has been welcome at argumentation conferences in Amsterdam, Windsor, Tokyo, and Santiago, and there is reason for optimism that culture-specific theories and findings will begin to appear more often and be written with the authority of non-Americans.

4. Final Thoughts

Interpersonal argumentation work, while never shunned, has also never characterized the larger share of argumentation research. In the monumental *Handbook of Argumentation Theory*, the chapter on American argumentation work devotes only about a third of its pages to the sort of research I have reviewed here, and the rest concerns rhetoric and debate (van Eemeren, Garssen, Krabbe, Snoeck Henkemans, Verheij, & Wagemans, 2014, ch. 8). In other domains, such as handbooks of interpersonal communication, argumentation never receives its own chapter. It is far less prominent in mainstream communication research than it is in argumentation studies.

Much of our research was originally stimulated by external work done in psychology, but it remains distinct from the studies done in social and cognitive psychology (e.g., Adler & Rips, 2008; Fishbein & Ajzen, 2010). Cognitive scientists are interested in how people approach deductive (or inductive, abductive, and causal) reasoning problems, but those problems are ordinarily presented in schematic ways without the idiosyncrasy or flourish that give color to actual interactions. These researchers sanitize the people out of the arguments. Reaction times and rational accuracy might be studied in that community, whereas we would want to know how participating in an interpersonal argument (rather than decoding a textual one) could affect one's emotional or relational life. The social psychological approach to persuasion focuses largely on the internal cognitive or attitudinal processes of a single message recipient. Questions surrounding the producer (and even the production) of persuasive messages are rarely raised, and even the descriptions of the persuasive messages are often frustratingly vague. It is quite rare for a psychologist to describe the evidence and arguments that were used in persuasive stimuli, or to do any systematic manipulation of these argument features. Nonetheless, we continue to study this work with profit, and have recently added evolutionary psychology to our reading lists (e.g., Mercier & Sperber, 2011). Our distinctiveness derives from our commitment to the idea that "arguments are in people," to quote Brockriede. This leads to our simultaneous study of content and person, supplemented by our understanding that both are contextualized within situations and phases of both personal and relational development.

I have tried to trace the development of research and theory regarding interpersonal arguing from its beginnings at the University of Illinois to its current standing across the globe. The repeated names in this brief summary should stimulate a reader's suspicion that this research area is not well populated. Enough good work has been done to advance our understandings in many respects: conversational structure, personality traits, development of skills throughout childhood, central arguments within close relationships, persuasive effects, situational considerations, arguers' understandings of what they are doing when they argue, and cultural variability in all of this. Still, more work and more workers are needed, in the U.S. and elsewhere.

References

Aakhus, M. (2007). Communication as design. *Communication Monographs, 74*, 112-117.
Adler, J. E., & Rips, L. J. (Eds.) (2008). *Reasoning: Studies of human inference and its foundations*. Cambridge: Cambridge University Press.
Arnett, J. J. (2000). Emerging adulthood: A theory of development from the late teens through the twenties. *American Psychologist, 55,* 469-480.
Benoit, P. J., & Benoit, W. L. (1990). To argue or not to argue. In R. Trapp & J. Schuetz (Eds.), *Perspectives on argumentation: Essays in honor of Wayne Brockriede* (pp. 55-72). Prospect Heights, IL: Waveland.
Bevan, J. L., Finan, A., & Kaminsky, A. (2008). Modeling serial arguments in close relationships: The serial argument process model. Human Communication Research, 34, 600-624.
Bevan, J. L., Hale, J. L., & Williams, S. L. (2004). Identifying and characterizing goals of dating partners engaging in serial argumentation. Argumentation and Advocacy, 41, 28-40.
Bitzer, L. F., & Black, E. (1971). The prospect of rhetoric. Englwood Cliffs, NJ: Prentice-Hall.
Brockriede, W. (1975). Where is argument? *Journal of the American Forensic Association, 11,* 179-182.
Brockriede, W., & Ehninger, D. (1960). Toulmin on argument: An interpretation and application. *Quarterly journal of speech, 46,* 44-53.
Burleson, B. R., & Kline, S. L. (1979). Habermas' theory of communication: A critical explication, *Quarterly Journal of Speech, 65*, 412-428.
Canary, D. J., & Seibold, D. R. (2010). Origins and development of the conversational argument coding scheme. *Communication Methods and Measures, 4,* 7-26.
Cohen, H. (1985). The deveopment of research in speech communication: A historical persecptive. In T. W. Benson (Ed.), *Speech communication in the 20th century* (pp. 282-298). Carbondale, IL: Southern Illinois University Press.
Cohen, H. (1994). *The history of speech communication: The emergence of a discipline, 1914-1945*. Annandale, VA: Speech Communication Association.
Congalton, K. J., & Olson, C. D. (1993). Argumentation and the older person: An analysis of unplanned discourse. In R. E. McKerrow (Ed.), *Argument and the postmodern challenge* (pp. 246-251). Annandale, VA: Speech Communication Association.
Cox, J. R., & Willard, C. A. (1982). Introduction: The field of argumentation. In J. R. Cox & C. A. Willard (Eds.), *Advances in argumentation theory and research* (pp. xiii-xlvii). Carbondale, IL: Southern Illinois University Press.
Croucher, S. M. (2013). The difference in verbal aggressiveness between the United States and Thailand. *Communication Research Reports, 30,* 264-269.
Croucher, S.M., Braziunaite, R., Homsey, D., Pillai, G., Saxena, J., Saldanha, A., Joshi, V., Jafri, I., Choudhary, P., Bose, L., & Agarwal, K. (2009). Organizational dissent and argumentativeness: A comparative analysis between American and Indian organizations. *Journal of Intercultural Communication Research, 38,* 175-191.

Dearin, R. D. (1969). The philosophical basis of Chaim Perelman's theory of rhetoric. *Quarterly Journal of Speech, 55*, 213-224.

Delia, J. G. (1977). Constructivism and the study of human communication. *Quarterly Journal of Speech, 63*, 66-83.

Eemeren, F. H. van, Garssen, B., Krabbe, E. C. W., Snoeck Henkemans, A. F., Verheij, B., & Wagemans, J. H. M. (Eds.) (2014). *Handbook of argumentation theory*. Dordrecht: Springer.

Eemeren, F. H. van, Garssen, B., & Meuffels, B. (2009). *Fallacies and judgments of reasonableness: Empirical research concerning the pragma-dialectical discussion rules* (Vol. 16). Springer Science & Business Media.

Eemeren, F. H. van, & Grootendorst, R. (1982). Unexpressed premises: Part I. *Journal of the American Forensic Association, 19*, 97-106.

Eemeren, F. H. van, & Grootendorst, R. (1983). *Speech acts in argumentative discussions*. Dordrecht, Holland: Foris.

Eemeren, F. H. van, Grootendorst, R., Jackson, S., & Jacobs, S. (1993). *Reconstructing argumentative discourse*. Tuscaloosa, AL: University of Alabama Press.

Ellis, D. G. (2012). *Deliberative communication and ethnopolitical conflict*. New York: Peter Lang.

Fishbein, M., & Ajzen, I. (2010). *Predicting and changing behavior: The reasoned action approach*. New York: Psychology Press.

Fitch, K. L. (1998). *Speaking relationally: Culture, communication, and interpersonal connection*. New York: Guilford.

Hample, D. (1978). Predicting immediate belief change and adherence to argument claims. *Communication Monographs, 45*, 219-228.

Hample, D. (1985). A third perspective on argument. *Philosophy and Rhetoric, 18*, 1-22.

Hample, D. (2005). *Arguing: Exchanging reasons face to face*. Mahwah, NJ: Lawrence Erlbaum Associates.

Hample, D., & Allen, S. (2012). Serial arguments in organizations. *Journal of Argumentation in Context, 1*, 312-330.

Hample, D., & Anagondahalli, D. (2015). Understandings of arguing in India and the United States: Argument frames, personalization of conflict, argumentativeness, and verbal aggressiveness. *Journal of Intercultural Communication Research, 44*, 1-26.

Hample, D., & Cionea, I. (2012). Serial arguments in inter-ethnic relationships. *International Journal of Intercultural Relations, 36*, 430-445.

Hample, D., & Dallinger, J. M. (1987). Cognitive editing of argument strategies. *Human Communication Research, 14*, 123-144.

Hample, D., & Irions, A. (in press; 2015). Arguing to display identity. *Argumentation*.

Hample, D., & Krueger, B. (2011). Serial arguments in classrooms. *Communication Studies, 62*, 597-617.

Hample, D., Han, B., & Payne, D. (2010). The aggressiveness of playful arguments. *Argumentation, 24*, 405-421.

Hample, D., Paglieri, F., & Na, L. (2012). The costs and benefits of arguing: Predicting the decision whether to engage or not. In F. H. van Eemeren & B. Garssen (Eds.), *Topical themes in argumentation theory: Twenty exploratory studies* (pp. 307-322). New York NY: Springer.

Hample, D., & Richards, A. S. (in press; 2015). A cognitive model of argument, with application to the base-rate phenomenon and cognitive-experiential self-theory. *Communication Research*.

Hample, D., Richards, A. S., & Skubisz, C. (2013). Blurting. *Communication Monographs, 80*, 503-532.

Hazen, M. D. (2007). Dissensus as value and practice in cultural argument. The tangled web of argument, con/dis-sensus, values and cultural variations. In H. V. Hansen, et al. (Eds.), *Dissensus and the search for common ground*, CD-ROM (pp. 1-43). Windsor ON: Ontario Society for the Study of Argumentation.

Henle, M. (1962). On the relation between logic and thinking. *Psychological review, 69*, 366-378.

Henrich, J., Heine, S. J., & Norenzayan, A. (2010). The weirdest people in the world? *Behavioral and brain sciences, 33*, 61-135.

Hochmuth, M. (1958). I. A. Richards and the 'new rhetoric.' *Quarterly Journal of Speech, 44*, 1–16.

Hochmuth, M. (1952). Kenneth Burke and the "new rhetoric". *Quarterly Journal of Speech, 38*, 133-144.

Infante, D. A. (1987). Enhancing the prediction of response to a communication situation from communication traits. *Communication Quarterly, 35*, 308-316.

Infante, D. A., & Rancer, A. S. (1982). A conceptualization and measure of argumentativeness. *Journal of Personality Assessment, 46*, 72-80.

Infante, D. A., & Wigley, C. J. (1986). Verbal aggressiveness: An interpersonal model and measure. *Communication Monographs, 53*, 61-69.

Jackson, S. (1998). Disputation by design. *Argumentation, 12*, 183-198.

Jackson, S., & Jacobs, S. (1980). Structure of conversational argument: Pragmatic bases for the enthymeme. *Quarterly Journal of Speech, 66*, 251-265.

Jackson, S., Jacobs, S., Burrell, N., & Allen, M. (1986). Characterizing ordinary argument: Substantive and methodological issues. *Journal of the American Forensic Association, 23*, 42-57.

Jacobs, S. (1999). Argumentation as normative pragmatics. In F. H. van Eemeren, R. Grootendorst, J. A. Blair, & C. A. Willard (Eds.), *Proceedings of the fourth international conference of the International Society for the Study of Argumentation* (pp. 397-403). Amsterdam, the Netherlands: SICSAT.

Jacobs, S., & Jackson, S. (1989). Building a model of conversational argument. In B. Dervin, L. Grossberg, B. J. O'Keefe, & E. Wartella (Eds.), *Rethinking communication: Volume 2, paradigm exemplars* (pp. 153-171). Newbury Park, CA: Sage.

Jeffrey, R. C. (1964). A history of the Speech Association of America, 1914–1964. *Quarterly Journal of Speech, 50*, 432-444.

Johnson, A. J. (2002). Beliefs about arguing: A comparison of public-issue and personal-issue arguments. *Communication Reports, 15*, 99-112.

Johnson, A. J., Hample, D., & Cionea, I. A. (2014). Understanding argumentation in interpersonal communication: The implications of distinguishing between public and personal topics. *Communication Yearbook, 38*, 145-173.

Johnson, K. L., & Roloff, M. E. (1998). Serial arguing and relational quality: Determinants and consequences of perceived resolvability. *Communication Research, 25*, 327-343.

Katriel, T. (1986). *Talking straight:* Dugri *speech in Israeli Sabra culture*. Cambridge: Cambridge University Press.

Kline, S. L., & Carlson, K. (1989). Children's rhetorical skill. In B. E. Gronbeck (Ed.), *Spheres of argument* (pp. 489-496). Annandale, VA: Speech Communication Association.

Kline, S. L., & Oseroff-Varnell, D. (1993). The development of argument analysis skills in children. *Argumentation and Advocacy, 30*, 1-16.

Kuhn, D. (2005). *Education for thinking*. Cambridge, MA: Harvard University Press.

Kuhn, D., Katz, J. B., & Dean, D., Jr. (2004). Developing reason. *Thinking & Reasoning, 10*, 197-219.

Kuhn, D., & Udell, W. (2003). The development of argument skills. *Child Development, 74*, 1245-1260.

Malis, R. S., & Roloff, M. E. (2006). Demand/withdraw patterns in serial arguments: Implications for well-being. *Human Communication Research, 32*, 198-216.

Martin, R. W., & Scheerhorn, D. R. (1985). What are conversational arguments? Toward a natural language user's perspective. In J. R. Cox, M. O. Sillars, & G. B. Walker (Eds.), *Argument and social practice* (pp. 705-722). Annandale, VA: Speech Communication Association.

McCroskey, J. C. (1969). A summary of experimental research on the effects of evidence in persuasive communication. *Quarterly Journal of Speech, 55*, 169-176.

Mercier, H., & Sperber, D. (2011). Why do humans reason? Arguments for an argumentation theory. *Behavioral and Brain Sciences, 34*, 57-111.

Miller, G. R., & Nilsen, T. R. (1966). *Perspectives on argumentation*. Chicago, IL: Scott, Foresman.

O'Keefe, B. J., & Benoit, P. J. (1982). Children's arguments. In J. R. Cox & C. A. Willard (Eds.), *Advances in argumentation theory and research* (pp. 154-183). Carbondale, IL: Southern Illinois University Press.

O'Keefe, D. J. (1975). Logical empiricism and the study of human communication. *Communications Monographs, 42*, 169-183.

O'Keefe, D. J. (1977). Two concepts of argument. *Journal of the American Forensic Association, 13*, 121-128.

O'Keefe, D. J. (1997). Standpoint explicitness and persuasive effect: A meta-analytic review of the effects of varying conclusion articulation in persuasive messages. *Argumentation and Advocacy, 34*, 1-12.

O'Keefe, D. J. (1998). Justification explicitness and persuasive effect: A meta-analytic review of the effects of varying support articulation in persuasive messages. *Argumentation and Advocacy, 35*, 61-75.

O'Keefe, D. J. (2015). *Persuasion: Theory and research*, 3d. ed. Los Angeles, CA: Sage.

Paglieri, F. (2009). Ruinous arguments: Escalation of disagreement and the dangers of arguing. In J. Ritola (Ed.), *Argument cultures: Proceedings of OSSA 2009*. http://scholar.uwindsor.ca/ossaarchive/OSSA8/papersandcommentaries/121/

Paglieri, F., & Castelfranchi, C. (2010). Why argue? Towards a cost-benefit analysis of argumentation. *Argument and Computation, 1*, 71-91.

Park, H. S., Levine, T. R., Kingsley Westerman, C. Y., Orfgen, T., & Foregger, S. (2007). The effects of argument quality and involvement type on attitude formation and attitude change: A test of dual-process and social judgment predictions. *Human Communication Research, 33*, 81-102.

Rancer, A. S., & Avtgis, T. A. (2014). *Argumentative and aggressive communication: Theory, research, and application*, 2d. ed. New York: Peter Lang.

Rancer, A. S., Baukus, R. A., & Infante, D. A. (1985). Relations between argumentativeness and belief structures about arguing. *Communication Education, 34*, 37-47.

Reynolds, R. A., & Reynolds, J. L. (2002). Evidence. In J. P. Dillard & M. Pfau (Eds.), *The persuasion handbook* (pp. 427-444). Thousand Oaks, CA: Sage.

Thompson, W. N. (1967). *Quantitative research in public address and communication*. New York: Random House.

Trapp, R., & Hoff, N. (1985). A model of serial argument in interpersonal relationships. *Journal of the American Forensic Association, 22*, 1-11.

Weaver, A. T. (1959). Seventeen who made history—the founders of the association. *Quarterly Journal of Speech, 45*, 195-199.

Weaver, R. M. (1953). *Ethics of rhetoric*. New York: Regnery.

Weger, H., Jr. (2001). Pragma-dialectical theory and interpersonal interaction outcomes: Unproductive interpersonal behavior as violations of rules for critical discussion.

Argumentation, 15, 313-329.
Wenzel, J. W. (1979). Jürgen Habermas and the dialectical perspective on argumentation. *Journal of the American Forensic Association, 16,* 83-94.
Wenzel, J. W. (1974). Rhetoric and anti-rhetoric in early American scientific societies. *Quarterly Journal of Speech, 60,* 328-336.
Wenzel, J. W. (1980). Perspectives on argument. In J. Rhodes & S. Newell (Eds.), *Proceedings of the Summer Conference on Argumentation* (pp. 112-133). Alta, UT: Speech Communication Association.
Willard, C. A. (1983). *Argumentation and the social grounds of knowledge.* University, AL: University of Alabama Press.
Willard, C. A. (1989). *A theory of argumentation.* Tuscaloosa, AL: University of Alabama Press.
Xie, Y., Hample, D., & Wang, X. (in press). A cross-cultural analysis of argument predispositions in China: Argumentativeness, verbal aggressiveness, argument frames, and personalization of conflict. *Argumentation.*
Yost, M. (1917). Argument from the point-of-view of sociology. *Quarterly Journal of Speech, 3,* 109-124.

Chapter 16

Ethos, Familiars and Micro-Cultures

Michael A. Gilbert

Department of Philosophy, York University, Toronto, Canada

1. Knowledgeability, Trustworthiness and Likability

In this chapter I want to examine the nature of personal ethotic standings that we, as individual arguers, apply to others and seek to have applied to us. According to Persuasion Theory, there are several concepts that are regularly applied to public individuals with regard to how much we trust them and accept their statements, views and opinions. I want to suggest that these can also be applied on a micro-level with respect to individual relationships between what Gilbert calls "familiars" (Gilbert, 2014a; 2014b). These are people ranging from close friends to tradespeople to teachers and doctors with whom one has regular or occasional interactions, some of which are bound to be argumentative in the extended sense of the term. The core concepts upon which Persuasion Theory rests are first knowledgeability, secondly trustworthiness, and thirdly liking (Baudhuin and Davis, 1972; Davis & Hadkis, 1995; O'Keefe, 2002; Rosenthal, 1966). These ideas are generally used to evaluate a public speaker or referenced expert. The higher one assesses an individual on the three scales the more likely one is to grant them a strong ethotic rating and, ergo, take what they say as true and/or correct.

By "ethotic rating," I mean a matrix of the feelings and attitudes one has to another individual. Simplistically, the higher the rating, the more highly valued the individual with respect to their judgment and other personality factors. In what follows I will show that a variety of approaches including Persuasion Theory, Facework, Politeness Theory and Gricean Theory can work together to help explain the nature and importance of ethos and ethotic rating. It should be noted that ethos is at work in all interactions, but never more important than when arguing.

There are two items that need to be pointed out. The first is that most of the research I have found concerns the assessment of non-familiars, of people whose opinions one might read or come across via media or reports. This means, secondly, that the assessment is *from* the individual assessor regarding the presenter. This latter is very important because it eliminates a great deal of the force of facework (Arundale, 2006). That is, the nature of the relationship does not involve such factors as the opinion of the speaker of me, or the maintenance of our relationship for the simple reason that there is none. These are normally the jobs of "facework," and that is why I say it is mostly eliminated. The main factor is that in the most usual situations studied in Persuasion Theory the assessment is one way, while in interactions with familiars it is two ways. This means that while I evaluate you, you are evaluating me. When Canadian Prime Minister Stephen Harper makes a speech he is not at the same time evaluating me, even though I am evaluating him. And, as far as facework is concerned we do not have a relationship to maintain, so I can be as rude to his face on the screen, send a mean screed via Twitter, or begin an anti-Harper Facebook page with impunity. With familiars, on the other hand, what they say impacts their ethotic rating, and what I reply, show, or react will impact mine. For example, if I tweet something mean about a colleague, that is very likely to worsen our relationship.

Familiars come in a wide variety of shapes and sizes and there is a great range of degrees of familiarity within the larger set of familiars. There are some people I see only occasionally such as, perhaps, my physician, auto mechanic or accountant, and others such as friends, family, and co-workers with whom I interact very frequently. It is also important to notice that within each range there are differences. So my auto mechanic might be a recent addition to my life, but my doctor may have been my physician for many years. Each has an ethotic rating based on those years. My assessment of their knowledgeability, trustworthiness, and likability changes and varies as interactions accrete. That is, ethotic ratings do not spring up full blown. Most of us, when meeting someone for the first time give them pretty much a *tabula rasa* though tending to the positive. We say that someone who trusts no one is cynical, and someone who trusts too much is gullible (Gilbert, 2007). The idea is to begin, *ceteris paribus*, with a more or less positive inclination, and then see what happens[1]. Moreover, as I have pointed out elsewhere (Gilbert, 2014a) we often know something, even if not much: e.g., who made the introduction, where was the introduction made, what was the context of the introduction, and so on.

Given all this, we create their ethotic ratings of people by judging them along the three main parameters discussed above. O'Keefe (2014) describes what he calls "expertise" and I refer to as knowledgeability as follows:

> The expertise dimension (sometimes called "competence", "expertness," "authoritativeness," or "qualification") is commonly represented by scales such as experienced-inexperienced, informed-uninformed, trained-untrained, qualified-unqualified, skilled-unskilled, intelligent-unintelligent, and expert-not expert. These items all seem directed at the assessment of (roughly) whether the communicator is in a position to *know* the truth, to know what is right or correct. (O'Keefe, 2014).

Without getting into issues regarding truth, we shall allow that when one is in a position of requiring information one does not have, then an expert, i.e., someone with more knowledge than oneself is called upon. As far back as 1979 I was claiming that the rough and tumble understanding of an expert is relative: "An expert is anyone who knows more than we do when we want to know it" (Gilbert, 2008, p. 79). What will change is that the more personal and pressing the information, the more important the source credibility will be. O'Keefe points out that likability becomes less important as the topic becomes more personally relevant (*ivi*, p. 10). In other words, I might take my friend's advice on what cell phone to purchase, but not so much when it comes to which surgery I should have.

Even trustworthiness is liable to become less important when expertise is crucial. "The trustworthiness dimension (sometimes called 'character,' 'safety,' or 'personal integrity') is commonly represented by scales such as honest-dishonest, trustworthy-untrustworthy, open-minded-closed-minded, just-unjust, fair-unfair, and unselfish-selfish. These items all appear to be related to the assessment of (roughly) whether the communicator will likely be inclined to *tell* the truth as he or she sees it" (O'Keefe, 2014, pp. 10-12). For our purposes

---

[1] In all of this I am envisioning the most common situations. Meeting someone who has a decidedly creepy look or threatening demeanor may alter the most common reaction.

we want our familiars to be honest, hear our wants and goals, and address our questions and interests with our values in mind. So I might want to consult someone who I do not trust a great deal if I think they will provide the necessary information I need because they are knowledgeable. Clearly, this can be difficult if I am concerned that they may not tell the truth.

This is compensated for, I believe, by the degree of intimate knowledge we often have with familiars. The better we know someone the more fine-tuned their ethotic rating becomes. We learn to rate *by context* rather than holus-bolus, and this is an important difference between public figures and familiars. We tend to judge public figures in a holistic way as trustworthy or not because we usually hear them focused on one area. Trying to extend a public figure's knowledgeability *as a result of their fame* incurs the fallacy of *ad verecundiam*. One recent example is Stephen Harper's touting that the famed hockey play Wayne Gretzky supports him. But someone's knowledge is rarely universal whether they are famous people or familiars. It is the context that determines both what we need and to whom we turn. My friend Andy knows everything there is to know about music, but I would never go to a baseball game with him – all he does is prattle on about the music being played between batters. My friend Emma, on the other hand, has an average knowledge of music, but knows everything there is to know about baseball. She's the one I call when I have an extra ticket. Knowledgeability, then is context dependent. If we consider an individual's knowledgeability indicated by some numeric KN, then we must add a context to that rating: KN_{ci}. So Andy's knowledge rating might be KN_m, where $_m$ is the appropriate music context, might be quite high, but Emma's KN_s is, when $_s$ is sports, higher than Andy's.

The question then arises as to whether trustworthiness and likability are also context dependent. Here one must consider if these two characteristics, (let us indicate them by TR and LK,) are taken in a broad sense or a narrow sense. If the latter, then I trust Emma around sports matters more than Andy, and prefer going to a concert with Andy more than Emma. This has most to do with the KN of each of them with respect to the particular subject matter. The context of a baseball game is a very different one from a concert, and I know that each has a different expertise. Andy knows a great deal about music and, so, it is more interesting to go to a concert with him. The narrow sense is subject and context specific and has its base in knowledgeability. However, it strikes me that this comes close to reducing them to knowledgeability insofar as the reasons for the changes in TR and LK are directly correlated to changes in KN. So, I believe this would be an error and not nearly so revealing as if we use the broad sense of the terms. By this I mean that my evaluation of Andy's LK is a function of many situations and cuts across various contexts, and ipso facto for TR.

Our ethotic ratings of familiars rests on a series of interactions stretching over time. Our initial rating may be neutral, but will also be affected by a number of factors including, for example, who made the introduction (McCroskey & Dunham, 1966; Tompkins, 1967). If someone whose financial acumen I respect recommends an accountant to me, then my initial ethotic rating may be positive rather than neutral. But the rating will change as the interactions we have continue. The more contact and longer the relationship the more finely tuned will be the ethotic rating, and the more it will depend on my own assessments of TR and LK. With friends our expectations and regard for them has been set, and while subject to change, will not do so easily. It is likely the case that, with some exceptions the stronger the relation the more dramatic must be the event precipitating a change, though such a cataclysmic event can certainly happen.

It also seems to me that these three characteristics are at least somewhat independent variables. While on the one hand they may well impact each other, on the other it is not hard to imagine liking someone who is not very trustworthy. The loveable rascal falls into this category, and the fact that I like Jude does not always mean I feel he is dependable. Similarly, I may not enjoy the company of Olivia but find she is the person whom I most trust. It is easy to see that which ethotic function is paramount can depend on the role an individual plays in one's life. A surgeon lacking in bedside manner might still be the first choice for your operation insofar as her KN and TR functions are high, though her LK is low. A friend may be a bit of a fool and have low KN and TR ratings, but still a high LK. A close friend might be high in TR and LK, but really have no KN status in anything significant. The *goals* one has in connecting with different persons determine which of the characteristics will be paramount. Each of the three can itself be a goal.

2. Grice, Cooperation and Politeness

The preceding discussion concerns how we organize ethotic ratings by breaking them down into component parts. The question remains how we arrive at them. Here I believe it behooves us to introduce Grice's maxims with the addition of tenets of politeness (Grice, 1975; Leech, 1983; 2007). Grice's masterly "Logic and Conversation" first appeared as a William James Lecture in 1967. There he spoke of the difference between what a speaker S literally says and what S expects a hearer H to understand. The meaning, the conversational implicature, depends on knowledge of the context as well as the assumption that S is following the rules of conversation. S expects H to take the *intended* meaning of the statement rather than the literal meaning of the words. So, for example, when Sophie leaves the exam room and bumps into her friend James, and James asks, was the exam hard? When Sophie responds: Professor Gilbert set it, she assumes that James will understand. Notice that we will not unless we have the information about the difficulty of Prof. Gilbert's exams that they share. This is carried even further when Willard talks about co-orientation, since that refers to the sharing of language and not just background information (Willard, 1978, p. 127).

To avoid confusion, I want to be clear that I am inserting Grice into the framework of KN, TR and LK. Our Gricean expectations will first be a function of how well we know someone and the extent to which we share the context. In addition, the extent to which I expect someone to "pick up" or understand a conversational implicature may also be a function of degree of familiarity or LK. Further, KN and TR can well be a function of our beliefs regarding our partner's inclination to follow the maxims.

Grice provides us with four maxims which are generally followed in conversation. It is the assumption that S, the speaker, is following these maxims that allows the hearer, H, to infer the correct meaning. The maxims are as follows. (The following is all from Grice, 1989, pp. 26-29).

QUANTITY
1. Make your contribution as informative as is required (for the current purposes of the exchange).
2. Do not make your contribution more informative than is required.

QUALITY
1. Do not say what you believe to be false.
2. Do not say that for which you lack adequate evidence.

RELATION
Be relevant.

MANNER
1. Avoid obscurity of expression.
2. Avoid ambiguity.
3. Be brief (avoid unnecessary prolixity).
4. Be orderly.

Presumably, the following of these maxims will lead to a reasonable ethotic rating. This, however, is more complicated than it might at first seem. One reason has to do with principles of politeness.

The nicer, wiser and more agreeable one is, the better one's *overall* rating, especially when it comes to credibility, i.e., TR. According to Burgoon, "the ideal source is highly… responsible, reliable, honest, just, kind, cooperative, nice, pleasant, sociable, cheerful, friendly, and good-natured and only slightly … expert, virtuous, refined, calm, composed, verbal, mild, extroverted, bold and talkative" (Burgoon, 1976, p. 205). Note that these do, to a fair extent, fit with Grice's maxims, though, as we shall see, not always. The agreement is especially so when it comes to Quality, though many seem to apply quite clearly to likability which Grice does not address. Rather it is others who add certain social aspects to Grice's maxims, primarily through the device of politeness. While there is a huge font of research regarding politeness, I will indicate only one stream, viz., that of Leech. He originally followed Grice in using the term 'maxims' for his rules (Leech, 1983), but later (Leech, 2007) decided that the word 'tenets' was preferable. They are as follows. (O = other person, S = speaker)

A. Place a high value on O's wants (GENEROSITY)
B. Place a low value on S's wants (TACT)
C. Place a high value on O's qualities (APPROBATION)
D. Place a low value on S's qualities (MODESTY)
E. Place a high value on S's obligation to O (OBLIGATION of S to O)
F. Place a low value on O's obligation to S (OBLIGATION of O to S)
G. Place a high value on O's opinions (AGREEMENT)
H. Place a low value on S's opinions (OPINION-RETICENCE)
I. Place a high value on O's feelings (SYMPATHY)
J. Place a low value on S's feelings (FEELING-RETICENCE)
(Leech, 2007, pp. 182-188)

So we now have concrete, in fact measurable, qualities that can inform the ethotic rating of an individual. Someone who undertakes these politeness tenets will stand a good chance of raising her ethotic rating in O's eyes. Someone who does not will *ceteris paribus* lower it. However, note that knowledgeability can still override various of these depending on context. Indeed, O'Keefe, citing several researchers, explains:

> The effects of liking on persuasive outcomes are minimized as the topic becomes more personally relevant to the receiver. Thus although better-liked sources may enjoy some general persuasive advantage, that advantage is reduced when the issue is personally relevant to the receiver. […] When receivers find the topic personally relevant, they are more likely to engage in systematic active processing of message contents and to minimize reliance on peripheral cues such as whether they happen to like the communication source. But when personal relevance is low, receivers are

more likely to rely on simplifying heuristics emphasizing cues such as liking ("I like this person, so I'll agree"). (O'Keefe, 2014, p. 10)

So the more important something is to us, the more we care less about likability and more about knowledgeability. Of course, the greater the interpersonal relations, the more likability plays a role. So Leech's tenets would appear to be most important in ongoing and routine familiar situations.

2.1. Culture and Context

It is, as we have seen, important to understand how ethotic ratings are constructed and how they may change and vary, and in this investigation I am trying to bring together several approaches that can nicely combine to provide some explanatory insights. These rules can work together so as to permit us to compare the way public discussion by public figures differs from our daily and much more frequent interactions. The suggestions so far indicate that the ratings follow rules, whether those of Communication Theory, Gricean conversation or others. However, the rules governing such ratings are not universally stable. Both context and culture play roles in the alteration of maxims and tenets. One such alteration has been much discussed, and that is the difference between Eastern and Western cultural rules. There are many ways to evaluate the differences but one important way has to do with self-construal. Citing the work of Markus and Kitayama (1991), Cross et al., say:

> As they illustrated, Western cultures prioritize the individual over the group, and individuals seek independence, autonomy, and separateness from others. In East Asian cultures, the group is prioritized over the individual, and individuals seek to fit into the group and maintain harmony in the group. (Cross, Hardin, & Swing 2009, p. 143)

As a result, a characteristic that may improve ethos in a Western setting, may diminish it in an Eastern, and *vice versa*. This comes out, for example in the idea of directness. Westerners tend to be very direct and assert their theses at the outset of their argument and then put forward reasons in a direct dialectical way. In many Eastern cultures the opposite is true and a rhetorical model issued that is less direct and less confrontational (Kirkpatrick, 1995, p. 274). Citing Oliver (1971) Kirkpatrick writes:

> In a comparison of Asian and Western rhetorics, Oliver notes points of general difference between them. These include their function, which in Asia is, according to Oliver, to promote harmony while in the West their function is to promote the welfare of the individual. Thus the discourse of Asian rhetoric tends to avoid argument and persuasive fervor. What is persuasive for Asian rhetoric(s) "is appeal to established authority buttressed by analogical reasoning which sought to clarify the unfamiliar through comparison with the familiar" (1971, p. 263). (Kirkpatrick, 1995, p. 274)

Indeed, all forms of communication, so *ipso facto*, argument, are formed by the foundational values and beliefs in a culture. Facework, the usually inherent and implicit maintenance of relationships, is present in all cultures, but the manner in which it is undertaken may vary widely. Facework, which occurs in all interactions, includes my concern of what you think of me, what I think of you, and what we feel about each other (Gilbert, 2014a). This is extended in FNT: the fundamental assumption of the FNT [Face negotiation Theory] is that people in all cultures try to maintain and negotiate face in communication. Thus, conflict is essentially a face-negotiation process, but cultural value orientations and individual attributes shape one's self/other-oriented facework and conflict behavior. Key propositions of the FNT include that members of collectivistic cultures or individuals who are interdependent in self-construal tend to be more other/mutual-face oriented, avoiding, obliging, compromising, and integrating, whereas members of individualistic cultures or individuals who are independent in self-construal tend to be more self-face oriented and competing (Zhang, Ting-Toomey & Oetzel, 2014, p. 374).

If we refer back to Grice's maxim's and Leech's tenets, we can see that there may well be some relativity that is required when shifting from one culture to another. Being direct can be appropriate manner in one culture but not in another, and in many cultures speaking the truth, saying what you believe to be false, is often the appropriate and polite thing to do. One example with which I am quite familiar involves a conflict between the maxim of Quality and the Principle of Politeness. Gazdar, for example, uses Hintikka's formulation of the maxim: "'QUALITY: Say only what you know.' So, when S utters α then S implicates Kα, where K means that S knows α" (Gazdar, 1979, p. 46). But in Mexico, it is so rude not to be able to help someone that it is more polite, when someone asks directions, to make them up rather than say you don't know. As a result, Mexicans know that if someone hesitates when asked which way is it to the library, then ignore whatever they say. Moreover, there are many circumstances when lying occurs both to save one's face or to avoid hurting someone's feelings (DePaulo & Kashy, 1998, passim). Lies may be opportunistic as when you say that your essay is almost complete because you do not want to be thought of as a procrastinator, or altruistic as when you express sympathy that is not really felt.

The very idea of arguing and with whom one may argue is highly culturally and context bound. In many cultures, especially Eastern, disagreeing or arguing with an elder or hierarchical person is at least frowned upon and often forbidden. As explained elsewhere (Gilbert, 2014b, p. 7) such restrictions can even lead to disasters as those in an inferior position show deference to those above them in status. Moreover, cultural styles can overpower maxims and tenets. Jews and Italians, for example, typically elaborate their claims, add stories, examples, and appeal to others for affirmations. So, the idea that one should only say what is required would be considered rude in most Jewish and Italian interactions as opposed to say, Japanese where minimalism is valued. Schiffrin (1984) writes:

> Topics of talk not inherently defined as disputable, however, also became the focus of sustained disagreement. Questions which had been minimally answered by non-Jewish speakers in other Philadelphia neighborhoods with just the information requested prompted arguments from my informants: questions about the location of a family doctor, belief in fate, educational background, solutions for personal problems, childhood games, location of friends and family, evaluation of local restaurants, who to invite to a party. (Schiffrin, 1984, p. 319).

Even relevance can vary insofar as many cultures use analogy and narrative as a basic communicative and argumentative form, rather than rely on direct disagreement or contradiction. Face has different meanings in different cultures. In Eastern cultures it means that one has not lived up to one's responsibilities of one's social position, while in the West it may mean that one has not achieved the desired goals.

> Deep-rooted in the emphasis on social relationships is the theme of face. In the collectivistic Korean society (and in other East Asian societies, such as China and Japan), the individual is not inner-directed but is rather governed by a need for not losing face. "Face ... is lost when the individual fails to meet essential requirements placed on him or her by virtue of the social position he or she occupies (Ho, 1976).

> Some Korean phrases strongly indicate the importance of face... A Korean proverb says: "Better to die rather than to live in dishonor/disgrace" — an interesting contrast with the American, "Give me liberty or give me death." In collective cultures, in which saving face is a critical matter, face-supporting behavior (i.e., avoiding hurting the hearer's feelings, minimizing imposition, and avoiding negative evaluation by the hearer), rather than efficient and direct behavior, may lead to a desirable outcome in the long run. (Kim, 1994, pp. 134-135).

Another quote from Kim refers directly to Grice:

> One of Grice's maxims for cooperative conversation is the maxim of manner, which recommends that speakers should avoid ambiguity and obscurity of expression (Grice, 1975). Several authors have suggested that, although direct communication may be a norm in the United States (an individualistic society), Grice's maxim of manner is less applicable in cultures with different value orientations. Okabe (1987), for instance, has shown that in Japan the traditional rules of communication, which prohibit criticizing the listener directly, asserting oneself, demanding, or rejecting, are more important than Grice's maxim of manner. (Kim, 1994, p. 135).

Bringing this back to our concerns, it becomes apparent that ethos is a highly variable concept. What can give rise to a high ethotic rating in one culture or context can lower it in another. There is a disconnect between rules of behavior and rules of communication and ethotic ratings because we need to know more than just what the actions, linguistic or physical are.

3. Ethos

At the beginning of this chapter a very important point was made, viz., ethotic ratings of experts are made by a hearer H regarding a speaker S, and not vice versa. I have argued, however, that this is not the case in ongoing relationships with familiars. In those contexts, you may try to influence your ethotic rating by following various rules and performing face-saving acts as best you can, but you may or may not succeed. (Indeed, most of us are familiar with an attempt to repair actually making things worse.) However, one's ability to both judge and be judged is much greater in the case of familiars as opposed to public speakers and expert personalities. As individuals interacting with individuals we are always working within the limits and strictures of Facework (Arundale, 2006; Bargiela-Chiappini, 2003). However, this concept has itself been challenged as too narrow in some interesting recent work. Locher and Watts prefer to talk of "relational work" rather than the traditional "face work:

> [P]oliteness is only a relatively small part of relational work and must be seen in relation to other types of interpersonal meaning Relational work refers to the "work" individuals invest in negotiating relationships with others. Human beings rely crucially on others to be able to realize their life goals and aspirations, and as social beings they will naturally orient themselves towards others in pursuing these goals. (Locher & Watts, 2005, p. 10).

I believe this approach makes more sense, and is richer when considering ethos. The idea of face is one-sided with S attempting to mold herself into what she wants to be seen as.

Relational work, on the other hand, shows a greater emphasis on communicators working together on a relationship. It is not one-sided but interactive.

Following this line of thought, I hypothesize that the maintenance and creation of ethotic ratings is a part of relational work. By that I mean that S and H, *when they are close familiars,* work together on their interactional ethotic standings. In close groups: spouses, best friends, family, inner circle friends the ethotic rating has been long and well established. There may be bumps in the road, and even patches of black ice, but the core evaluations do not change easily. Yes, there can be a cataclysmic event that will change everything, but the very fact that it must be a major occurrence demonstrates the solidity of the ethotic relationship. Outer circle friends, neighbors, colleagues and professional and trade acquaintances can change more easily, but there too the length of time will be a factor in how easily one alters the ethotic rating.

The structure of an ethotic rating is not only a function of closeness and relation time, but the goals and roles involved as well. I might give my auto mechanic a high likability (LK) rating, but if I come to believe he is charging me for parts I did not really need, then it quickly becomes clear that trustworthiness (TR) carries greater weight than likability. On the other hand, LK is likely first when it comes to friends and family, though TR will also play a role.

Ethotic ratings among familiars exist in a micro-context and even micro-culture that has its own set of maxims of conversation and tenets of politeness. S's knowledge of this is part and parcel of the communicative and behavioral framework in which ethos works. Acceptable behavior in one setting is unacceptable in another. Recall the film *Annie Hall* in which two people, Alvy Singer (Woody Allen), a died in the wool New York Jewish Manhattanite, and Annie Hall (Diane Keaton), a mid-western Bible Belt Christian fall in love (Allen, 1977). There is a wonderful sequence of split scenes between Annie's family at Thanksgiving where conversation is extremely polite and moderated, and Alvy's where everyone is loudly and boisterously talking at once. Both families are operating within their own micro-culture and would be mystified and even appalled by the other's. Yet they manage to fall in love and create a separate and independent micro-context. This does happen with bumps, as when Annie orders a corned beef sandwich on white bread with a glass of milk in a NY deli. (That's just not done. It's always rye bread, and never milk with meat.)

The point of this is the factors KN, TR, and LK play different roles in different relationships. The more important someone's KN is, the less important is their LK and, often, even their TR. This is a different situation from that of public experts. Consequently, when dealing with the ethos of familiars we cannot rely solely on traditional judgments and models that apply primarily to experts. First, other factors including the micro-culture and emphasis on LK come into play, and secondly, so does the fact that the rating is more interactive – we rate each other, rather than just O rating S. So understanding ethos requires that we be very sensitive to the numerous distinct situations in which it takes place and the innumerable varieties of relationships into which we enter.

References

Allen, W. (1977). *Annie Hall*. United Artists.
Arundale, R. B. (2006). Face as relational and interactional: A communication framework for research on face, facework, and politeness. *Journal of Politeness Research. Language, Behaviour, Culture, 2*(2), 193-216.
Bargiela-Chiappini, F. (2003). Face and politeness: new (insights) for old (concepts). *Journal of Pragmatics, 35*, 10-11.
Baudhuin, E. S., & Davis, M. K.. (1972). Scales for the measurement of ethos: Another attempt. *Speech Monographs, 39*(4), 296-301.
Burgoon, J. K. (1976). The ideal source: A reexamination of source credibility measurement. *Central States Speech Journal, 27*(3), 200-206.
Cross, S. E., Hardin, E. E., & Swing, B. G. (2009). Independent, relational, and collective-interdependent self-construals. In Leary, M. R.. & Hoyle, R. H. (Eds.), *Handbook of individual differences in social behavior*, New York: Guilford Press.
Davis, M., & Hadkis, D. (1995). Demeanor and credibility. *Semiotica, 106*(1-2), 5-54.
DePaulo, B. M., & Kashy, D. A. (1998). Everyday lies in close and casual relationships. *Journal of Personality and Social Psychology, 74*(1), 63-79.
Gazdar, G. (1979). *Pragmatics: implicature, presupposition and logical form*. New York: Academic Press.
Gilbert, M. A. (2007). Natural Normativity: Argumentation Theory as an engaged discipline. *Informal Logic, 27*(2), 149-161.
Gilbert, M. A. (2008). *How to win an argument: surefire strategies for getting your point across*. (3rd ed.). Lanham, MD: University Press of America. Original edition: 1979.
Gilbert, M. A. (2014a). *Arguing with People*. Calgary, AB: Broadview Press.
Gilbert, M. A. (2014b). Rules Is Rules: Ethos and situational normativity. International Society for the Study of Argumentation 2014, Amsterdam.
Grice, P. H. (1975). Logic & Conversation. In *Studies in the way of words*. Cambridge, MA: Harvard University Press.
Grice, P. H. (1989). *Studies in the way of words*. Cambridge, MA: Harvard University Press.
Kim, Y. Y. (1994). Cross-Cultural Comparisons of the Perceived Importance of Conversational Constraints. *Human Communication Research, 21*(1), 128-151.
Kirkpatrick, A. (1995). Chinese rhetoric: Methods of argument. *Multilingua, 14*(3), 271-295.
Leech, G. (1983). *Principles of pragmatics, Longman linguistics library*. London ; New York: Longman.
Leech, G. (2005). Politeness: is there an East-West divide. *Journal of Foreign Languages, 6*(3), 1-30.
Locher, M. A., & Watts, R. J. (2005). Politeness theory and relational work. *Journal of Politeness Research. Language, Behaviour, Culture, 1*(1), 9-33.
McCroskey, J. C., & Dunham, R. E. (1966). Ethos: A confounding element in communication research. *Speech Monographs, 33*(4), 456-463.
O'Keefe, D. J. (2002). *Persuasion: theory & research*. (2nd ed). Thousand Oaks, CA: Sage Publications.
O'Keefe, D. J. (2014). *Persuasion : theory & research* (3rd ed). Thousand Oaks, CA: Sage Publications.
Rosenthal, P. I. (1966). The concept of ethos and the structure of persuasion. *Speech Monographs, 33*(2), 114-126.

Schiffrin, D. (1984). Jewish Argument as Sociability. *Language in Society, 13*(3), 311-335.

Tompkins, P. K. (1967). The McCroskey-Dunham and Holtzman reports on "Ethos: A confounding element in communication research". *Speech Monographs, 34*(2), 176-178.

Willard, C. A. (1978). A Reformulation of the Concept of Argument: The Constructivist/Interactionist Foundations of a Sociology of Argument. *Journal of the American Forensic Association, 14,* 121-140.

Zhang, Q., Ting-Toomey, S., & Oetzel, J. G. (2014). Linking Emotion to the Conflict Face-Negotiation Theory: A U.S.-China Investigation of the Mediating Effects of Anger, Compassion, and Guilt in Interpersonal Conflict. *Human Communication Research, 40*(3), 373-395.

Chapter 17

Multimodal Persuasion in Judicial Debates

Francesca D'Errico[1] & Antonella Bellon[2]

[1] Faculty of Psychology, Uninettuno University, Rome, Italy, f.derrico@uninettunouniversity.net; [2] TIM S.p.A., Legal Office, Milan, Italy, antonundici@gmail.com

Abstract. The present chapter deals with the role played by two opposite persuasion strategies in judicial debates: raising one's power or lowering others' by means of expressions of dominance and discredit. Dominance and discredit can be multimodally expressed (Poggi & D'Errico, 2010; D'Errico & Poggi, 2014). Therefore, this work will analyze three different Italian trials in terms of both verbal (argumentation) and non-verbal signals (e.g., face, gesture, postures and prosodic features). Taking into account the seriousness of the verdicts in these trials, observational analyses point out how argumentative, discrediting and dominance moves can contribute in persuading the judge or the audience.

1. Introduction

Whereas classic research in social psychology has focused on the source, the message and the audience's characteristics as determinants of the process of social influence in a "neutral" context, the current studies of public communication are strongly oriented towards the forms of conflictual interactions (Hovland *et al.*, 1953; Mucchi Faina *et al.*, 2013).

Among these, a form that is effective in promoting interest but ineffective for political trust (Mutz & Reeves, 2005) is that in which the source of persuasion (e.g., a political leader in political debates or a lawyer in a judicial debate) aims at discrediting the opponent.

This contribution will present three observational studies on judicial debates, in which the acts of discredit are analyzed in depth by means of a "multimodal persuasion" approach , according to which every verbal and bodily signal is interpreted as a way to persuade an audience.

A judicial debate is a communicative interaction where a speaker A tries to persuade both the interlocutor (i.e., the Judge) and a third party (i.e., the audience) that s/he is right, while opponent B is wrong, and that B presented a distorted image of things (e.g. by making a wrong reconstruction of the misdeed) and proposed goals (i.e., condemning or acquitting the defendant) that are not sound or effective. The debate is therefore a persuasive interaction, presenting every kind of move of persuasion. Here, persuasion is intended either in its stronger sense of convincing someone to carry out some action, or in its weaker sense of convincing others that something is true. In both cases, one must bear arguments in support of one's plea, i.e., good reasons for the other to believe that something is really the way the persuader puts it, or that doing a certain action is right or convenient.

In addition to discredit, another way for the attorneys to persuade a judge and convey credibility is to communicate dominance. In doing so, attorneys have to manage a "tricky dominant position", as they need to display their power and self-confidence and at the same time to respect the judge's dominance (Higdon, 2008).

In this paper we analyze multimodal moves employed during the debates of three Italian trials (involving Fabrizio Corona, Luciano Moggi and Stefano Cucchi), in which contestants aim to gain credibility and power by acting in opposite ways: either by discrediting the opponent or by expressing their own dominance. Both strategies can be performed by verbal (at the linguistic and argumentative level; see Walton 2002) and body signals (facial expressions, gestures, gaze, posture, may all be exploited to directly or indirectly convey one's dominance or negative evaluations about the opponent). The analysis is based our model of persuasion (Poggi, 2005).

2. Dominance and Discredit in Multimodal Persuasion

In addition to the literature in argumentation theory, there are studies on persuasion in social psychology that point at the importance of the characteristics of credibility, reliability and attractiveness of the persuader in increasing the effectiveness of the persuasive message. Thus, not only the moral image of the persuader, as Aristotle claimed, counts in conveying persuasion; a more powerful source is more persuasive (Burgoon, 1990; Poggi & D'Errico, 2010). In this regard, two notions can be considered relevant: dominance and discredit. We will define them according to a theoretical model of mind, social relations and social interaction based on the notions of goals and beliefs. Then we will focus on the notion of dominance, by illustrating the possible moves in a multimodal perspective, and finally we will concentrate on discredit. Let us first overview the basic assumptions of this model.

2.1. Goals, Social Interaction and Social Evaluation

According to Conte and Castelfranchi (1995), any action performed by a system, social interaction, and communication between systems is regulated by goals. A system pursues its goals through plans, i.e., hierarchically organized sequences of actions. But when it does not have the power to achieve its goals, because of a lack of the necessary resources or skills, it may need another system to adopt its goal, and help him achieving it. Social interaction is the result of reciprocal goal adoption, instrumental, like commerce, or disinterested, like altruism. To gain goal adoptions, social influence (i.e., having another person pursuing the goal in its stead) and image (i.e., eliciting particular evaluations by others) may be exploited. In order to decide whether to adopt others' goals, people usually evaluate them – in terms of their adequacy – regarding the various goals (criteria): beautiful/ugly, good/bad, intelligent/stupid... Since our image determines our relationships with others, to have a positive image becomes a crucial goal for people, who do not want to elicit negative evaluations. There are two possible kinds of negative evaluations: a first kind of inadequacy, or lack of the power necessary to achieve some goals; and a second kind of noxiousness: a negative power that may hurt others, or thwart someone's goals.

People can evaluate others for many reasons, but especially when they must decide to trust a person, for example when being subject to a persuasive attempt. In this case, there are two important criteria that are used in evaluating people. The first is *benevolence*, or the willingness of the others to adopt our goals, and to not harm us or act aiming only at their own interests. The second criterion is *competence*, that is, the possession of knowledge, reasoning and planning skills. But while these criteria may be sufficient conditions in everyday persuasion, in other cases (e.g., in political discourse, see D'Errico & Poggi, 2012) the speaker must also show *dominance*, or the capacity of influencing other people, in order to impose his/her will, and to win in contests.

2.2. Raising One's Own Power: Displaying Dominance

Dominance is a relational construct (Castelfranchi, 2003; Burgoon *et al.*, 1990) implying a power comparison: to have "more power than" another. Strangely enough, it may be enhanced through communication: to communicate that one has power is a way to maintain power or even to acquire more. This is why, in public debates, people send each other dominance signals.

D'Errico and Poggi (2010) carried out a multimodal analysis of twelve dominance strategies, i.e., sets of communicative acts that provide a peculiar idea of how the Sender is dominant with respect to the Addressee. These range from the more aggressive strategy of imperiousness (or "blatant strategies"), judgement, invasion, norm violation and defiance, to the more "subtle" ones like touchiness and victimhood; ending up with the distancing strategies of haughtiness and other ways to show superiority, like irony and ridicule, easiness, carelessness, and assertiveness. In all these ways people try to appear better – stronger, more intelligent, noble, or important – than others.

Taking into consideration all these multimodal dominance moves within political debates is necessary to contextualize them within the judicial framework.

Higdon (2008), for example, acknowledges that the persuadee (the judge) is in a clearly dominant position with respect to the attorney: he is in the «tricky position of having to employ non-verbal behavior that connotes dominance (as dominant behaviors are more likely to persuade) yet, at the same time, display nonverbal cues to indicate both an awareness of and a respect for the judge's ultimate authority» (p. 656). To this purpose, Benus *et al.* (2014) investigated the production of conversational fillers and acoustic intensity in the judicial context, along with the patterns of turn-taking and the linguistic style markers as communicative social signals related to power relations, conflict, and voting behavior. They found some "successful" acoustic patterns and turn taking violations in relation to particular phases of the trial.

2.3. Lowering the Other's Power: Discredit

Another way to persuade others during a debate is by lowering the other's power and devaluating their image by accusation, criticism, or even insult: in short, speech acts that aim at discrediting the other. But what does it mean to cast discredit over someone?

Discredit is not a specific communicative act, but a perlocutionary effect of some communicative acts (Austin, 1962), that is in some cases created on purpose by the Sender. Discredit can be defined as the spoiling of the image of a person (B) in the eyes of other people (C), caused, either deliberately or not, by a person (A), by means of communicative acts that mention or point at actions or qualities of (B) that are considered negative by the third party (C) (Poggi & D'Errico, 2012). In political debates, politicians often attack each other by performing discrediting acts, that can be distinguished on the basis of their target feature. The target of discrediting, i.e., the criterion in terms of which the A casts a negative evaluation over B, can be either Competence, Dominance, or Benevolence. On the Competence side, one may cast doubts on the opponent as being ignorant or stupid; on the

Dominance side, the opponent could be attacked concerning his/her being helpless, inconsequential or ridiculous. Finally, on the Benevolence side, the opponent could be accused of being dishonest, cheating, or immoral.

3. Argumentations and Multimodal Persuasion

In addition to the multimodal communication approach, the analysis of multimodal persuasion in legal debates can also rely on rhetoric and argumentation theory.

In the field of argumentation, Walton (2008) provided a systematic classification of "argumentation schemes", in which he defined and classified more than 60 argumentations within the pragma-dialectic theory, based on a "presumptive" way of reasoning. In the pragma-dialectical perspective, unlike in logical, rhetorical, and dialogical approaches (Feteris, 2012), judicial argumentation is considered part of a rational critical discussion aimed at the resolution of a dispute, thus combining a pragma/rhetoric and a dialectic approach. In this sense an argumentation must aim at the dialectical goal of resolving a difference of opinion in a reasonable way, and at the rhetorical goal of persuading the judge/audience.

Arguers, in Walton's (2008) description, can choose from different types of argumentation schemes. Since arguments are especially needed when there are no evidences or facts to present, arguers need to set out their plea by making persuasive inferences. As a consequence, most of the schemes are based on presumptive ways of reasoning. Walton then acknowledges a number of *descriptive argumentation schemes* that give some hints, either based on the source of information, as in the case of the *argumentation from evidences, memoriam, perception, from ignorance,* or based on emotions, as in the case of the *argumentation schemes from threat, fear appeal, danger appeal, distress*.

Walton also lists a number of *normative argumentation schemes,* where sentences or arguments are based on norms, rules or other positive or negative examples, analogies, authorities or experts. Among these, there are the *argumentation schemes from rules, values, examples, analogies, expert opinion, commitment, popular opinion, witness testimony* or *position to know,* and *ethotic argument*. Particular violations of normative schemes are the so-called *ad hominem, generic ad hominem* and *circumstantial ad hominem.*

Normative argumentation can be distinguished from the *presumptive argumentation schemes,* which are based instead on possible inferences, as in the case of *abductive argumentation, practical reasoning* or *pragmatic inconsistency, from consequences, from sign to consequences, slippery slope argument, from alternatives, from oppositions, from evidences to a hypothesis, cause to effects, from bias, from correlation to cause*. Other argumentation schemes are based on the classification and specification of verbal features used in the argumentation, as in the case of the *vagueness* or *arbitrariness of verbal classification,* that allows a proper contextualization of the arguments.

In particular, the act of discrediting the other to prove him wrong may sometimes be a case of *ad hominem* fallacy. It violates the "freedom rule" according to which individuals engaged in a discussion must be free to provide arguments without fearing of being attacked.

What are the arguments commonly used in forensic persuasion? According to Walton, the forms of argument used in court are based on three types of thinking: in addition to the inductive and deductive, he acknowledges a third type of argument, called "presumptive". This particular argument, instead of being based on certain premises, hinges on abduction, defining the judicial context as oriented to the plausibility of the evidence.

Then, what are the arguments commonly used in legal persuasion? Walton (2002) finds the most common ones in the theory of argumentation, ranging from those based on empirical inference, as in the case of *abductive argument* or *from position to know*; those based on evidence, as in the case of *best explanation* of an expert or a witness, and those rule-based, as in *analogy* or *established rule*.

Besides the argumentational perspective, it seems useful to consider a multimodal persuasion perspective for which not only the verbal or argumentative aspects, but also the bodily ones are relevant. Namely, both discrediting moves (D'Errico *et al.*, 2013) and the "management of dominance communication" (Poggi & D'Errico, 2010) have a persuasive import. As shown by Higdon (2008), in the judicial context it is strategic, as well as required, to comply with the judge's dominance.

4. Multimodal Persuasion. Three Case Studies: Corona, Moggi and Cucchi

In this work, three cases of discredit in judicial communication will be taken into account, and analyzed with a multimodal persuasion approach. The communicative acts aimed at persuasion will be described both in terms of speech and in terms of body signals. The purpose of this work is to show which multimodal maneuvers of discredit and dominance are successful in persuading judges. Also, the work aims to state whether an excessive use of discrediting can lead to failure.

The analysis of three criminal trials described below is based on the theoretical model of discredit as a strategy of persuasion (Poggi & D'Errico, 2013), and focuses on the moves aimed at exhibiting dominance, that can be made by all those involved or having a role in the trial (Judge, PM[1], Defense, Accused, and Witness). These three trials had a high mediatic resonance in Italian media, and enjoyed a great following both when mentioned in newspapers or TV news and when they were broadcasted by *"Un giorno in pretura"* ("A day in a district court"), a TV programme that deals with judicial reporting.

Besides having been central in the public opinion, Corona and Moggi's trials involved two persons belonging to the worlds of sports and entertainment, who are well-known to the Italian public. The third case involved some branches of penitentiary police and doctors accused of having caused the death of a young man, Stefano Cucchi, jailed for drug dealing and died during detention. Corona and Moggi's trials ended with two sentences of imprisonment, respectively of seven and two years (lapsed in 2015). After Cucchi's trial, some of the defendants were condemned for a few months (and their sentences later declared null), while many of them were acquitted.

For each of the three case studies, we will present observations of the witnesses' testimonies, and will analyze the multimodal behaviors of defense, public prosecutor, accused person, judge and witness. The trials will be discussed following a descending order based on the severity of the judgment with respect to the alleged crime: first Corona's, then Moggi's, and finally Cucchi's trial.

5. Procedure, Methodological Choices and Description of Cases

For each of the three trials, we collected video material from the TV programme *"Un giorno in pretura"* ("A day in a district court"), with a total duration of 3 hours and 15

---

[1] "PM", stands for *"Pubblico Ministero"* (Public Prosecutor).

minutes (1 hour and 24 minutes for Corona's trial, 51 minutes for Moggi's and 1 hour for Cucchi's). We found examples of both verbal and bodily dominance, and of discrediting moves.

The fragments were transcribed, analyzed and classified by two independent expert coders. The verbal communication of some passages of the debates was transcribed, using Walton's classification (2008) for codifying the argumentation schemes. This methodology is particularly suitable, as it is based on a "presumptive" way of reasoning, including potential biases, but also probabilistic ways of reasoning. For trials in which the discrediting moves were mainly or exclusively performed by means of body signals, the communication was annotated for all the relevant modalities. An annotation scheme was constructed, based on the principles of the *"musical score of multimodal communication"* (Poggi, 2007), where each signal is analyzed in terms of its physical features (e.g., for gesture: handshape, location, orientation and movement, *expressivity parameters* of temporal extent, spatial extent, fluidity, power and repetition; for gaze: eyes direction, eyebrows and eyelids position and movements) and its literal and possible indirect meaning, which was expressed through a verbal paraphrase. Based on these meanings, a set of discrediting moves was built up. Here is an example of annotation (Table 1).

1. Sender and time	Defense 3.22	Defense 3.26
2. Speech	*"Ricorda quale arbitro e il perché?"*	*"Allora glielo dico io… Racalbuto!"*
3. Argumentation scheme	*Argumentum ad memoriam*	*Argumentum ad memoriam*
4. Meaning	"Do you remember which referee, and why?"	"Then I'll tell you… Racalbuto!"
5. Body signals	**Gaze:** *Direct and fixed toward interlocutor* **Posture:** *stretched forward*	**Prosody:** *Sudden interruption*
6. Meaning of body signals	I dare you. You can remember.	You are unable to tell me the truth.
7. Indirect meaning	You are trying to conceal the truth.	You are a liar.
8. Dominance or discrediting moves	Discredit on benevolence	Discredit on benevolence

Table 1. An annotation scheme of discrediting acts

In this fragment of Moggi's trial, Moggi's defense attorney interrogates Nucini, a referee of the Italian Football League who stated that Luciano Moggi, manager of the Juventus football team, influenced the referees' choices in exchange of favors.
Moggi's defense, at minute 3.22 (col.1) asks Nucini: *"Ricorda quale arbitro e il perché?"* ("Do you remember which referee, and why?") using an argumentation from memory (col.2-3) while gazing at him directly and staring, his body stretching forward (col.4), with a direct meaning of defeat (col.5) and an indirect meaning of absence of truth (col.6). Without giving him the chance to answer, at time 3.26 Defense says (col.1): *"Glielo dico io… Racalbuto!"* ("Then I'll tell you… Racalbuto!") (col.2). The interruption of the interlocutor (col.5) is a direct sign of impossibility to tell the truth (col.6), and this kind of discrediting moves can be classified as benevolence discredit.

5.1. Corona's Trial

Fabrizio Corona has been the object of numerous criminal proceedings between 2007 and 2013; here only the trial in which he was charged of extortion is taken into account.

The analysis was carried out on the trial held at the Court of Milan (*Tribunale di Milano*) in which the PM Francesco Di Maio claimed a condemnation of seven years and two months in prison for Fabrizio Corona, accused of attempted extortion and blackmailing to the detriment of some celebrities using scandalous photos. Among the blackmailed were Lapo Elkann, company manager and nephew of the famous industrialist Gianni Agnelli, and Adriano, a well-known soccer player of an Italian team. In appeal, the judge sentenced Corona to one year and five months in prison. Recently the Supreme Court cancelled this sentence and restored the previous judgment of 3 years and 2 months in prison.

5.2. Moggi's Trial

Luciano Moggi, former manager of the Italian football teams Roma, Lazio, Torino, Napoli and Juventus, in May 2006 was involved in the scandal known as *"Calciopoli"*[2] for criminal association aimed at fraud in football games. The trial against him was held in Naples. Moggi used to entertained relationships with people who managed football matches, or designated referees and journalists who judged the referees' work. Thanks to their compliance, he put pressure to determine the victories of his team. After the appeal process held in Naples, the Court confirmed the judgment of criminal association for Moggi, Pairetto and Mazzini, and condemned them to two years and four months in jail (less than the previous judgment of five years and four months), stating that the fraud in sport was extinct due to excessive judgment delay.

Many witnesses were heard by the Court, including Danilo Nucini, who was referee from 1995 to 2005, and Zdenek Zeman, a team coach who had been an opponent of Moggi before trial.

5.3. Cucchi's Trial

The death of Stefano Cucchi occurred in October 2009 during his detention for drug dealing. Some prison officers and some doctors of a hospital were accused of having caused his death.
After being arrested, Cucchi was brought to the police station, and there he was found in possession of various drugs and a medication (he suffered from epilepsy). Detention was decided for him; at the time he had no physical trauma and weighted 43 kilograms. The following day he was put on trial for summary. While entering the courtroom, he had difficulty walking and talking, and he showed evident bruising around his eyes. Talking to his father just before the hearing, he did not tell him he had been beaten.
Despite his precarious conditions, the judge set a new hearing for him to be celebrated a few weeks later, and stated that he had to remain at the Regina Coeli prison. After the hearing, Cucchi's physical conditions further worsened. He was visited at the hospital and the scoresheet reported injuries and bruising on his legs, face (including a broken jaw),

---

[2] The term *"Calciopoli"* (loosely translatable as "Soccer Gate") refers to an Italian football scandal that involved Italy's top professional football leagues, Serie A and Serie B. The scandal was uncovered in May 2006 by Italian police, when a number of telephone interceptions showed a thick network of relations between team managers and referee organizations. The team managers of some major teams have then been accused of rigging games by selecting favorable referees.

abdomen (including bleeding bladder) and chest (including two fractures to the spine). No hospitalization was granted to him.

In prison, his conditions got even worse, and he died on October 22, 2009: at the time of death, Cucchi weighed only 37 kilograms. During the investigation of the causes of death, a Ghanaian witness declared that Stefano Cucchi had actually revealed to him he had been beaten. Among the suspects, in addition to the prison officers, the doctors Aldo Fierro, Stefania Corbi and Rosita Caponnetti, were put on trial for leaving Cucchi without proper medical assistance and therefore die for starvation.

They defended themselves by saying it was the young man who refused treatment. The Third Court of Assizes of Rome (*III Corte d'Assise di Roma*) sentenced four doctors working at the hospital Sandro Pertini to one year and four months of prison, whereas the head physician was sentenced to two years in prison for manslaughter (suspended sentence). Another doctor was sentenced to eight months of prison due to false declaration, while six nurses and prison guards were acquitted because they would not have in any way contributed to the death of Cucchi.

6. Results. Trials' Argumentative Moves

The analysis of the verbal arguments used in the three trials helps us to deepen the analysis of the strategies used, in a perspective of multimodal persuasion. With their persuasive aims – in addition to the discrediting and the dominance moves – the actors involved in the trials present information that will affect the court's final decision. Specifically, in the following we will analyze the argumentative choices made by Public Prosecutor, Defense and Witnesses of the three trials.

6.1. Arguments in Corona's Trial

Corona's trial has a particular structure, as it is the Accused who mostly argues (see later in discrediting moves), often aggressively and violating the turn-taking rules. From an argumentative point of view, Corona makes a series of *ad hominem* attacks towards both the Witnesses and the PM, to the point of attributing lack of sexual attributes as a (catachrestic) metaphor to indicate lack of courage.

Corona, as a responsible of a talent agency, was in possession of several scandalous pictures, and "suggested" celebrities that appeared in those pictures to buy them, in change of large sums of money. During the trials, he frequently uses *arguments from rules* with the aim of explaining the unknown rules and dynamics of the show business, and the thin line between the specific tasks of his "work" and the crime of extortion. What follows is an example of how Corona explains the importance of possessing a newspaper in order to cause a real offense to a public person:

> Let's assume I wanted to extort money from someone, ok? I could go and tell someone I have these pictures here, and if I publish them I can spoil you, give me 50,000 euros or I will publish them. I could never make such a statement because I cannot publish the pictures, I can just take them to the newspaper[3].

---

[3] *"Metti caso io decido di fare un'estorsione, no, vado da una persona e gli dico io ho queste fotografie qua se le pubblico ti rovino o mi dai 50mila euro o io le pubblico. Io non potrei mai fare una dichiarazione del genere perché io non posso pubblicare le fotografie, io posso solo portarle al giornale."*

In this type of normative argument, Witnesses respond with some other presumptive argument, intended to help the audience to understand similar cases: for instance, by means of an *argument from analogy*, an *argument from consequences* (i.e., assuming very negative consequences descending from given scenarios), or an *argument from evidence to hypothesis* (i.e., assuming very negative consequences descending from real facts).

This is the case of Alberto Vergani, the manager of Marco Melandri, a motorcyclist to whom Corona wanted to sell incriminating photos, that could have distanced his fans: "*I fotografi avrebbero potuto farla passare come una sorta di... – avrebbero commentato – ecco perché Melandri è andato fuori*" – ("The photographers could have presented it as a sort of ...- they would have commented – this is why Melandri went off"). Vergani, in his role of Witness, suggests that the pictures in Corona's hands could actually have damaged Melandri's public image, and therefore this argument becomes a way to explain the point of view of the injured part with respect to laws possibly unknown to the Judge.

6.2. Arguments in Moggi's Trial

The arguments in Moggi's trial are mainly "descriptive" and "norm violation-based", since the Defense uses topics particularly related to *evidences* or *ad hominem*. However, the emotional orientation of this trial is peculiar, as it is positive for the Defense, while very negative for the Witnesses. In fact, the Defense uses *ad hominem* arguments in an ironic way by making fun of the Witnesses, while the Witnesses (in the case below, Nucini, the former Premier League referee) react with *arguments from distress* (1), *need for help* (2), and *values* (3):

> (1) Yep. I have seen some amazing things. I believe that there is intelligence... let me put it this way, but there must be a limit to making fun of people. You cannot, in front of 36 adults, with wives and children at home, show an episode that is not a criminal offence and say it is. You cannot make fun of a person who commits all week to earn the Sunday football game, and is denied this only because he does not understand certain logics. It is not possible. Well it is not possible[4].

> (2) Well I would have liked to see you there. Red turning into black, yellow turning into white. I see here in the courtroom some referees, fuck, a referee[5].

> (3) But I wonder, do those individuals have dignity? But what dignity![6]

Moreover, Witnesses reply with other arguments that specify the terms or rules linked to Moggi's behaviors by using *vagueness or arbitrariness of verbal classification*, or *rules*, as in the case of the coach Zeman that tries to defend himself by highlighting how his dismissals were not related to a lack of competence but to the orders given by Moggi.

After the clarifications and the indignation of the Witnesses, both Moggi's Defense and Moggi himself as the Accused reply with arguments *from abduction* and *ad hominem*, at the same time demonstrating the absurdity of the evidence and attacking the Witnesses

---

[4] "*Sì. Ho visto delle cose sorprendenti. Io credo che all'intelligenza c'è... passatemi il termine, ma la presa per i fondelli ci deve essere un limite. Non si può, davanti a 36 persone adulte, con a casa mogli e figli, fargli vedere un episodio che non è calcio di rigore e dire che è calcio di rigore. Non si può prendere per i fondelli una persona che tutta la settimana si fa un mazzo per meritarsi la serie A, e le viene negata solo perché non capisce certe logiche. Non è possibile. Ma non è possibile.*"
[5] "*Ma io vi avrei voluto vedervi lì. Rosso che diventava nero, giallo che diventava bianco. Io vedo qua in aula alcuni arbitri, cazzo, un arbitro.*"
[6] "*Ma io mi domando, ma questi individui hanno una dignità. Ma che dignità!*"

regarding their professional skills. At minute 15.57, Moggi says about Zeman, concerning his testimony:

> The problem is that Zeman said many things, he said that he is the best (coach) in Europe but he is the only one claiming it; he said that he was never exonerated, but he was many times. So he said a lot of false things, so many that I will ask my lawyers to sue him for calumny. (...) He goes to Naples in 2001 where he was exonerated after making two points out of 12 he had at his disposal. Why? Because he is not able to be a coach, the same way he speaks does he go on the football pitch, he is slow and awkward in speaking and the players never understand him.[7]

6.3. Arguments in Cucchi's trial

The prevailing arguments in Cucchi's trial are mostly associated with the evidence, the memories and perception of the death scenario; however, interesting differences emerge with respect to the roles. The PM (one of the lawyers representing Cucchi's family), uses the *arguments from evidence, ad memoriam* and *from practical reasoning* when trying to bring out facts or real memories from the witnesses. He also frequently exploits the *argument from vagueness of verbal classification* when trying to specify the term "crushed", that can unequivocally testify the accused's guilt. Even the Witnesses argue about evidence, memories, perceptions, but also *from ignorantiam*, and similarly does the public defender of Cucchi, when he is testifying as a Witness about Cucchi's swollen body and head:

> Given his thinness, in my opinion, slightly, I am not a doctor but I had not seen slightly swollen, I do not know if it's the right word, with respect to this congenital thinness.[8]

Similarly, the Defense frequently uses argument *from ignorantiam* and *from pragmatic inconsistency*, to diminish the importance of the Witness' statements, who accuse prison guards and medical staff.
Like in the following:

> Samura Jaja (the Witness) says that he attends the discussion, but cannot understand, since he cannot understand Italian well, what is discussed. He does not... says that at a certain point he heard a thud. He hears someone crying, he hears kicks, he hears him while being dragged and the only thing he hears, quite honestly is Stefano's door opened.[9]

6.4. Arguments in the Three Trials

---

[7] *"Il problema è che Zeman ha detto tante cose: ha detto che è il migliore d'Europa ma lo dice solo lui, ha detto che non è mai stato esonerato, ma è stato esonerato tante volte. Quindi ha detto un sacco di cose false tant'è che io chiederò ai miei avvocati di denunciarlo per calunnia. (..) Va al Napoli nel 2001 dove viene esonerato dopo aver fatto due punti su 12 che ne aveva a disposizione. Perché? Perché questo non sa allenare, così come parla va in campo, è lento nel parlare e impacciato e i giocatori non lo capiscono mai."*
[8] *"Data la magrezza, secondo me, leggermente, non sono un medico però non avevo visto leggermente gonfio, non so se è l'aggettivo giusto, rispetto a questa magrezza congenita."*
[9] *"Samura Jaja (il testimone) dice che assiste alla discussione, ma non riesce a capire perché non riesce a capire bene l'italiano, di che cosa si discute. Non... dice che ad un certo punto ha sentito un tonfo. Sente piangere sente dei calci lo sente mentre viene trascinato e l'unica cosa che sente, molto onestamente è la porta della di Stefano aperta."*

In order to differentiate the three trials even more, the following can be said: Corona's trial is mainly subject to aggressive devaluation of the Witnesses referring to their lack of "masculine attributes". Corona also casts doubts on the Judge and PM's competence, by means of explanations about show business' implicit rules. Witnesses react to this "aggressive-didactic" defensive approach by pointing at the effects resulting from Corona's "work" in terms of the damages to integrity and image, by using the *argument from important consequences*. The conduct of cross-examination by Moggi's Defense, with its use of *ad hominem* arguments, is equally aggressive. However, differently from Corona's trial, it is much more "emotionally driven": while using *ad hominem* argument, Defense often ridicules the Witnesses, conveying positive emotions of amusement (Poggi, 2010), while Witnesses respond with *distress*, i.e., with negative emotions of anger and sadness, but also with appeals to morality and "values". Finally, Cucchi's trial presents an argumentative approach that is mostly based on facts (evidence, memoriam or perception) and denials of the key Witnesses' statements - *from ignorance* or *pragmatic inconsistency* relating to the facts or narratives.

7. Discrediting Moves

Discrediting moves can be found in all three trials. They are computed from the quantitative point of view in Table 2. The multimodal coding following the annotation scheme previously described allows to compute the total amount of discredit moves while distinguishing them in terms of their target (dominance, competence, or benevolence) and of the different roles of the discrediter. As we can see from the results below the three trials significantly differ as to the discrediting strategies exploited (chi square; $p<0.001$).

7.1. Corona's Discrediting Moves

On Corona's trial, we can clearly see that the main discrediter is the Accused person (41,8%), followed by the public prosecutor (PM: 27,6%) and the Witness (18,4%). Corona's trial is a sort of "discrediting war" between the Accused (Corona) and the PM, where the Defense has the task of lowering the tones (we reported just 5% acts of discredit from the Defense). In particular, Corona discredits the PM mostly in terms of dominance (66,7%) and competence (45,8%) and more the PM compared to other roles. On the other hand, the PM tries to capture the Judge's attention on Corona's offensive way of answering, yet in 40% of cases he discredits Corona's benevolence (40%), attacking his unfairness, but in a somehow resigned way.

Let us now see a qualitative analysis of the cases.

7.1.1. Corona (Accused) Discredits the PM's Dominance

The analysis of the hearing sequence highlights the strategy used by Corona - the accused person - to discredit his opponents in breach of the procedural and communicative rules.

At minute 54 of the analyzed video - broadcasted by the TV programme *"Un giorno in pretura"* – Corona tries to describe the scandalous photos that depicted the football player Adriano in morally execrable poses by *overlapping on the public prosecutor's speaking turn* and *displaying an angry face* to him, who is casting doubts on his deductions. Corona interrupts him by saying *"No, allora lei mi prende per scemo! Se lei guarda queste foto le prende per normali?"* ("No, then you are taking me for a fool. If you look at these pictures, do you consider them normal?").

	Roles	DOMINANCE	BENEVOLENCE	COMPETENCE	tot
CORONA	JUDGE	2,6%	14,3%	4,2%	7,1%
	PUBLIC PROSECUTOR	17,9%	40,0%	25,0%	27,6%
	DEFENSE	7,7%	2,9%	4,2%	5,1%
	ACCUSED	66,7%	11,4%	45,8%	41,8%
	WITNESS	5,1%	31,4%	20,8%	18,4%
total within roles		100,0%	100,0%	100,0%	100,0%
total within discredits		39,8%	35,7%	24,5%	100,0%

Table 2. Corona Case. Discredit type*Role in the trial

He speaks with *fast speech rate* and *high tone of voice* when saying: "*È l'idea che si farebbe chiunque! Se lei non ce la fa a capire, vuol dire che ha qualcosa che non...*" ("It's the idea that everybody would have of it. If you can't understand, this means you have something that doesn't...").

The use of *argumentum from popular opinion* and *ad hominem* is accompanied by a *sarcastic tone of voice*, a *slight smile* and the *eye gaze turned toward the audience in the courtroom*, in search for complicity and approval of the sexual act represented in the picture. Then Corona goes on quietly alluding to the sexual inability of PM, by making *symbolic gestures* referred to a lack of sexual competence.

7.2. Moggi's Discrediting Moves

Moggi's trial presents a different pattern of discrediting moves from the ones described above (see Table 3). It seems to follow a more predictable trend, since the Defense counts 35.5% of the total amount of discrediting acts, followed by the Witnesses (22.6%) and finally by the Accused (Moggi: 19.4%). Also in this case, the most frequent targets of discredit are dominance (41.9%) and competence (39.8%). As to the role, we can clearly see how the Defense more than anyone else discredits others' dominance (43,6%) and competence (32,4%) while the Witness mostly discredits the lawyer's benevolence (23,5%) and competence (24,3%).
In fact, these results exemplify what occurs more frequently during this trial, i.e., that the Defense behaves very aggressively against the Witnesses who testify against Moggi. This also emerges from the observational analysis below.

7.2.1. Moggi's Defense Discredits Witness' Competence

In the fragment analyzed, at minute 4.35, Moggi's defense is questioning the Witness - the referee Nucini, who accuses the manager of having prevented him from refereeing the Premier League. According to the referee's testimony, Moggi, that was at the time manager of the Juventus football team, disadvantaged him because he did not favor Juventus.

	roles	DOMINANCE	BENEVOLENCE	COMPETENCE	tot
MOGGI	JUDGE	10,3%	17,6%	16,2%	14,0%
	PUBLIC PROSECUTOR	5,1%	17,6%	8,1%	8,6%
	DEFENSE	43,6%	23,5%	32,4%	35,5%
	ACCUSED	20,5%	17,6%	18,9%	19,4%
	WITNESS	20,5%	23,5%	24,3%	22,6%
total within roles		100,0%	100,0%	100,0%	100,0%
total within discredits		41,9%	18,3%	39,8%	100,0%

Table 3. Moggi's case. Discredit type*Role in the trial

During the interrogation the defense lawyer, *with his torso stretched forward, gazing into the eyes of the witness, fixing defiantly at him, with a low speech rate and half open eyelids expressing boredom*, asks a series of questions implying an *argument from memory*: "*Ricorda quale arbitro durante la gestione Bergamo-Pairetto ebbe a riportare la sospensione più lunga e si ricorda il perché*" ("Do you remember which referee received the longest suspension, and do you remember why?"). As the Witness hesitates in trying to answer, the Defense *abruptly says*: "*Ricorda male! allora glielo dico io... Racalbuto!*" ("You remember it wrong! Then I'll tell you... Racalbuto!").

This use of the argument from memory actually implies a discredit of the Witness' competence – of his memory, in fact. But more than that, a discrediting attitude is also present in the Defense's lack of compliance with turn-taking rules, consisting in his overlapping with the Witness' speech and in an aggressive exploitation of the question-answer pair. In the sentence above, he actually answers the question he himself posed. Moreover, immediately after that sentence, he repeats a similar question, and orders the Witness to answer:

> Do you remember the year of the Bologna-Juventus match, about which you repeatedly beckoned, who won the championship? Do not make comments, be polite and please simply answer the question. Period![10]

7.2.2. Moggi's Defense Discredits the Witness' Dominance

At minute 4:14 Zeman (a famous football coach, called to testify against Moggi) claims that he used to be considered one of the best coaches, but that, when he publicly attacked the Juventus team, his career was virtually over, due to Moggi's interference.
After that, the lawyer begins questioning Zeman: "*Lei aveva fatto solo un punto. Perché si meraviglia dell'esonero?*" (You only had gained one point. Why are you surprised with your exemption?), still *laughing at the Accused*. Then the Defense lawyer *looks away from Zeman* and starts talking quietly with the other lawyer while *giggling*, implying how ridiculous is Zeman's statement.

At minute 8:50, the public prosecutor says: "*Quanti esoneri ha subito nella sua carriera? Non il perché, ma il numero...*" ("How many exemptions have you received in your career? Not the reasons why, but their number..."). While the Judge invites to pay attention to the question, the PM shows resignation (*mouth closed with corner slightly downward*, meaning dissatisfaction), and then says: "*Vabbè presidente, lasciamo perdere...*" ("Well, president, never mind..."). This multimodal behavior conveys an ironic resignation discrediting Zeman on its linguistic ability. Everyone laughs.

---
[10] "*Ricorda nell'anno del Bologna-Juventus di cui lei più volte ha fatto cenno chi vinse il campionato? Non faccia commenti, abbia la cortesia di rispondere alla domanda. Punto!*"

At minute 9:59 the lawyer continues: *"Che lei fosse il migliore del mondo lo abbiamo capito, ma risponda sì o no, le giustificazioni non interessano"* ("That you were the best coach in the world we understand, but please answer with a yes or no, we are not interested in justifications"). From an argumentative point of view the lawyer bases his defense on an "argumentum from evidence"; the number of exemptions demonstrates Zeman's incompetence, but the claim "You're the best in the world" accompanied by an *ample gesture* that amplifies such a supposed "talent", his *laughters* and *shared smiles*, the words spoken in a *low voice with other bystanders* during the interrogation all show a mockery of the witness on his ability to face the trial, and discredit Zeman as having no credibility and being helplessness.

7.3. Discrediting Moves in Cucchi's trial

The analysis of the different kinds of discrediting strategies in Cucchi's trial (Table 4) reveals a significant (*chi-square*, $p < 0.001$) prevalence of discredit based on competence (45.9%), as compared to discredit on dominance (31.1%) and benevolence (23%). In particular, with respect to the role, the strategy of damaging the image of others is used primarily and to the same extent by PM and Defense (31.1%), followed by the Witnesses (28.4%). The targets mainly used by the PM are dominance and competence while in the case of the Witness is benevolence (41.2%).

The Defense lawyers of doctors and penitentiary police aim at discrediting the competence (35.3%) and benevolence (29.4%) of the Witnesses. In fact, during the trial, Defense asks questions to the Witnesses, with the intention of recusing them on the basis of either prior statements or their impossibility to see or hear what they claim to have seen or heard. In the latter case, the Defense claims that the Witnesses were in prison at the time of the facts, and thus their reliability is questioned (see below).

	roles	DOMINANCE	BENEVOLENCE	COMPETENCE	tot
CUCCHI	JUDGE	13,0%	0,0%	5,9%	6,8%
	PUBLIC PROSECUTOR	34,8%	23,5%	32,4%	31,1%
	DEFENSE	26,1%	29,4%	35,3%	31,1%
	ACCUSED	4,3%	5,9%	0,0%	2,7%
	WITNESS	21,7%	41,2%	26,5%	28,4%
total within roles		100,0%	100,0%	100,0%	100,0%
total within discredits		31,1%	23,0%	45,9%	100,0%

Table 4. Cucchi's Case. Discredit type*Role in the trial

7.3.1. The Defense Attacks the Witness on His Benevolence

At minute 24.43 a young man who was in the cell next to Cucchi while in prison, is heard as a witness of the events occurred during his detention.

The Witness claims to have seen through the peephole of the cell two guards that led Cucchi out of the cell. In saying *"due guardie"* (two guards) he emphasizes the number "two" placing the index and middle fingers on the table. The Defense of the prison guards says there are no windows in the cell, then the Witness, using an "argument from evidence", and with an incredulous face (*open mouth* and *wide open eyelids*) declares: "*Come non ci sono le finestrelle. Ci sta un buchetto così... è così se vuole si può mettere nella mia cella!*" ("Why, there are no windows? There is a little hole like this... it's the truth, if you want you can enter my cell!"), while miming the size of the small "window" with an *iconic gesture*. The Defense attacks him by denying the evidence "*Io ci sono stato... se uno sta seduto lì non si vede. Poi la corte lo verificherà*" (I have been there... if you are sitting there, you cannot see. Then the court will verify it), with *jerky gestures* and *half interrupted voice*. Irritated, the attorney attacks the validity and reliability of the words of the Witness, who replies with an "argumentum from evidence": "*Ho capito. Per me ci possiamo andare ora, anche adesso*" ("I understand. If it were up to me, we could go there even now").

8. Dominance Moves

Since lawyers are in the so-called "tricky position" (Higdon, 2008), dominance communication is of great importance in the processes of legal persuasion. We therefore also focused on the analysis of the different strategies of persuasion aimed at communicating the dominance of the counterpart, using "multimodal dominance moves" (Poggi & D'Errico, 2010). In both verbal and bodily communication, we took into account the so-called "subtle" dominance strategies (carelessness, self-confidence, touchiness, ridicule, victimhood) and the "blatant" strategies (judgment, imperiousness, invasion, norm violation and defiance).

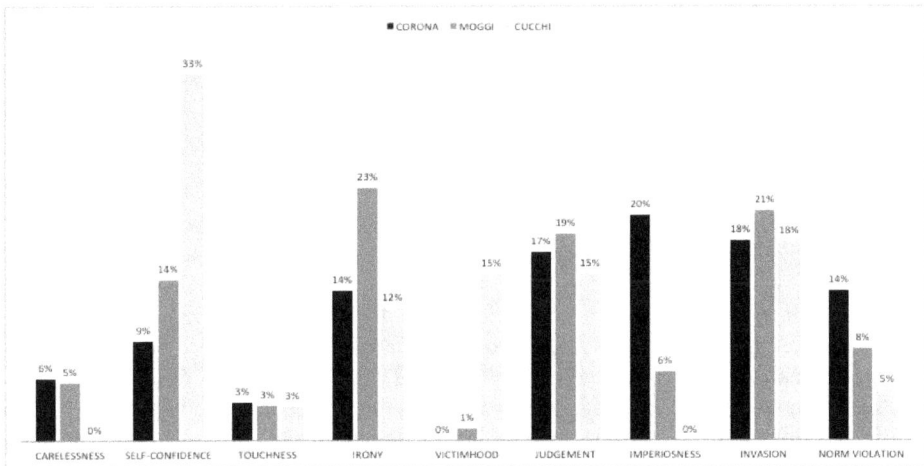

Table 5. Dominance moves*types of trials.

The results in Table 5 significantly show how the different dominance moves are used in the three trials. In Cucchi's trial there is a prevalence of subtle dominance moves based on self-confidence (33%) and victimhood (15%), while the blatant are less frequent (invasion: 18%, and judgement: 15%). The prevalence in Cucchi's trial of dominance moves based on self-confidence, including strategies of assertiveness and sometimes easiness (Poggi & D'Errico, 2010), emerged mainly from depositions gave by Cucchi's relatives and fellow detainees; they often respond very calmly and carrying arguments based on "evidence" and on the "memories" of what happened. *Victimhood* is instead a frequent strategy of both the

PM and the Defense lawyers of prison officers. They expressed it with a calm tone and sad, serious faces, when they were asked whether Cucchi was or was not already malnourished and suffering ("Did he walk slowly? Did he complain of pain?"). In doing so, they tried to diminish or strengthen the violence of the guards, and therefore of their guilt. The use of a victimhood strategy varies according to the debate phase: while Defense uses victimhood immediately after the description of Cucchi's imprisonment, the PM uses it throughout the debate, in order to stress how Cucchi was already infirm.

In Moggi's case, mixed types of dominance moves are recurrent: the blatant moves are *invasion* (21%) and *judgement* (19%), but *irony and ridicule* are certainly prevailing (23%). Differently from what happens is Cucchi's trial, Moggi's lawyers diminish the relevance and credibility of witnesses by means of teasing, highlighting how absurd accusations are, mainly in an indirect and funny way. They invade the interlocutor's speech, frequently overlapping with it, and formulate questions in such a way as to elicit laughter in bystanders.

Corona's case presents instead an absolute prevalence of blatant dominance strategies (69% of the total). There is a significant frequency of *imperiousness* (20%), *invasion* (18%), *judgement* (17%) and *defiance-norm violation* (14%) strategies, especially on the side of the Accused, who proves himself very arrogant in going against all legal procedures. *Easiness* is also frequent, as displayed by Corona's postures, sometimes relaxed and nonchalant, sometimes definitely arrogant (14%).

Corona imposes his conversational laws and regulates the debate prescribing his standards and criteria to PM, and when he does not like insistence, he loudly protests, as in the analyzed case: *"Mi porti rispetto, esigo educazione nei miei confronti, mi deve guardare in faccia quando mi parla"* ("Be respectful to me, I demand education towards me, you must look at me when you speak to me"). The Prosecutor promptly replies: *"Quando avrà finito di fare il suo show le farò la domanda"* ("Once you have finished doing your show, I'll ask my question"). Corona then stops and talks with witnesses (min. 7.50, deposition given by Melandri), he raises his voice, corrects the PM, shows annoyance, and attacks him on the basis of his the skill to conduct the process and explain the facts. He insults, uses profanity and vulgarity, expresses extremely negative evaluations on the people involved in the events, to such an extent that the Judge and the PM repeatedly reproached him.

9. Discussion

"Multimodal persuasion" is a theoretical and methodological framework within the multimodal communication research, that aims at analyzing persuasive processes by means of a multimodal approach (Poggi, 2007). In this chapter we have observed a particular setting of persuasion, the judicial debate, that puts the persuaders (the lawyers in particular) in the position of being credible and dominant persuaders while also respecting the judge's dominance. In order to explore the different levels of judicial multimodal persuasion, we chose to observe three frequent types of communicative acts: discrediting, argumentative and dominant moves. Discrediting and dominant moves require to deepen both the speech and the body level of persuasion by looking at its "peripheral" and superficial aspects that consist in how lawyers convey their Defense; while the argumentative moves are mostly focused on the verbal and rhetorical side and on how persuaders think and reason: these are the "central" aspects of persuasion (Petty & Cacioppo, 1986).

To carry on our research on multimodal persuasion we selected three very renowned trials in Italy: Corona's, Moggi's and Cucchi's. Corona and Moggi are two famous persons from entertainment and sport businesses, who in a certain sense have dramatized their trials and made entertainment out of them. This is not the case for police agents and doctors accused of having caused the death of Stefano Cucchi, a common person imprisoned for drug possession. Aware of these differences, we have used our model of persuasion and communication based on goals and beliefs, focusing our attention on discredits, argumentations and expressions of dominance (Castelfranchi, 2003; Poggi, 2005; Paglieri, 2015; Poggi & D'Errico, 2010, D'Errico & Poggi, 2014).

The analysis of the three trials have pointed out how discredits, argumentations and dominance moves can explain the persuasiveness of legal strategies. If we discuss the cases by ordering them in relation to the severity of penalties (starting from the most to the less severe) – respectively Corona, Moggi and Cucchi – we can also note how they are more and more aggressive and then less persuasive.

In Corona's trial, discredit is recurrent mainly on the opponent's dominance (and on his sexual possibilities and experiences), and cast by the Accused person toward the Public Prosecutor and the Judge. Coherently with this, we find that frequent arguments are *ad hominem* but also *from rules*, because Corona clearly decides to explain to the audience the implicit norms within show business, in order to demonstrate how his faults can be shared also by the injured part. This maneuvering is conveyed by blatant dominance moves, as Corona uses *invasion, imperiousness, judgement and norm violation,* thus showing extreme aggressiveness.

In Moggi's trial, first of all, the higher number of discrediting acts is performed by the Defense (following the trial's procedure), directed mostly on dominance but also on the injured part's competence, avoiding to attack the Defense lawyer directly, as in Corona's case. As regards argumentations, the most frequent are *ad hominem,* but also arguments from *evidence,* aimed to attack the Witnesses but also giving some elements to prove how Moggi can't have so much power in affecting the referees' decisions. To this purpose, Moggi's lawyer conveys this content by means of *irony* and by *ridiculing witnesses* but also *invading* and expressing *judgments,* thus showing both subtle and blatant dominance moves.

Finally, the trial that gives the lower numbers of condemnations - toward police agents and doctors of public hospital – is Cucchi's. In this case, the Defense attacks the Witnesses just on their *competence*, and on their possibility to have seen or heard violence on Cucchi in prison. Here we found a lower number of discrediting acts, mostly performed by the defense lawyers. The argumentations used by the Defense are based on *evidence, memories* and on *pragmatic reasoning*, and thus mainly based on facts and their potential inconsistencies. Particularly interesting in this case are the witnesses' arguments that range from *ignorantiam* to *vagueness of verbal classification*, when they indirectly affirm the Accused persons' responsibility or, on the contrary, they specify words in order to accuse by giving the right weight to words. As to dominance moves, in this case subtle strategies are more frequent, PM and Defense are quite apparently respectful of each other turn, and *victimhood* and *self-confidence* prevail. The moves of victimhood are functional to the phases of the debate: PM presents Cucchi as a victim mostly after the imprisonment, while Defense does so for all the duration of his life in jail, in order to show how Cucchi was already sick before that.

References

Austin, J. L. (1962). *How to do Things with Words*. London: Clarendon Press (1962).
Burgoon, J. K., Birk, T., & Pfau, M. (1990). Nonverbal Behaviors, Persuasion, and Credibility. *Human Communication Research*, 17, 140-169.
Castelfranchi, C. (1988). *Che figura. Emozioni e Immagine Sociale*. Bologna: Il Mulino

Castelfranchi, C. (2003). Micro-Macro Constitution of Power. *ProtoSociology, Internaional. ournal of Interdisciplinary Research*, 18-19, 208-265.
Conte, R., Castelfranchi, C. (1995). *Cognitive and Social Action*. London: University College.
D'Errico F., Poggi I. (2014). Acidity. The hidden face of conflictual and stressful situations. *Cognitive Computation*, 6(4), 661-676.
D'Errico F., & Poggi I. (2013) Discrediting Body. A Multimodal Strategy to Spoil the Other's Image. In *Multimodal Communication in Political Speech. Shaping Minds and Social Action*. Springer, Lncs, 7688, 181-206.
D'Errico F., & Poggi I. (2012) Blame the Opponent! Effects of Multimodal Discrediting Moves in Public Debates. *Cognitive Computation*, 4(4), 460-476.
Feteris, E.T. (2012). Strategic maneuvoeuvring with linguistic arguments in legal decisions: a disputable literal reading of the law. *International Journal of Law, Language & Discourse*, 2 (1), 106-125.
Higdon, M. J. (2008) Oral Argument and Impression Management: Harnessing the Power of Nonverbal Persuasion for a Judicial Audience. *Kansas Law Review*, 57, 3.
Hovland, C. I., & Weiss, W. (1951). The Influence of Source Credibility on Communication Effectiveness. *Public Opinion Quarterly*, 15, 635-650.
Kennedy, G. A. (2006). *On rhetoric: A theory of civic discourse*. Oxford University Press.
Mucchi Faina, A. (2013). *L'influenza sociale*. Bologna: Il Mulino.
Petty, R., & Cacioppo J. (1986) *Communication and persuasion: central and peripheral routes to attitude change*. New York: Springer-Verlag..
Paglieri, F. (2015). Arguments, conflicts and decisions. In F. D'Errico, I. Poggi, A. Vinciarelli, & L. Vincze (Eds.), *Conflict and negotiation: social research and machine intelligence*, pp. 117-136, Berlin: Springer.
Poggi, I, D'Errico F., & Vincze L. (2012). Ridiculization in Public Debates: Making Fun of the Other as a Discrediting Move. In *Proceedings of the 8th conference on (LREC'12). Istanbul 21-27 May (ELRA)*, pp. 44-50.
Poggi, I., & D'Errico F. (2010). Dominance in Political Debates. In A. A. Salah et al. (Eds.) HBU, LNCS 6219, pp. 163--174. Heidelberg: Springer.
Poggi, I., Cavicchio F., & Magno Caldognetto, E. (2007). Irony in Judicial Debates: Analyzing the Subtleties of Irony while Testing the Subtleties of an Annotation Scheme. Language Resources and Evaluation, *41(3-4)*, 215-232.
Poggi, I. (2007) *Mind, Hands, Face and Body. A goal and belief view of multimodal communication*. Berlin: Weidler Buchverlag.
Poggi, I. (2005) The Goals of Persuasion. *Pragmatics and Cognition, 13(2)*, 297-336.
Walton, D., Reed, C., & Macagno, F. (2008). *Argumentation Schemes*. Cambridge: Cambridge University Press.
Walton, D. (2002). *Legal Argumentation and Evidence*. University Park (PA).: Penn State Press.

Chapter 18

Finding Mussolini's Charisma in His Multimodal Discourse

Isabella Poggi[1] & Francesca D'Errico[2]

[1] Dipartimento di Filosofia Comunicazione e Spettacolo, Università Roma Tre, isabella.poggi@uniroma3.it;
[2] Facoltà di Psicologia, Università Internazionale Telematica UNINETTUNO, f.derrico@uninettunouniversity.net

1. Introduction

The term "charisma" in modern literature was first launched by Max Weber in 1912 (Weber, 1920). He claimed that during periods of crisis for a people or a nation, a "charismatic" leader may emerge, responding to the necessity for a strong leadership to come out of the crisis. Weber called this quality "charisma" (from Greek "charis", grace), thus considering it a grace, a divine gift that only some people may be enlightened with, and defined it as an "extraordinary quality" of a person who is believed to be endowed with superhuman properties, in such a way as to induce people to acknowledge him as a leader, to the point of making a cult of him. Though he did not describe this gift at length in his work, he was nonetheless somehow prophetic, since the century just rising at his time was the century of charismatic leaders, like Benito Mussolini, Hitler, Stalin, Mao Dse Dong, Fidel Castro, and maybe others…

Benito Mussolini based the fortunes of his dictatorship not only on his totalitarian imperium but also on his force of persuasion. His discourses are not particularly impressive from a literary point of view, nor particularly strong or convincing on the argumentative side – he had a bombastic, somehow trivial rhetoric, and according to some theater men, like Dario Fo, he was a bad actor, too – but he definitely did have charisma, as witnessed by the strong influence that he exerted on the Italian people, not only thanks to the violence of his dictatorship.

In this work we present a theoretical model of charisma and analyze Mussolini's multimodal discourses to find out how the persuasive mechanisms of his "actio" make his communication an example of charismatic communication.

2. Charisma

According to a sociocognitive model of charisma (D'Errico *et al.*, 2012; Signorello *et al.*, 2012), charisma may be seen as a set of internal features of a person that, when manifested by some external displays – some traits displayed or behaviors performed in various modalities – trigger a set of emotions in other people, that have the effect of influencing them in a peculiar way. In particular, the way in which a "charismatic" person influences others is that s/he induces them to pursue some goals not through coercion but voluntarily and willingly, while feeling involvement and enthusiasm in what they do together or on behalf of that person. Studying charisma therefore means on the one side to specify what are those internal features and on the other to track what external displays in the multimodal traits or behaviors of a person – words, prosody, voice, gesture, posture, face, gaze, body – manifest the internal features.

In the model above, the internal features of charisma in a great part correspond to aspects of persuasive discourse. According to Aristotle, to persuade the Audience the Orator may exploit three strategies: *logos* (rational argument), *pathos* (the appeal to the Audiences' emotions), and *ethos* (the Orator's personality). Within *ethos*, based on studies on how politicians are evaluated by laypeople, and on how they discredit each other, Poggi *et al.* (2012) single out three criteria of evaluation: benevolence – a politician's tendency to care the others' interests, more than his own; competence – his intelligence, expertise, planning capacity, creativity; and dominance – the capacity to win in competition and to impose his will.

From a qualitative study (Signorello *et al.*, 2012) asking participants to generate adjectives describing charismatic and non-charismatic persons, a list was drawn of 68 adjectives that cluster around the following dimensions:

1. *emotional intelligence*: the charismatic leader has a high tendency, and a high skill to feel emotions himself, to manifest them, and to be empathic with others' emotions (stressed by adjectives like *enthusiastic, passionate, empathetic*).
2. *sociability and inclusiveness*. The charismatic leader is people-oriented, inclusive, and makes followers feel "similar" to and "together" with him (he is *extraverted, sociable*).
3. *competence*: the charismatic leader has various physical and mental skills, and is endowed with them at a surprising and admirable extent. He may be physically strong, possibly skilled in sports, but mainly he has notable cognitive and communicative skills, among which visionary, creativity, foresight, strategic intelligence (he is *visionary, creative, enterprising, clear, persuasive*).
4. *dominance*: the charismatic leader is dominant, he often challenges traditions and dares other leaders, while not submitting to others (he is *active, dynamic, courageous, vigorous*).
5. *emotional induction*: the leader's charisma causes emotions in people (he is *attractive, charming, seducing*): they are charmed, subjugated by him, infected with his enthusiasm, hence willing to comply with his will.

These are in general the internal features of a charismatic leader. Subsequent research on how they are manifested and perceived in political leaders' voice found out that different dimensions of charisma may be more evident in different leaders, or in the same leader in subsequent phases of his leadership. This was evidenced, for example, in a study about Umberto Bossi, the founder and first leader of Lega Nord (North league), a racist party aiming at the secession of some regions of North Italy (Padania) from the rest of Italy. During his long leadership across fifteen years, Umberto Bossi had a stroke that impaired his speech and changed his voice; from acoustic analysis and perceptual studies on his speech acts and his voice quality some years before and some years after the stroke, it emerged (Signorello *et al.*, 2012; D'Errico *et al.*, 2013) that the dimensions above can be grouped around three factors, proactive-attractive (mainly described by adjectives like *lively, dynamic, charming, convincing*), calm-benevolent (*wise, calm, just, intelligent, easy, sincere*), authoritarian – threatening (*determined, threatening, disturbing, individualist, authoritarian*), and Bossi's charisma, as perceived from his voice, shifted from a more proactive-attractive and authoritarian threatening type before the stroke to a calm-benevolent type after the stroke.

3. Tracking Mussolini's Charisma from his Multimodal Discourses

It is empirically proved that Benito Mussolini was a charismatic leader. The empirical evidence is in the whole Italian history since 1922 – the year of his "March on Rome" – through 1945 – the year of his death. That so many millions of people for so many years had no suspicion that the content of his communication might be fake, misleading, a source of illusion, to the point of enthusiastically following him with trust and faith straight to ruin, is a deadly demonstration that he did have charisma.

The goal of this paper is to find out what are the specific features of Mussolini's charisma by tracking how they leak from his words and his multimodal communicative behaviors. To do so we adopted a two-step procedure. First we conducted a qualitative analysis on a small corpus of Mussolini's discourses, to find out the communicative aspects (words, speech acts, gestures, postures...) that most typically display his charisma. After singling out these aspects, we run a quantitative analysis on a much wider corpus of his discourses to test if the charismatic aspects found are actually represented with a significantly high frequency in a wider sample, thus allowing us to demonstrate the bulk of Mussolini's communicative charisma.

4. Corpora and Procedure

The first corpus we took into exam – corpus A – is a set of fragments from 10 discourses by Mussolini[1] since 1931 through 1940, ending with the discourse delivered immediately after the declaration of war on June 10th, 1940. These discourses are given in various Italian and foreign cities (Roma, Milano, Bari, Forlì, Aprilia, Berlin, Washington) and in various contexts: from the inauguration of a town to a salute to Hitler and his followers, from war declaration to feast for victory in Ethiopia.

On this corpus we run a qualitative analysis. In all fragments, we analyzed the verbal and the body behavior (after Poggi, 2007; Poggi *et al.*, 2013), tracking on the one side the type of communicative acts performed, on the other side their style, and the general exploitation of psychosocial strategies, in order to find out how the features of charisma listed above leak from these fragments.

More specifically, to analyze Mussolini's multimodal discourses, we take into exam all communicative modalities and their levels of analysis: we start from the lexical and syntactic aspects, then we move to the phonetic features of his speech, to his gestures, postures, his behaviors in general, and finally to his most frequent communicative acts. For all signals and aspects of signals we wonder which of them convey meanings that are related to the features of charisma posited by our model above, and which ones specifically. What meanings for instance are a cue to the expression of dominance, which to empathic communication, and so on.

The second corpus – corpus B – is a wider selection of entire discourses by Mussolini. We collected 15 of his discourses, all in the same time lapse as the fragments above, but

---

[1] We are grateful to TV Journalist Emanuele Colarossi for proposing an interview on Mussolini's charisma and for providing precious excerpts of his discourses.

different for the context, audience and communicative situation in which they were delivered: namely, 5 discourses to the crowd (the "oceanic crowds", as he himself called them) (discourse type 1); 5 to the Italian Parliament, in an institutional context (discourse type 2); and 5 to more restricted audiences in specific situations, like inauguration of a Surgery Conference, a discourse at the Opera of Rome or at a children summer camp. Our hypothesis is that the most typical aspects of charismatic communication are more frequently represented in the discourses to the crowd than in the other contexts.

This corpus was processed through an automatic quanti-qualitative analysis by TalTac (*Trattamento Automatico Lessicale e Testuale per l'Analisi del Contenuto*, i.e. "Lexical and Textual Automatic Processing for Content Analysis", Bolasco, 2013), a software for textual data analysis based on a lexicometric approach: an application of statistical principles to textual corpora. Textual statistics aims to extract the semantic level in a text starting from the list of words obtained by statistical analysis; for example, in the specificities analysis (namely "characteristic lexicon"), the software extracts a list of significant words obtained by a statistical comparison between sub-parts of text according to selected variables.

The lexical analysis includes some descriptive information, particularly interesting for understanding significant trends in Mussolini's discourse, like the *imprinting of the text*, that includes *adjective analysis* and *time analysis*.

Two more types of analysis are those of *peculiar and characteristic lexicon*. The peculiar lexicon is composed by the words that result over-represented in the text under analysis by comparing the corpus to an external frequency lexicon, taken as a reference model (in this case we used the standard Italian, a resident resource in Taltac). The measure of the variance from the reference lexicon is represented by the standard deviation (s.d.), which is the deviation between the frequencies of lexical forms in the analyzed text and those in the frequency lexicon (Bolasco, 2013).

The characteristic lexicon is created by dividing a corpus into sub-texts (so called sub-occurrences) according to the different levels of a chosen variable, in our case the discourse type. To find out the characteristics of charismatic discourse we divided the corpus into 3 subtexts corresponding to the three types of discourse because we needed to compare Mussolini's discourses to "oceanic crowds" (type 1), that we predicted to be more charismatic, with those in institutional contexts (type 2) and small specific contexts (type 3).

Then the different sub-texts are compared, by a t-test analysis, to extract a list of words overrepresented or underrepresented with respect to a normal distribution (the characteristic element index is calculated for all the units with a frequency of more than 5, with a probability threshold set at 5% through T-test; Bolasco, 2013).

Our corpus B counts 51034 occurrences (N) and 9699 different words (V), thus obtaining a medium lexical richness index [(V/N)*100], equal to 19,05%, that according to Bolasco (2013) is good enough to perform a lexicographic analysis.

5. Charisma in Mussolini's Words and Body

To illustrate the results of our analysis, in the following we overview those concerning Mussolini's words, types of communicative acts, voice, and body behavior. For each of these aspects, we on the one side illustrate our qualitative analysis of the signals conveying charismatic features, highlighting its psycho-social implications, on the other, the results of the quantitative lexicometric analysis.

5.1. The Quanti-qualitative Analysis of Mussolini's Discourses.

Adjective analysis. In order to perform our quantitative analysis we used the dictionary of positive and negative adjectives present in TalTac2 by analyzing the negative index[2] to identify polarization into positive and negative lexicon. The index reveals that the characteristic of negative polarity of words in the corpus is 36% (lower than 40% according to a research based on Italian corpora; Bolasco and Della Ratta, 2004) and it is oriented mostly in a maculine way (55% of adjectives are masculine).

Among these adjectives, Mussolini uses a large numbers of superlatives which represents 0.7% of the entire corpus of adjectives.

Time analysis. Time analysis reveals a very strong orientation to the present, because out of all verb frequencies Mussolini expresses time information most frequently as present (74%, as opposed to 19% past and 7% future). Future increases when Mussolini's discourses are directed to the crowd, where verbs in the future tense are 11%, significantly more frequent than in discourses to parliament and small groups.

5.1.1. Words

Let us start from the most trivial aspect of Mussolini's oratory: his words. His *lexicon* is characterized by two features, that we call "nicewordism" and "strongwordism": on the one side, he uses a high number of words that mention or evoke positive emotions and positive evaluations, on the other, his words very frequently refer to the highest possible quantity or intensity of things and properties.

Nicewordism. Mussolini makes a very frequent use of "hot words": ones that mention or refer to beautiful things or positive emotions, or that express positive evaluations likely to evoke them.

1. *l'eroismo individuale e collettivo del popolo italiano durante la guerra è stato sublime*
   (The individual and collective *heroism* of the Italian people during war has been *sublime*)
2. *Siamo fieri dell'intervento, fieri della guerra, fierissimi della nostra vittoria*
   (We are *proud* of our intervention, *proud* of the war, *most proud* of our victory)
3. *È questa l'Italia che essi [i martiri della rivoluzione fascista] volevano. L'Italia forte, ordinata, potente, tenace nei suoi sforzi e nelle sue fatiche*
   (This is the Italy that they [the martyres of fascist revolution] wanted. The *strong, ordered, powerful, tenacious* Italy in its efforts and its fatigues)
4. *Il grido della vostra esultanza pienamente legittima si fonde con quello che sale da tutte le città della Spagna.*
   (The cry of your fully legitimate elation merges with the one raising from all the cities of Spain)

These "nice" and "hot" words, mentioning and then inducing positive emotions in the audience, convey an element of pathos in Mussolini's discourse. As pointed by several authors (Stevenson, 1944; Macagno and Walton, 2014) the use of emotive words bears on

---

[2] The index is obtained by calculating the ratio between the total of negative occurrences and the total of positive ones (tot. Occ. Neg/tot. Occ. Pos·100).

an "emotion appeal"; and Mussolini does know how evoking emotions is a way to trigger the followers' will, and he leverages on this powerful weapon.

"Nice words" are in fact significantly represented in corpus B. Let us see some, mentioned with their standard deviation: *civiltà umana* (human civilization; 18,26), *eroi* (heroes; 4,3), *eroismo* (heroism; 15,15), *eroico* (heroic; 6,22), *eroicamente* (heroically; 16,33), *eroiche* (heroic; 6,14) *sublime* (sublime; 11,6), *fierissima* (proudest; 28,16), *fierezza* (pride, 20,25) *fieri* (proud; 6,26), *riconoscenza* (gratitude; 12,5), *gloriosa* (glorious; 23,24), *immemori* (forgetful; 35,88).

In addition, beside verbs like *consacrare* (consecrate, 22, 50), as predicted the "characteristic lexicon" in the speeches to the crowds shows a significant presence ($p<0.05$) of words strongly associated with very positive emotions: *cuore, pace, benessere, coraggio, magnifica, entusiasmo, orgoglio, commozione, degno, devozione* (heart, peace, wellbeing, courage, wonderful, enthusiasm, pride, meltage, worth, devotion).

The most frequent adjectives are focused on a very strong religious dimension, and the positive adjectivation becomes a way to target things and institutions as *sommo* (supreme, 81,9), *solenne* (solemn, 28,03), *degno* (worth, 14,52), *religioso* (religious, 11,92) by also losing all time and space references: *eterna* (eternal, 7,03), *eterno* (eternal, 5,03), *perpetuo* (perpetual, 4,96). On the contrary the negative adjectivation is defined by the death dimension: *putrefatto* (putrefied, 19,86), *funerei* (funereal, 18, 17), *derelitti* (derelict, 8, 16). In addition, in the corpus there are very frequent adjectives that cover labour and willingness: *industriosa* (industrious, 17,61), *feconda* (fertile, 11,60), *assidua* (assiduous, 9,59), *validissima* (most valid, 9,46).

Another aspect resulting from the quantitative analysis is the strong tendency of Mussolini to use "universal" and "inclusive" words when he speaks to the crowd, compared to parliament and small groups; when discourses are directed to the crowd, words like the following are very frequent: *Popolo* (= people: the most characteristic word; $p<0.000$), *rivoluzione, Terra, Popoli, Destino, Pace, Benessere, Avvenire, Volontà, Umanità, Famiglia, Patria, Diritti, Società* (revolution, Land, Peoples, Destino, Peace, Wellbeing, Future, Will, Mankind) ($p<0.05$). The characteristic words directed to the Parliament are more institutional: *Patto, Papa, trattative, concordato, sovranità, onorevoli, Parlamento, discussioni, ministro, convenzione, esercito, Stato, Governo, Re, Legge, Stati* (Pact, Pope, negotiation, Agreement, sovereignty, Lords, Parliament, discussions, miister, convention, army, State, Government, King, Law, States) ($p<0.05$); while the characteristic words addressed to small forums are focused on societal and economic aspects: *partito, economia, lavoro, università, sciopero, società, manifestazioni, organizzazione, cittadini, borghesia, politica, lavoratori* (party, economy, work, university, strike, society, manifestations, organization, citizens, bourgeoisie, politics, workers) ($p<0.025$).

Strongwordism. Another set of words contribute to provide an image of strength, power, dominance: words referred to high intensity of entities or qualities, mainly adjectives (*supremo* = supreme 23,30; *assoluto* = absolute 17,29) or adverbs in a morphologically or semantically superlative form (*giammai* = never 28,52; *solennemente* = solemnly 21,91; *perfettamente* = perfectly 18,89; *assolutamente* = absolutely 18,61; *eternamente* = eternally 16,49; *magnificamente* = magnificently 9,5; *irrevocabilmente* = irrevocably 8,5) Morphologically superlative words are those in which the highest degree of intensity is expressed (in Italian) by the suffix "*–issim*", like in *benissimo* (very well) or *tormentatissima Europa* (most tormented Europe), *questa ardentissima e fascistissima Milano* (this most burning and most fascist Milan), *noi lo diciamo nettissimamente* (we say it most sharply), *fierissimi della nostra vittoria* (most proud of our victory).

But in many cases, though intensity is not morphologically coded, the words used in any case convey the highest intensity, or the most extreme case of something: for instance *moltitudine immensa* (immense multitude), *ciò è matematicamente avvenuto* (this has mathematically occurred), *piramidale ignoranza* (pyramidal ignorance), *imperturbabile calma* (imperturbable calmness), *noi possiamo guardare con sovrumano disprezzo* (we can look with a superhuman contempt), *la più gigantesca dimostrazione che la storia del genere umano ricordi* (it is the most gigantic demonstration the history of human gender may remember). Everything in Mussolini's words is high, great, multiple, at the highest possible level of intensity, strength, power: exaggerated.

These "strong" words give an idea of dominance, of either Mussolini himself or his ingroup, either directly or indirectly. Directly when they refer, for instance, to the power or strength of the ingroup (like in *moltitudine immensa* = immense multitude, *sovrumano disprezzo* = superhuman contempt); indirectly when they mention something very strong or hard that the ingroup must face or confront (e.g. *l'ora delle decisioni irrevocabili* = the time of irrevocable decisions).

Such "strongwordism", though, has another important effect: it raises the level of certainty of Mussolini's statements. If I say "an immense multitude" of people instead of saying "there were many people", by raising the quantity of people I also imply that I am completely certain that people *were there* in fact.

Another way of conveying certainty is the use of "certainty words" like *indubbiamente* (undoubtedly), *promessa* (promise), *manterrò* (I will keep it), or *come sempre* (as always): all words containing a strong commitment (certainty of will) in their meaning.

Conveying high levels of certainty is important for the charismatic leader in two ways. First, a leader's job is to convince the audience to do what he wants. But being "convinced" can be defined as believing some beliefs with a high level of certainty (Poggi, 2005); and since the more a Sender communicates he himself is certain of the beliefs he is conveying, the more these beliefs tend to be considered certain (i.e., strongly believed) by the Addressee, then the charismatic leader must show very confident in what he says: he must convey a high level of certainty. Thus "strongwordism", besides directly conveying intensity, by indirectly conveying certainty also bears on the leader's *competence*: his knowing things, and his knowing them in a straightforward and not doubtful way. Moreover, this in some sense also bears on the dimension of *dominance*, in that it makes the leader appear particularly "categorical". Being "categorical" is a feature totally opposed to being uncertain or doubtful: one who is categorical is totally devoid of vagueness or uncertainty and gives the impression that not only is he totally certain of what he says, but what he says *is* the only truth, and he is totally committed to argue for it and to make it true in action. Thus "strongwordism", along with its aura of certainty and categoricity, finally also contributes to a job of *pathos*, because it induces self-confidence (a positive and pro-active emotion) in the Audience.

5.1.2. Communicative Acts

After single words, let us now consider Mussolini's communicative acts, a class including both speech acts and body communicative acts, performed by gesture, face, head, voice, posture. Our question is what types of communicative acts more typically contain the features of his charisma.

Incitations. A quite trivial communicative act in political speech are incitations: requesting the audience's action and encouraging it to take it. The most prototypical and straightforward are those during Mussolini's discourse of war declaration (Roma, June 10th, 1940):

5. *Popolo italiano! Corri alle armi, e dimostra la tua tenacia, il tuo coraggio, il tuo valore!*
(Italian people! Rush to arms, and demonstrate your tenacity, your courage, your value!)

Besides directly in the imperative mode, Incitations may also be phrased indirectly, for instance simply by a future tense:

6. *Io sento dalla vostra altissima temperatura ideale che se domani la Rivoluzione chiamerà, voi **risponderete** come un sol uomo.*
(I feel, from your hottest ideal temperature, that if tomorrow the Revolution calls, **you will respond** as a single man)

Other cases are also represented in Corpus B, like *risponderete* = you will respond 23,30; *darete* = you will give 22,90; *corri* = run 3,90)

5.1.3. Orders, Requests for Commitment

In some cases, Mussolini asks his people some commitment; but does not do so in a completely explicit way; rather, he presupposes that people ARE committed to do something, and by this he actually commits them to do so.

7. *Questa consegna io sono sicuro che diventa per voi immediatamente, nell'ora stessa in cui la pronuncio, un imperioso dovere. Dovete mettervi, come vi metterete, all'avanguardia per la valorizzazione dell'Impero.*
(This commitment I am sure becomes for you immediately, at the very same time I utter it, an imperious duty. You must put yourselves, as you will, at the forefront for the development of the Empire)

8. *Quale dunque è la parola d'ordine per il nuovo decennio, verso il quale noi andiamo incontro con l'animo dei vent'anni? La parola è questa: Camminare, costruire, e, se è necessario, combattere e vincere!*
(What is then the password for the new decennial, toward which we go with the spirit of a 20 years old? This is the password: to walk, to build, and, if necessary, to struggle and to win!)

Or even, he interprets an action by them as a true commitment by itself.

9. *Questo grido è come un giuramento sacro che vi impegna dinanzi a Dio e dinanzi agli uomini per la vita e per la morte.*
(This cry is like a sacred swear that commits you before God and before men by life and death)

Other words linked to these communicative acts are *request and action verbs*, like *spezzare* (to break, 41,47); *aspetterete* (you will wait, 29,22); *obbedire* (obey, 22,34); *ordine* (order 7,85).

This is the reasoning underlying this type of communicative acts.

>You must do X → I am sure you will do X → I do trust you
>But if
>I trust you (I, your Dux, make this gift to you of trusting you) →
>Then
>You must
>1. trust me
>2. do what I want you to do (X)

In a sense, Mussolini is relying on a sort of "cognitive dissonance" effect (Festinger, 1957): a well-known socio-cognitive mechanism, investigated by Social Psychology long after his death. If you perform some behavior, you are in some way inevitably induced to accept it and believe in that behavior; therefore, leading a person to do something is a way to convince him that doing this is right: action induction brings about internal persuasion. Actually, the typical (the right?) sequence should be: I lead you to believe X is right, and hence you finally decide to do X. But this is the other way around: I induce you to do X, hence you will start thinking that X is right. This way you are taken in the net, you cannot escape. If we define manipulation as a kind of social influence where I influence you without letting you know that I want to (and I do) influence you, well, this is a sort of manipulation; and the bulk of demagogy!

A similar mechanism underlies Mussolini's use of another type of communicative act: rhetorical questions.

5.1.4. Rhetorical Questions

Here are some of Mussolini's rhetorical questions, during his dialogue with the crowd:

>10. *Ne sarete voi degni?*
>   (Will you be worth thereof?)
>
>11. *Desiderate degli onori? Delle ricompense? La vita comoda? Esiste per voi l'impossibile?*
>   (Do you want honors? Rewards? A comfortable life? Does impossible exist for you?)

A rhetorical question puts two requirements on its Addressee. First, being a question, it requires an answer; second, being a rhetorical question, it asks for a specific wanted answer. Thus it forces the Addressee to say what the Sender wants, but through this it calls him to be responsible of what he says. Therefore, the effect aimed at, in both commitment requests and rhetorical questions, is to impose the other the responsibility for doing or saying something, though, actually, it is something that YOU want. By this device, Mussolini makes the crowd co-responsible with his own doings and sayings; or better, he makes them *believe* to be co-responsible, because in fact he is a dictator – the man of Providence – and he does not let anybody decide anything different from what he wants. By using rhetorical questions, Mussolini pretends to give the crowd the chance of deciding something, but in fact does not: one more case of covert induction; manipulation again.

5.1.5. Threats

In some cases, Mussolini's speech is threatening. Obviously, the target of his threats are most typically "others": enemies, rivals, in a word, the outgroup of Mussolini's present Audience. Yet, as we will see, sometimes he is threatening toward those, in the ingroup, that he suspects of not sure fidelity.

His language of threat, like generally does the language of dominance (Poggi & D'Errico, 2011), makes use of deontic words, like *deve* (must, 19, 33), *per forza* (compellingly, 10,32), *devono* (they must, 6,73), *necessariamente* (necessarily, 4,60).

In Bolzano, on August 31st, 1935, after Italy underwent the "sanctions" – an embargo to Italian military enhancement – he says:

> 12. *Il mondo deve sapere ancora una volta che sino a quando si parlerà in maniera assurda e provocatoria di sanzioni, noi non rinunceremo a un solo soldato, a un solo marinaio, a un solo aviere, ma porteremo al livello massimo possibile della potenza.*
> (The world must know, once again, that until people talk of sanctions in an absurd and provocative way, we will not give up any single soldier, any single marine, any single airman, but we will take to the maximum level of power)

And again:

> 13. *Il governo francese è perfettamente libero di rifiutarsi anche alla semplice distruzione di questi problemi, come ha fatto sin qui attraverso i suoi troppi reiterati e forse troppo categorici 'giammai'. Ma non avrà poi a dolersi se il solco che divide attualmente i due paesi, diventerà così profondo che sarà fatica ardua se non impossibile colmarlo.*
> (The French government is perfectly free to refuse even the bare destruction of these problems, as it has done so far through its too repeated and perhaps too categorical 'never'. But it will not have to complain if the furrow that presently divides the two countries becomes so deep that it will be hard if not impossible a fatigue to fill in it).

Il mondo deve sapere (the world must know) in 13) typically sounds as a warning to others, to the world. In 14), instead, the threaten is slightly more indirect: the French are free (not to do what we want), but then they have to take the burden of their choice. A way to make the others guilty, hence to refuse one's responsibility for possible retaliation.

Threats are conveyed by body signals even more frequently than by words.
In the speech after the end of the War in Spain, while saying

> 14. *Molti altri fra i nostri nemici mordono in questo momento la polvere.*
> (Many others among our enemies are biting the dust)
> He *shakes his right index finger up and down three times*; then he *shakes his right fist up and down twice.*

Shaking finger is a threatening gesture (Calbris, 1989; Streeck, 2007), while *shaking fist* is an aggressive gesture. Later, by alluding to the Spanish communists' motto *"no pasaràn"* (they will not pass through), he says:

> 15. *Siamo passati... e vi dico... e vi dico che pàsserémo.*
> (We have passed through... and I tell you... and I tell you that we will pass

through)
He utters *pàsserémo* by stressing the first and the third syllable, *pà-* and *ré-*, while the last syllable, *-mo*, though being the last utterance of this phrase does not have a descending intonation, but a rising one, as if leaving this promise/threat suspended in time. At the same time, he *shakes his right index finger*.

The intense stress on the two syllables of *passeremo* exhibits a quantum of aggressiveness, while its suspended ending, along with the meaning of the verb that is in the future tense, does not so much exult for the past but sounds as an incitation to Fascists and at the same time as a threat to Communists; as is confirmed by the threatening *index shaking*.

16. *Noi desideriamo che non si parli più di fratellanze, di sorellanze, di cuginanze e di altre tali parentele bastarde.*
    (We want people not to talk anymore of brotherhoods, sisterhoods, cousinhoods and of such other bastard kinships)

In the word *desideRiamo* (we want), the *"r"* is *uttered with particular intensity*, conveying strength and aggressiveness. At the same time, *right arm with hand palm cuts the air horizontally from left to right*: a gesture meaning "that's enough with this" (Kendon, 2004). The hand is not tense while uttering *fratellanze* (brotherhoods), but then, parallel to the words *parentele bastarde* (bastard kinships) becomes more *tense and rigid* and *turns up with the hand palm up and fingers curve* as a claw, a somehow iconic handshape that evokes grasping, scratching, hurting: again an aggressive gesture.

Threat is conveyed at the same time by words and voice in this example.

17. *Verso i popoli amici noi andiamo con un atteggiamento da amici. Contro popoli ostili noi avremo un chiaro, deciso, risoluto atteggiamento di ostilità.*
    (Towards friendly peoples we go with a friendly attitude. Against hostile peoples we will have a clear, determined, resolute attitude of hostility)

The *words are uttered very slowly*, which sounds disturbing and threatening, because it seems to imply: "I explain this very well, since if you do not understand it, it's bad for you". Then, after stressing the *"à"* of *ostilità* (hostility), he *protrudes his mouth while lowering his lips*, expressing anger but also dissatisfaction.

Actually, threatening messages are not addressed only to the "outgroup", strangers and enemies, but sometimes also to Italians: "them" among "us".

18. *Un governo che avesse disperso a frustrate la malagenia degli imboscati, se avesse punito severamente, con necessario piombo nella schiena, i disfattisti e i traditori*
    (A government that had dispersed the bad ilk of dodgers by means of lashes, if it had strictly punished, the defeatists and traitors by necessary lead in their back).

Here Mussolini refers to the "them" that are among us: those, among us, that are not reliable.

In another discourse, he uses a "carrot and stick" approach, alternating promise and threat.

19. *Gli ebrei di cittadinanza italiana i quali abbiano indiscutibili meriti militari o civili nei confronti dell'Italia e del Regime troveranno comprensione e giustizia. (...) il mondo dovrà forse stupirsi più della nostra generosità che del nostro rigore, a meno che, i semiti di oltre frontiera e quelli dell'interno, e soprattutto i loro improvvisati ed inattesi amici, che da troppe cattedre li difendono, non ci costringano a mutare radicalmente cammino .*
(The Italian citizen Jews who have indisputable military or civic merits towards Italy and Fascism will find comprehension and justice. (…) The world will have perhaps to be surprised more about our generosity than about our strictness, unless transboundary and within boundary Semites, and first of all their impromptu and unexpected friends, who from too many chairs defend them, force us to radically change our route)

On the one hand, he promises comprehension and justice to the Jews with particular merits, even arguing that he will surprise the world for his generosity; and only later does he launch his threat, introduced by the conjunction *a meno che* (unless).

The logical structure of threat is the following (Poggi, 2005; Guerini & Castelfranchi, 2006).

- A wants B not to do some X that A dislikes
  (for example, Mussolini wants Semites not to do X)
- to induce B not to do X, A commits himself to take a course of action Y aimed at thwarting B's goal Z
  (while A has not been aggressive to Semites and their friends so far, he will since now on change his route, if they do not comply with his will).

The conjunction *a meno che* (unless) introduces the case of B doing X (what A dislikes). Moroever, the threat has a double addressee (Goffman, 1981; Poggi, 2007): the Jews (definitely the outgroup), but also their friends in Italy (i.e., an outgroup within the ingroup). And the threat sounds even more aggressive towards these last (the Semites' friends) than toward the Semites themselves, as witnessed by the richer adjectivation devoted to them (*improvvisati ed inattesi amici* = impromptu and unexpected friends) and by the utterly negative adjective *troppe* (too many) in the modifier *che da troppe cattedre li difendono* (who from too many chairs defend them).

Finally, before saying *radicalmente cammino* (radically [change] route) Mussolini makes a long pause and utters *radicalmente* very speedily, in a hyper-articulated way, again while strongly stressing *r*, which expresses anger and hence threat. In fact, since anger causes aggression, if A expresses anger while mentioning the action Y he commits himself to do in order to thwart B's goal Z, A anticipates he will be particularly aggressive in doing Y.

5.1.6. Discrediting Acts

Mussolini's communicative acts seen so far are mainly aimed at stating his dominance: orders, incitations, rhetorical questions, threats. But to maintain one's power over the other, besides showing one is smarter or stronger than others, one can also show that others have less power, they are less smart or less strong than he. So, besides raising oneself and one's in-group, the Orator may lower others, the outgroup. One way to do so is to discredit them, that is, to spoil their image (D'Errico *et al.*, 2012) in front of the Audience.

A typical discrediting act is insult, that is, defining a Target as belonging to some degrading category (Poggi *et al.*, 2015), judged as lower than one the Target actually belongs, and such that anyone (and specifically the Target) would not like to belong to it.

Mussolini sometimes uses insulting nouns or adjectives:

20. *Per questi... residui... o residuati di tutte le logge*
(For these... residuals... or holdovers of all war surplus of all lodges)

The very word "residual" already conveys the idea of something discarded or not used for noble aims; but further, to call people "residual" is to reduce them to objects: a typical case of Bar-Tal's dehumanization devices (Bar-Tal, 1989; Volpato & Durante, 2003; Scardigno *et al.*, 2015).

From a quantitative point of view, dehumanization is frequently brought about by means of adjectives that evoke the idea of death (*putrefatto* = putrefied, 19, 86; *funerei* = funereal, 18, 17; *derelitti* = derelict, 8, 16); the negation of dominance (*miserevoli* = miserable, 11,21; *meschini* = mean, 7, 48; *patetici* = pathetic, 5, 42; *inutili* = useless, 4,16); or finally the opposite of benevolence (*sordidi* = sordid, 11,59; *infidi* = unreliable, 9,73; *traditori* = betrayers, 6,61; *barbaro* = barbarian, 4, 03).

A peculiar case of discredit is ridiculization (Poggi *et al.*, 2012): making fun of another is one of the cruelest ways to make him feel degraded, since it attributes one a negative evaluation of impotence; therefore, this is a very effective way to offend and abase the other, but at the same time a way to raise oneself by lowering the other.

Ridiculization is exploited by Mussolini, for instance, in his mocking laughter when he says:

21. *E quanto tempo dovrà ancora passare per convincersi che nell'apparato economico del mondo contemporaneo c'è qualche cosa che s'è incagliato e forse spezzato?*
(And how long will it take to get convinced that in the economic apparatus of today worlds there is something that got stuck and possibly broken?)

Here he considers the whole world as the outgroup and, by *opening his mouth* and *showing his teeth*, with *lip corners raised* and *squinting eyes*, he *laughs* at it.

Again, a peculiar way to make fun of others is to make a parody of them (Poggi & D'Errico, 2013), that is, to perform an exaggerated and distorted imitation of their possible behaviors. Mussolini makes a parody of "the others" in this example:

22. *Oltre le frontiere ci sono dei farneticanti, i quali non perdonano all'Italia fascista di essere in piedi. Per questi residui o residuati di tutte le logge è veramente uno scandalo inaudito che ci sia l'Italia fascista.*
(Beyond borders there are some raving people, who do not forgive Fascist Italy for standing up. For these residuals or war surplus of all lodges, it is really an unbearable scandal that Fascist Italy be there)

While uttering *è veramente uno scandalo inaudito* (it is really an unbearable scandal), he *raises both arms up* and then *draws half a circle* with them (like a rainbow), *finally dropping hands down*: he is mimicking a gesture of indignation of those "raving people", as if meaning: look how indignant they show about seeing us stand up! He thus makes a parody of his critics, to take vengeance of their criticism.

An important effect of discredit, here probably deliberately aimed at by Mussolini, is that speaking badly of others re-affirms values and hence strengthens in-group relationships: thus he creates complicity with the crowd.

5.1.7. Expressions of Affect

Mussolini sometimes explicitly expresses his affect to the audience:

23. *Io vi confesso di nutrire una sfumatura di simpatia per Aprilia (....) Potete contare sulla mia simpatia.*
(I confess that I feel a nuance of sympathy for Aprilia... you can count on my sympathy) (Aprilia is a town founded during the reclamation of Pontine Marshes, a country previously infested by malaria)

Sometimes his emotion expressions are phrased in the typical Fascist rhetoric, bombastic and full of adjectives. Like in

24. *...e il nostro amore per il popolo, amore armato e severo, è tutto vibrante di una profonda e consapevole umanità.*
(And our love for the people, an armed and strict love, all vibrates with a deep and aware humanness)

Expressing his emotions is a way for Mussolini to show his being human, similar to the people, and feeling emotions like his followers. And like the following communicative acts, expression of empathy and praise, it has a (perhaps wanted?) effect of seduction.

5.1.8. Expressions of Empathy

25. *C'è qualcuno che pensa che noi ci preoccupiamo dell'inverno dal punto di vista politico. È falso. [...] È dal punto di vista umano... Perché il pensiero che una famiglia soffra dà a me stesso una sofferenza fisica. Perché io so, so per averlo provato, che cosa vuol dire la casa deserta e il desco nudo.*
(There is someone who thinks that we worry about winter from a political point of view. This is false. [...] It is from the human point of view... because the very thought that a family suffers gives me physical pain. Because I do know, I know since I felt it myself, what does it mean a desert home and a nude table)

26. *Per noi Fascisti il popolo non è un'astrazione della politica, ma una realtà viva e concreta. Io soffro dei dolori del popolo.*
(For us Fascists the people is not an abstraction of politics, but a living and concrete truth. I suffer from the people's pains)

Empathy is definitely a central quality of a leader. But an effective leader must not only feel empathy towards his followers, he must express it. And Mussolini does know how to express his empathy, whether faked or sincere.

Yet, saying he suffers from the very same pains of his people has another goal too: to show he is similar and close to them. Like in this case:

27. *Camerati rurali, di Aprilia, di Pontinia, di Littoria e di Sabaudia! Voi potete contare sulla mia simpatia. È la simpatia di un uomo che ha l'orgoglio di dirvi che nelle sue vene scorre il sangue di autentici rurali.*
   (Rural camarades, from Aprilia, Pontinia, Littoria and Sabaudia! You can count on my sympathy. The sympathy of a man who has the pride to tell you that the blood of true rural people flows in his veins)

Here he does not simply say he has the same blood of his Audience (he is similar to them), but also that he is proud of this. The charismatic leader enhances group cohesion because, stating his closeness and similarity to each member of the group makes them feel close and similar to each other; further, by showing his appreciation of his and the group's quality, enhances their pride, enthusiasm, and motivation to follow him and do what he wants.

5.1.9. Praise

In various cases, Mussolini praises his audience.

28. *Mi accorgo da questo vostro urlo che avete buona memoria*
    (I realize from this shout of yours that you have a good memory)

29. *Ora che col vostro coraggio, col vostro sacrificio, con la vostra fede, avete dato un impulse potente alla ruota della Storia.*
    (Now that by your courage, your sacrifice, your faith, you have given a powerful impulse to the wheel of History)

30. *Voi mi avete atteso per sedici anni dando prova di quella discrezione che è un segno distintivo dei popoli di antica civiltà quali voi siete.*
    (You have been waiting for me for 16 years giving evidence of that discretion that is a distinctive sign of peoples of ancient civilization as you are)

31. *Questa Milano generosa, operosa, infaticabile.*
    (This generous, industrious, tireless Milan)

Praise is a typical act of seduction. By praising your followers, you declare your love to them; but besides this, you also satisfy their desire to be judged well by the leader. A charismatic leader is typically strongly admired by his followers, and admiration leads to the desire to be close to the the admired person, and to be appreciated by him (Poggi and Zuccaro, 2007). So the leader who expresses his appreciation of the followers' actions and qualities is further making them close to himself.

5.2. Communicative Creativity

A relevant aspect of Mussolini's charisma is his creative language. But what do we mean by creativity in general, and by communicative creativity?

5.2.1. Creativity and Charisma

According to our model, a characterizing feature of the charismatic leader, within his skills of competence, is his creativity. By creativity we mean the application of a divergent thinking, that is, a way of reasoning and connecting concepts different from the most common and widespread ones.

After the socio-cognitive framework adopted here, all people generate inferences, i.e., new beliefs, by applying inference rules, that is, connections among beliefs (types of reasoning) of different kinds: time, space, class-example, total-part, cause-effect, means-end, condition (Castelfranchi & Parisi, 1980). Inferences by definition are not totally certain: their degree of certainty is generally lower than that of beliefs acquired through perception, and varies a lot: depending on the certainty of the premises and the flexibility or necessity of the inference rules (Castelfranchi & Poggi, 1998), the belief resulting from the application of an inference may have a very low probability of being true – and yet be possible. We then define a creative belief as a belief that is very unexpected, one with a very low probability, a very low degree of certainty, drawn by inferences that many others would not have made.

A charismatic leader on the one side necessarily has a way of reasoning that must be similar to his followers' (or else he could not comprehend them nor be comprehensible by them, in a word, be "popular"); but at least to some extent, he must be endowed with a divergent and creative thinking, that he can apply at various levels of his leadership. First, he must build up a vision, that is, creatively imagine a future that is very different from present, and such that cannot be easily imagined without applying unexpected inference rules, or reasoning on premises that are not generally considered by other people. Second, to bring about his vision he may have to find out creative strategies (take for example the creativity of Gandhi in devising non-violence as a weapon to free India from violent Great Britain). Finally, this creative thinking may display in communication, to convey the fact that the leader is anyway different and better than the followers: someone to trust, to follow, to submit to.

Communicative creativity is expressed, for example, in finding unexpected surprising links between words, which allows one to make puns, cluster words that generally do not go together, create new words or new verbal imagines, using rhetorical figures, and so on.

5.2.2. *Mussolini's Creative Language*

Mussolini's language is full of neologisms and rhetorical figures.

First, being a follower of Futurism and a fan of the Italian poet D'Annunzio, he frequently used to create new words. His lexicon is often neologistic both for the creation of utterly new words and for the creative use of old words.

Here is an example of the latter

32. *Per questi residui o residuati di tutte le logge...*
    (For these residuals or war surplus of all lodges...)

In this sentence, the word *residui* (residuals) is used in its normal meaning; but then he adds the word *residuati*, that evokes (and precedes) the idiom *residuati bellici* (war surplus), a meaning not warranted in this context.

A case of utterly new coined word is the following:

33. *Nessun nemico peggiore della pace di colui che fa di professione il panciafichista o il paciafondaio.*
    (No worse enemy of peace than one who is a bellyfigist or a peacemonger by profession)

While *paciafondaio* (peacemonger) is a neologism that simply adds the suffix *–fondaio* (monger) to the root *pace* (peace) instead of the root *guerra-* (war), *panciafichista* (that could be roughly translated as "bellyfigist") is formed by *pancia* (belly) and *fico* (fig). A truly new creation.

In the lexicometric analysis, Mussolini's tendency to linguistic creation is also tested by a very large use of particular words that are not recognized by the system because not present in the Italian vocabulary: *anfrattuosità* ("anfrattuosità del Vaticano": something like *ravinity, fiordity* of the Vatican), *vociferatoria* ("la critica vociferatoria" = *rumorific* critic), *decrepitudine* ("paralitica decrepitudine" = paralytic *decrepitude*), necroforica (necroforica rendigote = *undertakerous* rendigote), *socialpussisti* (*socialpussists*), legazie, scioperaiolo ("sindacalismo scioperaiolo" = *strikeous trade-unionism*), *vigoreggiare* ("gioventù che vigoreggia" = *forcefuling* youth).

Mussolini's creativity is also expressed by his frequent use of rhetorical figures. He often uses metaphors: some already idiomatic (catachreses), like

34. *se [...] la grande campana suonerà a martello*
    if [...] the great bell rings for the funeral

Others creative, like:

35. *Pilotata energicamente dal primo ministro inglese, la navicella delle riparazioni e dei debiti è oggi nel porto di Losanna.*
    (Energically piloted by the English Prime Minister, the *nacelle* of repairs and debts is today in the harbour of Lausanne)

He also uses oxymorons, like *tragica contabilità* (tragic computation), or *il nostro amore per il popolo, amore armato e severo* (our love for the people an armed and strict love), *pace armata* (armed peace); and, coherently with his tendency to "grandeur", hyperboles: both idiomatized ones (*il calvario della guerra* = the calvary of war), and creative, metaphoric hyperboles (*questa Versaglia che agonizza* = This agonizing Versailles).

5.3. Voice

Within the phonetic aspects of Mussolini's speech, here we focus on three features. One is his wide use of rhetorical pauses (see for example: Duez, 1999). A pause is "rhetorical" when it is a silence not due to hesitation problems, that is, to the Speaker's need to think over what he is going to say, but to the deliberate goal of creating suspension, expectation, surprise, or impressing a nuance of importance and solemnity. Mussolini makes frequent rhetorical pauses, but in his case creating suspension and expectation often aims at a display of dominance: making a pause forces the audience to pay more concentrated attention, and also, by eliciting curiosity for what comes later, stresses the dependence of the Audience from the Orator.

Another interesting feature of his speech is hyper-articulation: a very accurate scanning of words, performed thanks to a low Speech Rate: as Lindblom (1990) and Pellegrino *et al.* (2013) point out, hyper-articulated speech is a type of "listener-oriented" speech, since the lengthening of syllabic durations allows listeners to interpret the message more easily. This phonetic feature points at Mussolini's populism: he wants to be well understood by all his people. We must remind that Mussolini was a school teacher before being a politician, and his skills of clear communication proved very important in his political career.

In some cases, though, the low speed of Mussolini's speech is not intended to clarity or communicative friendliness. To the opposite, it takes up a nuance of threat: in fact, speaking very slowly, since it leaves you time to understand the message, metacommunicates that it is very important and, enhancing clarity, rules out incomprehension and gives you no excuse for not complying with it.

Another phonetic feature conveying aggressiveness in Mussolini's speech is his way of uttering *r*s. When the discourse becomes particularly threatening, these consonants are lengthened and stressed: as he does for instance in the speech of War Declaration:

> 36. *Noi vogliamo spezzare le catene di ordine territoriale e militare che ci soffocano nel nostro mare*
> (We want to break the territory and military chains that suffocate us in our sea)

where all the "*r*"s in the phrase *le catene di oRdine teRRitoRiale* are geminate and hyper-articulated. These observations on the phonetic aspects of Mussolini's speech are, for the time being, simply based on a simple auditory overview; subsequent research might proceed to their instrumental acoustic analysis, and to a quantitative assessment of their actual frequency.

5.4. Body

Some aspects of Mussolini's body communication are known all around the world, and characterize him in an unmistakable way: his posture with erected bust, chin up, fists on hips. Being typical displays of pride (Tracy, 2004), and conveying a meaning of arrogant power over others, they insist on the dominance side of charisma.

But other aspects of charisma too are deliberately highlighted by Mussolini's body behaviors. In the videos by "Istituto Luce", a movie institution created in order to Fascist propaganda, he often exhibits his physical skills, like skiing or riding the horse. During the "wheat battle", a project aimed at giving an impulse to Italian agriculture, he gives the start to the opening ceremony by harvesting himself. All of these displays are aimed at showing a general image of competence of the Leader: Mussolini can do lots of things – even some that are not at all required from a political leader – and he can do them even better than others. This bears on the competence side of the leader's ethos, but also, again, on the dominance side, since his having more power than others in so many fields, justifies his having power over them. At the same time, though, his display of multiple skills also has a pathos effect, in that it favors the followers' admiration: an emotion felt for someone who has some qualities we would like to have, and that leads us to take him as a model, to imitate him, to follow him, to comply with him (Poggi & Zuccaro, 2007).

6. Multimodal Cues of Mussolini's Charisma

The goal of this paper was to find out how Mussolini's charisma is expressed, and what features of charisma are more frequently represented, in his multimodal discourses. Through qualitative analysis we selected the charismatic aspects of his communication; then through quantitative lexicographic analysis we demonstrated that the aspects selected are in fact significantly more represented in his discourses to the crowd, those in which most typically charisma should be displayed, as opposed to other discourses, in institutional contexts or to small audiences.

Let us summarize how the internal features of charisma are displayed and distributed across modalities.

As predictable, the aspect of charisma most typically leaking from Mussolini's discourse and general behavior is dominance. Mussolini's behavior is a continuous display of power: he is dominant in his body postures, his "strongwords" lexicon and his most recurrent communicative acts. But among them, his acts of incitation, requests for commitment, rhetorical questions, and discrediting acts are more a cue to a "proactive-attractive" charisma, while his commands and threats point at a "dominant-threatening" charisma. Finally, threat leaks in his voice mainly in his strongly articulated rs and in the low speed of his words, that when not aimed at being particularly clear finally sound as a strict warning.

The dimension of competence is displayed in Mussolini, on the body side, by his display of skills, and on the intellectual side, by his adorned rhetoric and his creative lexicon.

He displays benevolence, more specifically sociability and inclusiveness, by communicative acts that show similarity and togetherness with his people, as well as his clarity of speech, including hyperarticulation; moreover, he enhances identification of the crowd with the leader by means of discrediting acts that stress the contraposition between "us" and "them".

His feature of emotional intelligence is witnessed by expressions of affect and of empathy towards the followers, while emotion induction is also present: his "nicewords" tend to enhance the followers' motivation by inducing enthusiasm. "Strongwords", by conveying certainty, along with certainty words and the body display of pride, contribute to enhance the people's self-confidence, while his praises to followers, along with discredit of "the others", tend to enhance the followers' pride.

The notion of charisma, introduced by Weber at the beginning of last century, has proved truthful all along that century: charismatic leaders do exist, and they have a strong influence over other people. But what precisely is, in the mind and the body of a leader, that makes his charisma, might be seen as a "magic", some elusive element. The multimodal analysis of Mussolini's discourses allowed us to show that it is possible to single out the internal components of charisma, and to track which traits and behaviors express each of them.

Acknowledgments

We are indebted with Emanuele Colarossi and Luigi Bizzarri, journalist and coordinator of *"La Grande Storia"*, Channel RAI3, Rai Radiotelevisione Italiana, for giving us the chance to analyze Mussolini's discourses.

References

Bar-Tal, D. (1989). Delegitimization: The Extreme Case of Stereotyping and Prejudice. In: Bar-Tal, D., Graumann, C., Kruglansky, A.W., & Stroebe, W. (Eds.), *Stereotyping and Prejudice: Changing Conceptions*, pp. 169-182. New York: Springer Verlag.

Bolasco, S., & De Mauro, T. (2013). *L'analisi automatica dei testi: Fare ricerca con il text mining*. Roma: Carocci.

Bolasco S., & Della Ratta-Rinaldi, F. (2004). Experiments on semantic categorisation of texts: analysis of positive and negative dimension. In: Purnelle G., Fairon C., & Dister

A. (Eds.), Le poids des mots. Actes des 7es journées Internationales d'Analyse Statistique des Données Textuelles, UCL, Presses Universitaires de Louvain, pp. 1–9.

Castelfranchi C., & Guerini M. (2007). Is it a Promise or a Threat? *Pragmatics & Cognition* 15(2): 277–311, 2007.

Castelfranchi, C. & Parisi, D. (1980). *Linguaggio, conoscenze e scopi*. Bologna: Il Mulino.

Castelfranchi, C., & Poggi I. (1998). *Bugie, finzioni, sotterfugi. Per una scienza dell'inganno*. Roma: Carocci.

D'Errico F., Poggi I. & Vincze L. (2012). Discrediting signals. A model of social evaluation to study discrediting moves in political debates. *Journal of Multimodal User Interfaces*. 6: 163-178.

D'Errico F. & Poggi I. (2013). The Parody of Politicians. Multimodal distorted imitation aimed at political discredit. In *Proceedings of 4th IEEE Intl' CogInfoCom 2013* (Cognitive Infocommunication) – Budapest, pp. 423-428.

D'Errico F., Signorello R., Demolin D. & Poggi I. (2013). The perception of charisma from voice. A cross-cultural study. *Proceedings of ACII Conference, IEEE*, pp. 552- 557.

Duez D. (1999). La function symbolique des pauses dans la parole de l'homme politique. Faits de langues Année 1999, Vol. 7, n. 13, pp. 91-97.

Festinger, L. (1957). *A theory of cognitive dissonance*. Evanston, IL: Row & Peterson.

Lindblom, B. (1990). Explaining Phonetic Variation: A Sketch of the H&H theory. In: Harcastle, W.J., Marchal, A. (eds.) *Speech Production Modelling*, pp. 403-439. Dordrecht: Kluwer.

Macagno, F. & Walton, D. (2014). *Emotive Language in Argumentation*. New York: Cambdridge University Press.

Poggi, I. (2007). *Mind, hands, face and body. A goal and belief view of multimodal communication*. Berlin: Weidler.

Poggi I, D'Errico F. & Vincze L. (2012) Ridiculization in public debates: making fun of the other as a discrediting move. In N. Calzolari *et al.* (Eds.), Proceedings of the 8th conference on International Language Resources and Evaluation (LREC'12). Istanbul 21-27 May. European Language Resources Association (ELRA), pp. 44-50. http://www.lrecconf.org/proceedings/lrec2012/workshops/18.Proceedings%20ES3%202012.pdf

Poggi, I., D'Errico, F., & Vincze, L. (2015). Direct and indirect verbal and bodily insults and other forms of aggressive communication. In D'Errico, F., Poggi, I., Vinciarelli, A., & Vincze, L. (Eds.), *Conflict and Multimodal Communication. Social Research and Machine Intelligence*, pp.243-264. Berlin: Springer Verlag.

Poggi, I., D'Errico, F., Vincze, L. & Vinciarelli, A. (2013). *Multimodal Communication in Political Speech. Shaping Minds and Social Action*. Berlin: Springer Verlag.

Pellegrino, E., Salvati, L., & De Meo, A. (2013). Racism and Immigration in Social Advertisings Promoted by Italian Government and Non-governmental Institutions. In Poggi, I., D'Errico, F., Vincze, L. & Vinciarelli, A. (Eds.), *Multimodal Communication in Political Speech. Shaping Minds and Social Action*, pp. 207-219. Berlin: Springer Verlag.

Scardigno, R., Giancaspro, M.L., Manuti, A., & Mininni, G. (2015). The rhetoric of conflict inside and outside the stadium: The case study of an Italian Football Cheer Group. In D'Errico, F., Poggi, I., Vinciarelli, A., & Vincze, L. (Eds.), *Conflict and Multimodal Communication. Social Research and Machine Intelligence*, pp. 207-221. Berlin: Springer Verlag.

Signorello R., D'Errico F., Poggi I. & Demolin D. (2012). How charisma is perceived from speech. A multidimensional approach. *Proceedings of IEEE Social Computation*, Amsterdam, 3-5 september IEEE Computer Society, pp. 435-440.

Signorello, R., D'Errico, F., Poggi, I., Demolin, D. & Mairano, P. (2013). Charisma perception in political speech: A case study. In Mello, H., Pettorino, M. & Raso, T. (Eds.), Proceedings of the VIIth GSCP *International Conference: Speech and Corpora*, pages 343–348. Firenze: Firenze University Press.

Stevenson, C. (1944). *Ethics and Language*. New Haven: Yale University Press.

Streeck, J. (2007). Gesture in Political Communication. A Case Study of the Democratic Presidential Candidates during the 2004 Primary Campaign. *Research on Language and Social Interaction*, 41(2):154-186.

Tracy, J.L. & Robins, R.W. (2004). Show your pride: Evidence for a discrete emotion expression. *Psychological Science*, 15:194–197.

Volpato C., & Durante, F (2003). Delegitimization and racism. The social construction of anti-semitism in Italy. *New Review of Social Psychology*, 2:286–296.

Weber, M. (1920). *The theory of social and economic organization*. Oxford: Oxford University Press.

Chapter 19

Persuasive Lexicon Extraction From Political Speeches

Marco Guerini[1], Gözde Özbal[2] & Carlo Strapparava[3]

FBK-Irst - Povo, I-38100 Trento; [1]guerini@fbk.eu, [2]g.ozbal@fbk.eu,[3]strappa@fbk.eu

Abstract. This paper presents a resource of political speeches tagged with audience reactions called CORPS and distributional semantic approaches to extract lexica of persuasive words, where each word is associated with a 'persuasive impact' score. After the introduction of the resource, we compare several methodologies for extracting these lexica and provide a series of experiments to explore and compare the feasibility of the different approaches. The experiments, which were carried out on pairs of persuasive/non persuasive sentences, confirm the soundness of our idea and demonstrate that some methodologies stand out amongst the others.

1. Introduction

In recent years political discourses have received growing attention for persuasive communication analysis. Multimodal features are of paramount importance in this context. Non-lexical audio cues - such as audience reaction or speaker prosody - have been investigated in (Guerini, Strapparava, & Stock, 2008a, Hu *et al.*, 2008), as well as visual cues - such as speaker gesture or gaze - in (Poggi & Vincze, 2009). However, the lexical aspects of persuasive communication have received very little attention.

In this paper, we first present CORPS[1], a corpus containing transcripts of political speeches tagged with special non-lexical audio cues, i.e. audience reactions, such as APPLAUSE or LAUGHTER. It should be noted that the corpus is composed of transcriptions of speeches mostly given at public mass gatherings, so – in general – the audience is favorable to the speakers and the context is one of support. Of course, by giving value to audience reactions, we do not mean that the audience is effectively persuaded of some ideas or is induced to do something that they previously did not believe in. On the contrary, the audience tends just to react to signals, including an expected theme, a name, an expression, or the tone of the voice. Therefore the audience, so to say, resonates to a fragment of speech, which is meant to be of a persuasive genre. To be successful, the speaker's expression that immediately leads to an audience reaction must have been

---

[1] Information on how to obtain it is provided at http://hlt.fbk.eu/corps

coherently composed. We therefore believe that there is a wealth of linguistic material[2] that, by virtue of the validation provided by the audience reaction, can be used in various scenarios. So, the corpus has been built with the goal of allowing automatic processing of the data and to this end tags have been converted (from the original transcripts) to make them homogeneous in formalism and labeling. Metadata regarding the speeches have also been collected for eventual further analysis.

Given these premises, we provide a set of computational methodologies to extract lexica of persuasive words from CORPS, where each word is associated with a 'persuasive impact' score. To build the persuasive lexica, we used public reaction tags as indicators of hot-spots where persuasion attempts succeeded or, at least, a persuasive attempt was recognized by the audience[3]. In our view these hot-spots can thus be used to understand the potential persuasive use of words, by analyzing the distributional patterns of words around such audience reaction tags. To model our idea, we used four statistical approaches, namely: (i) term frequency-inverse document frequency (tf-idf), (ii) a weighted variant of tf-idf, (iii) Latent Semantic Analysis (LSA) and (iv) Latent Dirichlet Allocation (LDA). By grouping tags in sets that represent three main persuasive effects (namely positive, negative and ironical), for each approach we built persuasive lexica specialized in the corresponding effect. To validate the soundness of our idea and to compare the effectiveness of these lexica, we conducted experiments on pairs of persuasive/non persuasive sentences. The results demonstrate that some methodologies stand out amongst the others and that computational methodologies can be successfully utilized for choosing the right wording in persuasive communication.

The paper is structured as follows. In Section 2 we introduce the related works on political communication analysis and corpora. In Section 3 we describe the new release of the corpus, its characteristics, and how it has been collected; a particular attention is paid to a quantitative description of the resource according to several dimensions (e.g., speakers, temporal distribution, tag density, etc.). Sections 4 briefly discusses possible uses of CORPS. Then, from Section 5 on, we present the experiments we carried out to build and evaluate the persuasive lexica. Finally in Section 8, we draw our conclusions and outline possible future directions.

2. Related Work

In this section we present related works according to a broad distinction, i.e., works that focus mainly on applying computational linguistics to political communication analysis and works that concern political communication corpora annotated with different phenomena (e.g., linguistic, cognitive and sociological aspects).

Persuasion and automatic analysis of political communication. While there is a huge theoretical and empirical research on the rhetoric used by politicians, only in recent years there has been a growing interest in bridging the gap between qualitative analysis of political communication and computational linguistics in order to automatize tasks that were usually carried out manually. A well-detailed discussion on the broader problem of integrating information technologies with social science research can be found in (Cousins & Mcintosh, 2005).

---

[2] Given the textual nature of the corpus, rhetorical artifices based on prosody and other speech features cannot be addressed. These artifices are used to highlight key passages of a speech, with the help of high impact words or concepts.
[3] On this point see (Bull & Noordhuizen, 2000) about mistimed applauses in political speeches

The automatic analysis of political communication is mainly focused on *text categorization*. Text categorization deals with the task of assigning a document to a pre-defined set of categories, such as determining party position in a text (e.g., Republican or Democratic) - see for example the work presented in (Purpura & Hillard, 2006, Purpura, Hillard, & Howard, 2006). In (Purpura & Hillard, 2006) a topic spotting classification algorithm was used for the task of coding legislative activities into subject areas; the algorithm used a traditional bag-of-words document representation. In (Purpura *et al.*, 2006), the authors presented a method based on Support Vector Machines for classifying political emails according to the party that sent them (either the Republicans or the Democrats).

Finally, an automatic analysis of the lexical aspects of political communication, similar to the work presented in (Guerini *et al.*, 2008a) (but not considering the persuasive impact of words, based on audience reaction analysis), can be found in (Benoit & Laver, 2003, Laver & Benoit, 2002, Laver & Garry, 2000, Laver, Benoit, & Garry, 2003, Bligh, Kohles, & Meindl, 2004).

Corpora for the analysis of political communication. In recent years large amounts of political digitized data have become available through blogs, government records, newspapers and dedicated web sites. These data, if properly structured, can be exploited as resources in different research fields. As a consequence, the interest in the creation of annotated corpora with both manual and automatic methods has grown. In particular, text corpora have been built aiming at different kinds of analysis of political communication. For example, in the field of political psychology, (Dyson, 2008) performed a comparison of the cognitive architecture of political leaders on a corpus of prime ministerial responses in the British House of Commons annotated using the Hermann's conceptual complexity scheme (Hermann, 2003); (Klebanov, Diermeier, & Beigman, 2008) applied automatic semantic annotation techniques to analyze Margaret Thatcher's political rhetoric; meanwhile (Bevitori, 2007) annotated a corpus of parliamentary events about the war in Iraq with socio-linguistic tags in order to verify a number of role features, such as gender, party and institutional function. In (Franzosi, 2004) the authors created a large scale corpus of annotated political news from the journals of opposing parties, using the PC-ACE tool to manually annotate them with narrative and semantic aspects. This study aims to understand the characteristics of social events during the fascist period. In addition, audio-video corpora of political speech were created in a multimodal perspective, such as in (Poggi & Vincze, 2009) where an annotation scheme was proposed to analyze the persuasive importance of gesture and gaze in electoral debates.

With respect to these studies, it is worth noting that our data of interest are political monologues in textual form annotated with no lexical or linguistic information, but with audience reaction tags (i.e., a kind of non-lexical audio cues).

3. Corpus Building and Annotation

The first release of CORPS (Guerini *et al.*, 2008a) was made available in 2008 and consisted of about 900 speeches for a total of about 2.2 million words. In January 2011 a new corpus was released adding more than 2,700 speeches and reaching a total of almost 8 million words and more than 3,600 speeches. The annotation of the new speeches involved three annotators and required 1.25 person/months. For the new release we followed the annotation scheme previously used, deploying a semi-automated procedure with an ad-hoc annotation tool.

197 speakers are represented in the corpus even though most speeches are given by a total of six politicians, namely Bill Clinton, George W. Bush, Ronald Reagan, Dick

Cheney, Barack Obama, and John F. Kennedy. The speeches are in English, primarily delivered by native speakers and they represent monological situations. We decided not to include dialogical situations like in political debates since they are not in our current focus of research and they pose further problems in labeling and analysis. The temporal distribution of the speeches spans from 1917 to 2010. The original speech transcriptions were taken from the Web, mostly from government portals (e.g. the White House portal) and web sites of personal foundations (e.g., the William J. Clinton Foundation and the Margaret Thatcher Foundation).

3.1. Annotation Scheme

The collected files come from various Web sources and contain audience reaction tags. The annotation is aimed at:

- *normalizing such tags*, converting synonymic tags to a specifically designed annotation scheme. For example, some original transcriptions contained the tag {BIG-APPLAUSE} while others had {LOUD-APPLAUSE}: all these have been converted to {SUSTAINED-APPLAUSE}. This facilitates machine tractability of the corpus. Table 1 demonstrates a summary of the audience reactions tags and their conversion.
- *extracting metadata* from the speeches to make them automatically searchable (e.g. title, speaker, event, date). See Table 2 for a complete description of the structure of the speeches.

Audience Reaction Tag	Note
{APPLAUSE}	Main tag in speech transcription.
{LAUGHTER}	Main tag in speech transcription.
{SPONTANEOUS-DEMONSTRATION}	Tags replaced: "reaction" "audience interruption"
{STANDING-OVATION}	It replaces the corresponding annotation in the original source
{SUSTAINED-APPLAUSE}	Tags replaced: "big applause", "loud applause", etc.
{CHEERS}	Cries or shouts of approval from the audience. Tags replaced: "cries", "shouts", "whistles", etc.
{BOOING}	The act of showing displeasure by loudly yelling "Boo", Tags replaced: "hissing"
{TAG1 ; TAG2 ; ...}	In case of multiple tagging, tags are divided by semicolon. Usually there are at most two tags.
{AUDIENCE} [text] {/AUDIENCE}	Tag used to signal an audience intervention either positively or negatively focused.
Special Tag	**Note**
{AUDIENCE-MEMBER} [text] {/AUDIENCE-MEMBER}	Tag used to signal an intervention of single audience member such as claques speaking.

`{OTHER-SPEAK} [text]` `{/OTHER-SPEAK}`	Tag used to signal speakers other than the subject (like journalists, chairmen, etc.)
`{COMMENT="[text]"}`	Tag used for parenthetical comments not mappable to other tags

Table 1. List of main tags

As for what concerns the typology of persuasive communication (audience reaction), we can further individuate three main groups of tags, for analysis purposes:

- *Positive-Focus*: this group indicates a persuasive attempt that sets a positive focus in the audience. Tags considered (~49,000): {AUDIENCE}, {CHEERS}, {APPLAUSE}, {SPONTANEOUS-DEMONSTRATION}, {STANDING-OVATION}, {SUSTAINED-APPLAUSE}.
- *Negative-Focus*: it indicates a persuasive attempt that sets a negative focus in the audience. It is worth mentioning that the negative focus is set towards the object of the speech and not on the speakers themselves (e.g., "Do we want more taxes?"). Tags considered (~1,100): {BOOING}, {AUDIENCE} No! {/AUDIENCE}.
- *Ironical*: it indicates the use of ironical devices in persuasion. Tag considered (~16,000): {LAUGHTER}.[4]

{title} [mandatory - describing the speech] {/title}
{event} [not mandatory - derivable from the title] {/event}
{speaker} [mandatory] {/speaker}
{date} [mandatory] {/date}
{source} [mandatory - Internet address] {/source}
{description} [if present in the source] {/description}
{speech} [speech transcription with audience reactions tags] {/speech}

Table 2. Structure of a speech entry in CORPS

It should be noted that, rhetorically, positive-focus reactions can also be obtained from (sub-)fragments of speech that set a temporary negative focus in the audience, or even from a complete focus on negative aspects (usually political opponents' behavior). In fact, about 30% of the times, the rhetorical device used in political speeches to evoke applause is CONTRAST (see (Atkinson, 1984) and (Heritage & Greatbatch, 1986)).

Let us consider the speech that John F. Kennedy gave in Berlin on the 26th of June 1963, and in particular the following fragment, that led to an {APPLAUSE ; CHEERS} reaction: "Freedom has many difficulties and democracy is not perfect. But we have never had to put a wall up to keep our people in – to prevent them from leaving us." This fragment sets a double negative focus. First, by means of a CONCESSION, Kennedy sets a negative focus on the limits of the American social model: "Freedom has many difficulties and democracy is not perfect," then by means of a CONTRAST he sets a stronger negative focus on the Soviet social model: "But we have never had to put a wall up to keep our people in – to prevent them from leaving us." Still, the overall effect of the fragment, based on an implicit CONCESSION and an explicit CONTRAST, is to set people to a positive point of view on the American social model.

---

[4] If LAUGHTER appears in a multiple tag (e.g. together with APPLAUSE) by default this tag is associated to the ironical group. Note that calling this group "ironical" is a just a convention, it can encompasses humorous situations as well.

3.2. Statistics

This section presents some quantitative analysis about CORPS, which can give an idea of the nature of the corpus and represent the starting point for further qualitative and quantitative analysis (possible approaches relying on CORPS features will be discussed in Section 4). Henceforth, whenever referring to "tags", we will consider only audience reaction tags and discard the special tags defined in Table 1.

Table 3 demonstrates a survey on the main statistics about the corpus, while in Table 4 main statistics about tag frequencies in the corpus are provided. Additionally, Table 5 and Table 6 provide samples of {AUDIENCE} and {COMMENT} tags respectively.

In the following tables, "Tag-Density" refers to a measure that indicates how often audience reaction tags appear in the speeches under scrutiny; similarly "PF-density" refers to positive-focus tag, "I-density" to ironical-focus tag and "NF-density" to negative-focus tag density.

Total number of speeches:	3,618
Total number of speakers:	197
Total number of words:	7,901,893
Total number of tags:	66,082
Tag density (μ):	0.0084
PF-density (μ):	0.0062
I-density (μ):	0.0020
NF-density (μ):	0.00015
Temporal range (μ):	from 18/05/1917 to 16/09/2010

Table 3. Corpus main statistics

SINGLE TAGS	
{APPLAUSE}	46310
{LAUGHTER}	14055
{AUDIENCE}	1803
{BOOING}	756
{SPONTANEOUS-DEMONSTRATION}	313
{CHEERS}	234
{SUSTAINED APPLAUSE}	97
{STANDING-OVATION}	51
MULTIPLE TAGS	
{LAUGHTER ; APPLAUSE}	1579
{CHEERS ; APPLAUSE}	837
OTHERS	47
SPECIAL TAGS	
{AUDIENCE-MEMBER}	999
{COMMENT}	787
{OTHER-SPEAK}	404
GROUPED TAGS	
POSITIVE-FOCUS TAGS	49275
IRONICAL TAGS	15660
NEGATIVE-FOCUS TAGS	1147

Table 4. Tag main statistics

{AUDIENCE} Yes! {/AUDIENCE}	482
{AUDIENCE} No! {/AUDIENCE}	390
{AUDIENCE} Four more years! Four more years! {/AUDIENCE}	346
{AUDIENCE} Yes, sir {/AUDIENCE}	87
{AUDIENCE} U.S.A.! U.S.A.! U.S.A.! {/AUDIENCE}	41
{AUDIENCE} All right {/AUDIENCE}	39
{AUDIENCE} Flip-flop! Flip-flop! Flip-flop! {/AUDIENCE}	39
{AUDIENCE} Hooah. {/AUDIENCE}	38
{AUDIENCE} Reagan! Reagan! Reagan! {/AUDIENCE}	37

Table 5. Audience tag samples.

{COMMENT="Inaudible"}	257
{COMMENT="A toast is offered"}	30
{COMMENT="The bill is signed"}	30
{COMMENT="The medal was presented"}	26
{COMMENT="The medal was awarded"}	24
{COMMENT="Recording interrupted"}	18
{COMMENT="The citation is read"}	18
{COMMENT="The citation was read"}	16
{COMMENT="Interruption"}	9
{COMMENT="A moment of silence was observed"}	8

Table 6. Comment tag samples

Mathematically, "Tag-Density" can be calculated with two different average measurements: micro-averaged tag density (μ) and macro-averaged tag density (M). The idea is that, given a set of speeches - e.g., the speeches of Democrats -, the density of the tags can be computed either by counting all tag occurrences in the set and then dividing the result by the total number of words contained in those speeches (μ), or by computing the tag density for each category (Democrat speakers in our example) and then averaging over the results of each category in the set (M).

More formally, given a set of n speeches S, where a single speech is represented with s_i (i.e. $s_i \in S$), $|t_i|$ represents the number of tags in a given speech s_i and $|w_i|$ represents the number of words in the same speech; we can define μ as:

$$\mu = \frac{\sum_{i=1}^{n} |t_i|}{\sum_{i=1}^{n} |w_i|} \qquad (1)$$

In a similar way M can be defined as:

$$M = \frac{\sum_{i=1}^{|C|} \frac{|t_i|}{|w_i|}}{|C|} \qquad (2)$$

where $|C|$ represent the number of categories (speakers) in the set of speeches, and $|t_j|$ and $|w_j|$ represent the total number of tags and words for the category.

In the rest of the paper we will mainly provide micro-averaged tag density, since it represents the more general density within the corpus, but macro-averaged values will be provided as well when necessary for further analysis.

Speaker	Total Speeches	Tag-Density	PF-density	I-density	NF-density
Bill Clinton	889	0.007	0.005	0.002	0.00001
George W. Bush	427	0.015	0.012	0.002	0.00005
Ronald Reagan	388	0.004	0.001	0.003	0.00044
Dick Cheney	356	0.011	0.008	0.002	0.00061
Barack Obama	347	0.010	0.008	0.003	0.00007
John F. Kennedy	316	0.009	0.008	0.001	0.00000
Michelle Obama	107	0.009	0.005	0.003	0.00001
Margaret Thatcher	102	0.005	0.004	0.001	0.00001
Laura Bush	93	0.015	0.014	0.001	0.00000
Richard M. Nixon	61	0.006	0.005	0.000	0.00008
Al Gore	53	0.007	0.005	0.002	0.00004
Alan Keyes	51	0.004	0.003	0.001	0.00007

Table 7. Main speakers statistics - Micro-averaged densities (μ)

In Table 7 statistics about the main speakers are provided. We will not discuss it in details, since the analysis of the characteristic of each speaker is out of scope of the present paper. Still, we will introduce some interesting insights after aggregating the speakers in the subsequent tables.

In Table 8 some statistics about tag-densities are provided according to two main categorizations: Democrats/Republicans and Male/Female speakers. For the first categorization we used a subset of the most prominent speaker (i.e., 12 speakers accounting for 3,190 speeches with a coverage of 88% of the whole corpus). For the second categorization we used the whole corpus instead.

Party	Subset-Coverage	Tag-Density	PF-density	I-density	NF-density
Democrats	0.45	0.0075	0,0055	0,0019	0,000027
Republicans	0.55	0.0097	0,0072	0,0022	0,000309
Gender	Corpus-Coverage	Tag-Density	PF-density	I-density	NF-density
Females	0.11	0.0085	0.0067	0.0018	0.000007
Males	0.89	0.0083	0.0062	0.0020	0.000158

Table 8. Democrats/Republicans and Males/Females, micro-averaged densities (μ)

Since the data can be biased by some speakers being over-represented and others being under-represented, in Table 9 we provide the same tag-densities, this time macro-averaged, so to give every speaker the same importance in the final results.

Party	Subset-Coverage	Tag-Density	PF-density	I-density	NF-density
Democrats	0.45	0.0076	0.0056	0.0019	0.000036
Republicans	0.55	0.0094	0.0076	0.0017	0.000199
Gender	Corpus-Coverage	Tag-Density	PF-density	I-density	NF-density
Females	0.11	0.0068	0.0055	0.0013	0.0000007
Males	0.89	0.0070	0.0052	0.0017	0.0000444

Table 9. Democrats/Republicans and Males/Females, macro-averaged densities (M)

As can be seen from the data, while the Democrats/Republicans partition is well balanced (0.45 vs. 0.55), the Males/Females partition is unbalanced (0.89 vs. 0.11). In more details we can see that the tag density is slightly higher for Republican speakers (and the same holds for the subset of positive-focus tags), while the ironical-focus tags have almost the same density in both groups. Interestingly, the density of negative-focus tags (that represent a more "aggressive" kind of rhetoric, even if rarely used) is 11 times higher in the Republican group than that in the Democrat group. A similar consideration can be drawn for the male/female distinction: while all other tag densities are almost the same, for the negative-focus tags we have a density 60 times higher for male speakers.

Finally, in Figure 1, in order to get an overlook of how the rhetoric of politicians changed trough the years, the temporal distribution of tag densities is provided, according to a 10 years grouping (for every group the corresponding number of speeches is provided as well). We believe it is an interesting research topic understanding why tag densities raise or decrease across time, still, this is out of the scope of the present paper.

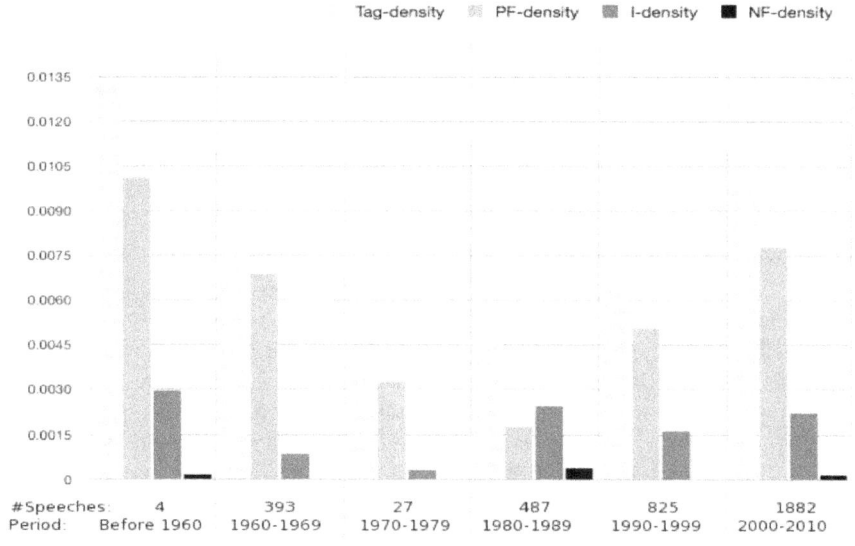

Figure 1. Temporal distribution of tag densities (μ)

4. Uses of the Corpus

CORPS and its new release allow the automation of several tasks in various theoretical and applied contexts. Given that CORPS is a relatively new resource, the full potential of approaches based on its characteristics is still to be extensively explored, nonetheless some examples of emerging trends are listed below.

- Persuasive expression mining. Some approaches hypothesize that the recognition and classification of phenomena such as applause, laughter, and speaker vocal effort can improve information retrieval (see, among others, (Bertoldi, Brugnara, Cettolo, Federico, & Giuliani, 2002) and (Hu et al., 2008)). On top of such features, approaches for extracting relevant linguistic material can be developed.
- Automatic analysis of political communication. By considering audience reactions, it is possible to individuate rhetorical phenomena that do not come into light with traditional approaches based on simple word usage (counting their occurrences). Several measures can be developed using different metrics like *z-score* - see for example (Conoscenti, 2011).
- Prediction of text impact. It is possible to use machine learning techniques for predicting the persuasive impact of novel speeches, in terms of audience reactions. In fact, with the huge amount of textual material that flows on the Web (news, discourses, blogs, etc.), it can be useful to have a measure for testing the persuasiveness of what we retrieve or possibly of what we want to publish on Web, as suggested in (Strapparava, Guerini, & Stock, 2010). Moreover, it is possible to answer such hypothetical questions as "How would a Democrat audience have reacted to this Republican speech?".
- Persuasive natural language generation. Just to mention the example of lexical choice: techniques that use domain information for choosing appropriate lemmata have been proposed, among others, in (Jing, 1998). In the Valentino prototype (Guerini, Strapparava, & Stock, 2008b, 2011) instead, lexical choice is performed on the basis of lemma impact rather than lemma use (i.e., the lemma with the highest *pi* is extracted). If the typology of persuasive communicative goal is specified (positive-focus, negative-focus, ironical), the choice can be further refined.

5. Pre-Processing CORPS

Among techniques at the root of traditional social science research, there are various text analysis methodologies, which can focus on the lexical, syntactic, or semantic level. For the analysis of CORPS the focus has been posed on the lexical level. If we want to properly analyze language use, we first have to "normalize" the text. In particular, to reduce data sparseness, we used a lemmatizer and a Part of Speech (PoS) tagger on the whole corpus, which gave for each token in the text the corresponding `lemma` and `pos`. So, at the lexical level we considered `lemmata` (e.g. the verb *to win*) rather than *tokens* (i.e., the form of the word, as it appears in the text: *win, wins, won*). In the following sections, if not differently stated, the term *word* indicates a `lemma#pos` - where `pos` can be `v` for verbs, `a` for adjectives, `r` for adverbs and `n` for nouns. So the word *to win* is represented as `win#v`. In the lexical analysis we further considered:

- the sentence immediately preceding audience reaction tags (under the hypothesis that it is exactly this sentence that triggered the audience reaction).

- the typology of persuasive communication (audience reaction: namely positive, negative or ironical focus).

As a next step, we split the corpus into sentences, keeping the information of whether a sentence is followed by an audience reaction or not. In this way we obtained a dataset of roughly 435,000 sentences including:

- ~49,000 sentences triggering (followed by) a positive focus tag
- ~1,000 sentences triggering a negative focus tag
- ~16,000 sentences triggering an ironical focus tag
- ~369,000 sentences which are not followed by any audience reaction tag.

An excerpt of the dataset is provided in Table 10. Each entry in this dataset consists of one sentence lemmatized and PoS-tagged. If a sentence is followed by an audience reaction, this information is provided at the end of that sentence.

...
be#v simple#a definition#n criminal#a
break#v law#n {IRONICAL_FOCUS}
so#r definition#n be#v criminal#n be#v where#r law#n be#v not#r very#r efficacious#a
really#r do#v not#r have#v effect#n your#a behavior#n {IRONICAL_FOCUS}
...
will#v not#r take#v stand#n know#v apparently#r do#v not#r know#v
life#n womb#n be#v not#r protect#v its#a mother#n choice#n
be#v God#n choice#n establish#v our#a right#n life#n {POSITIVE_FOCUS}
...

Table 10. Excerpt of the dataset in which each entry consists of one sentence lemmatized and PoS-tagged.

6. Methodologies for Extracting Persuasive Word Lexica From CORPS

Persuasive lexica can be extracted from CORPS by using different methodologies. The general idea we followed to build these lists is that a word is more persuasive if it appears frequently near audience reaction tags, and rarely far from them. To implement this idea several methodologies can be used, but most of them come from the realm of distributional semantics. Distributional semantics is a research area that implements methods for quantifying and categorizing semantic similarities between words, based on their distributional properties in large samples of language data (documents). The basic idea of distributional semantics can be summed up in the so-called Distributional hypothesis: linguistic items with similar distributions over documents have similar meanings. For our purposes we consider the single sentences as documents. From these we can either build virtual documents of sentences preceding audience reaction tags, or consider audience reaction tags as words that co-occur with other words within documents (sentences).

6.1. Tf-idf

Tf-idf, acronym for term frequency-inverse document frequency, is a statistical measure that is meant to indicate how important a word is to a document in a collection or corpus. The tf-idf value increases proportionally to the number of times a word appears in the document, but is offset by the frequency of the word in the corpus, which helps to adjust for the fact that some words appear more frequently in general. In our case, for each audience reaction tag (i.e., positive, negative and ironical) a virtual document including all the relevant persuasive sentences is created. Then, we calculate the tf-idf for each word present in the virtual documents by using the following formulae:

$$tf_{i,j} = \frac{n_{i,j}}{|d_j|} \qquad (3)$$

where $n_{i,j}$ is the number of occurrences of the lemma l_i in a virtual document d_j, while the denominator is simply the size (i.e., the number of lemmas) of the document d_j.

$$idf_i = \log \frac{|D|}{|\{d : l_i \in d\}|} \qquad (4)$$

where $|D|$ is the number of documents in the corpus (i.e. the 369,000 sentences which are not followed by any audience reaction tag plus the three virtual documents), while the denominator is the number of documents which contain the lemma l_i. The final tf-idf of a word is obtained by simply multiplying tf by idf. *The tf-idf for each word is then used as the pi score.*

6.2. Weighted Tf-idf

We also use a weighted variant of tf-idf, to capture the idea that words closer to a tag might be more persuasive than distant ones, even within the same sentence. In particular, the weight is calculated as the division of 1 by the distance of the word from the focus tag in terms of token count. This weight is included in the tf calculation as follows:

$$w\_tf_{i,j} = \frac{\sum w_i}{\sum_{l \in d_j} w_l} \qquad (5)$$

where $\sum w_i$ is the sum of the weights (i.e. distance from the focus tag) of the lemma l_i in a virtual document d_j and $\sum_{l \in d_j} w_i$ is the sum of the weights of all the lemmas existing in the document. idf_i is calculated in the same way as in Section 6.1. *The weighted tf-idf for each word is then used as the pi score.*

6.3. Latent Semantic Analysis

To get a similarity space with the required characteristics, we use Latent Semantic Analysis (LSA), a corpus-based measure of semantic similarity proposed by (T. Landauer, Foltz, & Laham, 1998). In LSA, term co-occurrences in a corpus are captured by means of a dimensionality reduction operated by a singular value decomposition (SVD) on the term-by-document matrix T representing the corpus. In LSA, similarity is computed in a vector space in which second-order relations among terms and texts are exploited using cosine similarity. LSA presents the advantage of yielding a vector space model that allows for a homogeneous representation (and hence comparison) of words, sentences, and texts. For representing a word set or a sentence in the LSA space we use the pseudo-document representation technique (Berry, 1992): each text segment is represented in the LSA space by summing up the normalized LSA vectors of all the constituent words, also using a tf.idf weighting scheme (Gliozzo & Strapparava, 2005). The methodology is unsupervised as we do not exploit any 'labeled' training material. According to (T. K. Landauer & Dumais, 1997), the power of the model comes from the optimal dimensionality reduction. Choosing the best r is a complex and still open problem. Empirically, it has been shown that NLP applications benefit from setting r in the range [50,400]. In particular, for the experiment in the present study, we employed $r = 200$. *For each word, the cosine similarity with the audience reaction tags is then used as the pi score.*

6.4. Latent Dirichlet Allocation

Latent Dirichlet allocation (LDA) is a flexible generative probabilistic model for collections of discrete data. LDA is an example of a topic model and was first presented as a graphical model for topic discovery by (Blei, Ng, & Jordan, 2003). LDA is based on a simple exchangeability assumption for the words and topics in a document; it is therefore realized by a straightforward application of de Finetti's representation theorem (Finetti, 1990). We can view LDA as a dimensionality reduction technique, in the spirit of LSA, but with proper underlying generative probabilistic semantics that make sense for the type of data that it models. In particular LDA is a three-level hierarchical Bayesian model, in which each item of a collection is modeled as a finite mixture over an underlying set of topics. Each topic is, in turn, modeled as an infinite mixture over an underlying set of topics. In natural language processing, LDA is useful for the following potential application areas: information retrieval (analyzing semantic/latent topic/concept structures of large text collections for a more intelligent information search; document classification/clustering, document summarization, and text/Web data mining community in general; and collaborative filtering. In the present paper, we run LDA with 50 topics. For LDA we selected for each focus tag the topic where the tag has the highest probability score. *Then we took the probability scores of all the words present in that topic and considered them as the pi of the corresponding words.*

7. CORPS Experiments

To run our experiments we partitioned our corpus of 435,000 sentences into a training and a test set. The training set consists of 424,590 sentences, while the test set contains 5,568 sentence pairs (one persuasive sentence paired with a non-persuasive one). In particular, the test set was built using a particular methodology that controls for confounding factors such as author and sentence length. The methodology we used to build these pairs is very similar to (Danescu-Niculescu-Mizil, Cheng, Kleinberg, & Lee, 2012, Tan, Lee, & Pang, 2014): for each P, where P is the sentence preceding an audience reaction (e.g., APPLAUSE, LAUGHTER), we selected a contrasting single-sentence $\neg P$ from the same speech. We required $\neg P$ to be close to P in the speech transcription, subject to the conditions that (i) P and $\neg P$ are uttered by the same speaker - which is trivial since these are monologues, where a single speaker is addressing the audience - (ii) P and $\neg P$ have the same number of words, and (iii) $\neg P$ is 5 to 15 sentences away from P. This last condition had to be imposed since, differently from (Danescu-Niculescu-Mizil et al., 2012), we do not have the evidence of which fragment of the speech exactly provoked the audience reaction (i.e., it could be the combination of more than one sentence). Then, we partitioned the test set into three subsets, namely $CORPS_{POS}$, $CORPS_{NEG}$ and $CORPS_{IRO}$, where each P is connected to the corresponding audience reaction focus tag.

The advantage of these kinds of test sets comparing two sentences of the same length is that it allows us to answer the following: is the difference in audience reactions – at least partially – due to lexical choices by the speaker or *just* to higher level and more abstract features (including for example argumentative schemes spanning over some sentences, or the tone of the voice)? Since we are focusing on the comparison of single sentences, where wording is more relevant than structure, if such an effect exists it should be modeled by some of our proposed approaches. We thus proceeded to built the persuasive lexica using the methodologies presented in the previous section.

To have a fair comparison of lexica quality (i.e., without any syntactic or compositional reasoning that can boost the performance) we used a naïve approach that averages over all the word scores in a sentence, similar, for example, to the approaches used in (Strapparava & Mihalcea, 2008) and (Staiano & Guerini, 2014).

	LSA_{IRO}	LSA_{NEG}	LSA_{POS}	LDA_{IRO}	LDA_{NEG}	LDA_{POS}
$CORPS_{POS}$	0.303	0.372	0.497	0.397	0.442	0.432
$CORPS_{NEG}$	0.551	0.580	0.551	0.522	0.551	0.449
$CORPS_{IRO}$	0.647	0.501	0.512	0.542	0.555	0.458
	$tfidf_{IRO}$	$tfidf_{NEG}$	$tfidf_{POS}$	w_tfidf_{IRO}	w_tfidf_{NEG}	w_tfidf_{POS}
$CORPS_{POS}$	0.511	0.523	0.578	0.522	0.537	**0.620**
$CORPS_{NEG}$	0.522	0.594	0.478	0.536	**0.681**	0.464
$CORPS_{IRO}$	0.642	0.582	0.547	**0.657**	0.579	0.534

Table 11. Experimental results in terms of the fraction of the pairs correctly categorized.

In particular, for our experiments we used the "average" of the corresponding persuasive scores – obtained from the lexicon under inspection – of all lemma-POS recognized in the text. Then, once we computed the scores for each P (S_p) and the corresponding score on $\neg P$ ($S_{\neg p}$), we considered a pair correctly classified if , incorrectly classified otherwise, including in the latter also the possibly undecidable sentences where $S_p = S_{\neg p}$. To understand if our persuasive lists are able to "grasp" the concept of persuasiveness, we should consider that a random baseline would correctly categorize 50% of the pairs. If our lists were able to outperform this baseline, this would mean that there is an underlying phenomenon of lexical persuasiveness that our lists are able to capture.

Table 11 demonstrates the experimental results on the three subsets of the test set by using the naïve approach with various lexica that we generated with the methodologies described in Section 6. The results are presented in terms of the fraction of the pairs correctly categorized. As can be seen, the weighted tf-idf variant is the one that obtains the best results (far above the random baseline) in a consistent way: i.e., the lexica built on the ironical virtual document performs better on the ironical test set, and the same holds for the positive and negative focus, indicating that not only we are able to recognize the persuasive impact of words, but also to differentiate among the various forms of lexical persuasion. W_tfidf is then followed, in terms of performances, by standard tf-idf based lexica, while LSA and LDA based lexica do not perform well. The better performance of the weighted variant of tf-idf in comparison to standard tf-idf can be explained by the fact that not all the words in a persuasive sentence have the same impact, but there is a sort of *crescendo* until the *apex*. A possible explanation for the poor performances of LSA and LDA, instead, can be the fact that they tend to "propagate" the persuasive relevance also to words that are similar among each other but not necessarily used in persuasive contexts. In other words, LDA and LSA tend to focus on the "topic" of the persuasive sentences, not on the actual linguistic realization of the topic itself: so, *in persuasion it is important to choose not only the right topic, but also the right words*.

8. Conclusions and Future Work

In this paper we have presented CORPS, a text resource containing political speeches annotated with audience reaction tags. After giving an overview of the related work, we have provided some analysis about the main features characterizing the corpus. Then, we have proposed various statistical methodologies, namely tf-idf, weighted tf-idf, LSA and LDA, to compile persuasive lexica from CORPS. In these lexica each word is associated with a 'persuasive impact' score, which was derived from the distributional patterns of that word around audience reaction tags. To compare the effectiveness of these lexica, we carried out classification experiments on pairs of persuasive/non persuasive sentences. The results demonstrate that some methodologies stand out amongst the others and that computational methodologies can be successfully utilized for choosing the right wording in persuasive communication. In particular, while weighted tf-idf outperformed all the other approaches, in many cases the lexica obtained with LSA and LDA were not effective. These findings suggest that in persuasion it is important to choose not only the right topic, but also the right words.

As regards future work, we plan to experiment with additional configurations of the methodologies we used (e.g., combining probability scores from LDA topics in a more complex way). We will also use machine learning approaches that combine persuasive impact of words with other lexical features. Finally, we will test our lexica on additional datasets from different domains – e.g., (Danescu-Niculescu-Mizil *et al.*, 2012, Tan *et al.*, 2014) – to understand if, and to what extent, the notion of lexical persuasiveness can be exported to contexts outside political communication.

References

Atkinson, J. (1984). *Structures of social action*. In J. Atkinson & J. Heritage (Eds.), (pp. 370–409). Cambridge, England: Cambridge University Press.
Benoit, K., & Laver, M. (2003). Estimating Irish party positions using computer wordscoring: The 2002 elections. *Irish Political Studies, 17(2)*.
Berry, M. (1992). Large-scale sparse singular value computations. *International Journal of Supercomputer Applications, 6*(1).
Bertoldi, N., Brugnara, F., Cettolo, M., Federico, M., & Giuliani, D. (2002). Cross-task portability of a broadcast news speech recognition system. *Speech Communication, 38*(3-4), 335–347.
Bevitori, C. (2007). Engendering conflict? A corpus-assisted analysis of women MPs positioning on the war in Iraq. *Textus, 20*(1), 137–158.
Blei, D., Ng, A., & Jordan, M. (2003, January). Latent dirichlet allocation. *Journal of Machine Learning Research, 3*, 993–1022.
Bligh, M. C., Kohles, J. C., & Meindl, J. R. (2004). Charisma under crisis: Presidential leadership, rhetoric, and media responses before and after the September 11th terrorist attacks. *The Leadership Quarterly, 15(2)*, 211–239.
Bull, P., & Noordhuizen, M. (2000). The mistiming of applause in political speeches. *Journal of Language and Social Psychology, 19*, 275–294.
Conoscenti, M. (2011). *The reframer: An analysis of barack obama's political discourse (2004-2010)*,. Roma: Bulzoni.
Cousins, K., & Mcintosh, W. (2005). More than typewriters, more than adding machines: Integrating information technology into political research. *Quality and Quantity, 39*, 581–614.
Danescu-Niculescu-Mizil, C., Cheng, J., Kleinberg, J., & Lee, L. (2012). You had me at hello: How phrasing affects memorability. In *Proceedings of the acl*.
Dyson, S. B. (2008). Text Annotation and the Cognitive Architecture of Political Leaders: British Prime Ministers from 1945-2008. *Journal of Information Technology & Politics, 5*(1), 7–18.
Finetti, B. de. (1990). *Theory of probability* (Vol. 1-2). Chichester: John Wiley & Sons Ltd.
Franzosi, R. (2004). *From words to numbers: Narrative, data, and social science*. Cambridge: Cambridge University Press.
Gliozzo, A., & Strapparava, C. (2005, June). Domains kernels for text categorization. In *Proc. of the ninth conference on computational natural language learning (CoNLL-2005)* (pp. 56–63). University of Michigan, Ann Arbor: .
Guerini, M., Strapparava, C., & Stock, O. (2008a). Corps: A corpus of tagged political speeches for persuasive communication processing. *Journal of Information Technology & Politics, 5*(1), 19–32.
Guerini, M., Strapparava, C., & Stock, O. (2008b). Valentino: A tool for valence shifting of natural language texts. In *Proceedings of LREC 2008*. Marrakech, Morocco: .
Guerini, M., Strapparava, C., & Stock, O. (2011). Slanting existing text with valentino. In *Proceedings of iui'11* (pp. 439–440). Palo Alto, USA.
Heritage, J., & Greatbatch, D. (1986). Generating applause: a study of rhetoric and response at party political conferences. *American Journal of Sociology, 92*, 110-157.
Hermann, M. G. (2003). The psychological assessment of political leaders. In J. Post (Ed.), (pp. 178–214). Lawrence Erlbaum Publishing Co.

Hu, Q., Goodman, F., Boykin, S., Fish, R., Greiff, W., Jones, S., . (2008). Automatic detection, indexing, and retrieval of multiple attributes from cross-lingual multimedia data.

Jing, H. (1998). Usage of wordnet in natural language generation. In S. Harabagiu (Ed.), *Proceedings of the Conference on the Use of WordNet in Natural Language Processing Systems"* (pp. 128–134). Somerset, New Jersey: Association for Computational Linguistics.

Klebanov, B. B., Diermeier, D., & Beigman, E. (2008). Automatic Annotation of Semantic Fields for Political Science Research. *Journal of Information Technology & Politics, 5*(1), 95–120.

Landauer, T., Foltz, P., & Laham, D. (1998). Introduction to latent semantic analysis. *Discourse Processes, 25*.

Landauer, T. K., & Dumais, S. T. (1997, April). A solution to plato's problem: The latent semantic analysis theory of acquisition, induction, and representation of knowledge. *Psychological Review, 104*(2), 211–240.

Laver, M., & Benoit, K. (2002). Locating tds in policy spaces: Wordscoring Dail speeches. *Irish Political Studies, 17(1)*.

Laver, M., Benoit, K., & Garry, J. (2003). Extracting policy positions from political texts using words as data. *American Political Science Review, 97(2)*, 311–331.

Laver, M., & Garry, J. (2000). Estimating policy positions from political texts. *American Journal of Political Science, 44(3)*, 619–634.

Poggi, I., & Vincze, L. (2009). Multimodal corpora. In M. Kipp, JC. Martin, P. Paggio, & D. Heylen (Eds.), (pp. 73–92). Berlin, Heidelberg: Springer-Verlag.

Purpura, S., & Hillard, D. (2006). Automated classification of congressional legislation. In *Proceedings of the Seventh International Conference on Digital Government Research.* San Diego, CA: .

Purpura, S., Hillard, D., & Howard, P. (2006). *A comparative study of human coding and context analysis against support vector machines (svm) to differentiate campaign emails by party and issues.*

Staiano, J., & Guerini, M. (2014). DepecheMood: a lexicon for emotion analysis from crowd-annotated news. *To appear in Proc. 52nd Ann. Meeting of the Assoc. for Computational Linguistics (ACL 14)*.

Strapparava, C., Guerini, M., & Stock, O. (2010). Predicting persuasiveness in political discourses. In *Proceedings of the seventh conference on international language resources and evaluation (lrec'10)*.

Strapparava, C., & Mihalcea, R. (2008). Learning to identify emotions in text. In *Proc. 23rd ann. acm symp. applied computing (sac 08)* (pp. 1556–1560).

Tan, C., Lee, L., & Pang, B. (2014). The effect of wording on message propagation: Topic-and author-controlled natural experiments on twitter. *ACL*.

Part V

Learning and Development

Chapter 20

The Trajectory of Argumentation and Its Multifaceted Functions

Matthew Fisher[1] & Frank C. Keil[2]

Yale University, New Haven, USA; [1] matthew.fisher@yale.edu; [2] frank.keil@yale.edu

Abstract. The manner, goal, and function of argumentation shift over the course of development. We suggest that early learners engage in proto-argumentation in a non-competitive "learning mode". The knowledge asymmetries in their social environment prompt argumentation to be used as a means of accumulating knowledge about the world. As early learners develop more sophisticated reasoning abilities, argumentation shifts to an "argue-to-win" style, which aims solely to promote one's own point of view and discredit others. The simplest form of this mode of argument is *ad hominem*-style attacks, which ignore the actual structure of the argument. Once the structure of the argument has been internalized, it is then possible to re-enter the learning mode equipped with more powerful tools for discovery. Though these modes at most times may be incompatible, certain strategies like "considering the opposite" (Lord, Lepper, & Preston, 1984) can be effective ways to leverage the features of reasoning to enable productive exchanges of ideas. Argumentation in this form can serve to expose gaps in understanding and probe the understanding of others.

1. The Trajectory of Argumentation and Its Multifaceted Functions

Arguments pervade a wide range of social interactions. They take place everywhere from classrooms (Chi, 2009) to courtrooms (Warren, Kuhn, & Weinstock, 2010) and occur across domains such as moral reasoning (Bloom, 2010; Haidt & Bjorklund, 2007) and scientific advancement (Latour & Woolgar, 1986; Osborne, 2010). Despite serving many important functions, there is no clear consensus on how to best characterize the concept of argumentation (Voss & Dyke, 2001). Argumentation has been described as containing multiple continua such as product-process, internal-external, and rhetoric-dialogical, where each instance of an argument can hold a different position on each dimension (Garcia-Mila & Andersen, 2007). Here, we suggest an additional dimension of argumentation that has not received attention. Arguments can be deployed with various goals and purposes and these characteristics create distinct modes of argument.

This chapter argues for distinguishable modes of argument that shift both across contexts and across the course of development. While philosophical and theoretical models of argumentation (i.e. Toulmin, 1958) are certainly important, those approaches may not describe the informal ways in which arguments are used by laypeople. Here, we focus on how arguments are construed psychologically. More specifically, we explore the possibility that arguments take on distinct and even incommensurable sets of characteristics depending on the context and period of development.

1.1. Modes of Argument

While there are many nuanced ways in which people argue in daily discourse, most broadly, two of the most common uses are winning and learning. Consider, for example, how a judge and a defense lawyer deploy arguments differently based on their role in the courtroom (Nickerson, 1991). The ideal judge aims to discern the truth of a matter and uses argumentation to evaluate evidence and probe for useful information. A defense lawyer, however, must carefully construct evidence to build a case supporting one particular outcome. These roles in the courtroom illustrate two distinct modes of argument. The judge exemplifies an argue-to-learn mindset — she freely follows the evidence and is ultimately concerned with finding the truth. The defense attorney exemplifies an argue-to-win mindset — he has a pre-determined position and is ultimately concerned with undermining the opposition and coming out on top. Note that both roles must use evidence to construct an argument to justify an assertion so that others see it as true (i.e. a suspect's innocence). The sense of argument that we trace in this chapter has this quality: it is truth-oriented in that the ultimate goal is to assert or evaluate a truth claim about a particular matter. Though the distinction is admittedly difficult to draw, this definition of argumentation excludes mere disputes, where the justifications offered are purely instrumental: people fighting to change a certain state of affairs to reach a desired end.

What is the evidence for these separate modes of argument? Experimental evidence suggests that these two approaches to argumentation constitute two distinct and potentially incommensurable mindsets. Social context is one factor that has been identified which can shift people into one mindset or the other (Fisher & Keil, under review). Specifically, when others are watching, winning becomes the goal of the argument, when others are not watching, the goal is to learn. In these studies, several indicators suggest that the cognitive construal of argumentation changes across social setting. First, participants chose to argue with those who are less knowledgeable about the topic of the argument when in a public setting, revealing their desire to outperform the opposition in the exchange. Second, the content of the arguments became more winning-oriented (as rated by independent judges) when in public. Finally, people produce lower quality arguments (again, as rated by independent raters) when they believe they are arguing in front of others. In a separate series of studies, when arguing cooperatively, people were more likely to hold a relativist understanding of truth, illustrating the downstream cognitive consequences of adopting a particular argumentative mindset (Fisher, Knobe, Strickland & Keil, under review). Thus, when people are arguing with the goal of learning, they shift their understanding of the underlying truth of the matter and see the truth as more subjective. These pieces of evidence support the notion that arguments can serve very different purposes at different times and in different contexts.

Although research has begun to unpack the different factors that influence the informal uses of argumentation, there remain many open questions. A developmental perspective offers a promising approach to better understanding argumentation by highlighting different components and precursors of how we argue. With much empirical work yet to be done, we offer a developmental roadmap for the various modes of argumentation. Given the novelty of the learn vs. win distinction, much of the available evidence is only indirectly related to our core distinction. Here, we organize the relevant evidence to sketch a plausible account for how argumentation develops. By examining the topic through this lens, we can isolate the cognitive underpinnings of learning and winning-oriented argumentation, better understand how they emerge, and see how they function.

1.2. Developing Modes of Argument

We propose that the ability to deploy arguments towards a specific goal develops in three distinct phases. In this chapter, we use this three-phase framework to organize the existing literature and suggest how observed patterns of argumentation relate to the modes of argument 1) As information seekers, children first exhibit a cooperative stance towards their interactions with others and use argumentation to promote learning. 2) As the ability to reason matures, trying to win becomes a major goal of argumentation as well. This mindset primarily manifests itself in one of two ways: people either discredit the source of argument through ad-hominem-style attacks or they deploy biased and self-serving reasoning mechanisms to construe the content of the argument in their favor. 3) Finally, the mechanisms used to win can be reoriented and leveraged to gain new insights from alternative perspectives through argumentation. When this final phase does occur, it allows the openness of Phase 1 to combine with the rigor of Phase 2 to create a powerful approach to learning.

2. Phase 1: Learning

2.1. Early Prosociality

The openness with which children learn from others may arise from the early emerging bias to cooperate. Converging evidence from a variety of fields of study suggests that human prosociality is intuitive (Rand, Greene, Nowak, 2012; Zaki, & Mitchell, 2013). While traditional models commonly assumed human cooperation requires deliberation and the control of selfish impulses, recent work suggests that in fact, cooperation is intuitive because it often creates benefits in social interaction. This theory is supported by the finding that 6-month-old infants exhibit prosociality, preferring helping agents to those who hinder another's action (Hamlin, Wynn, & Bloom, 2007). By 18-months, children engage in spontaneous helping behavior such as opening a cabinet or offering useful information without any expectation for a reward or repayment (Warneken & Tomasello, 2009). Only with increasing executive control can these intuitive responses consistently be overridden (Garon, Bryson, & Smith, 2008). These most basic forms of prosocial preferences and action help explain why children so willingly learn from others.

2.2. Trust in Testimony

Relying on the testimony of others is central to cognitive development (Wellman & Gelman, 1992). Children face vast knowledge asymmetries in relation to other human agents and need to rapidly absorb information from those who know more. Large swaths of human understanding require knowledge that the learner has never directly observed. Knowledge of this sort is often transmitted via testimony. Children are often willing to believe in the existence of unobserved causal agents, like those posited by science and religion (Harris, Pasquini, Duke, Asscher, & Pons, 2006). They will even override immediate sensory information in favor of another's testimony. Children with an understanding that appearance is not always reality accept an informant's incorrect labels for ordinary objects (i.e. calling a lemon a rattle; Lane, Harris, Gelman, & Wellman, 2014). In general, children display openness to acquiring information from the testimony of those around them. While part of this acceptance may reflect less cognitive capability to carefully scrutinize information, children are still able to utilize several heuristics as filters.

Children do not blindly accept any information from anyone; instead, early developing tendencies allow infants to identify trustworthy sources. These strategies may lay the foundation for the later development of the argue-to-win mentality. Eight-month-olds use the observed reliability of agents to inform future behavior (Tummeltshammer, Wu, Sobel, & Kirkham, 2014). Young children continue to track the performance of informants to select accurate sources (Koenig, & Harris, 2005; Koenig & Woodward, 2010). In these cases, children are using characteristics of the source to judge the trustworthiness of the message. Similar inferences characterize the argue-to-win mindset. Dismissing information based solely on the source may be a beneficial as children learn about the world, but may become less productive when externalized in the form of *ad hominem* attacks during argumentation.

In addition to evaluating sources, children also show early signs of scrutinizing the information they receive. Children who have difficulty with the appearance vs. reality distinction are surprised when adults make a claim that contradicts the child's perception (Pea, 1982) and will reject unexpected information (Harris, Pasquini, Duke, Asscher, & Pons, 2006; Robinson, Mitchell, & Nye, 1995). In these cases, children show a rudimentary ability to analyze the reliability of information based on the content, not merely the source. This capacity is far from adult-like competitive argumentation, but may be an important step in the development of the argue-to-win mindset. Despite children's selectivity, the formidable gaps in their knowledge require children to largely adopt a cooperative stance in knowledge acquisition.

2.3. Question Asking

It is no surprise that very young children ask questions. It is however, impressive how purpose driven their questioning can be. Even young preschoolers often ask questions with a clear goal of gaining explanatory insight. Thus children expect, and often receive, explanatory responses from adults. They can persist in asking questions and further adjusting them until they get explanations in response to their questions instead of mere facts or platitudes. (Callanan & Oakes, 1992; Chouinard, Harris, & Maratsos, 2007; Frazier, Gelman & Wellman, 2009). They are clearly capable of evaluating responses to their questions in terms of the explanatory insights they provide. They show pleasure and satisfaction with explanatory answers and often build on them; but they show impatience and dissatisfaction with scripted responses that offer no insight and often repeat the same why question or simply disagree with the non-explanatory answer (Frazier et. al., 2009). This is not the same as argumentation, but it does reveal an early ability to seek out information in a transactional manner and to evaluate the quality of the information received in real time and adjust one's queries accordingly.

2.4. Arguing to Learn: Acquiring Argumentation Skills

Children are cooperative and open to learning, but are they competent arguers? Children's argumentative skills may first develop out of interaction with family members (Stein & Albro, 2001). Children often engage in arguments and model the skill of offering justification for their positions after their parents (Dunn & Munn, 1987). The safety of the home environment offers children ample opportunity to practice producing and evaluating arguments as they resolve conflicts. Early in life, children display quite sophisticated skills in their arguments, demonstrating argumentative capabilities even before they enter the school setting. In disagreements between children at a nursery, 3 and 4-year-olds rarely end disputes in a simple refusal, but rather offer justifications for their position and ideally come to mutual understanding (Eisenberg & Garvey, 1981). Observations of children's interactions show that children use oral arguments to negotiate with others (Stein, Bernas, & Calicchia, 1997). These conflicts will focus on topics such as manners, possessions, and independence. Children will offer justification for their claims by appealing to sources such as their feelings, social rules and material consequences (Dunn & Munn, 1987). In one observational study, 18-month-olds would show signs of distress and anger when disputes arose over basic rights. By 24 months, children would respond similarly for conflicts regarding conventions. At 36 months however, children shifted strategies and began to offer justifications for their positions in these situations rather than simply expressing their discontent (*ibid*). These were the same topics in which mothers were most likely to offer justification as well, suggesting children readily pick up the skills of argumentation through observing their parents. Children show sensitivity to fallacious arguments: in appropriately simplified formats, 3-year-olds prefer non-circular arguments to circular arguments (Corriveau & Kurkul, 2014; Mercier, Bernard, & Clément, 2014). By 4 years of age, children are able to produce all of the essential components of an argument (Eisenberg, 1992) and show competence in understanding verbal structures like *modus tollens*, often used in formal arguments (Scholnick & Wing, 1991). Furthermore, children justify their assertions in arguments and understand that all views can be supported or challenged (Perlman & Ross, 2005).

Early in development, children's goals will conflict with others. For young children, the ultimate goal of deploying these sorts of arguments may be to act willfully on a set of beliefs (Dunn and Munn, 1987). In these situations, children will try to achieve their own goals and win the exchange. Children also offer justifications after breaking a norm or following a disputed or surprising assertion (Orsolini, 1993). Arguments like these can help children avoid embarrassment by providing justification for their actions. Children may display winning-oriented behavior, but in these cases they are not processing evidence in a truth-oriented argument; instead they are justifying impulses. We classify this behavior as a dispute rather than a genuine argument because children are not abstracting away from a particular topic and arguing for a specific view of the truth, but rather excusing their own behavior or attempting to overcome a personal obstacle. Though even in the case of disputes, children become less conciliatory over time (Tesla & Dunn, 1992).

2.5. *Arguing to Learn: In Practice*

The use of arguments in educational settings demonstrates the effectiveness of argumentation as a learning tool. Argumentation in the classroom can enable students to shift perspectives and change their initial thoughts through teacher-led classroom discussions (Leitão, 2000). They often utilize arguments to improve reasoning through group collaborations. When children are paired with someone who has a different perspective on a problem, they then improve when tested individually (Perret-Clermont, 1980). In general, these sorts of group interactions demonstrate the "two wrongs make a right" effect (Ames & Murray, 1982): the performance by group members often exceeds that of the group's top individual performers. Arguments amongst members of a group facilitate this improvement in performance (Schwarz, Neuman, & Biezuner, 2000).

When children in educational environments engage in collaboration they outperform children in the control groups. These benefits have been found across multiple disciplines

(Webb & Palinscar, 1996). A recent meta-analysis shows the benefits of interactive learning are superior to other methods like constructive activities or passive learning (Chi, 2009). Student groups do not produce higher quality scientific arguments than individuals, but those who generate arguments in a group setting show greater mastery over the material (Sampson & Clark, 2009). By high school, students are able to explicitly identify a wide variety of fallacies in arguments (Weinstock, Neuman, & Glassner, 2006; Weinstock, Neuman, & Tabak, 2004), suggesting that while young children are capable arguers, the ability to do so continues to develop throughout the school-age years.

The improvements observed in the group settings are due to genuine conceptual improvements, not simply increased motivation or being rewarded for group interactions. The encouragement students receive from family members and teachers does not increase academic performance; in fact, external rewards can sometimes hurt performance (Deci, Koestner, & Ryan, 2001). These considerations suggest group reasoning itself, and perhaps argumentation in particular, may be responsible for these positive effects of collaboration.

3. Phase 2: Winning

As the cognitive mechanisms required for adult-like reasoning coalesce, people's goals in arguments may often shift to winning. Working memory may play a key role in the shift to an argue-to-win mindset. This mode of argument requires a certain level of metacognitive awareness: instead of being embedded in a dispute, like young children, the arguer must have the ability to abstract out and view language as an object to which reasoning can be applied— connecting new arguments to the previous components of the exchange. Even into the high school years, students often fail to offer arguments relevant to the other elements of the argument (Jiménez-Aleixandre et al., 2000). Abstracting in this way while simultaneously using reasoning to produce and evaluate arguments is cognitively taxing, requiring a certain amount of working memory to function. It is part of a much larger pattern of increasing metacognitive awareness during the school years (Whitebread, Bingham, Grau, Pino Pasternak, & Sangster, 2007).

It may be that the biases found in human reasoning arise not for the function of achieving perfect rationality, but for winning arguments (Mercier & Sperber, 2011). Communication plays a vital role in human life and reasoning, construed as argumentative, would be adaptive in that it makes an individual less susceptible to misinformation. It is a way of ensuring epistemic vigilance (Sperber, Clément, Heintz, Mascaro, Mercier, Origgi, & Wilson, 2010). This approach helps explain why the argue-to-win mindset would only emerge once reasoning has more fully developed and can be applied to truth statements, as opposed to mere disputes.

Once the mechanisms of reasoning have matured, even if reasoning is for arguing, it does not entail that people's initial arguments will be particularly convincing. In fact, people often do not spontaneously produce sophisticated arguments (Kuhn, 1991). This deficit could be because arguers need the feedback of a live interaction to force them to find better arguments. Indeed, in argumentative contexts, people are found to be quite skilled arguers (for review see: Mercier, 2011).

We posit that in a competitive setting, such as a live interaction between two people who disagree about a topic important to them, people easily enter the argue-to-win mindset. This mindset can manifest itself in two distinct ways. These two types of arguing-to-win approximate the distinction between the analytical and heuristic routes to persuasion (Eagly & Chaiken, 1993; Petty & Cacioppo, 1984). The first argue-to-win strategy focuses only on the characteristics of the other arguer. The personal, *ad hominem* style attacks used in this mindset have no bearing on the actual content on the argument. The second, and more cognitively demanding argue-to-win strategy, addresses the substance of the argument, but in a biased, unconstructive manner. When engaged in this version of the winning mindset, people produce and evaluate information in a way that supports a previously determined position. Interactions of this sort are typically unproductive and end in a stalemate but do at least sharpen a sense of one's own position. It is a clear indication that one of these winning mindsets is active when people resist attempts at persuasion and become even more confident in their original point of view (Tormala & Petty, 2002). In the following sections, we examine the evidence for these two types of the argue-to-win mindset.

3.1. Discrediting the Source

Just as children adjust their trust in testimony based on information about the source, adults use information about a source as a heuristic for processing information from that source. In many cases, inferring the quality of information from the source may be an effective approach. For example, informants who are seen as trustworthy (Mills & Jellison, 1967) or as experts (Rhine & Severance, 1970) more effectively persuade others. In general, the degree to which a source garners credibility, the more like their message will be received (Petty & Wegener, 1998). In these cases, people use generally reliable cues about the source.

People also rely on less rational cues to determine their trust in a source. For example, people consider members of the in-group as more reliable informant than members of the out-group (Clark & Maass, 1988). When recipients perceive similarities with the source, the information they receive leads to more changes in behavior and attitudes (Feldman, 1984). Physical attractiveness is also used to evaluate sources of information— purchasing intentions increase as a function of attractiveness of the endorser (Kahle & Homer, 1985). These sorts of cues used to interpret information from a source, can be utilized to try and discredit an opponent in an argument.

From a logical perspective, the *ad hominem* argument is a logical fallacy (Walton, 1995; Woods, Irvine, & Walton, 2004). The characteristics of the arguer do not necessarily have any bearing on the actual argument. Despite recognizing clear cases of *ad hominem* as unreasonable in experimental settings, overt *ad hominem* it is still used in argumentative discourse, often going undetected by those involved in the argument (van Eemeren, Garssen, & Meuffels, 2012). In the argue-to-win mindset, *ad hominem* arguments are used as attempts to overpower the opponent and forcibly assert an alternative view. This approach is obviously unfruitful for making progress in any discussion, but it may be a way in which one person can exert dominance over another in front of a relevant social group. This may help explain why people are more likely to enter the argue-to-win mindset when in public (Fisher & Keil, under review).

3.2. Discrediting the Message

The second variety of arguing to win, focuses on the actual information presented in an argument. As arguers produce and evaluate arguments, they must deploy a set of reasoning mechanisms to process that information. The biases that have been found to be prevalent in human reasoning are consistently self-serving — they do not consider evidence in an optimally rational way. In the same vein as the argumentative theory of reasoning (Mercier & Sperber, 2011), we suggest that the function of reasoning can be to win arguments. That is, reasoning may be adaptive in that it helps protect against misinformation by producing and evaluating arguments.

If the function of human reasoning is to achieve pure rationality, one of its most puzzling features would be the confirmation bias: the tendency to find and interpret evidence in a manner that supports previously held beliefs (Nickerson, 1998). This phenomenon is nicely illustrated by the Wason selection task (Wason, 1966). People are presented with 4 two-sided cards: a 4, 7, E, and K are visible. They then must choose which two cards to turn over to determine if the following rule is true: if there is a vowel on one side then there is an even number of the other. People intuitively seek to confirm the rule and choose the cards mentioned in the rule (the 4 and E). The rarely chosen correct answer (the 7 and E card), is not made salient by the rule and thus reasoning does not generate justifications for why those cards would be the correct response (Evans, 1996). Relatedly, people find arguments that support a position they already hold as more convincing than opposing arguments (Lord, Ross, & Lepper, 1979). Accumulating evidence only to support one side will not be conducive to determining the actual truth of the matter. It is, however, an effective strategy if winning is the goal of the interaction. Constantly amassing evidence that one's own points of view are accurate makes those positions much easier to defend if ever challenged in an argument.

People also exhibit motivated reasoning: the tendency to arrive at desired conclusions by generating justifications (Kunda, 1990). Motivated reasoning pervades many aspects of thought, but features especially prominently in moral cognition. The intuitionist model of morality posits that moral justifications emerge after reasoning has been applied to intuitive moral judgments and generated rationalizations for why the intuitions are objectively true (Haidt, 2001). This habit of thought is exposed through a phenomenon called "moral dumbfounding". People consider moral dilemmas in which they unable to generate justifications for their intuition that an action is immoral (e.g. eating one's already dead dog). Even after failing to provide a cogent rationalization for their moral judgment, people refuse to change their initial judgment (Bjorklund, Haidt, & Murphy, 2000). In an argumentative setting, motivated reasoning makes even the most convincing evidence unpersuasive to those in the winning mindset. This strategy would not lead to true conclusions, but would be very helpful in winning an argument. It is difficult to lose an argument if one is unable to be convinced that their own view is mistaken.

Consider the ways in which reasoning leads to suboptimal decision making. People make decisions, not in a rational way, but in way that is easy to justify. For example, people tend to avoid changing a course of action once they already have committed time, energy, and resources (Arkes & Blumer, 1985). However, people avoid the sunk-cost fallacy when reasons are available to them to justify the waste created by switching (Soman & Cheema, 2001). Reasoning may explain the strength of framing effects as well: more effortful processing leads to a stronger effect of framing (Igou & Bless, 2007). Alternative frames may prompt people to generate reasons to justify decisions in those contexts. Making decisions can lead people to be overly active in searching for reasons to justify their choices, resulting in reliance on irrelevant reasons. In some instances, the same reason will be used to both accept and reject an item (Brown & Carpenter, 2000). In an argument, the tendency to make easy-to-justify choices, leads arguers to make arguments that will help them avoid losing, but may not ultimately lead to the truth.

These features suggest that a major function of reasoning is winning arguments. Here, we have reviewed only a small sample of the evidence to support this view (see Mercier & Sperber, 2011 for a comprehensive treatment). The development of reasoning enables the argue-to-win mindset, leading arguers to become biased as they process information. Activated by particular factors, such as social setting (Fisher & Keil, under review) the argue-to-win mindset is a counterproductive approach when exchanging ideas with others.

4. Phase 3: Using Winning to Learn

While the argue-to-win mindset presents a rather grim view of argumentation, the mechanisms supporting it can also be used for much more productive ends. Under certain conditions, the open and cooperative approach seen in early development can reemerge in adults. Several requirements must be met before one can successfully argue to learn as an adult. Cognitively, one must be able to engage in an additional level of abstraction. In the argue-to-win mindset arguments are applied to topics, but in the argue-to-learn mindset arguments are applied to arguments. In the argue-to-win mindset, one might say, "That position is wrong because of this evidence", but in the argue-to-learn mindset, one might say, "The argument from that evidence does not lead to that conclusion." At this level of abstraction, reasoning can scrutinize itself, leading to less biased conclusions. Additionally, to adopt the argue-to-learn mindset, one must be willing to "lose" the argument. If both arguers refuse to adjust their beliefs, no progress can take place. However, if they are willing to update their views as the argument develops, learning can be achieved. In this mode, arguers are like scientists testing hypotheses, willing to revise their theories as they encounter new evidence. The truth may not present itself immediately, but through the iterative testing of ideas, better answers will emerge. When the stakes of the argument are too high, this approach is unlikely to be used. In cases where there is less personal investment in the issue at hand, the argue-to-learn mindset is more likely.

4.1. Group Reasoning

Collaborative group reasoning, especially when there is a demonstrably correct answer, can help counteract individuals' biased reasoning and lead to learning. When groups reason about the Wason selection task, for example, no member holds a strong commitment to a particular solution. Group members are free to question others' intuitions and justifications. This leads to the construction of more sophisticated arguments than those generated by participants in isolation, leading groups to find the solution to the problem 75% of the time, while individuals find it only 9% of the time (Moshman & Geil, 1998). The transcripts of these sorts of group interactions verify that individuals must be convinced that the solution they initially proposed is incorrect before considering alternative solutions from other group members (Trognon, 1993). When group participation is high, more dissent within the group leads to more innovative solutions (De Dreu & West, 2001). The various points of view within a group help challenge assumptions that may have otherwise been taken for granted and could have been preventing progress towards the solution to a problem. Group diversity is more likely to cause emotional conflict, but if steps are taken to avoid such problems, diverse groups make better decisions (Priem, Harrison, & Muir, 1995; Williams, & O'Reilly, 1998). When groups reason together, using arguments to uncover alternative approaches, the correct solution tends to emerge (Laughlin & Ellis, 1986). When arguing in a competitive fashion, marked by a total unwillingness to change positions, the benefits of group reasoning are lost. However, when people are willing to be proven wrong, they end up arriving at better solutions collectively. This sort of collaborative reasoning promotes innovation and progress in society, including in science (Dunbar, 1997).

4.2. Considering the Opposite

During group reasoning, the others in the group counter the biases of individual reasoning. Similar benefits occur when individuals pit their own reasoning biases against themselves. Considering the opposite, a strategy similar to "playing devil's advocate", entails seeing the other side of an issue (Lord, Lepper, & Preston, 1984). In this mode of thinking, the biases of reasoning, such as the confirmation bias, are counteracted. An individual generates the strongest arguments in favor of one position, and then, ideally, also generates the best possible arguments for the opposing position. The individual is then able to fairly consider both perspectives and can then identify the best argument. This approach helps eliminate bias and create a more objective approach to the available evidence. For example, when people are deeply emotionally invested in topics, they tend to overestimate how well they can justify the underlying arguments. But by first considering opposing points of view, they are able counteract the influence of emotional investment and assess themselves more accurately (Fisher & Keil, 2014). Merely considering any other alternative, not just the opposite, can lead to similar debiasing effects (Hirt & Markman, 1995). This strategy is highly effective for individuals reasoning alone, but can also be beneficial when used in group settings (Schwenk & Valacich, 1994). Even if an interlocutor does not adopt the argue-to-learn mindset, one can still reap the benefits of a learning mentality by using the consider-the-opposite strategy.

5. Conclusion

Arguing with others and considering the opposite improves reasoning, but the situations in which these strategies work may be limited. The previous findings most often emerge in cases where no one in the group enters the interaction with a firmly held set of beliefs. In daily discourse, arguments often occur when this is not the case. Further research is needed to determine how to best promote learning-oriented arguments for firmly held topics, but there exist some promising preliminary solutions. Recent evidence suggests that the context in which arguments take place shifts the mindset of the arguers. When arguing in a public setting people tend to adopt a winning mindset, while in private they adopt a learning mindset (Fisher & Keil, under review). Several other factors, such as the stakes involved or priming of argument styles may also influence which goal is adopted in an argument, but the simple physical context of an argument, whether it is in public or private, can be surprisingly influential. Future work is needed to identify the conditions under which people willingly use arguments to learn and how they might resist influences that push them towards argue to win mindsets.

The development of argumentation may closely track the development of epistemological understanding (Kuhn, Cheney, & Weinstock, 2000). Throughout development, the understanding of truth shifts. Initially children have a "realist" view, where reality directly corresponds with statements made about it. This stance towards truth may contribute to children's openness to learning. As they seek to better understand reality, the assertions made by those around them can serve as effective guides to discovery. The argue-to-win mindset is likely to be related to the "absolutist" understanding of truth. In this phase, people have a binary view of truth: an assertion is either completely true or completely false. This view of truth helps set the stage for arguments to be viewed as battlegrounds where there will be a definite winner and a definite loser. In this context, arguers will do whatever it takes to outperform the opposition. Finally, the "evaluativist" conception of truth, where knowledge is uncertain and assertions and arguments must be evaluated, supports the argue-to-learn mindset in adults. In this phase, thinking critically about multiple positions generates the best possible understanding of the truth of the matter. Like arguing to learn, relatively few adults reach an evaluativist epistemic understanding.

Arguments are important throughout the lifespan, yet the role they play in a particular context can vary to a large degree. The development of argumentation follows a trajectory whereby people first engage in collaborative learning, then deploy the mechanisms of reasoning to win, and finally apply reasoning to arguments themselves in order to learn. These three phases of development clarify how arguments are used in everyday discourse and how the goal of an argument can shift dramatically. The developmental approach highlights the role of working memory, meta-representations, and reasoning in shifting from one phase of argumentation to the next. It may also be that primitive versions of all three phases can appear very early. For example, younger children can have greater metacognitive awareness in domains where they have massive expertise, such as child dinosaur experts (Gobbo & Chi, 1986) and it is possible then, when operating in such domains that they may also be better able to not only argue to win but even to argue to learn in the most sophisticated sense. This raises the possibility that, with appropriate interventions and training, it might be possible to foster the growth of even the most productive forms of argumentation even in young children. Arguments have the potential to lead to gridlock, demonstrating the worst of human reasoning, but they also have the potential to lead to powerful new insights.

References

Ames, G. J., & Murray, F. B. (1982). When two wrongs make a right: Promoting cognitive change by social conflict. *Developmental Psychology, 18*, 894–897.

Arkes, H. R., & Blumer, C. (1985). The psychology of sunk cost. *Organizational Behavior and Human Decision Processes, 35*(1), 124–140.

Bjorklund F., Haidt J., & Murphy S. (2000). Moral dumbfounding: When intuition finds no reason. *Lund Psychological Reports, 2*, 1–23.

Bloom, P. (2010). How do morals change? *Nature, 464*(7288), 490.

Brown, C. L., & Carpenter, G. S. (2000). Why is the trivial important? A reasons-based account for the effects of trivial attributes on choice. *Journal of Consumer Research, 26*(4), 372–385.

Callanan, M. A., & Oakes, L. M. (1992). Preschoolers' questions and parents' explanations: Causal thinking in everyday activity. *Cognitive Development, 7*(2), 213–233.

Chi, M. T. H. (2009). Active-Constructive-Interactive: A conceptual framework for differentiating learning activities. *Topics in Cognitive Science, 1*(1), 73–105.

Chouinard, M. M., Harris, P. L., & Maratsos, M. P. (2007). Children's questions: A mechanism for cognitive development. *Monographs of the Society for Research in Child Development*, 1–129.

Clark, R. A., & Delia, J. G. (1976). The development of functional persuasive skills in childhood and early adolescence. *Child Development, 47*, 1008–1014

Clark, R. D., & Maass, A. (1988). The role of social categorization and perceived source credibility in minority influence. *European Journal of Social Psychology, 18*(5), 381-394.

Corriveau, K. H., & Kurkul, K. E. (2014). "Why does rain fall?": Children prefer to learn from an informant who uses noncircular explanations. *Child Development, 85*(5), 1827–1835.

De Dreu, C. K. W., & West, M. A. (2001). Minority dissent and team innovation: The importance of participation in decision making. *Journal of Applied Psychology, 86*, 1191–1201.

Deci, E. L., Koestner, R., & Ryan, R. M. (2001). Extrinsic rewards and intrinsic motivation in education: Reconsidered once again. *Review of Educational Research, 71*(1), 1–27.

Dunbar, K. (1997). How scientists think: On-line creativity and conceptual change in science. In T. B. Ward, S. M. Smith, & J. Vaid (Eds.), Creative thought: An investigation of conceptual structures and processes (pp. 461–493). Washington, DC: American Psychological Association.

Eagly, A. H., & Chaiken, S. (1993). The psychology of attitudes. Belmont: Thompson/Wadsworth.

Eisenberg, A. R., & Garvey, C. (1981). Children's use of verbal strategies in resolving conflicts. *Discourse Processes, 4*, 149–170.

Eisenberg, A. R. (1992). Conflicts between mothers and their young children. *Merrill-Palmer Quarterly, 38*, 21–43.

Evans, J. S. B. (1996). Deciding before you think: Relevance and reasoning in the selection task. *British Journal of Psychology, 87*(2), 223–240.

Feldman, R. H. (1984). The influence of communicator characteristics on the nutrition attitudes and behavior of high school students. *Journal of School Health, 54*(4), 149–151.

Fisher, M., & Keil, F.C., (under review) Arguing to win or to learn: Situational constraints prompt contrasting mindsets. *Journal of Experimental Psychology: General.*

Fisher, M., Knobe, J., Strickland, B., & Keil, F.C., (under review) The influence of social interaction on intuitions of objectivity and subjectivity. *Psychonomic Bulletin and Review.*

Fisher, M., & Keil, F. C. (2014). The illusion of argument justification. *Journal of Experimental Psychology: General, 143*(1), 425–433.

Felton, M., & Kuhn, D. (2001). The development of argumentive discourse skill. *Discourse Processes, 32*(2–3), 135–153.

Frazier, B. N., Gelman, S. A., & Wellman, H. M. (2009). Preschoolers' search for explanatory information within adult–child conversation. *Child Development, 80*(6), 1592–1611.

Garcia-Mila, M., & Andersen, C. (2007). Cognitive foundations of learning argumentation. *Contemporary Trends and Issues in Science Education, 35*(1), 29–45.

Garon N., Bryson S. E., & Smith I. M. (2008). Executive function in preschoolers: A review using an integrative framework. *Psychological Bulletin, 134*, 31–60.

Gobbo, C., & Chi, M. (1986). How knowledge is structured and used by expert and novice children. *Cognitive Development, 1*(3), 221–237.

Haidt, J. (2001). The emotional dog and its rational tail: A social intuitionist approach to moral judgment. *Psychological Review, 108*(4), 814–834.

Haidt, J., & Bjorklund, F. (2007). Social intuitionists reason, in conversation. In W. Sinnott-Armstrong (Ed.), Moral Psychology, vol. 2: The cognitive science of morality: Intuition and diversity (pp. 241–254). Cambridge, MA: MIT Press.

Hamlin J. K., Wynn K., & Bloom P. (2007). Social evaluation by preverbal infants. *Nature, 450*, 557–559.

Harris, P. L., Pasquini, E. S., Duke, S., Asscher, J. J., & Pons, F. (2006). Germs and angels: The role of testimony in young children's ontology. *Developmental Science, 9*(1), 76–96.

Hirt, E. R., & Markman, K. D. (1995). Multiple explanation: A consider-an-alternative strategy for debiasing judgments. *Journal of Personality and Social Psychology, 69*(6), 1069–1083.

Igou, E. R., & Bless, H. (2007). On undesirable consequences of thinking: Framing effects as a function of substantive processing. *Journal of Behavioral Decision Making, 20*(2), 125–142.

Jiménez-Aleixandre, M. P., Bugallo Rodríguez, A., & Duschl, R. (2000). "Doing the lesson" or "doing science": Argument in high school genetics. *Science Education, 84*, 757–792.

Kuhn, D., Cheney, R., & Weinstock, M. (2000). The development of epistemological understanding. *Cognitive development, 15*(3), 309–328.

Kahle, L. R., & Homer, P. M. (1985). Physical attractiveness of the celebrity endorser: A social adaptation perspective. *Journal of Consumer Research, 11*(4), 954–961.

Koenig, M. A., & Harris, P. L. (2005). Preschoolers mistrust ignorant and inaccurate speakers. *Child Development, 76*, 1261–1277.

Koenig, M. A., & Woodward, A. L. (2010). Sensitivity of 24-month-olds to the prior inaccuracy of the source: Possible mechanisms. *Developmental Psychology, 46*, 815–826.

Kuhn, D. (1991). *The Skills of Arguments*. Cambridge: Cambridge University Press.

Kunda, Z. (1990). The case for motivated reasoning. *Psychological Bulletin, 108*(3), 480–498.

Lane, J. D., Harris, P. L., Gelman, S. A., & Wellman, H. M. (2014). More than meets the eye: Young children's trust in claims that defy their perceptions. *Developmental Psychology, 50*(3), 865–871.

Latour, B., & Woolgar, S. (1986). *Laboratory life: The construction of scientific facts*. Princeton, NJ: Princeton University Press.

Laughlin, P. R., & Ellis, A. L. (1986). Demonstrability and social combination processes on mathematical intellective tasks. *Journal of Experimental Social Psychology, 22*(3), 177–189.

Leitão, S. (2000). The potential of argument in knowledge building. *Human Development, 43*(6), 332–360.

Lord, C. G., Lepper, M. R., & Preston, E. (1984). Considering the opposite: A corrective strategy for social judgment. *Journal of Personality and Social Psychology, 47*(6), 1231–1343.

Lord, C. G., Ross, L., & Lepper, M. R. (1979). Biased assimilation and attitude polarization: The effects of prior theories on subsequently considered evidence. *Journal of Personality and Social Psychology, 37*(11), 2098–2109.

Mercier, H. (2011). Reasoning serves argumentation in children. *Cognitive Development, 26*(3), 177–191.

Mercier, H., & Sperber, D. (2011). Why do humans reason? Arguments for an argumentative theory. *Behavioral and Brain Sciences, 34*(02), 57–74.

Mercier, H., Bernard, S., & Clément, F. (2014). Early sensitivity to arguments: How preschoolers weight circular arguments. *Journal of Experimental Child Psychology, 125*, 102–109.

Mills, J., & Jellison, J. M. (1967). Effect on opinion change of how desirable the communication is to the audience the communicator addressed. *Journal of Personality and Social Psychology, 6*(1), 98–101.

Moshman, D., & Geil, M. (1998). Collaborative reasoning: Evidence for collective rationality. *Thinking & Reasoning, 4*(3), 231–248.

Nickerson, R. S. (1991). Modes and models of informal reasoning: A commentary. In J. F. Voss, D. N. Perkins, & J. W. Segal (Eds.), Informal reasoning and education (pp. 291–309). Hillsdale, NJ: Erlbaum

Nickerson, R. S. (1998). Confirmation bias: A ubiquitous phenomenon in many guises. *Review of General Psychology, 2*(2), 175–220.

Orsolini, M. (1993). Dwarfs do not shoot": An analysis of children's justifications.

Cognition and Instruction, 11(3&4), 281–297.
Osborne, J. (2010). Arguing to learn in science: The role of collaborative, critical discourse. *Science, 328*(5977), 463–466.
Pea, R. D. (1982). Origins of verbal logic: Spontaneous denials by two- and three-year-olds. *Journal of Child Language, 9*, 597–626.
Perret-Clermont, A.-N. (1980). *Social interaction and cognitive development in children.* London: Academic Press.
Petty, R. E., & Cacioppo, J. T. (1984). Source factors and the elaboration likelihood model of persuasion. *Advances in Consumer Research, 11*, 668–672.
Petty, R. E., & Wegener, D. T. (1998). Matching versus mismatching attitude functions: Implications for scrutiny of persuasive messages. *Personality and Social Psychology Bulletin, 24*(3), 227–240.
Priem, R. L., Harrison, D. A., & Muir, N. K. (1995). Structured conflict and consensus outcomes in group decision making. *Journal of Management, 21*(4), 691–710.
Rand D. G., Greene J. D., & Nowak M. A. (2012). Spontaneous giving and calculated greed. *Nature, 489*, 427–430.
Rhine, R. J., & Severance, L. J. (1970). Ego-involvement, discrepancy, source credibility, and attitude change. *Journal of Personality and Social Psychology, 16*(2), 175–190.
Robinson, E. J., Mitchell, P., & Nye, R. (1995). Young children's treating of utterances as unreliable sources of knowledge. *Journal of Child Language, 22*, 663–685.
Sampson, V., & Clark, D. (2009). The impact of collaboration on the outcomes of scientific argumentation. *Science Education, 93*(3), 448–484.
Scholnick, E. & Wing, C. (1991). Speaking deductively: Preschoolers' use of if in conversation and in conditional inference. *Developmental Psychology, 27*(2), 249–258.
Schwarz, B. B., Neuman, Y., & Biezuner, S. (2000). Two wrongs may make a right ... If they argue together!. *Cognition and Instruction, 18*(4), 461–494.
Schwenk, C., & Valacich, J. S. (1994). Effects of devil's advocacy and dialectical inquiry on individuals versus groups. *Organizational Behavior and Human Decision Processes, 59*(2), 210–222.
Shaw A., & Olson K. R. (2012). Children discard a resource to avoid inequity. *Journal of Experimental Psychology: General, 141*, 382–395.
Soman, D., & Cheema, A. (2001). The effect of windfall gains on the sunk-cost effect. *Marketing Letters, 12*(1), 51–62.
Sperber, D., Clément, F., Heintz, C., Mascaro, O., Mercier, H., Origgi, G., & Wilson, D. (2010). Epistemic vigilance. *Mind & Language, 25*(4), 359–393.
Stein, N. & Albro, E. (2001). The origins and nature of arguments: Studies in conflict understanding, emotion, and negotiation. *Discourse Processes. 32*(2&3), 113–133.
Stein, N. L., Bernas, R. S., & Calicchia, D. J. (1997). Conflict talk: Understanding and resolving arguments. In T. Givon (Ed.), Conversation: Cognitive, communicative and social perspectives: Typo-logical studies in language (Vol. 34, pp. 233–267). Amsterdam: John Benjamins.
Tesla, C., & Dunn, J. (1992). Getting along or getting your own way: The development of young children's use of argument in conflicts with mother and sibling. *Social Development, 1*(2), 107–121.
Tormala, Z. L., & Petty, R. E. (2002). What doesn't kill me makes me stronger: The effects of resisting persuasion on attitude certainty. *Journal of Personality and Social Psychology, 83*, 1298–1313.
Toulmin, S. (1958). *The uses of argument.* Cambridge: Cambridge University Press.
Trognon, A. (1993) How does the process of interaction work when two interlo- cutors try to resolve a logical problem? *Cognition and Instruction, 11*(3–4), 325–345.

Tummeltshammer, K. S., Wu, R., Sobel, D. M., & Kirkham, N. Z. (2014). Infants track the reliability of potential informants. *Psychological Science, 25*(9), 1730–1738.

van Eemeren, F. H., Garssen, B., & Meuffels, B. (2012). The disguised abusive ad hominem empirically investigated: Strategic manoeuvring with direct personal attacks. *Thinking & Reasoning, 18*(3), 344–364.

Voss, J. & Dyke, J. (2001). Argumentation in psychology: Background comments. *Discourse Processes, 32* (2&3), 89–111.

Walton, D. (1995). A Pragmatic Theory of Fallacy. Tuscaloosa, AL: University of Alabama Press.

Warneken F., & Tomasello M. (2009). Varieties of altruism in children and chimpanzees. *Trends in Cognitive Sciences*, 13, 397–402.

Warren, J., Kuhn, D., & Weinstock, M. (2010). How do jurors argue with one another? *Judgment and Decision Making, 5*(1), 64–71.

Wason, P.C. (1966). Reasoning. In Foss, B. (Ed.), New Horizons in Psychology. (pp. 106–137). Hardmondsworth: Penguin.

Webb, N. M., & Palinscar, A. S. (1996). Group processes in the classroom. In D. C. Berliner, & R. C. Calfee (Eds.), Handbook of educational psychology (pp. 841–873). New York: Prentice Hall.

Weinstock, M. P., Neuman, Y., & Glassner, A. (2006). Identification of informal reasoning fallacies as a function of epistemological level, grade level, and cognitive ability. *Journal of Educational Psychology, 98*(2), 327–341.

Weinstock, M., Neuman, Y., & Tabak, I. (2004). Missing the point or missing the norms? Epistemological norms as predictors of students' ability to identify fallacious arguments. *Contemporary Educational Psychology, 29*(1), 77–94.

Wellman, H. M., & Gelman, S. A. (1992). Cognitive development: Foundational theories of core domains. *Annual Review of Psychology, 43*(1), 337–375.

Whitebread, D., Bingham, S., Grau, V., Pino Pasternak, D., & Sangster, C. (2007). Development of metacognition and self-regulated learning in young children: Role of collaborative and peer-assisted learning. *Journal of Cognitive Education and Psychology, 6*(3), 433–455.

Williams, K. Y., & O'Reilly, C. A. III. (1998). Demography and diversity in organizations: A review of 40 years of research. In B. M. Staw & L. L.Cummings (Eds.), Research in organizational behavior (Vol. 20, pp.77–140). Greenwich, CT: JAI Press.

Woods, J., Irvine, A., & Walton, D. (2004). Critical Thinking: Logic & The Fallacies. Toronto: Prentice Hall.

Zaki, J., & Mitchell, J. P. (2013). Intuitive prosociality. *Current Directions in Psychological Science, 22*(6), 466–470.

Chapter 21

Arguing Your Way Out of Confusion

Blair Lehman[1] & Art Graesser[2]

[1] Educational Testing Service, Princeton, USA, blehman@ets.org; [2] University of Memphis, Memphis, USA, graesser@memphis.edu

Abstract. Confusion is an inevitable experience during the learning process. However, it is not the case that confusion is always detrimental to learning; in fact, research has shown that confusion is related to increased learning. This benefit stems from the effortful cognitive activities (e.g., problem solving, deliberation, reflection) that occur during confusion resolution, as opposed to the mere occurrence of confusion. There is still, however, a paucity of research as to what types of activities or interventions will facilitate confusion resolution and motivate students to engage in this challenging and effortful process. One activity that shows a great deal of promise is the construction of an argument. Argument construction provides students with the opportunity to compare and contrast opposing views as well as consider why one perspective is correct. The present chapter discusses a learning environment that experimentally induced confusion via the presentation of contradictory information and then utilized argument construction as a confusion resolution intervention. Contradictory information was presented by two animated pedagogical agents during the discussion of the scientific merits of a research case study. Students were then asked to give their opinion of the case study and construct an argument to convince the agent with which they disagreed that their opinion was correct. Students constructed arguments either with or without the presence of additional information about the research methods concept being discussed. The quality and structure of student arguments were assessed in three ways: (1) presence of claims and evidence, (2) semantic match score to ideal arguments, and (3) linguistic features such as cohesion and syntactic complexity. The impact of argument construction on both confusion resolution and learning are discussed.

1. Introduction

Confusion is inevitable during the learning process. As new information is encountered, learners must determine how to integrate it with previously held mental models. Research over the last decade has shown that experiences of confusion during learning are related to increased learning, particularly at deeper levels (Craig, Graesser, Sullins, & Gholson, 2004; D'Mello, Lehman, Pekrun, & Graesser, 2014; Graesser & D'Mello, 2012; Graesser et al., 2008; Lehman, D'Mello, & Graesser, 2012; Lehman et al., 2013). The assumption, according to cognitive conflict (see Limón, 2001, for review), cognitive disequilibrium (Festinger, 1957; Graesser, Lu, Olde, Cooper-Pye, & Whitten, 2005; Piaget, 1952), and impasse-driven theories of learning (Brown & VanLehn, 1980; VanLehn, Siler, Murray, Yamauchi, & Baggett, 2003), is that it is not the mere occurrence of confusion that promotes learning. These theories posit that it is the effortful process of resolving confusion that leads to learning.

There has been evidence from recent research that non-persistent, but not necessarily resolved, confusion is related to increased learning compared to persistent, unresolvable confusion (D'Mello & Graesser, 2011, 2012, 2014; D'Mello, Taylor, & Graesser, 2007; Lee, Rodrigo, Baker, Sugay, & Coronel, 2011; Lehman & Graesser, 2015; Liu, Pataranutaporn, Ocumpaugh, & Baker, 2013; Rodrigo, Baker, & Nabos, 2010). However, there is a paucity of research investigating methods to facilitate confusion resolution. Given that confusion often stems from a conflict between competing explanations (Brown & VanLehn, 1980; Caroll & Kay, 1988; D'Mello *et al.*, 2014; VanLehn *et al.*, 2003), argument generation may be a viable method to facilitate confusion resolution. Under these circumstances, generating an argument forces students not only to pick one explanation, but also to explain why that explanation is correct and the other explanation is incorrect while providing evidence to support those claims (Scheuer, Loll, Pinkwart, & McLaren, 2010). There are, of course, other models of argumentation. However, in the context of confusion resolution, the model in which students must select one option and defend that selection seems to be the most appropriate. This process of examining both explanations and reviewing the support for each can provide students the opportunity to resolve their confusion and the opportunity to reach a better understanding.

The remainder of this chapter is organized as follows. First, we discuss confusion and its relationship to learning. Second, the viability of argument generation as a method to facilitate confusion resolution is discussed. Third, the findings from a study that experimentally induces confusion via contradictory information and utilized argument generation as a confusion resolution method are discussed. Finally, the implications of integrating argumentation into learning environments are discussed.

1.1. Confusion and Learning

Confusion has been defined as an emotion (Rozin & Cohen, 2003), a knowledge emotion (Silvia, 2010), an epistemic emotion (Pekrun & Stephens, 2012), an affective state that is not an emotion (Keltner & Shiota, 2003), and a cognitive state (Clore & Huntsinger, 2007). We are defining confusion as an epistemic or knowledge emotion (Pekrun & Stephens, 2012; Silvia, 2010). With this definition taken into consideration, confusion is triggered by contradictions, anomalies, system breakdowns, errors, impasses, or when there is uncertainty about how to proceed (Brown & VanLehn, 1980; Caroll & Kay, 1988; VanLehn *et al.*, 2003). These events trigger confusion because they present competing explanations or, at the least, indicate to students that there is a problem with their current understanding (Piaget, 1952). However, it is important to note that these events do not always trigger confusion. Chinn and Brewer (1993) identified seven frequent responses to anomalous data. The seven responses are to (1) ignore new information, (2) reject new information, (3) view new information as irrelevant to the current mental model, (4) hold new information in abeyance, (5) reinterpret new information to fit with the current mental model, (6) minimally adjust the current mental model to integrate new information, and (7) adjust or replace the current mental model to integrate new information. Unfortunately, five out of these seven responses involve ignoring or manipulating evidence such that confusion resolution would not be needed. If confusion is not addressed and students do not engage in the process of confusion resolution, it is likely that understanding would still remain at a shallow level or could be potentially erroneous. This highlights the importance of creating a clear contrast between competing explanations such that only one explanation can be correct.

This theoretical benefit of confusion has been supported by empirical evidence. Some of the first evidence showing that confusion can be beneficial came from an initial investigation into student emotions during interactions with AutoTutor (Graesser *et al.*, 2004), a mixed-initiative natural language intelligent tutoring system (ITS, Craig *et al.*, 2004). Craig *et al.* found that not only was confusion a prominently occurring emotion, but it was also positively correlated with learning. This can be contrasted with another frequently occurring, negatively-valenced emotion, boredom, which was found to be negatively correlated with learning. The prevalence of confusion has been found in other ITSs such as Crystal Island (McQuiggan, Robison, & Lester, 2010), The Incredible Machine (Baker, D'Mello, Rodrigo, & Graesser, 2010), and Aplusix (Baker *et al.*, 2010). There has also been more recent evidence of a causal link between confusion and learning (D'Mello *et al.*, 2014; Lehman *et al.*, 2012; Lehman *et al.*, 2013).

The previously discussed research has focused on individual occurrences (i.e., at a critical juncture, students either were or were not confused) or the overall frequency of confusion during learning (i.e., confusion occurred for 20% of all emotion occurrences). However, it is also important to consider the temporal aspects of confusion during learning. D'Mello and colleagues, for example, have investigated the persistence of confusion (D'Mello *et al.*, 2007) and transitions between confusion and other emotions (D'Mello & Graesser, 2012) during interactions with AutoTutor. This work revealed the presence of vicious and virtuous cycles, which involve transitions between confusion, engagement/flow, and frustration (D'Mello & Graesser, 2012). The vicious cycle consists of transitions from confusion to frustration or persistent confusion (confusion to confusion transition), whereas the virtuous cycle consists of transitions from engagement/flow to confusion and back to engagement/flow. Put simply, the vicious cycle represents persistent, unresolvable (or hopeless) confusion and the virtuous cycle represents confusion that has been successfully resolved. As previously mentioned, cognitive conflict (Limón, 2001), cognitive disequilibrium (Festinger, 1957; Graesser *et al.*, 2005; Piaget, 1952), and impasse-driven theories of learning (Brown & VanLehn, 1980; VanLehn *et al.*, 2003) would predict that learning is most likely to occur when students have (1) attended to their confusion, (2) engaged in effortful cognitive activities such as problem solving, deliberation, and reflection, and (3) successfully resolved their confusion.

Empirical evidence has partially supported the prediction that fully resolved confusion will be more beneficial to learning than persistent, unresolved confusion. In general, these findings have shown that persistent, unresolved confusion is detrimental (Lee *et al.*, 2011; Rodrigo *et al.*, 2010) or not related to learning (Liu *et al.*, 2013). However, there is limited support for fully resolved confusion (Lehman & Graesser, 2015) and more support for non-persistent, but not necessarily resolved, confusion being beneficial for learning (D'Mello & Graesser, 2014; Lee *et al.*, 2011; Liu *et al.*, 2013; Rodrigo *et al.*, 2010). D'Mello and Graesser (2014), for example, found that participants with partially resolved confusion outperformed those with unresolved confusion. Liu *et al.* (2013) and Rodrigo *et al.* (2010) examined multi-emotion occurrences (e.g., confused-not confused-confused, confused-confused-confused) and also found that non-persistent, but not necessarily resolved, confusion was positively correlated with learning. It appears that engaging in confusion resolution activities may be more critical for learning than actually achieving successful resolution.

Although the findings that experiences of non-persistent confusion can be beneficial for learning, it is still important to remember the findings from Chinn and Brewer (1993). Specifically, recall that in many instances stimuli that should induce confusion and trigger effortful cognitive activities are ignored. It is then important to consider ways in which ITSs can facilitate the confusion resolution process. There are three primary ways that an ITS can facilitate the resolution process.

One method to facilitate confusion resolution is to simply draw attention to the state of confusion itself. If students do not acknowledge their confusion, or attempt to manipulate information such that confusion is no longer present (i.e., ignore the fact that two explanations are mutually exclusive and allow for both to explain the same event), then it is impossible for the beneficial cognitive activities, and ultimately learning, to occur. There are currently two ITSs that make use of this technique in conjunction with other techniques. Affective AutoTutor (D'Mello, Lehman, & Graesser, 2011) and UNC-ITSpoke (Forbes-Riley & Litman, 2011) detect confusion and uncertainty, respectively, and draw students' attention to their confusion during tutorial dialogue. By drawing students' attention to their confusion, these ITSs attempt to avoid the stumbling block of students ignoring their confusion.

The second method to facilitate confusion resolution is to motivate students to engage in or persist through this effortful and sometimes challenging process. This is the other technique utilized by Affective AutoTutor (D'Mello et al., 2011). Specifically, Affective AutoTutor provides students with an encouraging motivational statement (e.g., "Just keep going and I'm sure you'll get it.") and attributes the confusion to either the tutor agent or the material itself (e.g., "Some of this material can be confusing."). This attribution shift is particularly important because students may interpret confusion to be an indication of a lack of skills or intelligence and an inability to learn the material (D'Mello et al., 2014). Thus, this attribution shift can provide students with an increase in their confidence and boost their belief that they can resolve their confusion.

The third method to facilitate confusion resolution is to scaffold the confusion resolution process. Whereas the prior two methods involve interventions to initiate the confusion resolution process, this third method involves supporting students through the actual effort needed to resolve their confusion. UNC-ITSpoke (Forbes-Riley & Litman, 2011) utilizes this technique by providing additional scaffolding when uncertainty is detected. This additional scaffolding involves asking probing questions and providing additional information to facilitate resolution. However, this is not the only type of support that can be provided. More research investigating a variety of confusion resolution supports and their effectiveness is needed. We propose argument generation as one potential support to facilitate confusion resolution.

1.2. Argument Generation as a Method to Facilitate Confusion Resolution

The discussion of confusion during learning has highlighted four important factors to facilitate confusion resolution: (1) draw attention to the confusion, (2) create a contrast that is mutually exclusive (i.e., competing propositions cannot both be true in the current context), (3) motivate students to engage in confusion resolution, and (4) scaffold the confusion resolution process. We believe that argument generation can address each of these factors. Next we discuss the nature of argumentation in the present context.

We are focusing on collaborative argumentation, as opposed to debate-style argumentation, due to the greater emphasis on student learning of critical thinking, elaboration, and reasoning (Andriesson, 2006; Bransford, Brown, & Cocking, 1999). In the present context argument generation occurs when two competing perspectives are present and there is confusion about which perspective is correct. Table 1 shows an example conversation between a (human) student and two animated pedagogical agents. The two agents present competing perspectives, that are mutually exclusive, on the quality of the research case study being discussed (turns 5 and 6). After the agents present their opinions, the student is asked to select one of the two perspectives and generate an argument that defends why the perspective they have selected is correct and the other perspective is incorrect (turn 9). This argument generation task serves as a method to facilitate confusion resolution. This type of conversational argument generation is in line with a more collaborative orientation for argumentation.

Turn	Speaker	Dialogue
		A study showed that people who drank 6 cups of non-alcoholic beer self-reported levels of intoxication close to that of people who drank 6 cups of actual beer. A professor ran this study with one of his classes that had 100 students over 21, so he randomly assigned 50 to drink the real beer and 50 to drink the non-alcoholic beer. The professor decided to replicate the study two weeks later. He did the same exact study with the same class, making sure to randomly assign people to conditions. He found the same results again and concluded that people can feel just as drunk from drinking non-alcoholic beer as drinking alcoholic beer. <Research case study>
1	Dr. Williams [tutor agent]	Chris and I have been talking and we both think that there's a problem with this study. <Agreement>
2	Dr. Williams	Mary, do you think that there is there a problem with this study? <Forced-choice question>
3	Mary [human learner]	problem <Response>
4	Dr. Williams	I think we should look at the second study to see if that's the problem. <Advance dialogue>
5	Dr. Williams	This was a bad replication because he used the exact same population of students from the first study and it makes his conclusions weaker. Right Chris? <Perspective A>
6	Chris [peer student agent]	Completely disagree. The researcher has to use the same population to figure out if the first findings were just a fluke, so it was a good replication. <Perspective B>
7	Dr. Williams	Mary, should the researcher have used the same group of students in both studies or a different group? <Forced-choice question>
8	Mary	different group <Response align with Perspective A>
9	Chris	I still disagree. I'm not sure why the researcher should've used the same students. But I disagree with you, Mary. You need to prove to me that you're right. Put together a convincing argument about why he should've used different students. <Construct argument prompt>
10	Mary	I think that the researcher should use a different group of students for the replication study because he wants to know if drinking non-alcoholic beer can make people feel drunk. If he used the same group of students, he would only know that the effect is there for that specific group of students. If he finds the effect with a different group of students, then he can be more confident that the effect occurs generally. <Convincing argument>
11	Dr. Williams	Alright, thanks. <Acknowledge argument>

Table 1. Example collaborative argumentation between (human) student and two animated pedagogical agents

In the context of a conversation, it is unlikely that students would construct a lengthy, essay type of response to defend their position. It is more appropriate in this context for students to take a more informal approach to argumentation (turn 10); however, there are still certain structural components that would be needed for the argument to be effective. This argument would need to include claims, supporting evidence, and evidence against counter claims

(Scheuer et al., 2010). In the example argument in Table 1, the student, Mary, addresses each of these points by stating that she believes that the research should use a different group for the replication study (claim) because this will allow the researcher to understand the generalizability of the group (evidence), whereas using the same group will only allow the researcher to determine if the finding is repeatable with those specific students (evidence against counter claims). This ideal argument would involve weighing the pros and cons of each perspective. It is important to note that this is an ideal argument that would tap into the central route identified by Petty and Cacioppo (2012). It is also possible that students will take the peripheral route and will focus on less substantive reasons for their argument generation. Between these two extremes there are many variations in the quality of generated arguments.

Another task that could potentially facilitate confusion resolution is prompting students to explain why they hold a particular perspective when they are presented with the two mutually exclusive options. However, prior research suggests that explanation generation may not trigger students to fully engage in the cognitive activities that would lead to resolved confusion. Specifically, argumentation involves both constructing and defending explanations, whereas prompts to explain may only involve constructing explanations (Berland & Reiser, 2009). When only constructing explanations, students have been found to attend to claims but not the evidence that supports those claims (Brem & Rips, 2000; Kuhn & Katz, 2009). Defending, on the other hand, inherently involves acknowledging both one's own perspective and the competing perspective (Kuhn, 2010). Next, we address each factor to facilitate confusion resolution with respect to argument generation.

1.2.1. Does Argument Generation Draw Attention to Students' Confusion?

The previously mentioned interventions in Affective AutoTutor (D'Mello et al., 2011) and UNC-ITSpoke (Forbes-Riley & Litman, 2011) explicitly draw attention to confusion. Argument generation would not explicitly draw attention to the state of confusion itself (e.g., "It looks like you are confused" or "This material can be confusing"), but may do so implicitly. Argument generation requires students to consider and find support for why one position is correct and why another position is incorrect. This process may reveal to students gaps in their knowledge or uncertainties about which position is correct. This process has been described as "arguing to learn," which can be contrasted with "learning to argue" (Scheuer et al., 2010). Some researchers have even suggested that science and the process of learning science is argumentation, the "science as argument" conception (Kuhn, 2010). Thus, an inability to create a coherent argument may force students to acknowledge their confusion and hopefully trigger confusion resolution activities.

1.2.2. Does Argument Generation Create a Contrast that is Mutually Exclusive?

This issue is dependent upon how the competing positions are framed in the learning environment. In addition to framing the positions as mutually exclusive, it is also important to frame the scenario such that both positions are deemed as potentially valid. If one position is deemed irrelevant or categorically impossible, then the issue of mutual exclusivity is no longer important. However, if two relevant positions are presented to students in such a way that they must select one position, and by selecting one position students are in opposition to the other position, then this issue is addressed. The conversation in Table 1, for example, poses two opposing opinions about how to conduct a replication study. It is not possible in this example conversation for both opinions to be correct; however, both opinions are presented as valid within the current context. This process of having to select one position and defend it, regardless of the quality of the argument generated, will highlight to students that only one position can be kept.

1.2.3. Does Argument Generation Motivate Students to Engage in Confusion Resolution?

To consider this issue we can compare argument generation to explanation of one's position. While an explanation of one's opinion does involve getting to justify oneself, argument generation also includes an element of challenge that may make the task more engaging for students. The challenge of getting to prove oneself correct may motivate students to explain more fully not only their own position, but also why the other position is incorrect. This desire to win the argument or be correct may give students an increased sense of control in the learning environment, which can increase overall engagement and motivation (Malone & Lepper, 1987).

1.2.4. Does Argument Generation Scaffold the Confusion Resolution Process?

Argument generation scaffolds confusion resolution by promoting the cognitive activities that can lead to successful confusion resolution, such as deliberation and reflection. However, for students low in prior knowledge about the concepts being discussed or argument generation (i.e., provide support for claims, discussing both positions) this may not be enough scaffolding. This situation can be remedied with two additions to the argument generation task. First, low prior knowledge can be addressed by providing students with more information, similar to the technique used in UNC-ITSpoke (Forbes-Riley & Litman, 2011). Second, low argumentation ability can be addressed by scaffolding the argument generation process itself. There are already several ITSs designed for this very purpose (see Scheuer et al., 2010, for a review). Currently these ITSs involve constructing an argument with a diagram; however, this type of instruction could be adapted to a more conversational interaction. We focus on providing students with additional information to further scaffold the confusion resolution process in the present research.

Given this potential for argument generation as a method to facilitate confusion resolution, we have conducted an experiment that investigates its effectiveness. In the experiment, students learn about research methods concepts in a learning environment that involves the discussion of the scientific merits of research case studies (see Table 1 for an example conversation). The discussion involves a trialogue (i.e., three-party conversation) between the student and two animated pedagogical agents (tutor agent and peer student agent). Students attempt to resolve their confusion by arguing their position against one of the agents.

2. Argument Generation as a Confusion Resolution Intervention

We conducted a study in which 147 undergraduate students (105 Female, 42 Male) learned about research methods concepts in a learning environment that experimentally induced confusion and facilitated confusion resolution with argument generation. Participants ranged in age from 56 to 18 years old ($M = 21.6$, $SD = 6.62$). Fifty-seven percent of participants were African-American, 37% were Caucasian, 3% were Asian, 2% were Hispanic, and 1% reported other for ethnicity. Participants completed this study for course credit, although the learning activity was not part of their course material. Ninety percent of participants had not taken a course that discussed research methods concepts.

In the learning environment students engaged in a trialogue with two animated pedagogical agents to discuss the scientific merits of research case studies. The two animated pedagogical agents served the roles of tutor and peer student. Figure 1 shows the learning environment interface. There are six main components of the interface: (A) tutor agent, (B) peer student agent, (C) case study presentation, (D) additional information presentation, (E) trialogue transcript, and (F) text box for students to enter typed responses.

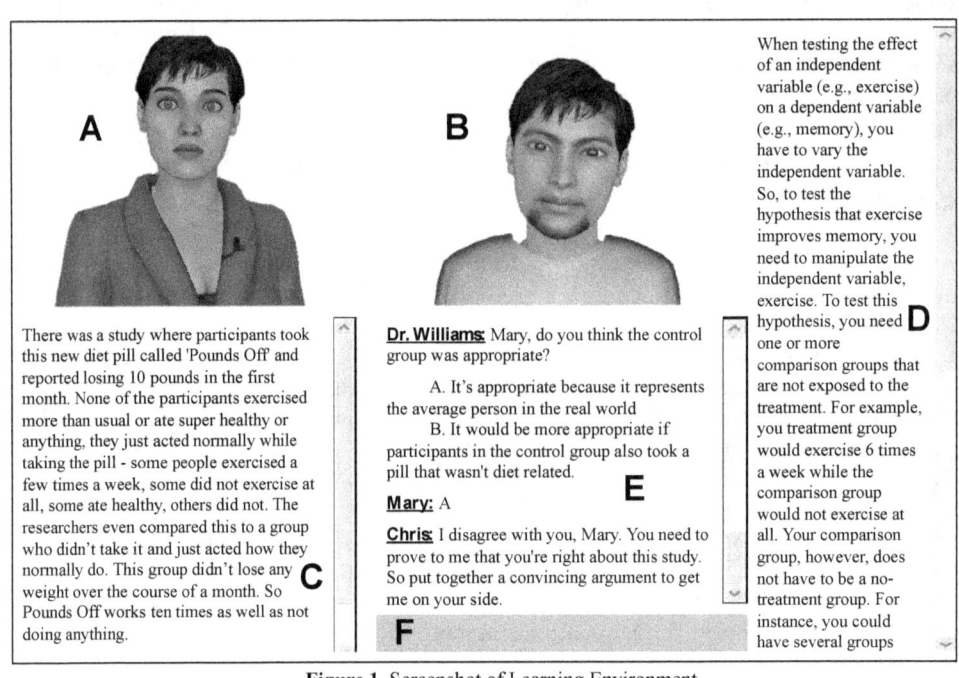

Figure 1. Screenshot of Learning Environment

Each trialogue involved the discussion of one case study. Students first read the case study and then began the discussion. Figure 2 shows the overall study procedure and the order of events in each of the six trialogues that students completed. At the beginning of the discussion the two agents present their shared opinion that the case study does have some flaw, as can be seen in the trialogue excerpt shown in Table 2. Each case study discussed did contain one flaw such as inappropriate control group or experimenter bias. The study being discussed in the trialogue in Table 2 is flawed because an inappropriate control group was used. After the (human) student (Mary) has given an opinion about the potential for a flaw in the case study, the confusion induction manipulation occurs. The agents present contradictory opinions on the nature of the flaw in the current study. In the excerpt in Table 2, for example, the tutor agent (Dr. Williams) presented a correct opinion (turn 5) and the peer student agent (Chris) disagreed and presented an incorrect opinion (turn 6). This was the True-False condition, whereas in the False-True condition the peer student agent presented the correct opinion in the disagreement. Students completed two trialogue interactions in the True-False condition and two trialogue interactions in the False-True condition (4 total). After the agents presented their opinions, the students were asked to give their opinion (turn 9). However, the student's opinion was restricted to agreeing either with Dr. Williams or Chris. This method of confusion induction has been found to induce confusion effectively (D'Mello *et al.*, 2014; Lehman *et al.*, 2013).

Turn	Speaker	Dialogue
		There was a study where participants took this new diet pill called 'Pounds Off' and reported losing 10 pounds in the first month. None of the participants exercised more than usual or ate super healthy or anything, they just acted normally while taking the pill - some people exercised a few times a week, some did not exercise at all, some ate healthy, others did not. The researchers even compared this to a group who didn't take it and just acted how they normally do. This group didn't lose any weight over the course of a month. So Pounds Off works ten times as well as not doing anything. <Research case study>
1	Dr. Williams [tutor agent]	Well Chris and I talked it over and the two of us think that there's something problematic about this study. <Trial 1 opinions>
2	Dr. Williams	Mary, do you think there might be a problem with this study? <Forced-choice question 1>
3	Mary [human learner]	problem <Response 1>
4	Dr. Williams	Let's talk about the control and the experimental groups they used. <Advance dialogue>
5	Dr. Williams	The control and experimental groups were too different. The control group should do something to make the two groups more similar. <Trial 2 opinion>
6	Chris [peer student agent]	I don't agree with you at all, Dr. Williams. The control group was good. There weren't any differences between the two groups besides the diet pill. <Trial 2 opinion>
7	Dr. Williams	Mary, do you think your classmates would be confused? <Confusion prompt>
8	Mary	confused <Confusion judgment>
9	Dr. Williams	Mary, do you think the control group was appropriate? <Forced-choice question 2>
		A. The control group is appropriate because it represents the average person in the real world and what he or she would do if they weren't taking the diet pill (i.e., acting normally). <Incorrect response>
		B. The control group would be more appropriate if it were changed so that participants in the control group also took a pill, a pill that wasn't diet related. <Correct response>
10	Mary	A <Response 2>
11	Chris	I disagree with you, Mary. You need to prove to me that you're right about this study. So put together a convincing argument to get me on your side. <Construct argument prompt>
12	Mary	The control group was appropriate because... <Convincing argument>
		To test this hypothesis, you need one or more comparison groups that are not exposed to the treatment... <Explanatory text>
14	Dr. Williams	Mary, do you think one of your classmates would be confused? <Confusion prompt>
15	Mary	not confused <Confusion judgment>

Table 2. Trialogue Excerpt from the True-False Condition

The confusion resolution intervention began after the induction manipulation. Students are prompted by the agent with which they had disagreed to generate a convincing argument. In Table 2, for example, Mary agreed with Dr. Williams (turn 10). Chris then asks Mary to convince him that she is correct about the study being discussed (turn 11). There were three intervention conditions: (1) Argument Only, (2) Argument then Information, and (3) Argument + Information. In the first condition, Argument Only, students are only prompted to generate an argument, as shown in Table 2. The second condition, Argument then Information, is identical to the first condition except that after students complete the argument they are presented with additional information (component D in Figure 1). The additional information presented was a brief explanatory text that addressed the research methods concept being discussed (e.g., control group), but did not address the actual case study being discussed. In the third condition, Argument + Information, students received the same argument prompt as the previous two conditions, but were provided with the additional information during argument generation. This final condition served as a more scaffolded argument generation task. Students completed four trialogue interactions with the same intervention condition for all interactions.

Next, we discuss how confusion, argument quality, and learning were assessed in the present experiment.

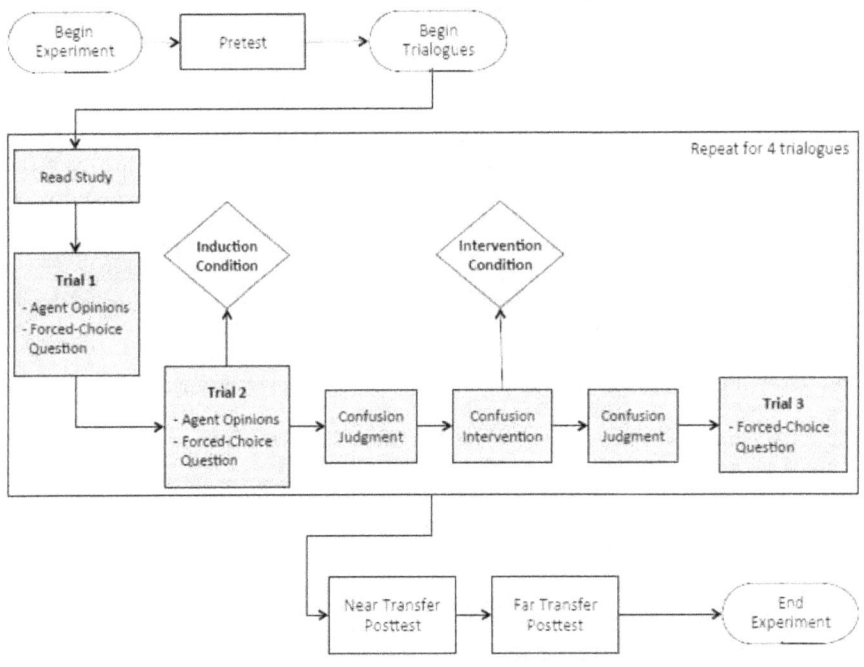

Figure 2. Study Procedure

2.1. Confusion Resolution

Confusion was assessed in each trialogue via two online confusion judgments. The confusion judgments occurred after the confusion induction manipulation (Time 1, turn 8 in Table 2) and after the confusion resolution intervention was completed (Time 2, turn 15 in Table 2). At each time the confusion judgment was phrased, "Do you think a classmate would be confused?" Students were then given the options confused or not confused to select. The confusion judgment was phrased in this manner in order to avoid students' potential bias when reporting confusion (D'Mello et al., 2014). Students may believe that experiences of confusion are indicative of an inability to succeed or a general lack of skills or intelligence, which could create a bias to not report their confusion. Previously this type of confusion judgment was found to be related to students' processing time, such that it appears that students are taking their own confusion into consideration when responding (Lehman et al., 2012).

There were four potential confusion resolution outcomes: (1) none (not confused at Time 1 and Time 2), (2) resolved (confused at Time 1, not confused at Time 2), (3) unresolved (confused at Time 1 and Time 2), and (4) created (not confused at Time 1, confused at Time 2). In the present investigation we focused only on those cases in which students were confused at Time 1. Thus, we focus on the comparison of cases in which students had resolved and unresolved confusion.

2.2. Argument Structure and Quality Measures

There are a variety of methods to assess the quality of an argument (Scheuer et al., 2010). In the present investigation we assessed argument quality and structure with four methods: length, presence of claims and evidence, linguistic complexity measures, and semantic match score to an ideal argument. Each of these measures is discussed below, with the exception of length, which was measured via word count.

2.2.1. Presence of Claims and Evidence

There are multiple methods to code scientific arguments (see Sampson & Clark, 2008); however, the most appropriate method for the present study was the scheme presented by Zohar and Nemet (2002), which incorporates both the content quality and the structure. This coding scheme involved identifying instances of claims and evidence in the argument. The scheme has been adapted to include the rating of claim and evidence quality (accurate, inaccurate), based on the suggestion of Sampson and Clark (2008). Two human raters coded the arguments for the presence of claims, evidence, the amount of evidence, and the quality of claims and evidence. A subset of the corpus was first coded to compute reliability (*kappa* = .842). The corpus was then divided evenly between the raters for coding. This coding scheme was then used for five argument quality dependent measures: claim quality, evidence quality, amount of evidence, overall presence score, and overall quality score. The present investigation only considers the overall presence score, which was derived from the coding scheme with scores of 0 (neither claim nor evidence was present), 1 (claim or evidence was present), and 2 (claim and evidence were present).

2.2.2. Linguistic Complexity Measures

The linguistic complexity measures were computed with an automated text analysis tool called Coh-Metrix (Graesser & McNamara, 2011). Although Coh-Metrix provides over 100 linguistic measures, the present investigation focused on the five easability components that were previously found (Graesser, McNamara, & Kulikowich, 2011) as well as an overall composite score of these five components. Easability refers to the degree of ease with which readers can both encode the text and extract the meaning of the text. The first component is *narrativity*, which refers to the degree to which a text tells a story with characters, events, places, and elements that are familiar to the reader. The second component is *referential cohesion*, which refers to the overlap in words and ideas across sentences within the text. The third component is *syntactic simplicity*, which refers to the simplicity of sentence structure and the number of words in sentences. The fourth component is *word concreteness*, which refers to the degree to which the words evoke mental images. The fifth component is *situation model cohesion*, which refers to the degree of coherence in the text, particularly the presence of causal, intentional, and temporal connectives. Finally, the overall composite score represents the degree of *formality*. For each of these components a higher score represents a text that is easier to understand, with the exception of formality. For formality a lower score represents a text that is easier to understand. However, the relationship between text ease and argument quality has not yet been determined. In other words, it is not clear how scores on each of these measures relate to an argument of higher or lower quality. Thus, the present investigation of linguistic features is exploratory in nature.

2.2.3. Semantic Match Score

Student arguments were compared to prototypical correct responses that were created by a content expert. Prototypical correct responses were unique to each of the four research methods concepts discussed during the trialogues. Student arguments were compared to prototypical correct responses using an inverse word frequency weighted overlap (IWFWO) algorithm. The IWFWO algorithm is a word-matching algorithm in which each overlapped word is weighted on a scale from 0 to 1, relative to its inverse frequency in the English language, using the CELEX corpus (Baayen, Piepenbrock, & Gulikers, 1995). The inverse frequency allows for higher weighting of lower frequency, more contextually relevant words (e.g., replication, bias), while higher frequency words (e.g., and, but) are given a lower weighting. Comparisons resulted in a match score between 0 and 1 (1 = perfect similarity). This match score served as the semantic match score dependent variable in subsequent analyses.

2.3. Learning Outcome Measures

Research methods knowledge was assessed with a flaw-identification task. The flaw-identification task consisted of a description of a previously unseen research study and students were asked to identify flaw(s) in the study by selecting as many items as they wanted from a list of six research methods concepts. The list included four concepts that could potentially be flawed (i.e., discussed in the trialogues) and two distractor concepts (i.e., not discussed in the trialogues). Students also had the option of selecting that there was no flaw in the research study, although each study contained at least one flaw.

The flaw-identification task involved near and far transfer studies. The near transfer studies differed from the studies discussed in the trialogues on surface features only. Each near transfer study contained one flaw. Each concept discussed during the trialogues had one near transfer study, resulting in four near transfer studies. The far transfer studies differed from the studies discussed in the trialogues on both surface and structural features. For example, a surface feature difference could be taking a diet pill (original study) versus an acne pill (near transfer study), whereas structural feature differences could be experimental and do-nothing control groups (original study) versus three or more comparison groups all receiving some type of treatment (far transfer study). Each far transfer study contained two flaws, resulting in two far transfer studies in all. All transfer studies were presented after all four of the trialogues were completed.

3. Findings

To investigate argument generation as a confusion resolution intervention we conducted two phases of analyses. First, we investigated differences in argument structure and quality based on confusion resolution outcome (resolved vs. unresolved) and intervention condition (Argument Only vs. Argument then Information vs. Argument + Information). Second, we investigated differences in learning outcomes based on argument structure/quality, the argument structure/quality × confusion resolution outcome interaction, and the argument structure/quality × intervention condition interaction.

A mixed-effects modeling approach was adopted for all analyses due to the repeated measurements and nested structure of the data (Pinheiro & Bates, 2000). Mixed-effects models include a combination of fixed and random effects and can be used to assess the influence of the fixed effects on dependent variables after accounting for any extraneous random effects. This approach allows each of the four trialogue interactions that students completed to be considered separately in the analyses. There were a total of 424 cases included in the present analyses. The *lme4* package in R (Bates & Maechler, 2010) was used to perform the requisite computations.

Linear or logistic models were constructed on the basis of whether the dependent variable was continuous or binary, respectively. The random effects in all analyses were participant and concept (i.e., the flawed research methods concept in the research case study being discussed). In addition, all models included order as a fixed effect (order of concept presentation) and time on task (for each trialogue) as a fixed effect. Time on task was included as a fixed effect because the confusion intervention conditions were not equivalent in time on task. The time difference was due to the fact that the Argument + Information and Argument then Information conditions involved two tasks (argument generation and reading the text), whereas the Argument Only condition involved only one task (argument generation). The random effects and fixed effects of order and time on task were consistent across all models (control). Confusion resolution outcome (resolved cases = 209, unresolved cases = 215), intervention condition (Argument Only cases = 135, Argument then Information cases = 150, Argument + Information cases = 139), confusion resolution outcome × argument quality, or intervention condition × argument quality were the categorical fixed effects.

As mentioned previously, the analyses of argument structure and quality were exploratory in nature. Due to the exploratory nature of these analyses, the significance value was set to $p < .1$ in order to allow for the opportunity to determine more general patterns that may not be observable at more stringent significance levels.

3.1. Argument Structure and Quality

To investigate the relationship between argument structure/quality and confusion resolution we investigated differences in each argument structure/quality measure based on confusion resolution outcome, intervention condition, and confusion resolution outcome × intervention condition. None of the models for the confusion resolution × intervention condition interaction was significant (p's > .10), thus we will focus on the main effects of confusion resolution outcome and intervention condition.

First, we investigated differences in argument length. Argument length did not significantly differ based on confusion resolution ($p > .10$). However, there was a significant difference based on intervention condition [$\chi^2(2) = 5.36, p = .003$]. The students in both the Argument Only ($M = 36.5, SD = 28.8$) and Argument + Information conditions ($M = 44.2, SD = 37.1$) wrote longer arguments than those in the Argument then Information condition ($M = 33.8$, $SD = 21.0$). These findings suggest that the scaffolding provided in the Argument + Information condition had the benefit of allowing students to generate longer arguments, but this finding alone does not provide any information about the quality of these longer arguments. The next three investigations address this issue along with argument structure.

Second, we investigated differences in the overall presence score. The overall presence score was dummy coded for analyses, such that individual scores were predicted (i.e., 0, 1, or 2). As a reminder, a score of 0 represented an argument that had neither a claim nor evidence, a score of 1 represented an argument that had a claim or evidence, and a score of 2 represented an argument that had both a claim and evidence. Figure 3 shows the proportional occurrence of each presence score based on the confusion resolution outcome and intervention condition. For confusion resolution outcome, the only significant difference was for a score of 2 (claim *and* evidence). This model revealed that when students resolved their confusion they were more likely to include both claims and supporting evidence than when their confusion was unresolved [$\chi^2(1) = 1.96$, $p = .081$]. This finding suggests that the creation of a more complete argument may be related to successful confusion resolution.

For intervention condition, the only significant difference was for a score of 1 (claim *or* evidence). This model revealed that students in the Argument then Information condition were more likely to have a score of 1 than those in the Argument Only and Argument + Information condition [$\chi^2(2) = 4.53$, $p = .052$]. This is an interesting finding given that differences in argument quality between the Argument Only and Argument then Information conditions would not be expected since the conditions are identical in terms of argument generation.

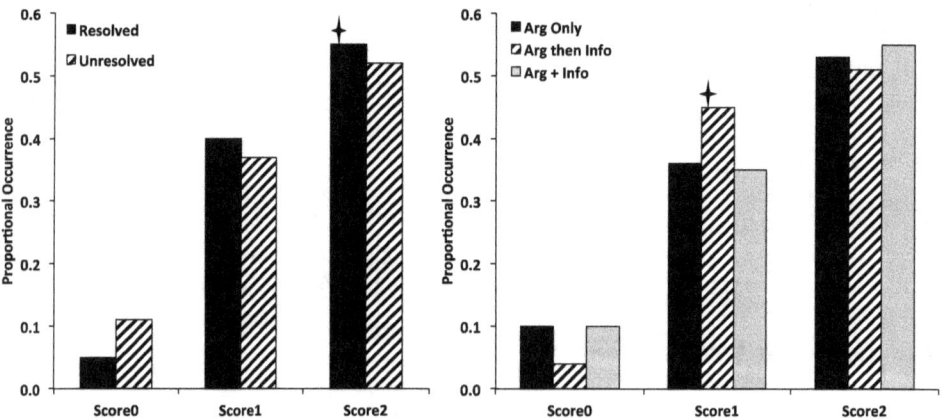

Figure 3. Proportional Occurrence of Overall Presence Scores and Average Semantic Match Score by Confusion Resolution Outcome and Intervention Condition

Third, we investigated the linguistic features of arguments. There were no significant differences based on confusion resolution outcome (p's > .10). Figure 4 shows the average percentile score for each linguistic feature. There were three significant models for intervention condition. Specifically, there were differences between intervention conditions for syntactic simplicity [$F(2) = 2.67$, $p = .036$], word concreteness [$F(2) = 4.26$, $p = .008$], and situation model cohesion [$F(2) = 1.98$, $p = .071$]. Students in the Argument Only condition had syntactically more complex arguments than those in both the Argument then

Information and Argument + Information conditions, used more concrete words than those in the Argument + Information condition, and had greater situation model cohesion than those in the Argument then Information condition. Students in the Argument Only condition also had more formal arguments [$F(2) = 3.01$, $p = .026$] based on the composite formality score from the five principle component Coh-Metrix measures of text difficulty. This finding shows that students in the Argument Only condition generated overall more complex arguments. Students in the Argument then Information condition used more concrete words than the Argument + Information condition.

Once again the differences between the Argument Only and Argument then Information conditions are interesting and surprising given the similarity between these two conditions. The differences for the Argument + Information condition suggest that perhaps providing the additional information does impact the linguistic nature of student arguments. The use of more abstract words, for example, could illustrate that students are making use of the text to increase the complexity of their vocabulary compared to students who are only tapping into their own knowledge in the Argument Only and Argument then Information conditions.

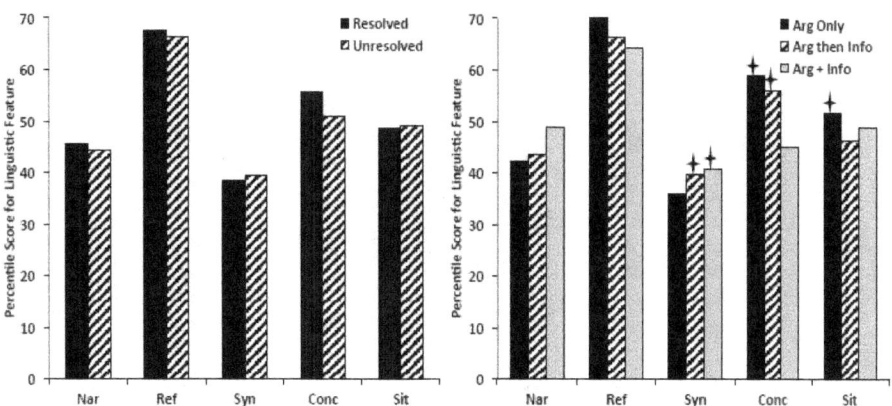

Figure 4. Linguistic Measures by Confusion Resolution Outcome and Intervention Condition (Nar = Narrativity, Ref = Referential Cohesion, Syn = Syntactic Simplicity, Conc = Word Concreteness, Sit = Situation Model Cohesion)

Fourth, we investigated the semantic match scores of arguments. There were not significant differences based on confusion resolution outcome (resolved: $M = .372$, $SD = .308$; unresolved: $M = .367$, $SD = .360$) or intervention condition (Argument Only: $M = .382$, $SD = .339$; Argument then Information: $M = .387$, $SD = .325$; Argument + Information: $M = .338$, $SD = .341$) (p's $> .10$). The lack of difference suggests that students' ability to mention the critical content did not vary based on confusion resolution outcome or intervention condition.

3.2. Learning Outcome

To investigate differences in learning outcomes we first split the argument quality measures into high and low cases via a median split. For formality, the high cases represent more complex arguments. For the five easability components, the high cases represent less complex arguments. Lastly, the high cases for semantic match score represent arguments with a higher semantic similarity to the ideal argument. We next investigated differences in learning outcome between the high and low cases for both the near and far transfer tasks. The models were not significant for the near transfer task (p's > .10); however, there were significant differences for the far transfer task. Figure 5 shows the proportion of correctly identified flaws on the far transfer task for the high and low cases for each argument quality measure. Students with syntactically more complex arguments [$\chi^2(1) = 1.57, p = .105$], more narrative arguments [$\chi^2(1) = 4.26, p = .020$], and higher semantic match scores [$\chi^2(1) = 7.51, p = .003$] performed better on the far transfer task. Together these findings suggest that generating a more complex argument that also addresses the appropriate content facilitates learning when confusion has occurred.

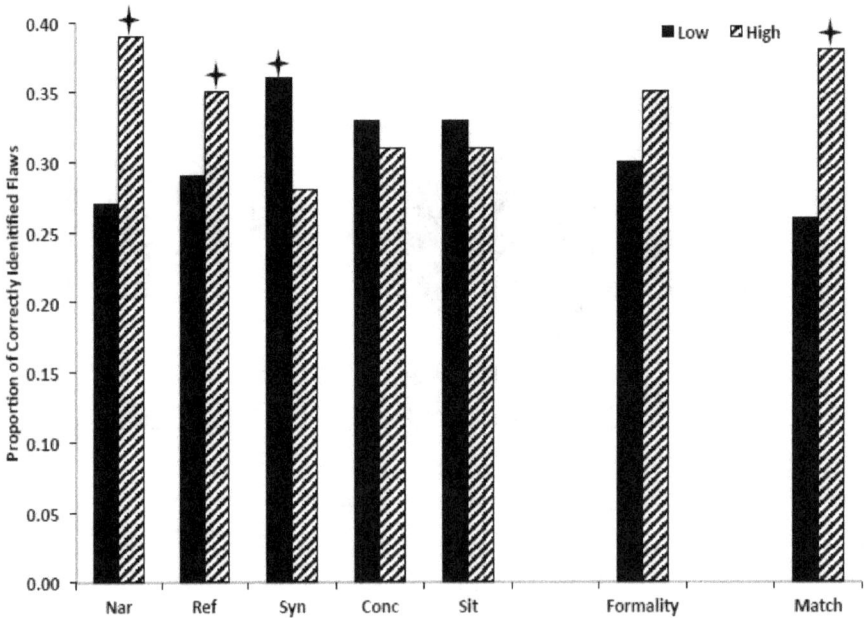

Figure 5. Far Transfer Learning Outcome by Argument Quality

Next, we investigated the argument structure/quality × confusion resolution outcome and argument structure/quality × intervention condition interactions with respect to performance on the transfer tasks. There were only two significant models and both involved the semantic match score on the far transfer task. Figure 6 shows the interaction pattern for both models. The semantic match score (high, low) × confusion resolution outcome (resolved, unresolved) interaction revealed that there was a significant difference in the low semantic match cases [$\chi^2(3) = 9.91, p = .019$], but there was not a significant difference between outcomes in the high semantic match cases. In the low semantic match cases, students performed better when they had unresolved confusion compared to when they had successfully resolved their confusion. This is an interesting finding because the low semantic match score cases indicate that students had not addressed the important keywords and concepts mentioned in the ideal argument. It may have been that the argument generation task highlighted to students the gaps in their knowledge and kept them in a confused state. Given the increased performance, it is unlikely that these unresolved confusion cases were instances of hopeless (unresolvable) confusion. This finding could be similar to findings in metacognitive research that better performing students are more attuned to what they do not know than their lower performing counterparts (Dunlosky & Lipko, 2007; Glenberg & Epstein, 1985; Graesser, D'Mello, & Person, 2009; Hacker, Dunlosky, & Graesser, 2009). In other words, after generating a shallow argument, higher performing students knew that they were missing some knowledge and reported continued confusion. Lower performing students, on the other hand, were over confident that they had addressed the necessary content and sufficiently understood the material to no longer be in a state of confusion.

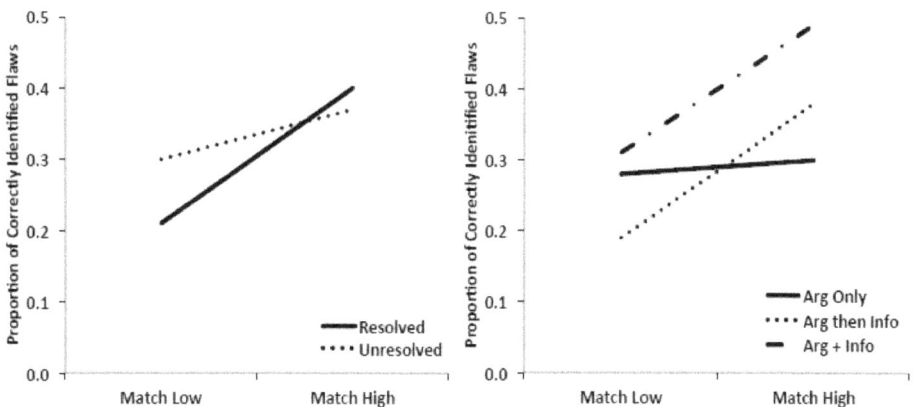

Figure 6. Far Transfer Learning Outcome by Argument Quality × Confusion Resolution Outcome and Argument Quality × Intervention Condition

The semantic match score (high, low) × intervention condition (Argument Only, Argument then Information, Argument + Information) interaction revealed that, when students had a low semantic match score, those in the Argument Only condition outperformed those in the Argument then Information condition, whereas when students had a high semantic match score those in the Argument + Information condition outperformed those in the Argument Only condition [$\chi^2(5) = 14.3, p = .014$]. This finding suggests that the scaffold of additional information may not have been universally useful. In the low semantic match score cases, for example, the student arguments did not address the important keywords and concepts in the ideal argument. This inability to address the important concepts could have highlighted knowledge gaps and triggered confusion. However, despite the fact that students in the Argument then Information condition had access to the additional information, these students did not perform as well on the far transfer task as students who never had access to this information in the Argument Only condition. In addition, the Argument then Information condition only outperformed the Argument Only condition when students had already included the important information in their arguments, prior to being presented with the additional information. It may be the case then that more adaptive scaffolding, such as highlighting important concepts in the text, is needed to increase the benefits of argument generation.

4. Discussion

Students frequently experience confusion during learning (Baker *et al.*, 2010; Craig *et al.*, 2004; Graesser & D'Mello, 2012; Graesser *et al.*, 2008; Lehman, D'Mello, & Person, 2008; Lehman, Matthews, D'Mello, & Person, 2008; McQuiggan *et al.*, 2010) and if the experience of confusion initiates effortful cognitive activities, they can reach a deeper level of understanding. However, students are often unable to engage in these effortful cognitive activities due to a lack of knowledge, skills, or motivation. It is then important that learning environments provide students with supports to initiate and persevere through confusion resolution. We have proposed argument generation as a potential intervention to facilitate confusion resolution and conducted a study to investigate the relationship between argument generation, confusion resolution, and learning. Earlier in this chapter we discussed four factors that a confusion resolution intervention would need to address. We now return to these factors to discuss how the argument generation task used in the present experiment met or did not meet each of these factors.

4.1. Did Argument Generation Draw Attention to Students' Confusion?

In the present investigation we did not directly address this point. However, when all confusion resolution outcomes were considered, we did find that students in the Argument then Information condition were more likely to have created confusion (not confused at Time 1 and confused at Time 2) than those in the two other conditions. This suggests that argument generation can draw attention to students' confusion, but it does not directly draw students' attention to their confusion. Perhaps the argument generation intervention can be augmented with the addition of a direct confusion recognition statement in a similar fashion to Affective AutoTutor (D'Mello *et al.*, 2011) and UNC-ITSpoke (Forbes-Riley & Litman, 2011).

4.2. Did Argument Generation Create a Contrast that is Mutually Exclusive?

In the present investigation we set up the scenario such that students had to agree with only one of the agents. Anecdotally, it was not the case that students attempted to adopt or reconcile both positions in their arguments. It appears, then, that associating the competing positions with agents in the discussion may facilitate creating mutually exclusive positions.

4.3. Did Argument Generation Motivate Students to Engage in Confusion Resolution?

In the present investigation we did not have a measure that assessed students' degree of engagement in the confusion resolution activities. However, one potential measure is the number of words that students wrote for their arguments compared to similar experiments in which students were only asked to explain their position. Three previous experiments utilized the same method of confusion induction and the same learning task (D'Mello *et al.*, 2014; Lehman *et al.*, 2013), but only asked students to explain why they had selected their position. In the present study the average argument word count across all conditions was 38.2 words, whereas in the three previous experiments the average word count for explanations was 15.4, 16.8, and 22.7 words, respectively. Although this is a very crude measure of engagement with confusion resolution activities, this comparison does show that students are willing to generate more text when prompted to argue as opposed to explain. Prior research has shown that longer contributions are correlated with higher learning gains (Chi, Siler, Yamauchi, Jeong, & Hausmann, 2001; Litman et al., 2006).

4.4. Did Argument Generation Scaffold the Confusion Resolution Process?

In the present investigation all conditions involved argument generation, so it is difficult to determine the degree to which this task scaffolded the confusion resolution process. However, it was the case that there were differences in learning outcomes based on the quality of the argument generated, which suggests that argument generation can facilitate the confusion resolution process. The Argument + Information condition contained an additional scaffold in that additional information was provided to students during argument generation. This additional information does seem to have aided students, but only when they integrated the information into their argument. It appears that a more adaptive type of scaffolding may be needed to see increased benefits from argument generation. The learning environment could, for example, automatically assess the quality of the argument, provide feedback to students, and then allow students to revise their arguments based on that feedback. Another option could be to have a discussion about the students' arguments, with agents asking probing questions that require the students to defend and revise their argument. This type of interactive argument generation could help students to both "learn to argue" and "argue to learn" (Scheuer et al., 2010).

It is also interesting to note that the argument generation does not appear to need to fully resolve confusion to benefit learning. There were in fact cases in which unresolved cases performed better on learning outcome measures than resolved cases. This is consistent with current research findings (D'Mello & Graesser, 2011, 2012, 2014; D'Mello et al., 2007; Lee et al., 2011; Liu et al., 2013; Rodrigo et al., 2010). It may be the case that the cognitive activities triggered by argument generation (e.g., deliberation, reflection) are sufficient to increase learning without necessarily requiring that the confusion be resolved.

4.5. Assessing Argument Quality

In addition to investigating argument generation as an intervention for confusion resolution, we have also investigated multiple methods for assessing argument quality in the present chapter. Traditionally, argument quality has been assessed in a variety of ways that often involve an analysis of the structure and quality of the information presented (Petty & Cacioppo, 2012; Sampson & Clark, 2008; Scheuer et al., 2010). We incorporated some of these methods as well as several automated analyses that addressed the linguistic features of the argument. The findings presented in this chapter showed that there were differences across a variety of potential argument quality measures with respect to confusion resolution and learning.

In addition to the previously discussed analyses, we can also investigate differences between the more traditional argument quality measures and the linguistic features from Coh-Metrix. When investigating overall presence score, for example, we found that arguments with a score of 2 (inclusion of both claims and evidence) were more narrative, syntactically simpler, used less concrete words, and had higher situation model cohesion than arguments with a score of 1 (inclusion of claims or evidence). Similarly, when investigating differences in high and low cases of semantic match score, we found that arguments with a high semantic match score had higher referential cohesion, used less concrete words, and had higher situation model cohesion. These patterns, along with the previously discussed findings, provide preliminary evidence to the utility of analyzing linguistic features as a method of evaluating the quality of arguments. In addition, these linguistic features can be automatically derived and built into conversational learning environments in order to provide feedback and scaffolding to students.

5. Concluding Remarks

In this chapter we have presented argument generation as a potential intervention for confusion resolution. Although the evidence is only preliminary, it appears that argument generation can be beneficial for learning. Argument generation requires students to consider both the positions presented and the reasons *why* one is correct and the other is incorrect, while also potentially engaging students to a greater degree than only explaining their position. Further research is needed to examine the way in which argument generation can benefit confusion resolution and learning as well as the potential methods for scaffolding the argument generation process. Overall, "arguing to learn" appears to be a beneficial approach to promote learning at deeper levels.

Acknowledgements.

This research was supported by the National Science Foundation (NSF) (ITR 0325428, HCC 0834847, DRL 1235958). Any opinions, findings and conclusions, or recommendations expressed in this paper are those of the authors and do not necessarily reflect the views of NSF.

References

Andriesson, J. (2006). Arguing to learn. In R. K. Sawyer (Ed.), *The Cambridge handbook of the learning sciences* (pp. 443-460). New York, NY: Cambridge University Press.

Baayen, R., Piepenbrock, R., & Gulikers, L. (1995). *The CELEX lexical database* (Release 2) [CD-ROM]. University of Pennsylvania, Linguistic Data Consortium, Philadelphia, PA.

Baker, R. S., D'Mello, S. K., Rodrigo, M. T., & Graesser, A. C. (2010). Better to be frustrated than bored: The incidence, persistence, and impact of learners' cognitive-affective states during interactions with three different computer-based learning environments. *International Journal of Human-Computer Studies, 68*, 223-241.

Bates, D. M., & Maechler, M. (2010). *Lme4: Linear mixed-effects models using S4 classes*. Retrieved from http://CRAN.R-project.org/package=lme4

Berland, L., & Reiser, B. (2009). Making sense of argumentation and explanation. *Science Education, 93*, 26-55.

Bransford, J. D., Brown, A. L., & Cocking, R. R. (1999). *How people learn: Brain, mind, experience, and school.* Washington, DC: National Academy Press.

Brem, S. K., & Rips, L. J. (2000). Explanation and evidence in informal argument. *Cognitive Science, 24*, 573-605.

Brown, J., & VanLehn, K. (1980). Repair theory: A generative theory of bugs in procedural skills. *Cognitive Science, 4*, 379-426.

Caroll, J., & Kay, D. (1988). Prompting, feedback and error correction in the design of a scenario machine. *International Journal of Man-Machine Studies, 28*, 11-27.

Chi, M. T. H., Siler, S., Yamauchi, T., Jeong, H., & Hausmann, R. (2001). Learning from tutoring. *Cognitive Science, 25*, 471-534.

Chinn, C. A., & Brewer, W. F. (1993). The role of anomalous data in knowledge acquisition: A theoretical framework and implications for science education. *Review of Educational Research, 63*, 1-49.

Clore, G. L., & Huntsinger, J. R. (2007). How emotions inform judgment and regulate thought. *Trends in Cognitive Sciences, 11*, 393-399.

Craig, S. D., Graesser, A. C., Sullins, J., & Gholson, B. (2004). Affect and learning: An exploratory look into the role of affect in learning. *Journal of Educational Media, 29*, 241-250.

D'Mello, S. K., & Graesser, A. C. (2011). The half-life of cognitive-affective states during complex learning. *Cognition & Emotion, 25*, 1299-1308.

D'Mello, S. K., & Graesser, A. C. (2012). Dynamics of affective states during complex learning. *Learning & Instruction, 22*, 145-157.

D'Mello, S. K., & Graesser, A. C. (2014). Confusion and its dynamics during device comprehension with breakdown scenarios. *Acta Psychologica, 151*, 106-116.

D'Mello, S. K., Lehman, B. A., & Graesser, A. C. (2011). A motivationally supportive affect-sensitive AutoTutor. In R. A. Calvo & S. K. D'Mello (Eds.), *New perspectives on affect and learning technologies* (pp. 113-126). New York, NY: Springer.

D'Mello, S. K., Lehman, B., Pekrun, R., & Graesser, A. (2014). Confusion can be beneficial for learning. *Learning & Instruction, 29*, 153-170.

D'Mello, S., Taylor, R., & Graesser, A. (2007). Monitoring affective trajectories during complex learning. In D. McNamara & J. Trafton (Eds.), *Proceedings of the Cognitive Science Society 2007 Conference* (pp. 203-208). Austin, TX: Cognitive Science Society.

Dunlosky, J., & Lipko, A. (2007). Metacomprehension: A brief history and how to improve its accuracy. *Current Directions in Psychological Science, 16*, 228-232.

Festinger, L. (1957). *A theory of cognitive dissonance*. Stanford, CA: Stanford University Press.

Forbes-Riley, K., & Litman, D. (2011). Benefits and challenges of real-time uncertainty detection and adaptation in a spoken dialogue computer tutor. *Speech Communication, 53*, 1115-1136.

Glenberg, A. M., & Epstein, W. (1985). Calibration of comprehension. *Journal of Experimental Psychology: Learning, Memory, and Cognition, 11*, 702-718.

Graesser, A. C., & D'Mello, S. K. (2012). Emotions during the learning of difficult material. In B. Ross (Ed.), *The psychology of learning and motivation*, vol. 57 (pp. 183-225). Waltham, MA: Elsevier.

Graesser, A. C., D'Mello, S. K., Craig, S. D., Witherspoon, A., Sullins, J., McDaniel, B., & Gholson, B. (2008). The relationship between affect states and dialogue patterns during interactions with AutoTutor. *Journal of Interactive Learning Research, 19*, 293-312.

Graesser, A. C., D'Mello, S. K., & Person, N. K. (2009). Meta-knowledge in tutoring. In D. J. Hacker, J. Dunlosky, & A. C. Graesser (Eds.), *Handbook of metacognition in education* (pp. 361-382). New York, NY: Routledge.

Graesser, A., Lu, S. L., Jackson, G., Mitchell, H., Ventura, M., Olney, A., ... Louwerse, M. (2004). AutoTutor: A tutor with dialogue in natural language. *Behavioral Research Methods, Instruments, and Computers, 36*, 180-193.

Graesser, A., Lu, S., Olde, B. A., Cooper-Pye, E., & Whitten, S. (2005). Question asking and eye tracking during cognitive disequilibrium: Comprehending illustrated texts on devices when the devices breakdown. *Memory & Cognition, 33*, 1235-1247.

Graesser, A. C., & McNamara, D. S. (2011). Computational analyses of multilevel discourse comprehension. *Topics in Cognitive Science, 3*, 371-398.

Graesser, A. C., McNamara, D. S., & Kulikowich, J. (2011). Coh-Metrix: Providing multilevel analyses of text characteristics. *Educational Researcher, 40*, 223-234.

Hacker, D. J., Dunlosky, J., & Graesser, A. C. (2009). *Handbook of metacognition in education*. New York, NY: Routledge.

Keltner, D., & Shiota, M. (2003). New displays and new emotions: A commentary on Rozin and Cohen (2003). *Emotion, 3*, 86-91.

Kuhn, D. (2010). Teaching and learning science as argument. *Science Education, 94*, 810-824.

Kuhn, D., & Katz, J. (2009). Are self-explanations always beneficial? *Journal of Experimental Child Psychology, 103*, 386-394.

Lee, D., Rodrigo, M., Baker, R., Sugay, J., & Coronel, A. (2011). Exploring the relationship between novice programmer confusion and achievement. In S. D'Mello, A. Graesser, B. Schuller, & J. Martin (Eds.), *Affective computing and intelligent interaction 2011* (pp. 175-184). Berlin/Heidelberg, Germany: Springer.

Lehman, B., D'Mello, S. K., & Graesser, A. C. (2012). False feedback can improve learning when you're productively confused (Manuscript in preparation).

Lehman, B., D'Mello, S. K., & Person, N. (2008, June). *All alone with your emotions: An analysis of student emotions during effortful problem solving activities*. Workshop on Emotional and Cognitive issues in ITS held in conjunction with Ninth International Conference on Intelligent Tutoring Systems, Montreal, Canada.

Lehman, B., D'Mello, S. K., Strain, A., Mills, C., Gross, M., Dobbins, A., ... Graesser, A. C. (2013). Inducing and tracking confusion with contradictions during complex learning. *International Journal of Artificial Intelligence in Education, 22*, 85-105.

Lehman, B., & Graesser, A. (2015). To resolve or not to resolve? That is the big question about confusion. In C. Conati & N. Heffernan (Eds.), *Proceedings of 17th International Conference on Artificial Intelligence in Education* (pp. 216-225). Hiedelberg, Germany: Springer.

Lehman, B. A., Matthews, M., D'Mello, S. K., & Person, N. (2008). What are you feeling? Investigating student affective states during expert human tutoring sessions. In B. Woolf, E. Aimeur, R. Nkambou, & S. Lajoie (Eds.), *Proceedings of the Ninth International Conference on Intelligent Tutoring Systems* (pp. 50-59). Berlin & Heidelberg, Germany: Springer-Verlag.

Limón, M. (2001). On the cognitive conflict as an instructional strategy for conceptual change: A critical appraisal. *Learning and Instruction, 11*, 357-380.

Litman, D. J., Rose, C. P., Forbes-Riley, K., VanLehn, K., Bhembe, D., & Silliman, S. (2006). Spoken versus typed human and computer dialogue tutoring. *International Journal in Education, 16*, 145-170.

Liu, Z., Pataranutaporn, V., Ocumpaugh, J., & Baker, R. (2013). Sequences of frustration and confusion, and learning. In S. D'Mello, R. Calvo, & A. Olney (Eds.), *Proceedings of the Educational Data Mining Conference 2013* (pp. 114-120). International Educational Data Mining Society.

Malone, T. W., & Lepper, M. R. (1987). Making learning fun: A taxonomy of intrinsic motivations for learning. In R. E. Snow & M. J. Farr (Eds.), *Aptitude learning and instruction*, vol. 3 (pp. 223-253). Hillsdale, NJ: Erlbaum.

McQuiggan, S. W., Robison, J. L., & Lester, J. C. (2010). Affective transitions in narrative-centered learning environments. *Educational Technology & Society, 13*, 40-53.

Pekrun, R., & Stephens, E. J. (2012). Academic emotions. In K. R. Harris, S. Graham, T. Urdan, S. Graham, J. M. Royer, & M. Zeidner (Eds.), *APA educational psychology handbook*, vol. 2 (pp. 3-31). Washington, DC: American Psychological Association.

Petty, R., & Cacioppo, J. T. (2012). *Communication and persuasion: Central and peripheral routes to attitude change*. New York, NY: Springer Science & Business Media.

Piaget, J. (1952). *The origins of intelligence*. New York, NY: International University Press.

Pinheiro, J. C., & Bates, D. M. (2000). *Mixed-effects models in S and S-PLUS*. New York, NY: Springer Verlag.

Rodrigo, M., Baker, R., & Nabos, J. (2010). The relationships between sequences of affect states and learner achievements. In S. Wong, S. Kong, & F.-Y. Yu (Eds.), *Proceedings of the Computers in Education Conference 2010* (pp. 56-60). Serdang, Malaysia: Faculty of Educational Studies, Universiti Putra, Malaysia.

Rozin, P., & Cohen, A. (2003). High frequency of facial expressions corresponding to confusion, concentration, and worry in an analysis of naturally occurring facial expressions of Americans. *Emotion, 3*, 68-75.

Sampson, V., & Clark, D. B. (2008). Assessment of the ways students generate arguments in science education: Current perspectives and recommendations for future directions. *Science Education, 92*, 447-472.

Scheuer, O., Loll, F., Pinkwart, N., & McLaren, B. M. (2010). Computer-supported argumentation: A review of the state of the art. *International Journal of Computer-Supported Collaborative Learning, 5*, 43-102.

Silvia, P. J. (2010). Confusion and interest: The role of knowledge emotions in aesthetic experience. *Psychology of Aesthetics Creativity and the Arts, 4*, 75-80.

VanLehn, K., Siler, S., Murray, C., Yamauchi, T., & Baggett, W. (2003). Why do only some events cause learning during human tutoring? *Cognition & Instruction, 21*, 209-249.

Zohar, A., & Nemet, F. (2002). Fostering students' knowledge and argumentation skills through dilemmas in human genetics. *Journal of Research in Science Teaching, 39*, 35-62.

Chapter 22

The Importance of Multi-Modality in Mathematical Argumentation

Baruch Schwarz[1] & Naomi Prusak[2]

The Hebrew University of Jerusalem, Israel, [1] baruch.schwarz@mail.huji.ac.il; [2] inlrap12@netvision.net.il

Abstract. Pioneering research indicates the importance of non-verbal channels in mathematical reasoning and mathematical argumentation. In this chapter, we go deeper in this direction to identify the role of gestures and other material actions may serve as argumentative moves on their own, not only as accompanying verbal utterances. We will present examples from a year-long course designed to foster creativity and reasoning in a context of collaborative mathematical problem-solving and argumentation among third-grade talented students. We will show how non-verbal actions – gestures as well as material actions (drawing, folding, cutting, etc.) interweave with children's verbal peer argumentation towards the construction of sound mathematical arguments. We will show that the multimodal argumentation afforded by the design led children to substantive conceptual insights in elementary geometry. We will suggest several roles of non-verbal actions in learning through mathematical argumentation. In particular, we will show the proleptic role of gestures and of inscriptions in argumentation in geometry.

1. Introduction and Theoretical Framework

Classical theories of Argumentation traditionally rely on words, written or oral: words seem a priori explicit in contrast with images, gestures or facial expressions. Indeed, van Eemeren and Grootendorst's pragma-dialectical approach (2004) regards verbal expressions in argumentative discourse as speech acts. At most, non-verbal channels are brought forward to strengthen the validity of the identification of an argument or of an argumentative move. However, this orthodox approach has been shattered during the last twenty years from several directions. First, non-verbal channels (e.g., visual channels) have been claimed to represent arguments in ways that are sometimes more convincing than verbal channels – like in the world of advertising (Birdsell & Groarke, 1996, 2007). Several criticisms have been raised towards an incautious identification of non-verbal channels of communication with argumentative moves. For example, Larvor (2013) has distinguished between the use of visual modes to express assertions and their use to express arguments in mathematics. This caveat being brought forward, Larvor convincingly showed that diagrams may serve visual argumentation in geometry when verbal arguments cannot express handily mathematical ideas.

A second direction for the necessity for replacing the verbal channel of argumentation by other channels comes from the context of *learning* – a context of efforts and of difficulties. Non-verbal channels are frequent not because they are more convincing but because they are less explicit, thus demand fewer efforts. Indeed, in learning contexts, the verbal production is often unclear and impoverished. As shown by Anderson and colleagues, the integrity of oral arguments expressed by children is very often lacking and researchers are obliged to complete children's productions according to reasonable assumptions, in order to analyze their argumentation (Anderson, Chinn, Chang, Waggoner, & Yi, 1997). Of course, this lack of clarity of arguments is around what can be understood from the oral productions only, but children's ability to communicate through social interactions suggests that intersubjectivity involves multiple modalities. The domain of mathematical reasoning provides a promising venue on this issue. In this domain, cognitive psychologists have achieved an impressive breakthrough to demonstrate people's tendency to use multimodal channels of reasoning in order to *lighten cognitive load*: Goldin-Meadow and colleagues (Goldin-Meadow, Nusbaum, Kelly, & Wagner, 2001) showed that both adults and children remember significantly more items when they gestured during their math explanations than when they did not gesture. Gesturing appeared to save the speakers' cognitive resources on the explanation task, permitting the speakers to allocate more resources to the memory task. It is widely accepted that gesturing reflects a speaker's cognitive state, but based on their observations, Goldin-Meadow and her colleagues suggested that, by reducing cognitive load, gesturing may also play a role in shaping that state. Socio-cultural psychologists go further in rejecting en bloc the purely verbal conception of thinking and reasoning to adopt a multimodal "material" conception of reasoning. For example, Radford (2009) follows the German social theorist Arnold Gehlen to adopt a "sensuous cognition" (Gehlen, 2009): The very texture of thinking is made up of speech, gestures, and our actual actions with cultural artifacts (signs, objects, etc.). Radford shows how reasoning in a mathematics classroom involves a sophisticated semiotic coordination of speech, body, gestures, symbols and tools: the very texture of thinking...cannot be reduced to that of impalpable ideas. It is instead made up of speech, gestures, and our actual actions with cultural artifacts (signs, objects, etc.)... Mathematical cognition is not only mediated by written symbols, but...is also mediated, in a genuine sense, by actions, gestures and other types of signs (pp. 111-112).

Duval, Ferrari, Høines and Morgan (2005) abound in this direction: the crucial properties of mathematical language cannot be thoroughly investigated without taking into account all the linguistic systems adopted in doing mathematics at any level, including written and spoken verbal language, symbolic notations, visual representations and even gestures (p. 790).

Arzarello and Sabena (2011) somehow conciliate between cognitive and socio-cultural perspectives to suggest that multimodal semiotic and theoretical control combine when students argue and prove in the graphical context of Elementary Calculus. These developments in socio-cultural psychology concur with new developments in theories of Argumentation that point at the importance of multimodality in different domains such as constructing metaphors or in advertising (Forceville, 2011).

It appears then that for theorists that hold different views, multimodality is omnipresent in mathematical reasoning. Goldin-Meadow has shown the role of gestures in reasoning in mathematics. She did not show whether they have an argumentative role, though. Radford adopts a very general approach to describe the construction of meaning when one interacts with signs, inscriptions and representations. Arzarello (2008) describes the *semiotic bundle* capitalized on when learners interact with each other and with signs, inscriptions and representations. However, beyond his general characterization of mathematical argumentation as being multimodal and as involving gestures, tools or instructions, he does not analyze the argumentative roles of different channels. The other researchers that have documented mathematical argumentation in a learning context do not focus on the multimodal aspect of argumentation. For example, Knipping and Reid (2013) uncover the argumentative structures of classroom talk in teacher-led proving activities. While their approach is interesting, it is not only limited to verbal production, but also focuses on proving – a context that can hardly been identified as argumentative.[1] Several researchers have observed mathematical argumentation as a dialectical activity that precedes the activity of proving and inscribing a formal proof, in which disagreements are settled (Boero et al. 2008; Douek, 2008; Schwarz, Hershkowitz & Prusak, 2010). However, they have not identified the functions of the non-verbal channels in multimodal argumentation yet.

To summarize, although the omnipresence of multimodality in mathematical argumentation is nowadays recognized, the functions of multimodality are not well understood yet, in the realm of reasoning and learning. We focus here on multimodal argumentation among unguided children engage in a learning situation, which invites them to argue in order to settle contradiction and disagreements. We present a task designed to create a situation of socio-cognitive conflict and consequently to trigger *productive argumentation* in the sense that the argumentation is designed to trigger learning processes in geometry. The socio-cognitive conflict could take place as a result of an arrangement of students in small groups and the design of a task that afforded the generation of multiple solutions, some of them being incompatible. We show that in this context, group discourse was argumentative and that it was multimodal. Our main theoretical contribution consists of identifying some functions of the non-verbal actions towards learning through argumentation. Argumentation is taken in a rich sense that includes the co-construction of arguments and critical discussions to settle disagreements. Let us first review the role of cognitive and socio-cognitive conflicts in knowledge construction.

1.1. The Role of Cognitive and Socio-Cognitive Conflicts in the Construction of Knowledge

Most of the models proposed to explain construction of knowledge have emphasized the role of cognitive conflict as a necessary condition for achieving it. *Cognitive conflict*, or *cognitive dissonance*, may occur whenever a learner is confronted by an event which varies from what is expected, where the event might be a result, fact, opinion, etc. Cognitive conflict is triggered by surprise, uncertainty, curiosity, or perplexity. When the newly assimilated information conflicts with previously formed mental structures, it may result in disequilibrium or a cognitive conflict (Piaget, 1975). Piaget claimed that this state of disequilibrium motivates the learner to seek equilibrium. Piaget outlined the importance of the imbalanced state for cognitive growth, where a balanced state is achieved through *accommodation/assimilation* towards equilibration by meeting the challenges of disequilibration. Empirical studies challenged Piaget's theory of development and the central role of cognitive conflict by showing that children adopt various ways to face a cognitive conflict without solving it (e.g., Chinn & Brewer, 1993).

Neo-Piagetians, such as Mugny and Doise (1978) did not consider the conflict as a psychological state but as a social setting. For neo-Piagetians (Mugny & Doise, 1978; Perret-Clermont, 1980), the *socio-cognitive conflict* is a setting in which different participants are involved and in which they face arguments or evidence that challenges their views. The challenge can originate from peers or from data that the participants handle. As

---

[1] Argumentation theorists such as Krabbe (2013) have claimed that proofs can be identified as arguments. We concur with this claim. However, we also claim that the elaboration of a proof in a learning context, especially when students are requested to do so without raising any conjecture about the problem at stake, is generally not an argumentative activity.

claimed by Mugny and Doise (1978): "socio-cognitive conflict is an important factor in all restructuration, whether collective or individual. Progress should therefore be most apparent when subjects of different cognitive levels actualize different approaches of the same task..." (p. 183).

The reallocation of the conflict from the cognitive to the design realm turns the issue of the socio-cognitive conflict to highly complex: the ways to resolve a socio-cognitive conflict are *a priori* extremely various. Intense research presently conducted on this domain indicates that the resolution depends on the discourse type and deeply affects learning. Schwarz and Asterhan (in press) integrate research findings to first recognize that argumentative discourse in general can foster conceptual learning (e.g., Asterhan & Schwarz, 2009). They go further to distinguish between disputative and deliberative argumentation with respect to further learning outcomes. A focus on the interpersonal, competitive dimension of social interaction may raise uncertainty and threaten self-competence (Butera & Mugny, 1995; Darnon, Butera & Harackiewicz, 2007), increase positive evaluations of the partner's competence (Darnon, Muller, Schrager, Pannuzzo & Butera, 2006; Gabriele & Montecinos, 2001), and raise concerns about group belonging or interpersonal relationships. It appears that learners adopting a deliberative discourse outperform those whose 145 discourse is disputative. These findings indicate that scripting collaborative argumentation may lead to conceptual learning. These research findings led us to observe the nature of argumentation in a situation of socio-cognitive conflict in a situation of collaboration on challenging tasks (Asterhan & Schwarz, 2016; Schwarz & Baker, 2016).

1.2. The Role of Argumentation and Multiple Channels of Communication in Learning Geometry

In this chapter, our ambition is (1) to exemplify that the resolution of socio-cognitive conflicts among young learners may occur through multimodal deliberative argumentation and (2) to identify the function of multimodality in learning through argumentation. While, as we just reported, the first insight is not totally new, the description of how multimodal argumentation in mathematics leads to conceptual learning is novel. It goes beyond what Radford, as well as Arzarello and Sabena showed so far – the fact that mathematical reasoning is multimodal. We will show the functions of the multiple modalities of argumentation in the resolution of the socio-cognitive conflict. Our adventure does not begin from scratch: Duval (2006) linked argumentation in elementary geometry to the mereological decomposition of shapes: division of the whole into parts with the aim of reconstructing another figure, allowing for the detection of geometrical properties. Duval's mereological decomposition is an excellent example of multiple solving strategies in geometry due to the fact that such decompositions can be executed materially (by cutting and reassembling), graphically (by drawing lines that reorganize the shapes), or by observing visually. Our working hypothesis was that it is important to encourage this strategy, namely the composing and decomposing of shapes, in students' mathematical activities. Following these theoretical considerations, we show in this chapter that gestures and material actions and the use of artifacts are important constituents of early geometrical reasoning, and that they are deployed in multimodal argumentation. And most importantly, we will attempt to identify some of the functions of multiple modalities in argumentation towards the learning of geometrical properties.

2. The Study

Three groups of 20 gifted and talented third-grade students participated in a special enrichment program in mathematics over three successive years. The students in each group attended 28 meetings over the course of one academic year. The course was designed to foster mathematical reasoning in a problem-solving context. The course combined problem solving in dyads or small groups, peer argumentation, and teacher-led discussion. The design of the activities was developed specifically for this course by the second author and relied on five design principles: (a) creation of problems with multiple solutions, (b) creation of collaborative learning situations, (c) stimulation of socio-cognitive conflict, (d) provision of tools for checking hypotheses, and (e) opportunity for reflection upon and evaluation of solutions.

About 25% of the activities dealt with issues related to the geometrical concepts of area and perimeter and their relationship. Each 75-minute lesson typically opened with a teacher-initiated 15-minute discussion to create a shared understanding of the activity. Next, the teacher distributed worksheets, and student groups (primarily dyads) worked collaboratively on the problem, completing a worksheet that scaffolds the reporting process (up to 40-50 minutes). The teacher circulated among the groups to answer questions and help when needed. At the end of the activity, the teacher orchestrated a reflective discussion on the activity. Socio-cognitive conflict was an integral part of many of the activities.

The second author designed an activity – "Sharing a cake", to facilitate the understanding of the concept of geometrical area. In particular, it was designed to realize that shapes may have equal area without being congruent. Figure 1 presents a shortened version of the activity. At the end of each section (1, 2a, 2b, and 2c) we collected the worksheets from the students so that they could not change their answers during the following task, when they might discover that they had been mistaken in a previous task. This allowed us to identify the emergence of new ideas about the area concept.

Although we do not report here on the ways students solved Task 1 (see Prusak, Hershkowitz & Schwarz, 2013), it is relevant to the focus of our analysis – the resolution of a socio-cognitive conflict, which occurred in Task 2. The goal of Task 1 was to encourage students to provide diverse solutions for the problem and diverse explanations to justify them. The students were explicitly required to explain and justify in writing each solution they drew. Nine grid squares, representing the cake, were provided to students on their worksheets in order to encourage them to find many diverse solutions; the grid provided a proper tool for checking hypotheses by comparing the area of shapes created. Therefore, Task 1 aimed at encouraging students to give verbal justifications to their solution to the partition of the cake.

In the students' worksheets, we identified four main types of solutions (see Fig. 2). In Type A, all four shapes were congruent and were created by simple partitions: drawing diagonals, perpendicular bisectors, or segments parallel to one side. In Type B, all four shapes were congruent but they were created with more sophisticated partitions. Type C solutions consisted of two different pairs of congruent shapes. And in Type D, there were no more than one pair of congruent shapes (quite often all four shapes were non-congruent – see Fig. 2). We found that 84% of the students produced at least three different types of solutions.

This means that they produced at least one solution in which not all of the four parts were congruent. We also identified three types of justifications, based on (1) congruency (46%), (2) a compose-and-decompose strategy (12%), and (3) counting (42%) (Prusak, Hershkowitz, & Schwarz, 2013).

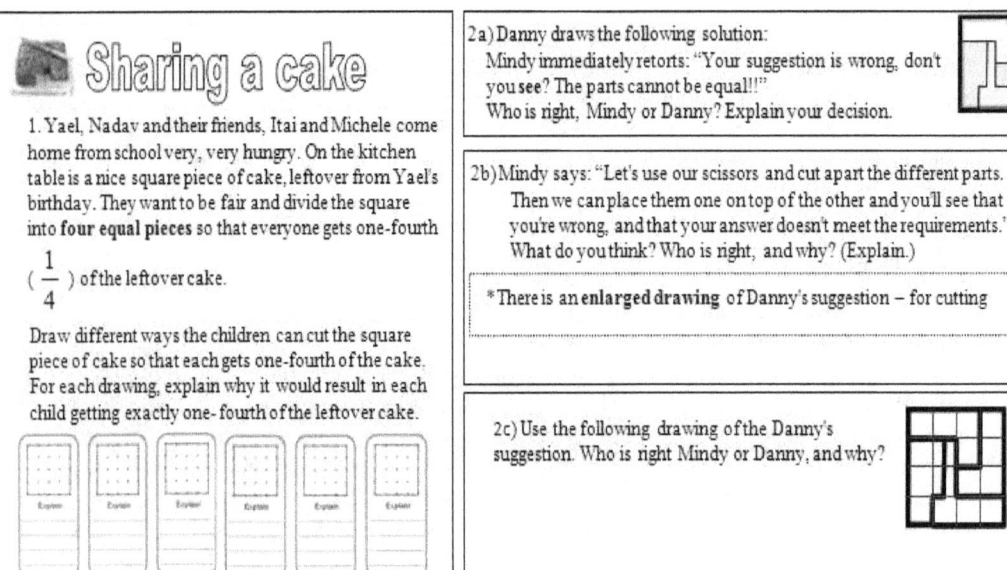

Figure 1. A shortened version of "Sharing a Cake"

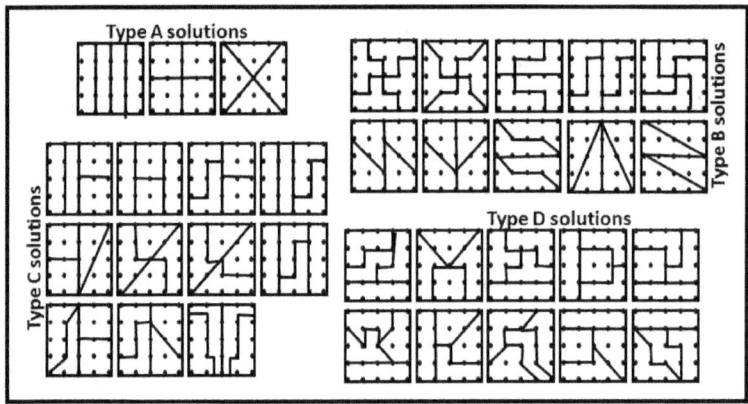

Figure 2. The four types of partition to task 1

The multiplicity of strategies for the partitions suggests that for the students, two geometrical shapes can have the same area without being congruent. However, this possibility is based on a visual sense supported by the dots provided to undertake the partition. We designed Task 2 to create a socio-cognitive conflict between students working in small groups: in Task 2a, the four parts do not "look" congruent. In Danny's solution, the area of part D (see Figure 3) looks bigger than the area of part C. Hence Mindy's claim (which appears on Task 2b), "Your suggestion is wrong, don't you see? The parts cannot be equal!!" stresses the apparent impossibility of congruence. Mindy's suggestion to use scissors is in fact an invitation to adopt compose-and-decompose strategies (Duval, 2006). Our working hypothesis was that in Task 2, students would agree that the areas are not equal and Mindy's invitation to use the scissors would create a socio-cognitive conflict to be solved in Task 2c.

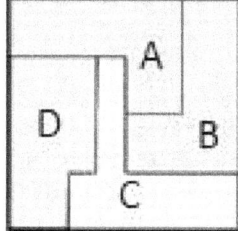

Figure 3. Task 2a: a task that affords socio-cognitive conflicts

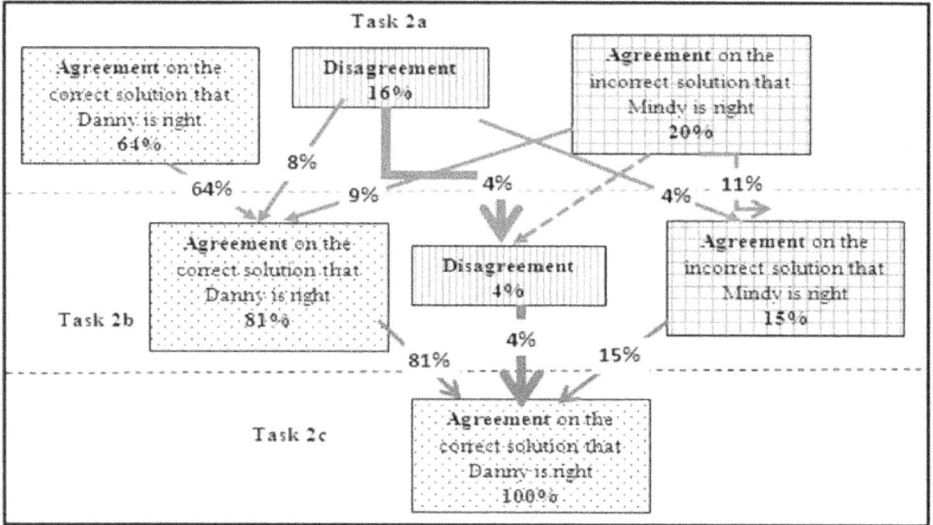

Figure 4. Interactions found in Tasks 2a, 2b, and 2c and their percentage distribution

Figure 4 displays the distribution of students' (dis)agreements in Task 2a, 2b, and 2c. We can see that already for Task 2a, 64% of the students agreed that Danny's solution is right. However, we found that almost 25% of the justifications given for the correct answer were incorrect or incomplete. For 36% of the students, the "Share the Cake" activity led to a socio-cognitive conflict: 16% initially disagreed and 20% agreed on the wrong solution, and subsequently faced together a challenge when they used their scissors. At the end after solving task 2c with the square grid, all agreed about Danny's right solution and most of the justifications were correct. In this chapter, we will follow the trajectories of some students who faced a socio-cognitive conflict. We present here three cases of solution of Task 2.

An important caveat should be stressed here. Our methodological approach relies on *case studies* in which learning was reached by resolving a socio-cognitive conflict through multimodal argumentation. We do not claim here that there are no other ways to learn geometrical properties in group/individual work. Rather, we focus on several cases in which argumentation were triggered. We should say, however, that in all the groups of the present study in which argumentation was triggered, argumentation was multimodal and led to the learning of a geometrical property. In this sense, the cases we present are representative.

2.1. Interaction 1: The Role of Encompassing Gestures in Solving a Socio-Cognitive Conflict

In Task 2a, Harry and Larry are in a situation of socio-cognitive conflict: Harry claimed that Danny was wrong and Larry claimed he was right. Larry's written justification given in Task 2a is that "Danny is right, the four parts are equal". Therefore, even though Larry's claim is correct, he does not justify it. We present in Table 1 an excerpt of the interaction between Harry and Larry in which they jointly reached the correct solution at the end of Task 2b. Nonverbal actions are marked in brackets. To understand the interactions, and especially the meaning of the gestures, we use the capital letters that name the different parts of the partition of Task 2a in Figure 3 (these letters do not appear in Task 2a).

Table 1 includes the first part of the protocol. Harry and Larry use four kinds of gestures. The first, a pointing gesture. The second is a measuring gesture: at least two fingers bind the length of a shape like in Harry5. As for the comparing gesture, it is done by using two fingers that refer in parallel to two shapes or by extending the measuring gesture by transferring the opening between two fingers that measure a length of one figure to another one. A variant of the measuring gesture is the well-known V-point gesture consisting of pointing simultaneously at two shapes to express an idea of equality (Novack et al, 2014). The encompassing kind of gesture consists of moving a finger around a geometrical figure, or encompassing the whole figure with all fingers (sometimes this gesture is accompanied by a shrinking or an expanding movement). It seems that the pointing gesture is used to refer to something evident: In Harry5, the pointing gesture is enacted to simply refer to A and B. In Harry8, the pointing gesture on D, is used to refer to something which seems "evident" to Harry, the fact that there are more square units in D than in A and B. The functions of the measuring and comparing gestures are clear. As for the encompassing gesture, it functions to compose parts A and B as one shape (in Larry13), and to decompose the square-cake in two equal parts (in Larry14). Table 1 also includes a material action – drawing square units. While this action is not verbal, it is not a gesture as gestures do not modify the material environment.

Let us now turn to the deployment of the discourse. Harry first opposed Larry's claim that Danny was correct (Larry4), by showing that Larry was correct for parts A and B only through a pointing gesture. Harry insisted that Danny was wrong and backed his claim visually with a measuring gesture (Harry5). He then silently drew perpendicular lines (Harry6) and used this inscription to count or estimate the number of the square units he drew (Harry8), and by such to check his argument that A and B have the same area but that the area of D is equal to neither of them. In Harry11, Harry reiterates his former argument. As the interviewer asks him about part C, he firmly declares "But part D is already not equal so they cannot be the same as the others. This inference points at a beginning of deductive reasoning. Larry, on the other hand, goes on claiming that Danny was correct by composing A and B as a geometrical figure whose area is half of the square, and composing C and D as another half (Larry13 and Larry14). Larry does not express this argument verbally but, rather by using an encompassing gesture to around the shapes composed and the square decomposed. This move is a challenge to Harry's argument; it is a weak challenge, though: Larry believed that showing that what he said was enough to claim that the area of each part was one-fourth of the square!

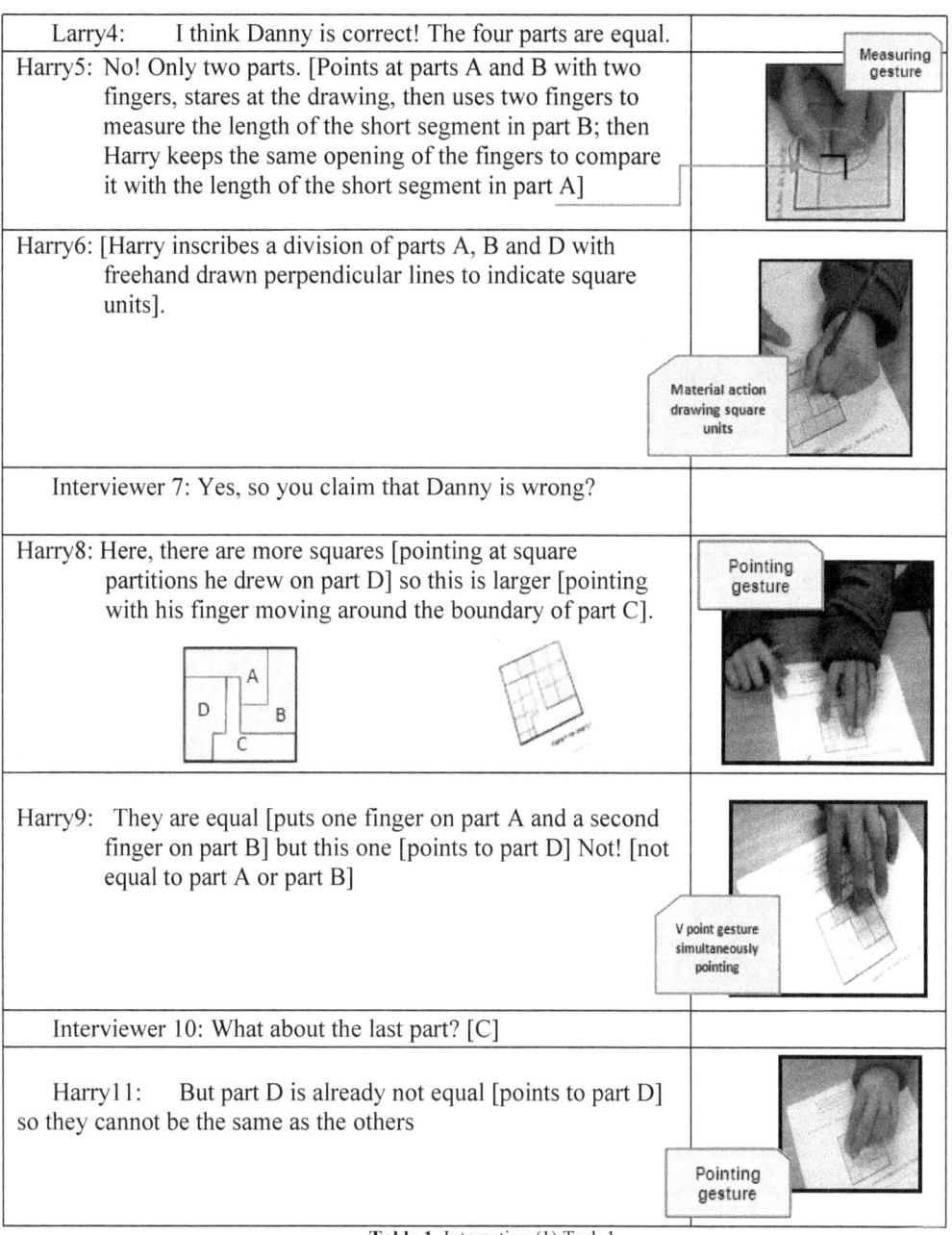

Table 1. Interaction (1) Task 1

This protocol shows that the discourse that deployed between Harry and Larry is argumentative from two perspectives: First Harry and Larry try to construct their own argumentations. Secondly, they confront their arguments critically. It appears that in this argumentation, gestures are enacted synchronously with verbal channels. However, the verbal is very often impoverished and the gestural very rich like in Harry5. Harry6 is an argumentative move – backing his claim that Danny is wrong, done non-verbally – by inscribing a division of parts A, B, and D. Let us look at the continuation of this interaction on Task 2b in which students are encouraged to use scissors.

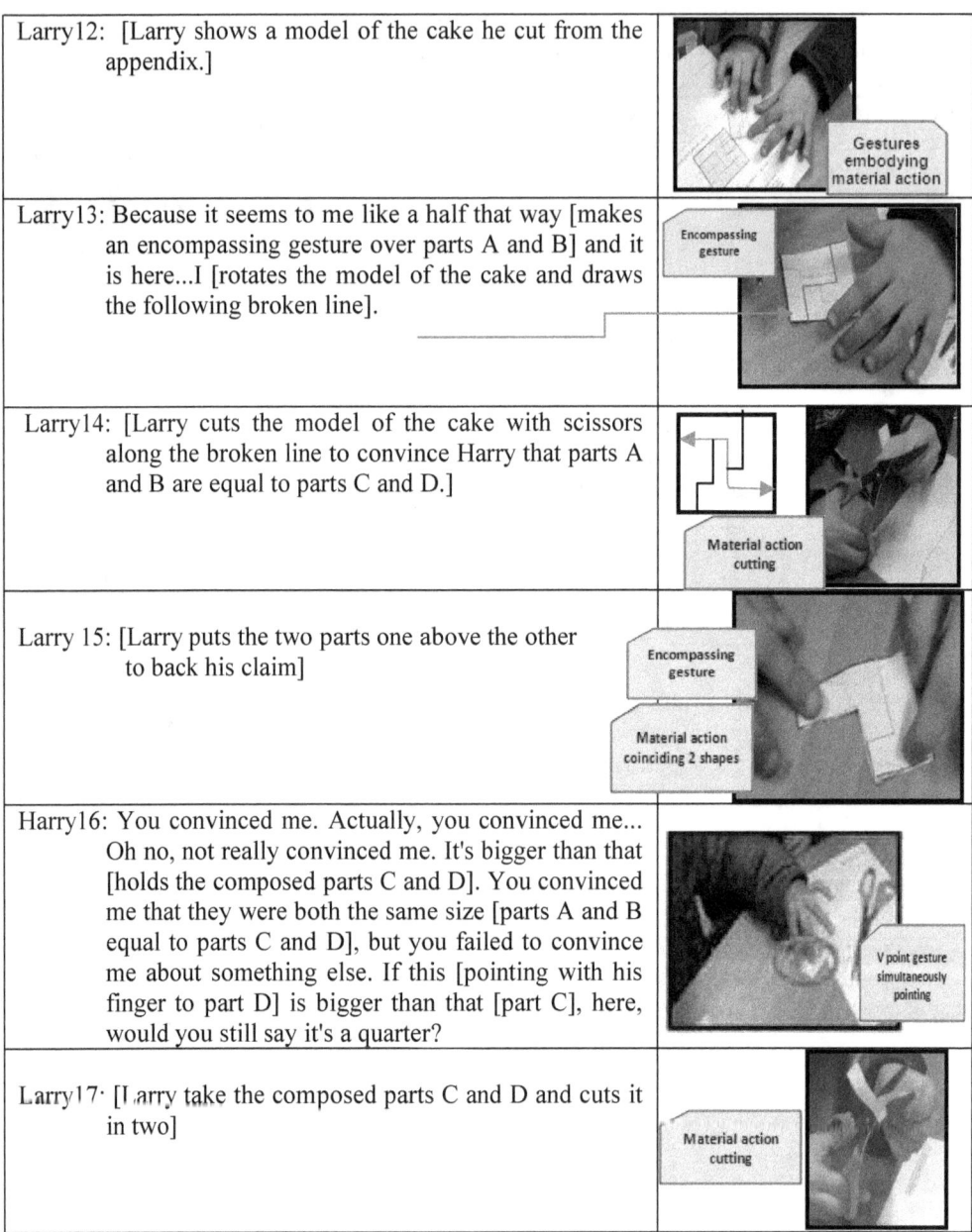

Larry12:	[Larry shows a model of the cake he cut from the appendix.]	Gestures embodying material action
Larry13:	Because it seems to me like a half that way [makes an encompassing gesture over parts A and B] and it is here...I [rotates the model of the cake and draws the following broken line].	Encompassing gesture
Larry14:	[Larry cuts the model of the cake with scissors along the broken line to convince Harry that parts A and B are equal to parts C and D.]	Material action cutting
Larry 15:	[Larry puts the two parts one above the other to back his claim]	Encompassing gesture / Material action coinciding 2 shapes
Harry16:	You convinced me. Actually, you convinced me... Oh no, not really convinced me. It's bigger than that [holds the composed parts C and D]. You convinced me that they were both the same size [parts A and B equal to parts C and D], but you failed to convince me about something else. If this [pointing with his finger to part D] is bigger than that [part C], here, would you still say it's a quarter?	V point gesture simultaneously pointing
Larry17:	[Larry take the composed parts C and D and cuts it in two]	Material action cutting

The importance of multimodality in mathematical argumentation

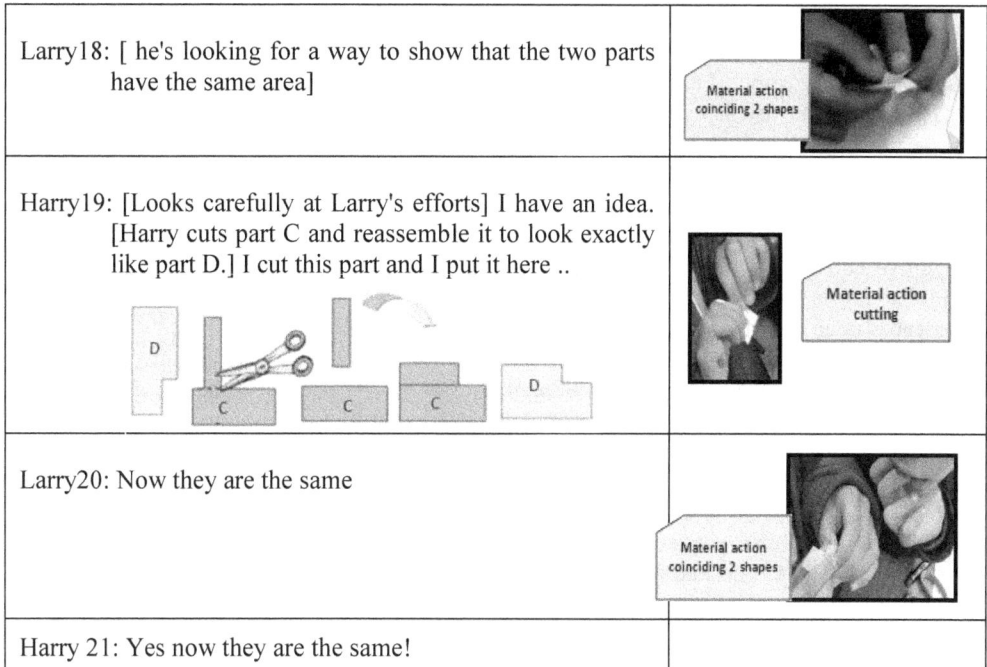

Table 2. Interaction (1) Task 2b

In Larry13, Larry continues his argument by enacting first a gesture embodying a material action: rotating a model, He then drawing a line and cuts the model of the cake. These gestures and material actions are done synchronously with verbal utterances. Here also the verbal is impoverished and the non-verbal is very rich. The material actions and the gesture have an argumentative role as they help bringing more evidence to Larry's argument. And indeed, Harry declares first he is convinced. But he retracts himself, again by using gestures (here a V-point gesture). However, he probably feel that Larry's argument is not logically valid, after he showed seeds of deductive considerations in Harry 11. It is probable that when in Harry16, Harry asserted "You convinced me that they were both the same size [parts A and B equal to parts C and D], but you failed to convince me about something else," he actually felt that the equality of areas between the shapes composed with A and B, and with C and D, and the equality of areas of A and B, does not necessarily imply that C and D have the same area. But this logical idea is too complicated to formulate verbally, and Harry expressed it by integrating verbal, gestural and material actions. Harry19 is a very interesting turn. Harry uses the scissors to cut C and reassemble the new shape coinciding with D. This non-verbal action completes Larry's argument. Multimodality here is diachronic and not synchronic: the gestural here does not occur at the same time as the verbal but the two occur one after the other. It is consecutive to it and to Larry's turns (13, 14, 15, 17, 18) in which Larry used the encompassing gesture several times. This gesture planted the idea of composing and decomposition. It seems reasonable that the encompassing gesture was proleptic to the material decomposition and composition done by Harry: It prepared the ground to this composition/decomposition.

Task 2c

Harry and Larry came to an agreement that Danny is correct at the end of their interaction in Task 2b. Their justification in 2c of the equality of the areas of C and D is different from in the interaction: They backed their claim with through a counting justification. They wrote in their worksheet: "Danny is right because in each part he drew, there are four squares."

In summary, this first interaction was initiated by a socio-cognitive conflict. The socio-cognitive conflict led to a disagreement between Harry and Larry. The design of the task afforded the emergence of a rich argumentative process: students not only constructed evidence-based arguments, but also challenged each other (and even refuted the arguments of the other). This rich argumentation was multimodal. The verbal argumentation was generally impoverished but was accompanied by non-verbal argumentation, gestural and material. The various kinds of gestures involved in multimodal argumentation included pointing, encompassing, measuring, embodied material actions, and comparison gestures. We saw that multimodality was mainly synchronic as the discourse very often deployed in parallel modalities. We identified two instances of diachronic non-verbal argumentation. We showed that this argumentation was productive in the sense that it led to a geometrical conceptual insight. The main function of the multimodality was to help expressing ideas that were too complex in a verbal mode. The compose-and-decompose actions are representative of this complexity. These actions were very often replaced by encompassing gestures, leaving an impoverished verbal discourse, and even a diachronic discourse (with segments that were totally non-verbal).

2.2. Interaction 2 Ofaz and Jonathan: Multimodal Argumentation Mediated by Inscriptions for Solving a Socio-Cognitive Conflict

Like Harry and Larry, Ofaz and Jonathan were in a situation of socio-cognitive conflict: Jonathan initially claimed that Danny was right and Ofaz claimed the opposite. He looked at Danny's drawing and wrote: No! Danny is not right! I measured with my ruler and it is wrong. As we will show it, this initial situation of disagreement led to a discourse which was argumentative and multimodal. However, as we will see, gestures played an additional function to the general function to express ideas that are too complex to be expressed verbally only.

This excerpt clearly shows that Ofaz and Jonathan were first in a situation of socio-cognitive conflict. In Jonathan2, Jonathan opposes Ofaz's claim by enacting a material action – inscribing segments on the given figure to decompose C and D. From his words "I'll show you that I'm right!", it is clear that he already sees the solution. His material actions are intended to convince his peer. In turns 2 to 6, Jonathan tries to be as clear as possible in his explanations. He interacts with material inscriptions (that change as a result of his material actions) demonstrates to Ofaz that he is right. In this ongoing process, Jonathan operates composing and decomposing transformations through the use of arrows on the inscription of the partitioned square and gestures referring to it. Ofaz remains puzzled as it seemed that he could not grasp the idea of decomposing and composing a figure (Ofaz7).

The importance of multimodality in mathematical argumentation

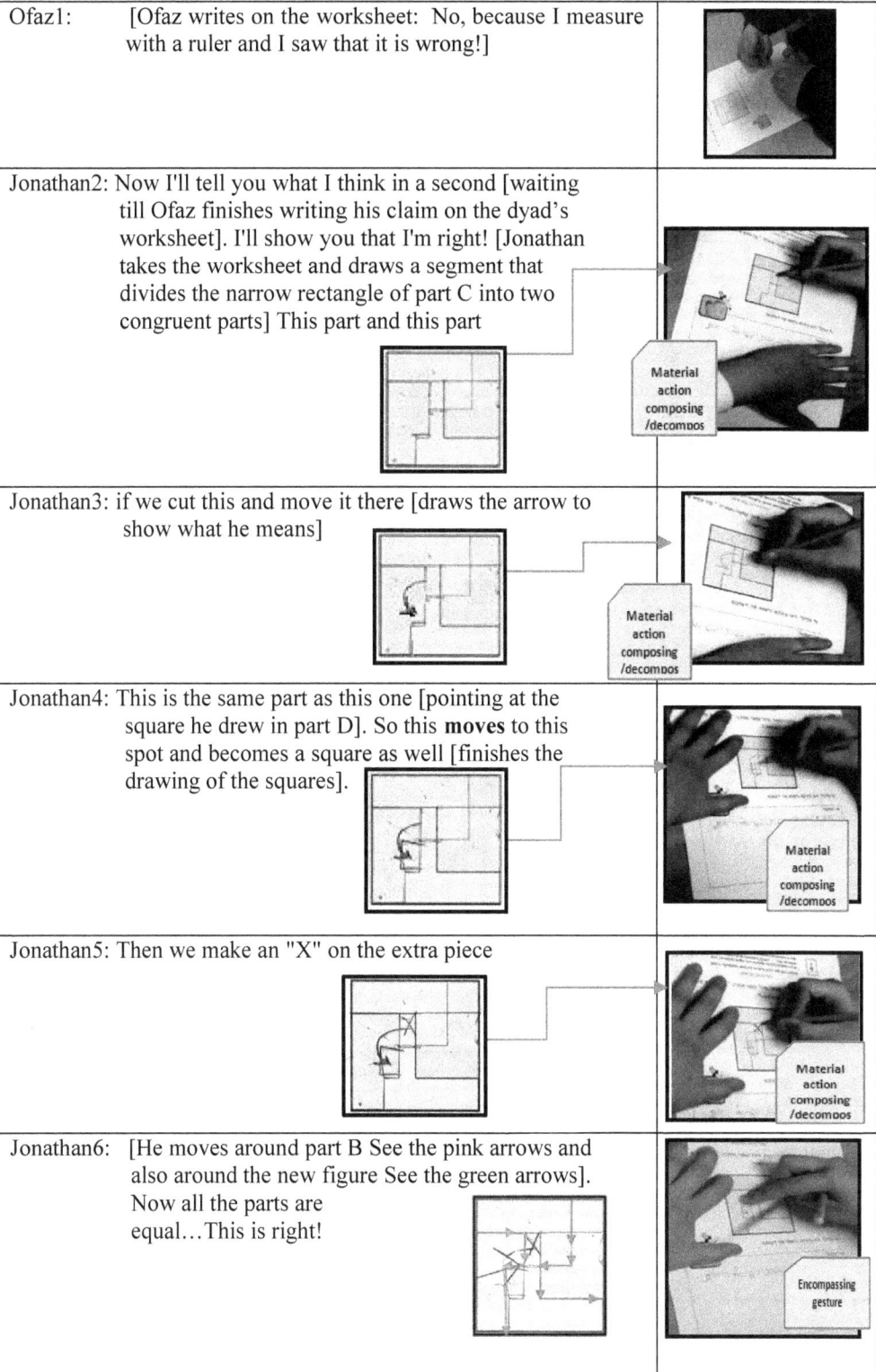

Ofaz7:	But it's impossible to **take away** parts and move them somewhere else. What if I cut the piece of the cake and **took it away**?
Jonathan8:	No! But the parts are equal! [Taps with his pencil on the drawing.] Let's *imagine* that this part was here. Let's say it is.
Ofaz9:	But now it is not somewhere else [pointing with the ruler on the drawing]. *Pointing Gesture*
Jonathan10:	No! I disagree and I'll write my claim now. [he adds his claim on the worksheet]
	Ofaz: No, because I measure with a ruler and I saw that it is wrong! Jonathan: In my opinion yes, if we say a piece is moved [indicated with the arrow] it was there

Table 3. Interaction (2) Task 2a

Task 2b gives the opportunity to our disagreeing dyad, Ofaz and Jonathan, to manipulate a real model while discussing an equitable division of the cake, instead of reasoning and arguing with an inscription. Table 4 displays the actions – verbal and non-verbal undertaken by the dyad.

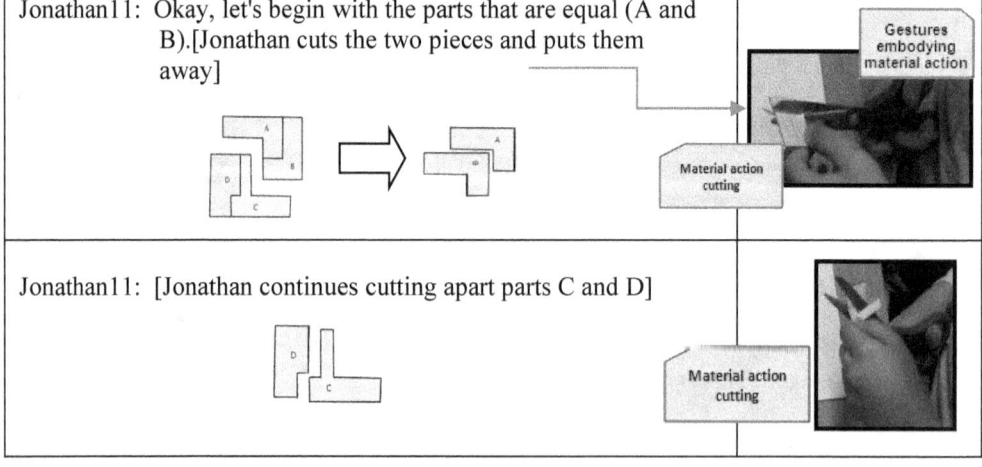

Jonathan11:	Okay, let's begin with the parts that are equal (A and B).[Jonathan cuts the two pieces and puts them away] *Gestures embodying material action* / *Material action cutting*
Jonathan11:	[Jonathan continues cutting apart parts C and D] *Material action cutting*

The importance of multimodality in mathematical argumentation

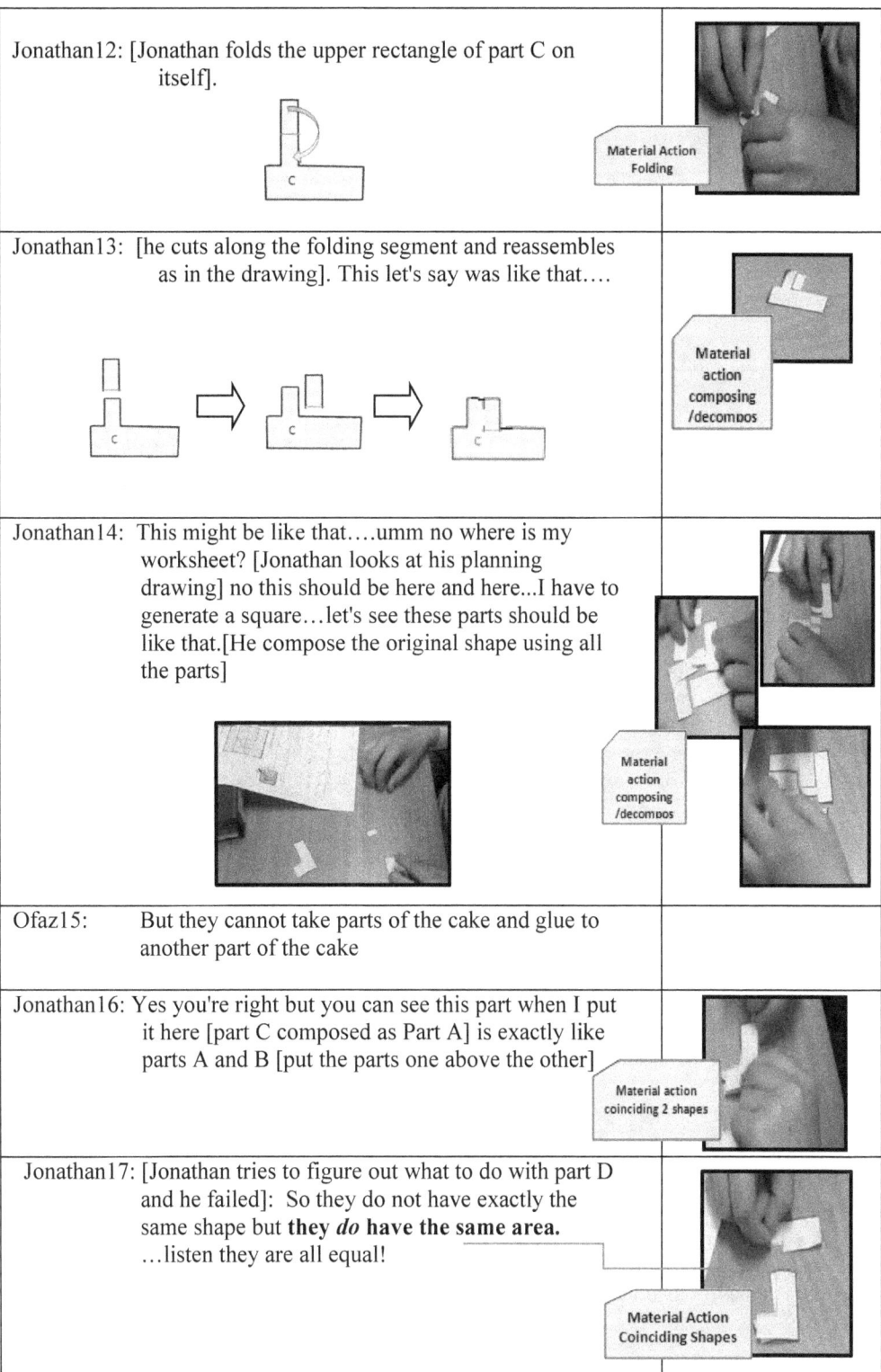

Ofaz18:	But you have here an extra piece…so I was right!! [Shows that the parts do not coincide.]	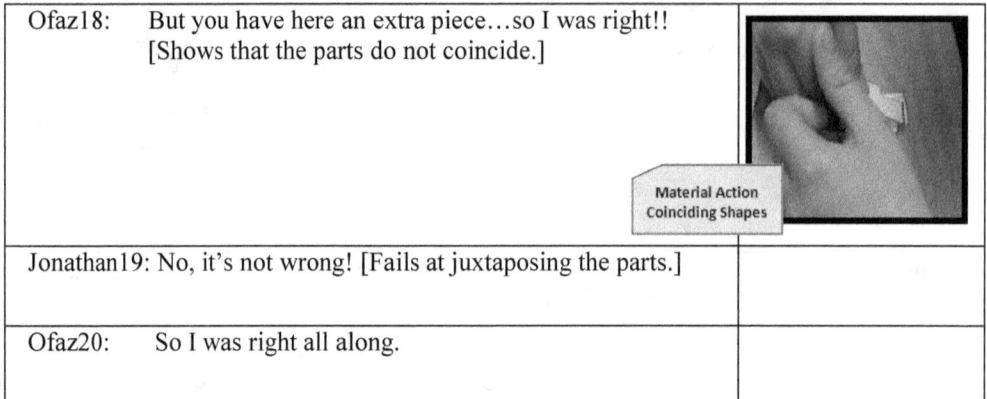 Material Action Coinciding Shapes
Jonathan19:	No, it's not wrong! [Fails at juxtaposing the parts.]	
Ofaz20:	So I was right all along.	

Table 4. Interaction (2) Task 2b

The continuation of the interaction between Jonathan and Ofaz shown in Table 4 consists more or less of successive efforts done by Jonathan to persuade Ofaz that Danny is right. He knows that if two shapes have the same area, operations of decomposition and (re-)composition lead one shape to coincide with the other. His reasoning in Jonathan11, 12 and 13 is abductive: his premise is that C and D have the same area (perhaps because this is what he feels intuitively. His swiftness to materially decompose C in Jonathan12 and recompose it in what he thinks to be D suggests that when being given the inscription representing the partition of the cake, he could reason and imagine this decomposition/composition. The turn 11-17 are actions Jonathan does in order to find evidence that his claim is correct. The verbal is impoverished or totally inexistent. Material actions and gestures provide the dominating modalities. Jonathan's failure to get the decomposed then recomposed shape made of D coincide with C leads him to declare in turn 17 "So they do not have exactly the same shape but they do have the same area. …listen they are all equal" reinforces our hypothesis that the inscriptions help him imagine the partition. Ofaz refutes Jonathan's argument by showing that two shapes (C recomposed, and D) do not coincide. Turn 20 where Ofaz declares "So I was right all along" confirms that this excerpt is of an argumentative nature from two perspectives: Jonathan's efforts to bring evidence to establish his argument, and the dialectic process between Ofaz and Jonathan during which Jonathan tries to strengthen his argument, and Ofaz refutes it by demonstrating that one shape obtained by transforming a shape through decompositions and compositions is not amenable to the other shape. But this refutation is ephemeral: In Task 2c, the dyad is given a proper inscription to reason with. With the help of this new inscription, the dyad will find the right way to see that C and D have the same area.

Task 2c

The provision of a grid in Task 2c helped the dyad reached an agreement on the correctness of Danny's solution; they succeeded in doing so by using the counting strategy. Ofaz wrote on their group worksheet (See figure 5): "Jonathan was right, but I changed my opinion because in each part there are 4 square units". Yet it is worth noting that they also indicate the composing and decomposing transformation that should be done on parts C and D. Here too, the interaction involved rich multichannel argumentation processes.

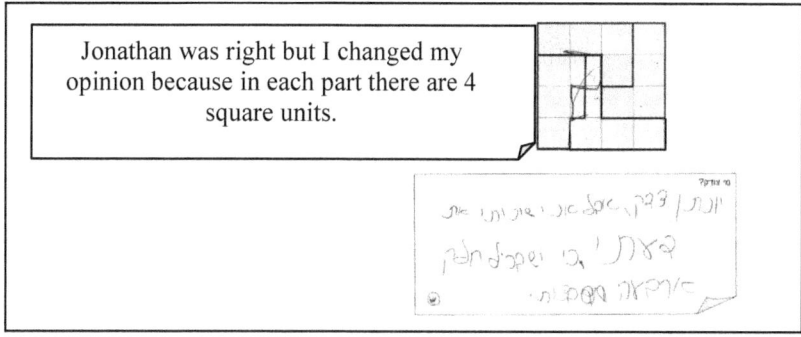

Figure 5. Final conclusion in task 2c

3. Discussion

This chapter was about multimodal argumentation in mathematics in learning contexts. We focused on two cases but those cases were representative in the sense that the educational design we provided enabled the creation of a socio-cognitive conflict: we arranged students in small groups (generally dyads); we knew that the geometric idea at stake (the fact that non-congruent shapes can have the same area) is not well understood among Grade 3 students. We designed a task affording multiple solutions and strategies. The two examples we brought are representative in the sense that many of groups found themselves in a situation of socio-cognitive conflict. Figure 4 shows that 16% of the students were in a situation of disagreement in the beginning of the task, and 20% agreed upon the wrong conclusion, which created another kind of socio-cognitive conflict. As aforementioned, for most of the 64% of the students who agreed that Danny is right by the beginning of Task 2, the justifications they provided were not satisfactory or unarticulated. This fact suggests that they were uncertain how to decide between the proposed opposing positions, that of Danny and that of Mindy. They were then also in a situation of socio-cognitive conflict. In this situation, we have presented two examples of interaction that show rich argumentation between peers. In other words, a proper design triggered the creation of a socio-cognitive conflict that led to rich argumentation. Argumentation was productive in the sense that in Task 2c, a written task, students' justifications showed that they learned the idea at stake, the fact that non-congruent shapes can have the same area.

We exemplified two instances of interactions in which argumentation were multimodal. As aforementioned, this phenomenon was ubiquitous in the design we elaborated. As we already mentioned in the introductory part, this phenomenon is not new in mathematical argumentation. In particular, we showed, as suggested by Goldin-Meadow and colleagues, that the enactment of certain gestures lessen cognitive load. This was the case of Harry's encompassing gesture that expressed the decomposition of a shape and the re-composition of another one, in order to demonstrate that the proposed partition was not equitable. What we showed, however, was that the verbal modality was generally impoverished in comparison with material actions and gestures. In fact, we showed that non-verbal modalities were generally synchronic with impoverished verbal modality and that non-verbal modalities were sometimes diachronic with the verbal.

One of the most important contributions of this paper is that we pointed out at the communicative functions of the non-verbal. As mentioned in the introductory part, multimodality in mathematical reasoning was explained by the cognitive needs of the children when engaging in demanding tasks. Argumentation in the present study was social, and argumentative moves were intended to show, to explain to the other, to persuade, to challenge or to refute. Of course, through these argumentative moves, which were very frequently dialectical, arguments were (co-)constructed, but the primary function of the argumentative moves was social. Needless to say, the function of the pointing gestures was generally much more basic than the other gestures: It primarily invited the other to follow the reasoning of his/her peer. Of course, our insights should not diminish the importance of gestures and other non-verbal actions for cognitive functioning.

The first example showed the importance of encompassing gestures to learn about geometry. The second example stresses the importance of inscriptions in reasoning. Again, this phenomenon is not new. In fact, it was discovered by Charles Sanders Peirce more than one hundred years ago (Peirce, 1931, the collected papers are posthumous). Peirce noticed that in the presence of inscriptions, the mathematician often adopts an abductive approach: The figure suggests the probable conclusion and the solver takes it for granted to ways that lead to this already know conclusion. This is exactly what Jonathan is doing when he tries to decompose part D and to re-compose it to obtain part A, which is congruent to part B. His failure to make the re-composed part coincide with C and his insistence on the fact that C and D have the same area suggest that Jonathan could imagine a partition based on the inscription. He understood that his partition does not make the job, but that the problem is not that he is wrong: He simply failed at finding the right cutting procedure. But he could imagine through the inscription given to him that such a partition exists. And although Ofaz is well aware that Jonathan failed at refuting his claim, he co-writes in Task 2c a solution based on Jonathan's imagination, probably because the grid provided helps Jonathan to propose a way to count the number of squares needed to cover D (which is in fact a new partition of D).

In conclusion, we showed that multimodal argumentation helped resolve a socio-cognitive conflict. The multimodality helped in handling the situation of disagreement on a challenging task. The two arguers tried to challenge each other or to persuade each other that they were right. The functions of gestures were multiple. First, their function was communicative: they helped solvers indicate their peers their problem-solving moves. Secondly, gestures helped solvers lessen their cognitive load. In some cases, gestures expressed abstract ideas – like the encompassing gesture that helped expressing the compose/decompose idea. The role of inscriptions was crucial for the intertwining of material non-verbal actions in mathematical argumentation. The inscriptions children were given and the inscriptions they themselves created served as new evidence strengthening their own positions. In this situation, whenever an argument was elaborated, it was not accepted by the peer until the end of the argumentation. The model of such an argumentation is pragma-dialectical. However, the central role of gestures, other non-verbal actions and inscriptions, bestow to mathematical argumentation in learning contexts a distinctive character.

References

Anderson, R. C., Chinn, C., Chang, J., Waggoner, M. & Yi, H. (1997). On the Logical Integrity of Children's Arguments. *Cognition and Instruction, 15*(2), 135-167.
Arzarello, F. (2008). The proof in the 20th century. In P. Boero (Ed.), *Theorems in schools: From history epistemology and cognition to classroom practices* (pp. 43–64). Rotterdam: Sense Publishers.

Arzarello, F. & Sabena, C. (2011). Semiotic and theoretic control in argumentation and proof activities. *Educational Studies in Mathematics, 77*, 189–206 DOI 10.1007/s10649-010-9280-3.

Asterhan, C. S. C. & Schwarz, B. B. (2016). Argumentation for Learning: Towards the development of a theoretical model. Educational Psychologist.

Asterhan, C. S. C., & Schwarz, B. B. (2009). Argumentation and explanation in conceptual change: Indications from protocol analyses of peer-to-peer dialogue. *Cognitive Science, 33*(3), 374-400.

Birdsell, D. S. & Groarke, L. (1996). Towards a theory of visual argument. *Argumentation and Advocacy, 33*(1), 1-10.

Birdsell, D. and Groarke, L. (2007). Outlines of a theory of visual argument. *Argumentation and Advocacy, 43*(3–4),103–113.

Boero, P., Douek, N., Morselli, F., & Pedemonte, B. (2010). Argumentation and proof: A contribution to theoretical perspectives and their classroom implementation. In M. M. F. Pinto & T. F. Kawasaki (Eds.), Proceedings of the 34th Conference of the International Group for the Psychology of Mathematics Education, vol. 1 (pp. 179–204). Belo Horizonte: PME.

Butera, F. & Mugny, G. (1995). Conflict between incompetences and influence of a low-expertise source in hypothesis testing. *European Journal of Social Psychology, 25*, 457-462.

Darnon, C., Butera, F., & Harackiewicz, J. M. (2007). Achievement goals in social interactions: Learning within mastery vs. performance goals. *Motivation & Emotion, 31*, 61-70. DOI 10.1007/s11031-006-9049-2.

Darnon, C., Muller, D., Schrager, S. M., Panuzzo, N., & Butera, F. (2006). Mastery and performance goals predict epistemic and relational conflict regulation. *Journal of Educational Psychology, 98*, 766-776.

Duval, R., Ferrari, P. L., Høines, M. J. & Morgan, C. (2005). Language and Mathematics. *CERME4/CERME4_WG8.*

Duval, R. (2006). Les conditions cognitives de l'apprentissage de la géométrie: développement de la visualisation, différenciation des raisonnements et coordination de leur fonctionnement. *Annales de Didactique et de Sciences Cognitives, 10*, 5-53.

van Eemeren, F. & Grootendorst, R. (2004). *A systematic theory of argumentation: The pragma-dialectical approach.* Cambridge University Press.

Forceville, C. (2008). "Metaphor in pictures and multimodal representations." In Raymond W. Gibbs (Ed.), *The Cambridge Handbook of Metaphor and Thought* (pp. 462-482). Cambridge: University Press.

Gehlen, A. (1988). *Man. His nature and place in the world.* New York: Columbia University Press.

Goldin-Meadow, S., Nusbaum, H., Kelly, S. D. & Wagner, S. (2001). Explaining Math: Gesturing Lightens the Load. *Psychological Science, 12*(6), 516-522.

Knipping, C. & Reid, D. (2013). Revealing structures of argumentation in classroom proving processes. In A. Aberdein and I. J. Dove (Eds.), *The Argument of Mathematics* (pp. 181–208). Dordrecht, Springer.

Krabbe, E. C. W. (2013). Arguments, Proofs and Dialogues. In A. Aberdein and I. J. Dove (Eds.), *The Argument of Mathematics* (pp. 21–34). Dordrecht, Springer.

Larvor, B. (2013). What philosophy of mathematical practice can teach argumentation theory about diagrams and pictures. In A. Aberdein and I. J. Dove (Eds.), *The Argument of Mathematics* (pp. 209–222). Dordrecht, Springer.

Mugny, G., & Doise, W. (1978). Socio-cognitive conflict and structure of individual and collective performances. *European Journal of Social Psychology,* 8(2), 181–192.

Novack, M. A., Congdon, E. L., Hemani-Lopez, & Goldin-Meadow, S. (2014). From action to abstraction: using the hands to learn math. *Psychological Science*, *25(4)*, 903-910. doi:10.1177/0956797613518351.

Pedemonte, B. (2007). How can the relationship between argumentation and proof be analysed? *Educational Studies in Mathematics,* 66(1), 23–41.

Peirce, C. S. (1931/1958). Collected Papers (Vol. I-VIII). In C. Hartshorne, P. Weiss, & A. Burks (Eds.), Cambridge, MA: Harvard University Press.

Perret-Clermont, A.-N. (1980). *Social interaction and cognitive development in children.* London: Academic.

Piaget, J. (1975). *The child's conception of the world.* Totowa, NJ: Littlefield, Adams (Originally published 1932).

Prusak, N. Hershkowitz, R. & Schwarz, B. B. (2013). Conceptual Learning in a principled design problem solving environment. *Research in Mathematics Education,* 15(3), 266-285.

Radford, L. (2009). Why do gestures matter? Sensuous cognition and the palpability of mathematical meanings. *Educational Studies in Mathematics,* 70(2), 111-126.

Schwarz, B. B. & Baker, M. J. (2016). Dialogue, Argumentation and Education: History, Theory and Practice. Cambridge University Press.

Schwarz, B. B., Hershkowitz, R. & Prusak, N. (2010). Argumentation and Mathematics. In C. Howe and K. Littleton (Eds.), *Educational Dialogues: Understanding and Promoting Productive Interaction* (pp. 115-141). Routledge.

Chapter 23

The Psychology of Far Transfer From Classroom Argumentation

E. Michael Nussbaum[1] & Christa S. C. Asterhan[2]

[1] University of Nevada, Las Vegas, Las Vegas, Nevada, USA, nussbaum@unlv.nevada.edu; [2] Hebrew University of Jerusalem, Jerusalem, Israel, asterhan@huji.ac.il

Abstract. Certain classroom programs that engage students in argumentive discourse over an extended period of time have been shown to result in far transfer effects in other disciplines. For example, argumentation-rich teaching in science classes or mathematics has resulted in higher student achievement in English Language Arts. In this chapter, we review previous explanations for these effects rooted in theories of development, argumentation schema, ACT-R theory, motivation, and situativity. We then extend these accounts by proposing that in these programs, students discover and practice "proactive executive control strategies". These strategies involve intentionally activating or inhibiting a certain cognitive process, such as protection from interference. The acquisition and strengthening of these strategies has been used to explain far transfer effects from working memory training to tests of fluid intelligence, based on a cognitive architecture proposed by Taatgen (2013). We propose that similar processes may be at work in argumentive learning environments. For example, when one is considering someone else's counterargument, one has to protect the mind from interference by one's own argument, and then switch attention back to one's argument to advocate or evaluate it. Our account is consistent with those explaining far transfer effects from the generation of general production rules (Koedinger & Stampfer, 2015) as well as the acquisition of conceptual agency through participation in conversations that matter (Greeno, 2006). Our theory also has the advantage, however, of uniting various levels of cognitive analysis, from the micro to the more molar.

1. Introduction

The identification of far transfer from long-term participation in argumentive discourse could be construed as revolutionary, challenging long-standing assumptions about the domain-specificity of learning. Our view, however, is that this progress is evolutionary, reflecting the refinement of theories of transfer that have been in existence for the last thirty years. There is, though, a new conceptualization of these theories, specifically by Taatgen (2013), that explains transfer effects more completely than previous theories.

This chapter reviews the literature on transfer effects from argumentive discourse, particularly that presented in a recently published volume edited by Resnick, Asterhan, and Clarke (2015). We then discuss previously proposed mechanisms and present a new proposal based on Taatgen's cognitive architecture. A central idea is that extended argumentive discourse develops proactive executive control strategies.

2. Argumentive Discourse and Transfer

Over the last four decades there has been rising interest among scholars from various disciplines in the role of discussion in learning and cognitive development. This research has been conducted by scholars from multiple theoretical perspectives, has employed a wide range of different methodologies, and has focused on different aspects of both learning and dialogue, and on different social settings. Even though each of these research communities has generated its own body of evidence, until recently these remained separate 'islands' of research. In a 2011 research conference and its subsequent edited volume (Resnick et al., 2015), these bodies of research and evidence were brought together to take stock of what we know about dialogue that promotes learning.

First, across research traditions there is consensus on the type of dialogue that promotes learning and development (Chi & Menekse, 2015; Howe & Abedin, 2013; Mercer & Littleton, 2007; Resnick et al, 2015). This kind of talk begins with inviting students to think aloud about a domain concept, and explain and reflect upon their own reasoning. They make public their half-formed ideas, questions, and explanations. Other students take up their classmates' statements: explaining or clarifying them, adding their own questions, reasoning about a proposed solution, or offering a counter claim or an alternate explanation. Discussion may be steered by a teacher, or peer-led. It may be conducted in whole classrooms, small groups or dyads, or even with an intelligent computer agent. The dialogue typically is not adversarial (Asterhan & Babichenko, 2015), hence our use of the term 'argumentive' rather than 'argumentative'.

Secondly, there is ample evidence from both descriptive and experimental research showing that argumentive discourse is associated with improved retention and understanding of subject matter. For example, students who discussed and challenged each other's explanations of different evolutionary phenomena improved their understanding of natural selection, as shown by improved performance on near transfer items a week later and compared to different control conditions (Asterhan & Schwarz, 2007). Likewise, Pauli and Reusser (2015) analyzed the quality of teacher-led classroom discussions in a three-lesson unit on the Pythagorean Theorem. Students' mathematics achievement on that topic was assessed before and after the unit. Even though overall classroom argumentation in these natural settings was not very frequent, it significantly predicted achievement scores, even after controlling for various variables (including student interest and coherence of mathematical content presentation). These retention and near transfer effects are commonly explained in terms of superior information processing (e.g., Chi & Menekse, 2015) and improved detection of misconceptions and errors (Asterhan & Babichenko, 2015; Chin & Osborne, 2009).

Third, argumentive discussions about multiple texts, moral dilemmas and other public concerns produces improvements in students' ability to write persuasive essays on that topic (e.g., Frijters, ten Dam & Gijlaartsdam, 2008; Kuhn, Shaw & Felton, 1997) and on new topics (e.g., Kuhn & Crowell, 2011; Reznitskaya, Anderson, Dong, Kim, & Kim, 2008), as well as to engage in argumentative discussions on novel topics (Crowell & Kuhn, 2013; Kuhn, 2010). The outcome measures of these interventions all target improvement in oral or written argumentation products, that is: the skills of argumentation (as measured, for example, by the number of counterarguments and rebuttals generated). This type of domain-specific skill transfer is explained by practice and skill refinement, and the internalization of the norms of argumentation (Kuhn, Zillmer, Crowell, & Zavala, 2013).

The majority of published research on argumentation-based learning has established a solid base of knowledge about how academic classroom discourse proceeds and how concepts are developed interactively. It has also been successful in showing that certain forms of discussion-based activities result in improved retention and understanding of the subject matter and improved capabilities to engage in oral and written argumentation.

In addition to these near transfer effects, however, a small, yet impressive set of studies shows that the benefits of structured classroom argumentation reaches beyond the discourse and the topic of discourse itself. It can lead to steep increases on standardized test scores years after the intervention program, transfer to other disciplines, and rises in cognitive skills and fluid intelligence. These are instances of far transfer.

When highly skilled teachers teach primary grade mathematics to previously underachieving students using a whole-class instruction method in which a problem is posed to the class and the teacher then leads a structured discussion of different proposed solutions (where students explain, challenge, and develop their own and others' ideas), steep changes in standardized test math scores occur, and there is transfer to reading test scores and retention of both reading and math gains for up to three years. Comparison students remained at low performance levels (O'Connor, Michaels & Chapin, 2015; Resnick, Bill, Lesgold & Leer, 1991). When middle school students participate for two years in a science education program where they are prompted to articulate, explore, and juxtapose their thoughts about various 'Thinking Science' activities, they perform significantly better than control groups on national achievement tests in science, taken three years later. Moreover, although the intervention was in science, large differences were also found in mathematics and English Language Arts (Adey & Shayer, 1993; Shayer, 1999). Several replications have been reported since (see Adey & Shayer, 2015 for a synopsis).

When upper elementary children are randomly assigned to participate in lessons of philosophical enquiry involving interactive dialogue in small groups for 1 hour a week over 16 months, they show improved non-verbal and verbal reasoning on standard tests of cognitive ability, compared to control subjects who participated in the regular curriculum. Moreover, these differences were retained two years later after they had transferred to secondary school (Topping & Trickey, 2007).

3. Some Possible Transfer Mechanism

Although the transfer effects of argumentive discourse have been well documented, the underlying causal mechanisms are unclear, particularly of the far transfer effects. Several mechanisms have been proposed, both cognitive and motivational.

3.1. Developmental Theories

One explanation, based on a Piagetian framework, is that dialogic instruction helps to develop formal operational thought. Adey and Shayer (2015) have used this account to explain the far transfer effects of their Cognitive Acceleration program (specifically from science to English Language Arts). Formal operations involve hypothetical and combinatorial reasoning (Inhelder & Piaget, 1955/1958), and dialogic instruction that involves argumentation and scientific reasoning would develop these general structures. Note that the term 'formal operations' was inspired by formal logic. Formal logic involves logical predicates (conditionals, disjunctions, negations) which are involved in argumentative reasoning.

It could be counterargued that argumentative reasoning is for the most part *informal*, but we find this term problematic. The term informal reflects, in logic, attention to issues of relevance, truth, and meaning, but these concepts can be formalized. In philosophy and computer science, various formal logics have been developed as alternatives to traditional, first-order logic (Stenning, 2002). Based on these logics, Douglas Walton and colleagues have specified various argument schemes (1996), such as 'argument from authority' or 'popular opinion', that are associated with informal fallacies but are also cogent forms of reasoning in many circumstances. These schemes include critical questions used to assess argument cogency. The schemes are not as general as formal logical operations, such as 'modus ponens', but nor are they domain specific. The schemes have a mid-level degree of generality. To the degree that individuals develop cognitive structures that are scheme based, some degree of transfer would be possible.

A larger problem with the formal operational account is that a Piagetian framework is inconsistent with much contemporary research in cognitive development. The Piagetian stages do describe general theoretical trends but admit of too many exceptions, anomalies, and intraindividual variation to serve as a convincing basis for an explanation of far transfer effects. More promising are neo-Piagetian theories that use development of functional working memory capacity, in combination with knowledge, as a driver of cognitive development (Fischer & Lazerson, 1984). For example, Case (1992) found that the ability to simultaneously attend to multiple dimensions of a problem (such as length and width in a conservation problem), and to coordinate and integrate these dimensions, contributes to higher cognitive performance.

Nussbaum and Edwards (2011) provide a relevant application. They documented how one seventh-grade student developed a cost-benefit schema through a gradual process of first differentiating and then integrating costs (i.e., disadvantages) and benefits through dialogic argumentation. The schema enabled transfer to a novel problem. This work was grounded in Nussbaum's (2008) argument-counterargument integration framework, which specifies cognitive operations such as weighing, generating and evaluating designs, and different refutational moves as mechanisms for integrating arguments and counterarguments into an overall position. The framework was inspired by neo-Piagetian notions of differentiation and integration, which are also found in work on 'integrative complexity' (Suedfeld, Tetlock, & Streufert, 1992). In addition, the argument-counterargument integration framework extends earlier work by Leitão on integrative replies (Leitão, 2000).

3.2. Argumentation Schema Theory (AST)

It is plausible that participants in argumentive discussions tend to construct various types of argument schemas that could serve as mid-level cognitive structures that support transfer. Argument schemas are different from schemes in that the latter are normative whereas the former are more psychologically based (and may be in various stages of development). Argument schemas are generic cognitive structures that include slots for various components of arguments, such as arguments, counterarguments, and integrations (Leitão, 2000), or grounds, warrants, backing, rebuttals, and qualifications (Toulmin, 1958). Bereiter and Scardamalia (1982) describe discourse schema for persuasion that include a statement of belief and slots for at least one reason, elaboration, example, statement on the other side, reason against, and conclusion.

Reznitskaya and Anderson (2002) proposed an expanded notion of argument schema as "a network that connects individual arguments, representing extended structures of argumentative discourse" (Chinn & Anderson, 1998, p. 321). They hypothesize that argument schemas function to direct attention to relevant information, facilitate comprehension of other's arguments, support retrieval of argument-relevant information from memory, organize argument-relevant information, and provide a basis for finding flaws. A key structural assumption is that argument schemas are composed of 'argument stratagems', which are tactics used in persuasion and reasoning. Anderson *et al.* (2001) provide examples of stratagems that 'snowballed' during collaborative reasoning discussions, i.e., were used with increasing frequency by different students. These examples included:

- I agree (or disagree) with [NAME].
- Hedge [PROPOSITION].
- What if [SCENARIO]?
- If [ACTION], then [BAD CONSEQUENCE], so [NOT ACTION].

Their notion of argument schema highlights the dialogic and pragmatic nature of such schemas, and not only their structural aspects. According to Anderson *et al.* (2001), "A complete argument stratagem is comprised of information about (a) the purpose or function of the stratagem, (b) the condition in which the stratagem is used, (c) the form the argument takes, (d) the consequences of using the stratagem, and (e) the possible objections to this form of argument" (p. 2).

Anderson and colleagues have used their Argumentation Schema Theory to explain the transfer effect from oral argumentation (specifically Collaborative Reasoning discussions) to written essays (Reznitskaya & Anderson, 2002). Essays from students participating in Collaborative Reasoning contained more arguments, counterarguments, and rebuttals than those from a comparison group (Reznitskaya *et al.*, 2008). The transfer task presumably activated an argument schema which contained these structural and rhetorical aspects. Although these would not be considered far transfer effects, as the essay writing was performed in the same class setting as the dialogue activities, argument schemas could plausibly be involved in far transfer effects.

3.3. ACT-R Theory

Another albeit complementary theory of transfer was proposed by Singley and John R. Anderson (1985). The theory is now rooted in Anderson's cognitive architecture, the Adaptive Control of Thought – Rational (ACT-R) and focuses on skill transfer. Procedural knowledge is mentally represented by if...then production rules which gain strength every time they are successfully used. When faced with a task, there may be competing production rules that are relevant, but the one with the greatest strength will fire and the others inhibited (Anderson & Lebiere, 1998).

According to ACT-R theory, the probability that learning on one task will transfer to another is a function of the number of production rules that the two tasks share in common. Also, in learning a new task, if there are no preexisting production rules directly relevant to a new task, an individual will search long-term, procedural memory for rules that share common goals or conditions and adapt the most relevant rule for the new situation by creating a more general rule. This process of generalization enables individuals to learn a new task quicker than would be the case otherwise; the reduction in time is known as 'savings'.

For example, Singley and Anderson (1985) studied whether training on the use of one text editor, such as ED or EDT, transferred to the learning of another, such as Emacs. There were in fact savings proportional to the number of shared goal conditions and subgoals. As an illustration of shared goals, deleting a line was a feature of all text editing systems but this goal was realized through different keystrokes (e.g., control-n vs. period-d). Participants in the experiment learned Emacs significantly faster if they had been previously trained on a line-based editor (ED or EDT).

Koedinger and Stampfer (2015) suggest that far transfer effects from argumentive discourse could be explained by the generation of rather general production rules, such as:

- If I'm considering a claim, search for evidence *for* it.
- If I'm considering a claim, search for evidence *against* it.
- If given evidence against a claim, find a counterargument.
- If a dialogue/argument norm is being violated, point it out (p. 275).

Although these rules are general, these authors emphasize that the transfer effect is not a general effect but the product of various skills, principles, and schemata (they call these 'knowledge components', or KCs). They postulate that KCs are not learned best through direct instruction but by engaging in discourse. Through discourse, students can observe and imitate talk moves made by others and induce KCs (including productions rules). Students also need to participate in dialogues frequently to strengthen and tune these rules and acquire fluency.

We note that similar claims are made by schema theorists, and that argument stratagems are likely mentally represented by production rules. Production rules specify both goals and conditions under which an action should be performed, which Reznitskaya and Anderson (2002) emphasize are important elements of argument stratagems.

3.4. Motivation

Argumentive reasoning typically involves effortful cognitive processes, for example, searching for relevant arguments and information, or coordinating and integrating arguments and counterarguments, etc. For transfer to occur, it is imperative that individuals be motivated to engage in these processes. Student interest (which we discussed previously as a control variable) is part of motivation, but motivation also involves various other psychological constructs (e.g., effort, value-expectancies, competence beliefs, etc.).

Engaging in controversy can be intrinsically motivating to students, especially if a student finds that some of their peers have different opinions. This can cause cognitive conflict or 'epistemic curiosity' (Berlyne, 1954). Students also have greater freedom of choice, which is intrinsically motivating (Ryan & Deci, 2000). Furthermore, Aristotle (trans. 2000) argued that using one's intellectual abilities is pleasurable. Increased engagement from argumentation can improve conceptual understanding (Asterhan & Schwarz, 2007) and can result in near transfer effects on tests that were the focus of instruction.

There may of course be individual differences in the degree of enjoyment individuals receive from engaging in effortful cognitive process, as reflected in such constructs as need for cognition (Cacioppo, Petty, & Kao, 1984) or the tendency to engage in active open-minded thinking (Baron, 1988). These reflect 'thinking dispositions' (Stanovich & West, 1997). These dispositions may be inculcated by recurrent engagement in argumentive discourse. Other motivational constructs that may be developed through argumentation in classrooms include positive beliefs about intelligence and learning (Dweck, 2006), persistence when faced with academic difficulties, and academic self-efficacy, each of which could contribute to better performance on transfer tasks, for example standardized tests (Resnick, 2015). The recurrent positioning of children as reasoners and as participants in collaborative problem solving and exploration may eventually be internalized, becoming an individual disposition (Gresalfi, 2009). This disposition can serve as a spark to activate argumentation schemas and related knowledge elements in new contexts. Conversely, activation of an argumentation schema in a new context can activate motivational tendencies, or, even more likely, the two may be coactivated. Regardless, the motivational mechanisms underlying far transfer are different from simply the increased conceptual engagement underlying near transfer.

Under a dispositional account, a middle school student participating in an argumentive program such as Cognitive Acceleration in Science Education (CASE) may learn not to shy away from intellectual challenges, would come to learn better from their errors, and would persist more often when faced with failure. This could translate into better achievement scores in different content domains during high school, as was found by Adey and Shayer (2015).

3.5. Situative Accounts

Rooted in situativity theory (Greeno & Moore, 1997), which focuses on individual-environment interactions, Greeno (2006) and Engle (2006) have argued that transfer effects likely involve 'conceptual agency' (Pickering, 1995), where the actor appropriates and modifies material or conceptual resources for the purpose of his/her activity and where the action is consequential. The notion of conceptual agency has some similarities to academic self-efficacy, but is broader in that the action can be framed as having consequences that go beyond the immediate school setting and contribute to an on-going dialogue about real-world issues that matter. Engle (2006) found that such framing promotes transfer, specifically learning framed expansively as connected to prior and future classroom activities, and to discourse and audiences beyond the classroom. Such discourse is metadiscursive, as "the contents of and participants in a particular episode are made relevant to other contents and settings to construct generality" (Greeno, 2006, p. 537). Individuals who exercise conceptual agency are creative and adaptive; they become authors of their own contributions, rather than merely reciting the contributions of others, and are accountable to the norms and standards of the classroom and broader intellectual communities in which they participate. They have authority to both problematize and resolve issues (Engle & Conant, 2002). In our view, transfer effects of participation in dialogic environments could be partially explained by the development of learner identities that involve conceptual agency and accountability. Although conceptual agency appears to be domain specific, what constitutes the boundaries of a domain are not fixed and could be quite broad; in fact, a learner could develop an identity as an intelligent novice (Bruer, 1993) and good problem solver who attempts to learn conceptual constraints and affordances of more specific domains.

We have presented conceptual agency as an individual characteristic, but it is also affected and supported by the activity system in which an individual is positioned. Conceptual agency can be analyzed at both the more molar, sociocultural level and at a more micro, individual level.

4. A More Refined Approach: Proactive Executive Control Strategies

In this section, we propose a more refined account than previous ones of far transfer effects from participation in argumentive discourse. The proposal offers mechanisms underlying dispositions, production rules, and argument schema, and complements situative concepts such as conceptual agency. At the heart of the proposal is the concept of 'proactive executive control strategies' (PECS) (Braver, Gray, & Burgess, 2007). In this section, we explain what PECS are and their place in a theory of far transfer developed by Niels Taatgen (2013), which we apply to the domain of argumentation. We first, however, apply the theory to working memory training and transfer, as this analysis will provide some insight into argumentive transfer.

4.1. Executive Control Strategies and Working Memory Training

Executive control strategies manage the content of working memory. These strategies set and maintain attention on goals and subgoals, activate relevant information, suppress irrelevant information, update information, and coordinate tasks and/or awareness of multiple constraints, switching between them when necessary (Taatgen, 2014). There is evidence that if executive control strategies are taught to individuals and practiced, the functional size of working memory is increased and performance on cognitive tasks enhanced, including transfer tasks.

There is a growing literature on working memory training (Morrison & Chein, 2011). For example, Karbach and Kray (2009) trained individuals of varying ages for several days on task switching. Task switching involves performing two simple tasks in mixed-task blocks, where subjects switch tasks on every second trial. One task involved deciding whether something was a vegetable or fruit, the second was whether a picture was small or large. In performing these tasks, individuals not only must practice task switching but also goal maintenance (remembering the goal for each trial) and inhibiting interference; thus several aspects of executive control were trained. Training involved four sessions spaced one week apart; in addition, there were two pretest and two posttest sessions. The training transferred to better performance on Stroop interference tasks, improved scores on the Raven progressive matrices test (a test of fluid intelligence), and increased working memory capacity (Taatgen, 2014). The far transfer effect sizes were largest for children (with most values greater than 0.70) and smallest for older adults (with most values greater than 0.40) (Karbach & Klay, 2009).

In another study, Mackey, Hill, Stone, and Bunge (2010) had upper-elementary students play various computerized and noncomputerized games for one hour, twice a week for eight weeks. In one condition, students played games requiring speed (such as Blink) and in the other games involving reasoning (such as Rush Hour). Games used for reasoning training "demanded the joint consideration of several task rules, relations, or steps" (p. 585). The reasoning training led to a significant increase from pre- to post-, specifically on the Test of Nonverbal Intelligence, which involves matrix reasoning (Cohen's $d = 1.51$), and on a test of spatial working memory (Forward Spatial Span, Cohen's $d = 0.65$). Improvements on these two outcome measures were not correlated, suggesting that the improvement in spatial working memory does not explain the improvements in fluid intelligence. Children in the speed condition did not improve much on these outcomes.

Other researchers finding transfer effect from working memory training on Stroop and/or matrix tasks included Klingberg *et al.* (2005), who worked with elementary students with ADHD for 25 days, and Jaeggi, Buschiuehl, Jonides, and Shah (2011), who trained elementary and middle-school students for 8 to 19 days on a demanding dual n-back task. Researchers only found transfer for those who scored above the median on the training task. This may shed light on why some other researchers have not found working memory training to transfer to fluid intelligence (Owen *et al.*, 2010; Redick *et al.*, 2013). In addition, Taatgen (2013) argues that transfer tasks used by Owens *et al.* were simple working memory tasks that most people may have already partially mastered rather than complex tasks involving conflict between rehearsal and a secondary task. Resolving such conflict was an important part of the training that was not adequately tested. On balance, we believe that the preponderance of evidence supports the claim that there are far transfer effects from working memory training.

4.2. Proactive Executive Control Strategies

According to Taatgen (2013), far transfer can occur when individuals learn and practice 'proactive executive control strategies'. PECS involves intentionally activating or inhibiting a certain cognitive process, such as protection from interference. For example, in a Stroop task, one might intentionally try to focus on the color of the ink and not the meaning of the words. Another example is finding some way to rehearse words during a complex working memory task designed to make rehearsal difficult. In a proactive strategy, certain control decisions are made before the task commences that sustain the goal context and prime certain attentional processes. In contrast, with a reactive strategy, one first processes the stimuli before making a control decision (e.g., focusing on ink color in the Stroop task and resolving any interference errors).

We propose that far transfer effects from instruction involving structured, academic dialogue and argumentation involve the acquisition of PECS. For example, when one is considering someone else's counterargument, one has to protect the mind from interference from one's own argument, and then switch attention back to one's argument to advocate or evaluate it. One needs to find a way to rehearse one's argument in the interim so as not to forget it. It is also desirable to remember the collective goals of the discourse (Walton, 1998) and one's individual goals. Goal maintenance, task switching, and protection from interference are all control functions that can be performed proactively, that is, without prompting, and which can be trained.

Other PECS that could be developed during argumentive discourse include discerning the similarities and differences between past, current, and possible future situations (Marton, 2006) as well as between arguments. Koedinger and Stamfer (2015) analyze how participation and discussion of experiments in the CASE program might result in variable and relationship abstraction, and that these knowledge components are also applicable to language arts. They describe an example task from their English language arts assessment that involves students in describing the views of one of the characters in a dialogue; "students must engage in variable abstraction to identify those views and how they differ from one character to another" (p. 280).

4.3. The Primitive Elements (PRIM) Theory

In this section, we briefly describe how PECS are part of a new theory of transfer, including far transfer, specifically Taatgen's PRIM theory (2013, 2014).

The PRIM theory specifies a cognitive architecture that is a major refinement of the ACT-R architecture. Production rules are composed of more primitive elements known as PRIMs that either compare, copy, or retrieve information (see Taatgen, 2014, for specific examples).

PRIMs are carried out one at a time before learning, but are compiled into larger units, including production rules, during learning. In contrast to ACT-R theory, more general rules are compiled first, before PRIMs involving more specific information are added. Skill learning proceeds from declarative to mixed declarative/procedural (intermediate stage) and then to a fully procedural stage. In the intermediate stage, rules contain declarative elements, known as operators, that control the order in which PRIMs are carried out. The operators can be either reactive or proactive; the latter are PECS. Once a skill is fully proceduralized and automated, the intermediate rules and associated operators, although no longer directly needed, are still present in memory and can be later reused in a transfer situation.

As an illustration (Taatgen, 2014), suppose one needs to maintain multiple items in working memory. A proactive strategy would be to store these items in declarative memory as a list, specifically by (1) initializing the list, (2) adding items to the list, and (3) recalling the list (which also strengthens memory of the list). This strategy can transfer from one working memory task to another, thus partially explaining working memory training effects. Such a strategy may be required during argumentive discourse in order to recall a chain of reasoning, although more complex structures than lists, e.g., networks or trees, may also be involved (Nussbaum, 2011).

4.4. What Explains the Transfer of PECS?

Just because knowledge or skill can transfer from one situation to another does not mean it will transfer. The history of learning theory is replete with transfer failures. We therefore need to explain why individuals may or may not transfer PECS from one situation to another. This is a key piece missing from most other explanations of argumentive far transfer (e.g., why should a disposition, schema, or metacognitive knowledge engendered in one context transfer to a vastly different one?). Activation mechanisms must be specified.

According to the PRIM theory, an operator is retrieved during the transfer task through spreading activation in declarative memory. Several operators with similar conditions may become activated even if the actions are different. In particular, there will be competition between proactive and reactive strategies. As in ACT-R, the PRIM theory assumes the rules/strategies with the most cognitive utility will be executed. Proactive strategies are more complex than reactive ones, and so will initially have less cognitive utility. Over time, however, with training, practice, and/or experience in a learning environment, a proactive strategy may gain strength as its utility increases. This occurs because the "cost" of implementing a proactive strategy declines as an individual becomes more fluid with it, and also because they may experience more benefits from using a proactive strategy.

Note that control strategies have to be learned, so there will be individual differences in strategy use and performance, including transfer performance. This fact might explain why in working memory training there is little correlation between improvements on working memory transfer tasks and fluid intelligence transfer tasks: individuals may have learned somewhat different strategies from the training, some of which are more applicable to different transfer tasks than others.

There is empirical support for the PRIM theory. Taatgen (2013) reports that in Singley and Anderson's (1985) text editing transfer study, their identical productions model only predicted half as much transfer as actually occurred. A computational model, known as Actransfer, based on the PRIM theory, successfully accounted for most of the additional variance. The model also explains some of the far transfer effects from working memory training to fluid intelligence tasks (e.g., Raven progressive matrices). Taatgen (2013) also successfully modelled the transfer of working memory training to the Stroop task, as reported by Chein and Morrison (2010), and the transfer of task switching training to Stroop, as reported by Karbach and Kray (2009). In addition, Taatgen (2013) used the theory to explain cases where there was no transfer, for example in Owens et al. (2010) on simple working memory tasks that did not involve a conflict with a secondary task (which was a characteristic of the training).

4.5. Theoretical and Practical Implications

Learning PECS takes time, several months at least, so that students can discover and practice PECS and use them successfully enough times so that a PEC strategy's cognitive utility becomes strong enough to compete with previously learned, suboptimal strategies. In a case study of how one student may have learned a particular argument stratagem, specifically asking "how much will something cost?" Nussbaum and Edwards (2011) showed that the student used the stratagem productively during a class discussion (because it resulted in further dialogue on the question) and then transferred and applied the stratagem when writing about a novel topic. Nussbaum and Edwards (2011) suggested that the cognitive utility of the stratagem for the student may have increased. In other cases, argument stratagems may be used inappropriately (Nussbaum, 2003), leading to a decrease in use.

Although an argument stratagem is not the same as a proactive control strategy, the latter is cognitively needed to implement the stratagem. In this example, the student would need, at a minimum, to maintain a goal to critique arguments, rehearse the position being critiqued, inhibit agreement with the position, recognize the relevancy of cost considerations, and produce the cost question at a relevant time. The student was most likely engaged with the argumentive discussion in a host of other ways as well.

Students learn argument stratagems often by observing other students use those stratagems (Anderson *et al.*, 2001), but the underlying PECS are not directly observed. So how are the PECS learned? It may be that students have to construct PECS to implement a stratagem, either by transferring and using a previously learned PEC strategy or by recruiting various element of a strategy (also previously learned) and combining them in a novel way. Students in argumentive learning environments are involved in a variety of cognitive activities, so students likely reuse certain PECS in various situations. Transfer within the specific learning environment may ultimately promote far transfer to other contexts. For example, accumulation of a variety of PECS might transfer to use in other subject domains and eventually to improved achievement on cognitive ability tests several months or years later.

Our account so far has stressed the critical importance of experience and practice. This is the learning side of the equation. Later, in a situation where transfer could potentially occur, spreading activation is used to retrieve a PEC strategy. The PRIM theory therefore highlights the critical importance of declarative memory. The emphasis is similar to that which Kuhn (2010) places on meta-knowledge. Kuhn and colleagues found that having students reflect on the quality of their arguments (and those of others) seemed to promote transfer between science and social studies tasks. Although it is hard to disentangle this feature of her learning environment from others, Morris, Miller, Anderson *et al.*, 2012, also found such reflection to promote transfer. Kuhn postulates that conscious knowledge about argumentation (and use of that knowledge) constitutes what she calls meta-procedural or meta-strategic knowledge and is critical for transfer. In our account, such knowledge is likely stored in declarative memory, but is also connected to intermediate task general rules that involve both declarative and procedural elements. The rules are "intermediate" because they are learned during the intermediate stage of skill learning.

In relation to argument schema theory, such schemas are also likely to be encoded by intermediate task general rules that become active through spreading activation and in turn, require proactive control strategies to help fill in slots of the schema. We argue that our account synthesizes these other theories into a more comprehensive one.

Another theoretical question concerns the relationship of PECS with conceptual agency. Proactivity involves goal-directed actions and therefore human agency, but not necessarily conceptual agency, which is a subset of human agency related to creative conceptual actions within a discipline. We think it is likely that conceptual agency promotes proactivity; in an environment where students are actively generating and evaluating ideas, PECS will be required. The converse is not necessarily the case; PECS may not necessarily promote conceptual agency, in the full sense described by Greeno (2006). There are a variety of PECS, some more complex than others. Even cognitive rehearsal is considered a PEC strategy, because an individual is actively trying to maintain information in working memory, but environments involving recitation (and not argumentive discourse or conceptual agency) could promote these types of PECS.

The importance of PECS (and the PRIM theory) is in our view theoretical, in explaining transfer effects. We are uncertain at this point whether it will be a useful framework for instructional design. We do, however, cautiously put forward three possible practical implications of the PRIM theory. First, instructional designs that promote conceptual agency will also promote proactive control. Second, learning environments should include a varied set of learning activities, so that primitive elements can be combined in a variety of ways, allowing appropriate PECS to be constructed when needed. As a result, more robust transfer could be expected on a variety of different tasks. Third, as mentioned previously, developing proactive control takes time, which could explain why students need to be engaged in argumentive discussions for at least 90 minutes per week over multiple months before far transfer effects are observed (Resnick, 2015).

5. Summary and Final Conclusions

In this chapter, we have reviewed some of the various mechanisms by which participation in argumentive discussion can result in far transfer effects in other domains. These include development of formal operational thought (an explanation that we do not favor), and the development of argument schemas, compiled from argument stratagems and somewhat general production rules. Argumentive discussions may also build thinking dispositions and student identities as active thinkers who exercise conceptual agency.

Drawing on Taatgen's PRIM theory of transfer, we have suggested that conceptual agency and participation in argumentive discussions may help build proactive executive control strategies (PECS), these being the "muscles of the mind". The PRIM theory holds that during learning, students develop and strengthen more general production rules before more specific ones; the more general ones involve linking condition-action primitives but without a control condition in the production rule. Rather, control is exercised by PECS, which are stored in declarative memory and can be triggered by spreading activation in a transfer situation. The representation of a PEC strategy linked to a general production rule form what Taatgen calls an intermediate task general rule. Students can learn, over time, that there are cognitive benefits to being proactive and exercising conceptual agency and that certain of these task general rules have cognitive utility. The fact that those rules have been learned and used before makes it easier for them to be used again, increasing their utility. It becomes easier to learn a proactive strategy for new tasks.

This account therefore involves cognitive and motivational mechanisms. We believe the PRIM theory is better specified than explaining far transfer through the acquisition of a general disposition, because dispositional accounts do little more than describe the behavior they attempt to explain. In contrast, we believe that concepts, such as expansive framing and student identity, are productive, as they help explain the transfer of conceptual agency from one discipline to another. While not part of the PRIM theory, these concepts complement it in an important way. Specifically, a student who thinks of herself as an intelligent novice may invest more cognitive effort in a transfer task, which can result in more cycles of spreading activation of increasing strength, increasing the chance that other relevant PEC strategies will be recalled.

Our account integrates cognitive accounts at different levels of analysis (sociocultural/situative, cognitive, and microcognitive/primitive). Even greater integration could be achieved with additional research. Such research could use the PRIM theory to explain why transfer may be found between some tasks and not others (considering the tasks demands of argumentation and of transfer tasks, and individual differences in strategy use), and to make and test predictions. Doing so will require analysis of the argument stratagems used by specific students, identification of the PECS and PRIMs involved, and analysis of transfer test items to make differential predictions about item performance based on the PECS and PRIMs involved in each task. In a similar vein, Koedinger and Stampfer, 2015, recommend that existing data sets contain more complete information on all the tasks used in instruction and assessment.

Because PECs are partially declarative and may often contain elements of self-talk (for example, to maintain focus on goals or rehearse items), think aloud protocols could be generated from students completing transfer item to further test our theory, specifically in regards to the type of 'meta-strategic' knowledge and 'identity conceptions' that transfer items may invoke. It predicts greater amounts of self-talk than would a general production rule model. In relation to other accounts, our approach synthesizes concepts from two major research programs (argument stratagems/schemas from Anderson *et al.*, 2001, and meta-strategic knowledge from Kuhn and Crowell, 2011), neither of which on their own has as much theoretical power to make differential predictions on how students would perform on transfer item (e.g., on standardized tests in a different domain). Although our account needs to be fleshed out in more detail, we believe it has the potential of moving the psychology of argumentation significantly forward. Or so it could be argued.

References

Adey, P., & Shayer, M. (1993). An exploration of long-term far-transfer effects following an extended intervention programme in the high school science curriculum. *Cognition and Instruction, 11*(1), 1 - 29.

Adey, P., & Shayer, M. (2015). The effects of cognitive acceleration. In L. B. Resnick, C. Asterhan, & S. N. Clarke (Eds.), *Socializing intelligence through academic talk and dialogue* (pp. 127-142). Washington, D.C.: AERA.

Anderson, J. R., & Lebiere, C. (1998). *The atomic components of thought*. Hillsdale, NJ: Lawrence Erlbaum Associates.

Anderson, R. C., Nguyen-Jahiel, K., McNurlen, B., Archodidou, A., Kim, S., Reznitskaya, A. ..., & Gilbert, L. (2001). The snowball phenomenon: Spread of ways of talking and ways of thinking across groups of children. *Cognition and Instruction, 19*, 1-46.

Aristotle (2000). *Nicomachean ethics* (R. Crisp, trans.). Cambridge, UK: Cambridge University Press.

Asterhan, C. S. C., & Babichenko, M. (2015). The social dimension of learning through argumentation: Effects of human presence and discourse style. *Journal of Educational Psychology, 107*(3), 740-755.

Asterhan, C. S., C., & Schwarz, B. B. (2007). The effects of monological and dialogical argumentation on concept learning in evolutionary theory. *Journal of Educational Psychology, 99*, 626-639.

Baron, J. (1988). *Thinking and deciding*. New York, NY: Cambridge University Press.

Bereiter, C., & Scardamalia, M. (1982). From conversation to composition: The role of instruction in a developmental process. In R. Glaser (Ed.), *Advances in instructional psychology* (Vol. 2, pp. 1-64). Hillsdale, NJ: Erlbaum.

Berlyne, D. E. (1954). A theory of human curiosity. *British Journal of Psychology, 45*, 180-191.

Braver, T. S., Gray, J. R., & Burgess, G. C. (2007). Explaining the many varieties of working memory variation: Dual mechanisms of cognitive control. In A. R. A. Conway (Ed.), *Variation in working memory* (pp. 76-106). New York, NY: Oxford University Press.

Bruer, J. T. (1993). *Schools for thought.* Cambridge, MA: MIT.

Cacioppo, J. T., Petty, R. E., & Kao C. F. (1984). The efficient assessment of need for cognition, *Journal of Personality Assessment, 48*, 306-307.

Case, R. (1992). *The mind's staircase: Exploring the conceptual underpinnings of children's thought and knowledge.* Hillsdale, NJ: Erlbaum.

Chein, J. M., & Morrison, A. B. (2010). Expanding the mind's workspace: Training and transfer effects with a complex working memory span task. *Psychonomic Bulletin & Review, 17*, 193-199. doi:10.3758/PBR.17.2.193

Chi, M. T. H., & Menekse, M. (2015). Dialogue patterns in peer collaboration that promote learning. In L. B. Resnick, C. S. C. Asterhan, & S. N. Clarke (Eds.), *Socializing intelligence through academic talk and dialogue* (pp. 263-274). Washington, D.C.: AERA.

Chin, C., & Osborne, J. (2010). Students' questions and discursive interaction: Their impact on argumentation during collaborative group discussions in science. *Journal of Research in Science Teaching, 47*(7), 883-908.

Dweck, C. (2006). *Mindset: The new psychology of success.* New York, NY: Random House.

Engle, R. A. (2006). Framing interactions to foster generative learning: A situative explanation of transfer in a community of learners classroom. *The Journal of the Learning Sciences, 15*(4), 451-498.

Engle, R. A., & Conant, F. R. (2002). Guiding principles for fostering productive disciplinary engagement: explaining an emergent argument in a community of learners classroom. *Cognition & Instruction, 20*(4), 399-483. doi:10.1207/S1532690XCI2004_1

Fischer, K. W., & Lazerson, A. (1984). *Human development: From conception through adolescence.* New York, NY: W. H. Freeman.

Frijters, S., ten Dam, G., & Rijlaarsdam, G. (2008). Effects of dialogic learning on value-loaded critical thinking. *Learning and Instruction, 18*(1), 66-82. doi:10.1016/j.learninstruc.2006.11.001

Greeno, J. G. (2006). Authoritative, accountable positioning and connected, general knowing: Progressive themes in understanding transfer. *The Journal of the Learning Sciences, 15*(4), 537-547.

Greeno, J. G., & Moore, J. (1993). Situativity and symbols: Response to Vera and Simon. *Cognitive Science, 17*, 49-59. doi:10.1016/S0364-0213(05)80009-6

Gresalfi, M. S. (2009). Taking up opportunities to learn: Constructing dispositions in mathematics classrooms. *Journal of the Learning Sciences, 18*(3), 327-369.

Howe, C., & Abedin, M. (2013). Classroom dialogue: a systematic review across four decades of research. *Cambridge Journal of Education, 43*(3), 325-356.

Inhelder B., & Piaget J. (1958). *The growth of logical thinking from childhood to adolescence* (A. Parsons & S. Milgram, Trans.). New York, NY: Basic Books (Original work published 1955).

Karbach, J., & Kray, J. (2009). How useful is executive control training? Age differences in near and far transfer of task-switching training. *Developmental Science, 12*(6), 978-990. doi:10.1111/j.1467-7687.2009.00846.x

Klingberg, D. K., Fernell, E., Olesen, P., Johnson, M., Gustafsson, P., Dahlström, K.,...Westerberg, H. (2005). Computerized training of working memory in children with ADHD – a randomized, controlled trial. *Journal of the American Academy of Child and Adolescent Psychiatry, 44,* 1767-186.

Koedinger, K. R., & Stampfer, E. (2015). Accounting for socializing intelligence with the Knowledge-Learning-Instruction Framework. In L. B. Resnick, C. S. C. Asterhan, & S. N. Clarke (Eds.), *Socializing intelligence through academic talk and dialogue* (275-288). Washington, D.C.: AERA.

Kuhn, D. (2010). Teaching and learning science as argument. *Science Education, 94*: 810-824.

Kuhn, D., & Crowell, A. (2011). Dialogic argumentation as a vehicle for developing young adolescents' thinking. *Psychological Science, 22,* 545-552. doi:10.1177/0956797611402512

Kuhn, D., Shaw, V., & Felton, M. (1997). Effects of dyadic interaction on argumentative reasoning. *Cognition and Instruction, 15,* 287-315.

Leitão, S. (2000). The potential of argument in knowledge building. *Human Development, 43,* 323-360.

Mackey, A. P., Hill, S. S., Stone, S. I., & Bunge, S. A. (2011). Differential effects of reasoning and speed training in children. *Developmental Science, 14,* 582–590. doi:10.1111/j.1467-7687/2010/01005.x

Marton, F. (2006). Sameness and difference in transfer. *The Journal of the Learning Sciences, 15*(4), 499-535.

Mercer, N., & Littleton, K. (2007). *Dialogue and the development of children's thinking: A sociocultural approach.* London, England: Taylor & Francis.

Morris, J., Miller, B., Anderson, R. C., et al. (2012). *Instructional discourse and argumentative writing.* University of Illinois, Urbana-Champaign.

Morrison, A. M., & Chein, J. M. (2011). Does working memory training work? The promise and challenges of enhancing cognition by training working memory. *Psychonomic Bulletin & Review, 18,* 46-60. doi:10.3758/s13423-010-0034-00

Nussbaum, E. M. (2003). Appropriate appropriation: Functionality of student arguments and support requests during small-group classroom discussions. *Journal of Literacy Research, 34,* 501-544.

Nussbaum, E. M. (2008). Using argumentation vee diagrams (AVDs) for promoting argument/counterargument integration in reflective writing. *Journal of Educational Psychology, 100,* 549–565. doi:10.1037/0022-0663.100.3.549

Nussbaum, E. M. (2011). Argumentation, dialogue theory, and probability modeling: Alternative frameworks for argumentation research in education. *Educational Psychologist, 46,* 84-106. doi:10.1080/00461520.2011.558816

Nussbaum, E. M., & Edwards, O. V. (2011). Argumentation, critical questions, and integrative stratagems: Enhancing young adolescents' reasoning about current events. *Journal of the Learning Sciences, 20,* 433-488. doi:10.1080/10508406.2011.564567

O'Connor, C., Michaels, S., & Chapin, S. (2015). "Scaling down" to explore the role of talk in learning: From district intervention to controlled classroom study. In L. B. Resnick, C. S. C. Asterhan, & S. N. Clarke (Eds.), *Socializing intelligence through academic talk and dialogue* (pp. 111-126). Washington, D.C.: AERA.

Owens, A. M., Hampshire, A., Grahn, J. A., Stentoon, R., Dajani, S., Burns, A.,...Ballard, C. G. (2010). Putting brain training to the test. *Nature, 465*(7299), 775-778.

Pauli, C., & Reusser, K. (2015). Discursive cultures of learning in (everyday) mathematics teaching: A video-based study on mathematics teaching in German and Swiss classrooms. In L. B. Resnick, C. S. C. Asterhan, and S. N. Clarke (Eds.), *Socializing intelligence through academic talk and dialogue* (pp. 181-193). Washington, D.C.: AERA.

Pickering, A. (1995). *The mangle of practice*. Chicago: University of Chicago Press.

Redick, T. S., Shipstead, Z., Harrison, T. L., Hicks, K. L., Fried, D. E., Hambrick, D. Z., Engle, R. W. (2013). No evidence of intelligence improvement after working memory training: A randomized, placebo-controlled study. *Journal of Experimental Psychology: General, 142*(2), 359-379. doi:10.1037/a0029082

Resnick, L. B., with Schantz, F. (2015). Talking to learn: The promise and challenge of dialogic teaching. In L.B., Resnick, C. S. C Asterhan, and S. N. Clarke (Eds.), *Socializing intelligence through academic talk and dialogue* (pp. 441-450). Washington, D.C.: AERA.

Resnick, L. B., Asterhan, C. S. C., and Clarke, S. N. (Eds.; 2015), *Socializing intelligence through academic talk and dialogue.* Washington, D.C.: AERA.

Resnick, L. B., Bill, V. L., Lesgold, S. B., & Leer, M. N. (1991). Thinking in arithmetic class. In B. Means, C. Chelemer, & M. S. Knapp (Eds.), *Teaching advanced skills to at-risk students* (pp. 27-53). San Francisco, CA: Jossey-Bass.

Reznitskaya, A., & Anderson, R. C. (2002). The argument schema and learning to reason. In C. C. Block & M. Pressley (Eds.), *Comprehension instruction: Research-based best practices* (pp. 319-334). New York, NY: Guilford.

Reznitskaya, A., Anderson, R. C., Dong, T., Kim, I.-H., & Kim, S.-Y. (2008). Learning to think well: Application of argument schema theory to literacy instruction. In C. C. Block & S. R. Paris (Eds.), *Comprehension instruction: Research-based best practices* (2nd ed.) (pp. 196-213). New York, NY: Guilford.

Ryan, R. M., & Deci, E. L. (2000). Intrinsic and extrinsic motivations: Classic definitions and new directions. *Contemporary Educational Psychology, 25,* 54–67. doi:10.1006/ceps.1999.1020

Shayer, M. (1999). *GCSE 1999: Added-value from schools adopting the CASE Intervention*. London, England: Centre for the Advancement of Thinking.

Singley, M. K., & Anderson, J. R. (1985). The transfer of text-editing skill. *International Journal of Man-Machine Studies, 22*, 403– 423. doi:10.1016/S0020—7373(85)80047-X

Stanovich, K. E., & West, R. F. (1997). Reasoning independently of prior belief and individual differences in actively open-minded thinking. *Journal of Educational Psychology, 89*(2), 342-357.

Stenning, K. (2002). *Seeing reason: Image and language in learning to think.* New York, NY: Oxford University Press.

Suedfeld, P., Tetlock, P. E., & Streufert, S. (1992). Conceptual/integrative complexity. In C. P. Smith (Ed.), in association with J. W. Atkinson, D. C. McClelland, & J. Veroff, *Motivation and personality: Handbook of thematic content analysis* (pp. 393-400). New York, NY: Cambridge University Press.

Taatgen, N. A. (2013). The nature and transfer of cognitive skills. *Psychological Review, 120*(3), 339-471.

Taatgen, N. A. (2014). Between architecture and model: Strategies for cognitive control. *Biologically Inspired Cognitive Architectures, 8,* 132-139. doi:10.1016/j.bica.2014.03.010

Topping, K. J., & Trickey, S. (2007). Collaborative philosophical enquiry for school children: Cognitive gains at two-year follow-up. *British Journal of Educational Psychology, 77*, 787-796.

Toulmin, S. (1958). *The uses of argument*. New York, NY: Cambridge University Press.
Walton, D. N. (1996). *Argumentation schemes for presumptive reasoning*. Mahwah, NJ: Erlbaum.
Walton, D. N. (1998). *The new dialectic*. University Park, PA: Pennsylvania State University Press.

www.ingramcontent.com/pod-product-compliance
Lightning Source LLC
Chambersburg PA
CBHW071941220426
43662CB00009B/939